Photosynthesis and Plant Light Requirements

Photosynthesis and Plant Light Requirements

Edited by:

Carlos Tello Lacal

Delve
PUBLISHING

www.delvepublishing.com

Photosynthesis and Plant Light Requirements

Carlos Tello Lacal

Delve Publishing

2010 Winston Park Drive,

2nd Floor

Oakville, ON L6H 5R7

Canada

www.delvepublishing.com

Tel: 001-289-291-7705
001-905-616-2116

Fax: 001-289-291-7601

Email: orders@arclereducation.com

ISBN: 978-1-77361-049-8 (Hardcover)

DECLARATION

Some content or chapters in this book are open access copyright free published research work, which is published under Creative Commons License and are indicated with the citation. We are thankful to the publishers and authors of the content and chapters as without them this book wouldn't have been possible.

ABOUT THE EDITOR

Carlos Tello Lacal

Carlos Tello (Valencia, Spain, 1981) studied Agricultural Engineering with specialization in Biotechnology at the Universidad Politécnica de Valencia, where he graduated with a BSc thesis on the characterization of an Arabidopsis mutant with increased tolerance to seed ageing and salt stress. He moved afterwards to Sevilla to complete his MSc and PhD at the Instituto de Recursos Naturales y Agrobiología, where he studied the mechanisms employed by plants in their adaptation to abiotic stress and more specifically the role of the Arabidopsis Na+/H+ antiporter SOS1 in the regulation of Na+ and K+ homeostasis. Since 2014, he is based in Zürich and develops several freelance science-related jobs. He has recently retrained himself in the field of cell culture in bioreactors by attending an advanced training course at the Zürcher Hochschule für Angewandte Wissenschaften.

TABLE OF CONTENTS

LIST OF CONTRIBUTORS

Carlos Tello
Instituto de Recursos Naturales y Agrobiología de Sevilla

Koichi Kobayashi
Graduate School of Arts and Sciences, The University of Tokyo, Tokyo, Japan

Tatsuru Masuda
Graduate School of Arts and Sciences, The University of Tokyo, Tokyo, Japan

Robert M. Larkin
Key Laboratory of Horticultural Plant Biology (Ministry of Education), College of Horticulture and Forestry Sciences, Huazhong Agricultural University, Wuhan, China

María del Rocío Gómez-García
Department of Plant Genetic Engineering, Center for Research and Advanced Studies-Irapuato Unit, National Polytechnic Institute, Km 9.6 libramiento norte carretera Irapuato-León, Irapuato, Guanajuato 36821, México

Neftalí Ochoa-Alejo
Department of Plant Genetic Engineering, Center for Research and Advanced Studies-Irapuato Unit, National Polytechnic Institute, Km 9.6 libramiento norte carretera Irapuato-León, Irapuato, Guanajuato 36821, México
Department of Biotechnology and Biochemistry, Center for Research and Advanced Studies-Irapuato Unit, National Polytechnic Institute, Km 9.6 libramiento norte carretera Irapuato-León, Irapuato, Guanajuato 36821, México

Henry Daniell
Department of Biochemistry, School of Dental Medicine, University of Pennsylvania

Choun-Sea Lin
Agricultural Biotechnology Research Center, Academia Sinica, Taipei, Taiwan.

Ming Yu
Department of Biochemistry, School of Dental Medicine, University of Pennsylvania

Wan-Jung Chang
Agricultural Biotechnology Research Center, Academia Sinica, Taipei, Taiwan.

Krzysztof Bobik
Department of Biochemistry, Cellular and Molecular Biology, University of Tennessee, Knoxville, TN, USA

Tessa M. Burch-Smith
Department of Biochemistry, Cellular and Molecular Biology, University of Tennessee, Knoxville, TN, USA

Roberta Croce
Department of Physics and Astronomy, Faculty of Sciences, VU University Amsterdam, Amsterdam, The Netherlands

Herbert van Amerongen
Laboratory of Biophysics, Wageningen University, Wageningen, The Netherlands

Sari Järvi
Molecular Plant Biology, Department of Biochemistry, University of Turku, Turku, Finland

Peter J. Gollan
Molecular Plant Biology, Department of Biochemistry, University of Turku, Turku, Finland

Eva-Mari Aro
Molecular Plant Biology, Department of Biochemistry, University of Turku, Turku, Finland

Miaomiao Liu
Institute of Medicinal Plant Development, Chinese Academy of Medical Sciences and Peking Union Medical College, Beijing, China

Shanfa Lu
Institute of Medicinal Plant Development, Chinese Academy of Medical Sciences and Peking Union Medical College, Beijing, China

PREFACE

Photosynthesis plays a central role in sustaining life on Earth, being its major outcomes the fixation of atmospheric CO_2 into carbohydrates and the production of O_2 as a result of the lysis of water. Numerous pigments, photoreceptors, light harvest complexes, electron transport chains and enzymes are involved in this process. Understanding the functioning of photosynthesis is the first step towards increasing its performance, which can ultimately increase crop yield. By compiling open access research papers, the present books reviews the state-of-the-art knowledge on the different aspects concerning photosynthesis, including the potential application of recent findings.

In Chapter number 1, the editor describes the photoreceptors employed by plants to convert the different wavelengths into signals, as well as the effects of light quality on plant growth and metabolism. Chapters 2 and 3 deal with chlorophylls and other plant tetrapyrroles by reviewing their synthesis and signaling, respectively. In turn, Chapter 4 describes the synthesis of the other major pigment type, carotenoids, in chili peppers. The next block of two chapters exposes the current knowledge on chloroplasts. Chapter 5 explores the phylogenetic diversity of chloroplast genomes and the potential of this knowledge in genetic engineering, while Chapter 6 reviews their inter-organellar and inter-cellular signaling. In Chapters 7 and 8, the respective functioning of photosystems I and II in light harvesting is detailed. Chapter 9 reviews the different reactions taking place at the thylakoid lumen, as well as the proteins involved. A detailed overview of the photosynthetic electron transporters plastoquinone and ubiquinone is presented in Chapter 10.

CHAPTER

1

INFLUENCE OF LIGHT WAVELENGTH ON PLANT GROWTH AND METABOLISM

Carlos Tello

Instituto de Recursos Naturales y Agrobiología de Sevilla

SPECTRUM OF SOLAR RADIATION

The spectrum of the solar radiation outside the Earth's atmosphere resembles that of a black body at 5900 K, with a peak at wavelengths of ~500 nm (Hulstron, 1989, Webster, 1999). After the most highly energetic radiations have been filtered by the atmosphere, over 99% of the solar radiation reaching the surface of the Earth is in the spectral region between 150 and 4000 nm, with approximately 41% within the visible and 51% within the infrared ranges. Based on its wavelength, this radiation can be classified into UV-C (100-280 nm), UV-B (280-315), UV-C (315-400), visible light (400-700) and near infrared (700-4000) (Mora et al., 2000; Godish, 2004; Eltbaakh et al., 2011). Because the atmosphere is more efficient at filtering the shortest wavelengths of solar radiation and its thickness is very variable through seasons, with altitude and latitude, the composition of the solar electromagnetic spectrum, especially the amount of UV-B radiation, is variable among locations (Paul and Gwyn-Jones, 2003; Pinedo et al., 2006).

Radiation within the 400-700 nm waveband is generally termed "photosynthetically active radiation" (PAR), since its absorption by chlorophyll and other photosynthetic pigments such as carotenoids drives CO_2 fixation through photosynthesis. As opposed to PAR, the fraction of wavelengths below 400 nm on the Earth's surface is very scarce, far-red radiation (700-780 nm) exhibits reduced efficiency at driving photosynthesis and radiation beyond these wavelength values is hardly absorbed by leaves (Emerson, 1958; McCree, 1972a, 1972b, 1981; Ort and Yocum, 1996). Although short-waved radiation (e.g. blue) has more energy than its long-waved (e.g. red) counterpart, all photons within the 400-700 nm range contribute equally to net photosynthesis upon absorption by photosynthetic pigments. However, not all the PAR wavelengths are absorbed with the same efficiency by leaves. The relative quantum efficiency (RQE) spectrum for chlorophyll shows that most plants reach major absorption peaks in the red (662 nm) and blue (430 nm) regions. While red light is very efficiently absorbed by chlorophyll and therefore plays an essential role in driving photosynthesis, not all blue photons can be used, since some of them are absorbed by inactive pigments such as anthocyanins or by carotenoids, which must transfer the excitation energy to chlorophylls with an efficiency varying from 90 to only 35% (McCree, 1971; Barnes et al., 1993; Atwell et al., 1999; Hogewoning et al., 2012). In turn, blue light fulfills a wide variety of roles in plant photomorphogenesis, such as regulation of stomatal movements, inhibition of hypocotyl growth and control of phototropism (Blaauw and Blaauw-Jansen, 1966; Cosgrove and Green, 1981; Schwartz and Zeiger, 1984; see sections 2 and 3). Although green light is not as efficiently absorbed as the two former, it can penetrate deeper into leaves and canopies, driving photosynthesis in places inaccessible to blue and red light (Sun et al., 1998; Terashima et al., 2009; Paradiso et al., 2011).

Radiation outside the PAR range does not drive photosynthesis, but plays some roles in plant morphogenesis and metabolism. Given the damaging effect of UV-B radiation on macromolecules, this radiation triggers the production of sunscreen pigments such as flavonoids and anthocyanins (Tevini et al., 1981; Beggs and Wellman, 1985; Chalker-Scott, 1999; Paul et al., 2006). Additionally, plants exposed to UV-B radiation grow more compact in morphology, produce tougher, more robust leaves and can increase their content in essential oils (Gardener et al., 2009; Kumari and Agrawal, 2009; Wargent et al., 2009; Hikosaka et al., 2010). The far-red to red light ratio is often referred to as "light quality" and regulates important processes such as flowering, shade avoidance and plant morphology (Eskins, 1992; Johnson et al., 1996; Jeong et al., 2012; Kubota et al., 2012; see sections 2 and 3).

PHOTORECEPTORS AND LIGHT PERCEPTION

Plants respond to changes in the quality, quantity and direction of light, which they use as signals to regulate their growth, development and metabolism. In order to detect these changes, plants are equipped with numerous proteins called photoreceptors, which detect different facets of the solar spectrum (Batschauer, 1999; Galvão and Fankhauser, 2015). Plant photoreceptors are present throughout the whole plant, are sensitive to different wavelengths, intensities and durations of light, and can trigger both local and long-distance responses (Kami et al., 2010; Chaves et al., 2011; Ito et al., 2012). In this section, the main light perception mechanisms involving phytochromes, cryptochromes, phototropins, photoreceptors from the Zeitlupe family and UVR8 will be described.

Phytochromes

Phytochromes were initially identified as red/far-red photochromic pigments with the ability to exist in two stable states: the dark-adapted, biologically unstable, red light-absorbing (l_{max} = 660nm) Pr and the excited, biologically active, far-red light-absorbing (l_{max} = 730 nm) Pfr. Both forms are mutually interconvertible upon excitation with red or far-red light, respectively, acting this mechanism as a switch for several signaling cascades (Butler et al., 1959). Phytochromes are dimers of 124 kDa proteins, each of which comprises an N-terminal photosensing (PSM) and a C-terminal HK-related domain (HKRD) that facilitates dimerization. The PSM region is formed by a PAS (Period/Arnt/Single-Minded) domain of unknown function, a GAF (cGMP phosphodiesterase/adenylyl cyclase/FhlA) region, which binds covalently to tetrapyrrole chromophores (bilins) such as phycocyanobilin, phytochromobilin or biliverdin, and a PHY (Phy-specific) domain, which stabilizes the activated state. The N-terminal extension (NTE) upstream of PAS, is required for Pfr stability and signaling (Cherry et al., 1992; Rockwell et al., 2006; Nito et al., 2013). X-ray crystallographic studies established that the GAF domain is intimately held together with PAS through a knot, while being widely separated from PHY and only connected with it by a long linker α-helix and a tongue-like hairpin. The bilin is attached to a conserved Cys residue at the GAF region via the C3 sidechain of its A-ring. (Wagner et al., 2005; Essen et al., 2008; Yan et al., 2008). Upon photoconversion from Pr into Pfr, the tongue refolds itself from an anti-parallel β-sheet structure into an α-helix and loops (Huges, 2010; Anders et al., 2011, 2013; Takala et al., 2014).

Plants typically contain multiple phytochromes (e.g. phyA to phyE in Arabidopsis), being phyA (light labile) and phyB (light stable) the predominant types. phyA is relatively abundant in etiolated seedlings and controls the very low fluence (10^{-9} $mmol \cdot m^{-2} \cdot s^{-1}$) and far-red high irradiance (0.1-50 $mmol \cdot m^{-2} \cdot s^{-1}$) responses, both of which promote cotyledon expansion and inhibit hypocotyl elongation, that is to say, cause seedling de-etiolation. Light exposure triggers both the rapid depletion of the protein and the reduction in PHYA transcript abundance. (Smith and Whitelam, 1990; Quail, 1994; Casal et al., 1997; Bae and Choi, 2008; Nagatani, 2010). The role of phyA in the control of flowering by day length, the induction by light of seed germination and the regulation of the circadian clock, especially under low photon irradiance, has also been demonstrated (Johnson et al., 1994; Botto et al., 1996; Shinomura et al., 1996; Weller et al., 1997; Somers et al., 1998; Devlin and Kay, 2000). In turn, phyB is more abundant in light-grown plants and is responsible for the red/far-red reversible response. This photoreceptor controls seedling de-etiolation and photoregulation of seed germination in response to red light. Additionally, it regulates gravitropic orientation in shoot and root, plant architecture, circadian period, stomatal development, flowering and freezing tolerance (Korneef et al., 1980; Reed et al., 1993, 1994; Shinomura et al., 1994, 1996; Smith, 1995; Poppe et al., 1996; Smith and Whitelam, 1997; Devlin and Kay, 2000; Halliday et al., 2003; Boccalandro et al., 2009). Although phyA and phyB are the most prominent phytochromes, Arabidopsis phyA phyB double mutants retain the capacity to respond to low red:far-red ratios and end-of-day far-red treatments, suggesting roles for the other phytochromes in light perception (Halliday et al., 1994; Devlin et al., 1996; Whitelam and Devlin, 1996;

Poppe and Schäfer, 1997). phyD shows approximately 80% sequence similarity with phyB and is clustered into a single group together with this protein and phyE. In turn, phyA is most closely related to phyC in spite of its unique properties. Phytochromes phyB-E are all responsible for the red/far-red reversible response (Clough and Vierstra, 1997; Goosey et al., 1997; Sharrock and Clack, 2002). phyC operates through modulation of phyB, with which it is obliged to form a heterodimer. Mutants deficient in PHYC exhibit reduced de-etiolation under red but not far-red light. Accelerated flowering was observed in phyC mutants in the Col background but not Ws, suggesting natural variation of its function (Franklin et al., 2003a; Monte et al., 2003; Clack et al., 2009). phyD has been observed to play a small role in light-induced germination and minor accessory functions to phyB in seedling de-etiolation, cotyledon expansion and anthocyanin accumulation, in addition to redundantly acting with phyB and E under low red:far-red ratios to mediate shade avoidance and repress flowering, or only with the former to repress the cold-acclimation CBF regulon at warm temperatures (Halliday et al., 1994; Aukerman et al., 1997; Hennig et al., 2002; Franklin et al., 2003; Franklin and Whitelam, 2007). Finally, phyE shows an essential role in promoting light-induced seed germination, especially under continuous far-red light or cold (<16°C) temperature, minor contributions to seedling de-etiolation and redundant functions with phyA and phyB in maintaining a compact rosette phenotype, in addition to its role in shade avoidance and flowering repression previously described (Devlin et al., 1998; Hennig et al., 2002; Franklin et al., 2003b; Heschel et al., 2007).

Cryptochromes

Cryptochromes were the first blue light receptors to be characterized in plants. They share sequence similarity with photolyases, which repair UV-induced DNA damage by removing pyrimidine dimers in the presence of this wavelength (Ahmad and Cashmore, 1993; Lin and Shalitin, 2003; Sancar, 2003). The cryptochrome/photolyase superfamily is comprised by cyclobuntane pyrimidine dimer photolyases, 6-4 pyrimidine-pyrimidone adduct photolyases, CRY-DASH proteins, animal cryptochromes and plant cryptochromes. Plant and animals have lost DNA-repair activity and instead regulate several growth and developmental processes, such as photomorphogenesis (in plants) or the circadian clock (in both) (Cashmore,

2003; Lin and Shalitin, 2003; Sancar, 2003; Chaves et al., 2011). Plant cryptochromes function as dimers, with each monomer consisting of two major domains: the N-terminal Photolyase Homologous Region (PHR) and the Cryptochrome C-terminal Extension (CCE) domain. The former is a highly conserved 500 amino acid region comprised by the N-terminal α/β and the C-terminal α subdomains. The primary cromophore flavin adenine dinucleotide (FAD) binds to the α subdomain to allow for blue light perception while a second chromophore, 5,10-methenyltetrahydrofolate, binds to the α/β region to transfer additional UV (370-390 nm) energy to FAD (Lin et al., 1995; Malhotra et al., 1995; Banerjee et al., 2007; Hoang et al., 2008). Upon irradiation with blue light, the semiquinone neutral FAD radical FADH· is generated, thereby triggering photoreceptor activation. FADH· reverts to FAD in the absence of blue light and its formation is antagonized by green light. The PHR domain is also required for cryptochrome dimerization (Sang et al., 2005; Rosenfeldt et al., 2008; Langenbacher et al., 2009). Conversely, the CCE domain, which functions in cryptochrome signaling, is highly divergent and poorly structured (Yang et al., 2000; Bräutigam et al., 2004; Yu et al., 2009). Both domains have been proposed to be in their inactive conformation in the darkness. Upon blue light perception by PHR, a conformational change in CCE would be triggered to allow for the initiation of signaling. Phosphorylation by upstream kinases on multiple serine residues could activate the CCE domain by causing its electrostatic repulsion from PHR (Chaves et al., 2011; Liu et al., 2011; Christie et al., 2015).

The genome of Arabidopsis encodes three cryptochromes, cry1-3. cry3 is a CRY-DASH protein localized in mitochondria and chloroplasts whose function is to repair UV-induced lesions in single-stranded DNA and loop structures of double-stranded DNA (Kleine et al., 2003; Selby and Sancar, 2006; Pokorny et al., 2008). cry1 an cry2 are located in the nucleus and function as major blue light receptors which regulate seedling de-etiolation and photoperiodic flowering, respectively, in response to this wavelength. Because cry2 undergoes blue light-dependent ubiquitination and degradation by the 26S proteasome in the nucleus, it carries out its functions under low light conditions (Ahmad and Cashmore, 1993; Cashmore et al., 1999; Kleiner et al., 1999; Liu et al., 2011; Zuo et al., 2012). Upon activation by blue light, cry1 and cry2 trigger changes in gene expression that alter plant development (Folta and Spalding, 2001; Ma et al., 2001; Ohgisi et al., 2004; Chaves et al., 2011). The bZIP transcription factor LONG HYPOCOTYL 5 (HY5) is targeted for proteolysis at the CUL4-DDB1 E3 ubiquitin ligase complex by the CONSTITUTIVE PHOTOMORPHOGENESIS 1 (COP1)/ SUPPRESSOR OF PHYA1 (SPA1) complex in etiolated seedlings. Upon binding to COP1/SPA1, cry1 and cry2 suppress the action of the complex and enhance the accumulation of HY5 and other factors promoting de-etiolation (Yang et al., 2000; Yu et al., 2009; Lau and Deng, 2012). In addition, cry2 controls the accumulation of transcription factors involved in the photoperiodic regulation of flowering. By binding to the COP1/SPA1 complex, it suppresses the proteolytic degradation of the zinc finger transcriptional regulator of flowering in response to long days CONSTANS (CO). Photoactivated cry2 also interacts with the basic helix-loop-helix (bHLH) transcription factor CRY-INTERACTING bHLH1 (CIB1), which promotes the expression of the transcriptional regulator of floral initiation FLOWERING LOCUS T (FT), as well as with other CIB family member of redundant function such as CIB2 and CIB3. (Liu et al., 2008, 2013; Kennedy et al., 2010; Zuo et al., 2011).

Phototropins

Phototropins are blue-light receptors that were named after their role in plant phototropism. They contain a Ser/Thr kinase domain at their C-terminus and two light, oxygen or voltage sensing (LOV) domains at their N-terminus. LOV domains are ~110 amino acid regions that bind non-oxidized flavin mononucleotide (FMN) in a non-covalent manner. Upon irradiation with blue light (l_{max} ~450 nm), the FMN is converted into its active signaling state by forming a covalent bond between its C(4a) carbon and the sulphur atom of a conserved LOV Cys residue, being the process reverted in the darkness (Christie et al., 1999, 2012a; Salomon et al., 2000; Swartz et al., 2001). The LOV2 domain has been demonstrated to be a major regulator of phototropin autophosphorylation, which is essential to its function, while LOV1 has been proposed to modulate LOV2 photoreactivity and contribute to protein dimerization (Christie et al., 2002; Cho et al., 2007; Jones et al., 2007; Nakasako et al., 2008; Sullivan et al., 2008; Okajima et al., 2014). Autophosphorylation occurs at multiple serine residues upon phototropin activation, being this modification essential for their signaling (Christie et al., 2015).

Arabidopsis contains two phototropins, pho1 and pho2, both of which are located at the plasma membrane and regulate blue light-induced photoresponses under weak light conditions (Sakamoto and Briggs, 2002; Takemiya et al., 2005; Kong et al., 2006). Genetic analyses showed overlapping functions for both proteins in the regulation of root and hypocotyl phototropism, chloroplast movement, stomatal opening, leaf positioning and leaf flattening. phot2 functions predominantly at high light intensities in the case of hypocotyl phototropism, while phot1 works over a broader range (Kagawa et al., 2001; Kinoshita et al., 2001; Sakai et al., 2001; Sakamoto and Briggs, 2002; Inoue et al., 2008). Specific roles for phot2 include chloroplast and nuclear avoidance in response to high intensity blue light and regulation of protein-mediated viral defense (Kagawa et al., 2001; Jeong et al., 2010; Higa et al., 2014). In turn, phot1 controls the rapid inhibition of hypocotyl growth upon transfer to light, transcript destabilization and suppression of lateral root growth (Folta and Kaufman, 2003; Folta et al., 2003; Moni et al., 2014). One of the best characterized phototropin signaling pathways is the activation of stomatal opening by blue light. Both phot1 and phot2 phosphorylate the guard cell-specific kinase BLUE LIGHT SIGNALING 1 (BLUS1) at the Ser348 within its C-terminus, which triggers a signaling cascade leading to H^+-ATPase phosphorylation by an unknown kinase followed by its activation through the binding of 14-3-3 protein to its C-terminus. As a result, the hyperpolarization of the plasma membrane activates K^+ inward-rectifying channels and the cell swelling resulting from the accumulation of K^+ causes stomatal opening (Kinoshita and Shimazaki, 1999; Shimazaki et al., 2007; Takemiya et al., 2013). Phosphorylation targets for phot1 involved in the regulation of phototropic growth include the auxin efflux carrier ABCB19, whose inhibition by phot1 seems to enhance phototropic responsiveness, and PHYTOCHROME KINASE SUBSTRATE 4 (PKS4), which has been shown to impact phototropism depending on its phosphorylation status, despite remaining its biochemical function unknown (Noh et al., 2003; Nagashima et al., 2008; Demarsy et al., 2012; Spalding, 2013; Kami et al., 2014).

Zeitlupe family of receptors

In addition to phototropins, three other photoreceptors containing LOV domains exist in Arabidopsis: ZEITLUPE (ztl), FLAVIN-BINDING

KELCH REPEAT F-BOX 1 (fkf1) and LOV KELCH PROTEIN 2 (lkp2). From N- to C- terminus, members of this family comprise a LOV domain, an F-box and six kelch repeats. Like that of phototropins, the LOV domain of this family binds to the chromophore FMN and is activated by blue light, resulting in the formation of an FMN-cysteinyl adduct. Owing to the presence of a nine amino acid insertion specific to this family in the loop between Eα and Fα, the photocycle is much slower and the decay of this state can occur in the order of hours or even days. Although this property makes the complete reversion of ZTL/FKF1/LKP2 proteins unachievable under a 24 hour day-night cycle, the instability and fluctuation in a circadian phase- and photoperiod-dependent manner shown by these proteins ensures a high turnover that promotes their recycling (Imaizumi et al., 2003; Kim et al., 2003, 2007; Zikihara et al., 2006; Sawa et al., 2007; Fornara et al., 2009; Ito et al., 2012; Pudasaini and Zoltowski, 2013; Suetsugu and Wada, 2013). F-boxes are associated with Skp Cullin F-box (SCF)-type E3 ubiquitin ligases, which target proteins for degradation, while kelch repeats mediate protein-protein interaction and heterodimerization among members of this family. Several proteins belonging to the Arabidopsis Skp1-like (ASK) family, such as ASK1, CULLIN 1 (CUL1) and AtRBX1, have been shown to interact with ZTL/FKF1/LKP2 and be necessary for their activity in vivo (Han et al., 2004; Kevei et al., 2006; Takase et al., 2011; Ito et al., 2012).

The functional divergence between these three proteins has been studied through the characterization of single and multiple mutants. While ztl mutants are primarily defective in circadian clock regulation and only slightly in flowering time, fkf1 show opposite phenotypes. Single lkp2 mutants exhibit nearly normal circadian and flowering phenotypes due to their low level of expression (approximately 4% of ZTL levels), but both the expression of this gene in the ztl background under the promotor of ZTL and its constitutive overexpression in wild type plants mimicked the effect of ZTL expression (Kiyosue and Wada, 2000; Nelson et al., 2000; Schultz et al., 2001; Imaizumi et al., 2003, 2005; Somers et al., 2004; Baudry et al., 2010; Takase et al., 2011). The ztl fkf1 double mutant, and more strongly the ztl fkf1 lkp2 triple mutant, but not the ztl lkp2 double mutant, showed longer period defects than the ztl single mutant. The triple mutant was slightly delayed under long-day conditions when compared to fkf1, while the ztl lkp2 line displayed an early flowering phenotype under short-day conditions that could be suppressed by introgressing FKF1. Taken together, these results indicate that ZTL/LKP2 primarily regulate the circadian clock, through which they indirectly control flowering time, and FKF1 predominantly controls flower photoperiodicity, showing all three receptors partially overlapping functions (Fornara et al., 2009; Baudry et al., 2010; Takase et al., 2011). ZTL has been shown to specifically interact with TIMING OF CAB EXPRESSION 1 (TOC1) and PSEUDO RESPONSE REGULATOR 5 (PRR5) in yeast two-hybrid assays. Both proteins rapidly decay in the darkness and are stabilized by blue light. These findings, together with their expression levels in ztl and ZTL-overexpressing lines, indicate that ZTL mediates the proteolytic degradation of TOC1 and PRR5 in a process inhibited by blue light (Más et al., 2003; Yasuhara et al., 2004; Kiba et al., 2007; Fujiwara et al., 2008; Baudry et al., 2010). FKF1 mediates photoperiodic flowering by regulating the activity of FT and CO in a long day-dependent manner (Sawa et al., 2007; Sawa and Kay, 2011, Song et al., 2012). The Dof transcription factors CYCLING DOF FACTOR (CDF) 1, -2 and -3 interact with the kelch repeats of FKF1 and LKP2, but not with those of ZTL. These transcription factors function as repressors of FT and CO transcription, as seen in studies with knockdown and overexpression mutants. FKF1 mediates photoperiodic degradation of CDFs in a proteasome-dependent manner, which activates CO and FT transcription. Additionally, it stabilizes CO proteins in the presence of blue light, thereby promoting FT expression (Imaizumi et al., 2005; Fornara et al., 2009; Song et al., 2012). ZTL and FKF1 are stabilized by binding to GIGANTEA (GI) through their LOV domains in an interaction enhanced by blue light. This radiation also contributes to the GI-mediated stabilization of ZTL and FKF1 by activating the degradation of the GI repressor COP1 by cryptochromes (Kim et al., 2007; Sawa et al., 2007; Yu et al., 2008; Lian et al., 2011; Liu et al., 2011; Zuo et al., 2011).

UVR8

Low doses of UV-B radiation have long been known to influence plant growth, photomorphogenesis and biochemical content, being these effects neither due to the action of the above-described photoreceptors nor the result of UV-triggered DNA damage (Wellman, 1976, 1983; Klein, 1978). The most well-known responses to UV-B include the suppression of both hypocotyl and root growth, the promotion of cotyledon opening and the synthesis of UV-absorbing flavonoid pigments (Ballaré et al., 1995; Kim et al., 1998; Boccalandro et al., 2001; Ryan et al., 2001; Whinkel-Shirley, 2002; Suesslin and Frohnmeyer, 2003; Tong et al., 2008; Conte et al., 2010). Kliebenstein and collaborators (2002) isolated an Arabidopsis mutant for the UV RESISTANCE LOCUS 8 (UVR8) in a screen for plants hypersensitive to UV-B radiation. The mutant also exhibited downregulation of the flavonoid biosynthesis enzyme CHS and increased expression of PATHOGEN-RELATD (PR) 1 and -5 proteins. The isolation of additional uvr8 alleles showed that CHS expression was only affected by UV-B, establishing that UVR8 acts in a UV-B-specific manner. Microarray analyses demonstrated that UVR8 regulates the expression of several genes in response to UV-B, such as those coding for chloroplast proteins, flavonoid biosynthesis enzymes, oxidative damage amelioration enzymes, transcription factors or DNA repair enzymes (Brown et al., 2005; Favory et al., 2009; Davey et al., 2012). UVR8 is a 440 amino acid and ~47 kDa protein consisting of a highly conserved seven-bladed β-propeller core similar to that of RCC1, a short N-terminal extension and a flexible 60 amino acid C-terminus (Renault et al., 1998; Makde et al., 2010; Christie et al., 2012b; Wu et al., 2012). The donut-shaped monomers come into contact via extensive face-to-face interactions to form a dimer in which blades 1 and 7 lie on the same side. This surface contains a high proportion of charged amino acids, including 12 UV-absorbing (7Trp + 2Tyr + 3Phe) and 31 salt bridge-forming (9Arg + 16Asp +6Glu) residues per monomer. Arg residues at the positions 228 and 338 are key to maintain the dimer structure (Christie et al., 2012b; Wu et al., 2012; Huang et al., 2014). Instead of binding to a chromophore, UVR8 senses UV-B radiation through specific Trp residues. Trp233 and Trp285, which are sufficiently close to each other and to Trp337 at the dimer interface to allow for the overlapping of their orbitals, have been shown to play an essential role in photoreception (Rizzini et al., 2011; Christie et al., 2012b; O'Hara and Jenkins, 2012; Wu et al., 2012).

Upon irradiation with UV-B, the monomers dissociate through yet unknown mechanisms, which initiates signal transduction cascades. The transcription of ELONGATED HYPOCOTYL 5 (HY5), which regulates many UVR8-target genes and sometimes acts redundantly with HY5

HOMOLOG (HYH), is rapidly stimulated after UV-B detection (Brown et al., 2005; Brown and Jenkins, 2008; Safrany et al., 2008; Favory et al., 2009). COP1, whose interaction with UVR8 is required for UV-B signaling, regulates the same set of genes as this receptor, including HY5 and HYH. UV-B exposure enhances the stabilization of UVR8 both through HY5 and the transcription factor FHY3, and by stabilizing COP1 (Oravecz et al., 2006; Favory et al., 2009; Heijde et al., 2013; Huang et al., 2012; 2014). COP1 binds to SUPPRESSOR OF PHYA-105 (SPA) proteins and acts as a substrate for the CUL4-DDB1 E3 ubiquitin-ligase complex, responsible for degrading proteins that promote de-etiolation such as HY5 in the dark. UV-B radiation reduces the interaction of COP1-SPA with CUL4-DDB1 and instead promotes its association with UVR8, in which the WD40 domain of COP1 interacts with a 27 amino acid region (C27) at the C-terminus of UVR8 (Osterlund et al., 2000; Zhu et al., 2008; Cloix et al., 20012; Lau and Deng, 2012; Huang et al., 2013). REPRESSOR OF UV-B PHOTOMORPHOGENESIS (RUP) proteins are negative regulators of UVR8 that mediate its dimerization. They bind to the C27 region of UVR8 and displace COP1 owing to the similarity of their sequence to the WD40 domain (Grüber et al., 2010; Cloix et al., 2012; Heijde and Ulm, 2013). Additional proteins involved in UV-B responses through the regulation of HY5 expression, such as BBX24/STO and RCD1, could also associate with complexes containing UVR8 (Indorf et al., 2007; Jiang et al., 2009; 2012; Jenkins, 2014).

PLANT GROWTH AND DEVELOPMENT UNDER DIFFERENT LIGHT WAVELENGTHS

Plant productivity

Early studies with light-emitting diodes (LEDs) proved that red light (600-700 nm) is the most efficient at driving photosynthesis. However, supplementation with blue light (400-500 nm) is required for normal plant growth (Barnes and Bugbee, 1991; Bula et al., 1991; Barta et al., 1992; Sager and McFarlane, 1997). Wheat and Arabidopsis produce more seeds, pepper, lettuce, radish and spinaches increase their biomass, Frigo strawberry plants yield more fruits with increased soluble sugar content, potato plants produce more tubers and cucumber, tomato, pepper, radish, lettuce and potato plants, as well as rose and chrysanthemum flowers, grow more compact when blue light is provided (Jose, 1977; Brown et al., 1995; Yanagi et al., 1996; Goins et al., 1997, 1998; Yorio et al., 2001; Chang et al., 2009; Samuoliene et al., 2010; Ouzounis et al., 2014). One reason for this is the promotion of stomatal opening by blue light, thereby allowing for an increased CO_2 uptake (Hogewoning et al, 2010; Nanya et al, 2012; Savvides et al, 2012; van Ieperen et al, 2012). Blue light also regulates plant metabolism and morphology. In cucumber, it enhances photosynthetic potential and transcription of genes required for CO_2 fixation. In rice, it causes higher stomatal conductance and photosynthetic rate owing to an increased content in chlorophyll and rubisco (Matsuda et al., 2004; Wang et al., 2009). However, not all plant species respond equally to the addition of blue light. While roses and chrysanthemums benefit greatly from blue light, campanulas show little increase in their stomatal conductance upon supplementation. Light quality also influences stomatal development, being their number and density increased upon supplementation with blue light in Whitania somnifera L. and decreased by UV-B treatment in soybean (Gitz et al., 2005; Lee et al., 2007; Ouzounis et al., 2014). In an experiment with pepper and spinach plants in which intra-leaf gradients of photoinhibition under different light wavelengths were measured, the severity of photoinhibition was in the descending order of blue, red and green near the adaxial surface and blue, green and red in the deeper tissue, which correlates with the chlorophyll and Mn absorption spectrums, respectively (Oguchi et al., 2011). Additionally, blue light increases plant transpiration and is not as efficient as red light in promoting rooting, which can be detrimental in plant propagation. Rooting has been shown to be improved by 100% red light in grapevine and Protea cynarodies. While this wavelength promotes the formation of the phenolic compound 3,4-dihydroxybenzoic acid at concentrations favorable to rooting, excessive amount of blue light can cause them to achieve inhibiting levels (Wu, 2006; Wu et al., 2007; Poudel et al., 2008; Wu and Lin, 2012). Nevertheless, the best rooting performance has been achieved with red and blue light mixtures in species such as banana, cotton, strawberry or gentian, varying the blue:red rations among species, developmental stages and background light conditions (Nhut et al., 2002, 2003; Moon et al., 2006; Li et al., 2010).

Green light can penetrate deeper into the leaf and plant canopy and therefore improve photosynthetic efficiency upon supplementation when compared to plants grown with only red and blue LEDs. In strawberries cultivated under polypropylene films, the green cover induced increased leaf area, root to shoot and leaf weight ratios, as well as specific leaf area, while it reduced harvest index and absolute and relative growth rates (Terashima et al., 2009; Casierra-Posada et al., 2012; Lu et al., 2012). Fresh and dry weight biomass were increased in lettuces grown under a mixture of red, blue and green light. However, high amounts (>50%) of green light reduced plant yield and altered stem and leaf elongation. Supplementation with green, yellow and orange light reduced growth rate in tomato plants. A possible explanation for the deleterious effects of high green light irradiance could be its negative impact in stomatal opening (Kim et al., 2004a, 2004b; Li and Kubota, 2009; Brazaityte et al., 2010).

Far-red light can have both positive and negative effects on crop production. While it increases leaf area in lettuce, thereby allowing greater light capture and growth rate, and promotes flowering in long-day ornamental species (Runkle and Heins, 2001; Li and Kubota, 2009; Stutte et al., 2009), it also inhibits germination in lettuce, causes stretching or bolting when supplemented at later stages of plant growth and reduces chlorophyll accumulation (Shinomura et al., 1996; Li and Kubota, 2009). Plant morphology can be manipulated through end-of-day (EOD) treatments with far-red or red light. EOD far-red causes plants to grow taller, while EOD red increases compactness (Kubota et al., 2012).

Accumulation of phytochemical substances

Because red and blue light are the wavelengths most efficiently driving photosynthesis and the latter promotes protein biosynthesis while preventing its degradation, these radiations are the most effective in enhancing the accumulation of soluble sugars and proteins, respectively, although the response varies among species and cultivars (Pan and Chen, 1991; Li an Pan, 1994; Bian et al., 2013). Red light was shown to increase soluble sugar accumulation in cucumber, tomato, radish, pea and strawberry (Cui et al., 2009; Zhang et al., 2009, 2010a; Liu et al., 2015), while blue light was found to increase the content of soluble sugars in tomato seedlings and the protein content in this species, radish, eggplant and strawberry, both alone and in combination with red light. In strawberry, red light also promoted the accumulation of vitamin

C, while blue light increased titrable acid. A red/blue/yellow ratio of 7/2/1 maximized the sugar-acid ratio, leave pigment content and fruit production (Chang et al., 2010; Zhang et al., 2010b; Li et al., 2015; Liu et al., 2015). A mixture of red, blue and white light was the most effective in driving the accumulation of soluble sugars in lettuce, while only green light was required in beet (Lin et al., 2013; Stagnari et al., 2014).

The biosynthesis of pigments is also influenced by light quality. Chlorophyll levels are enhanced by red and blue light, while high exposure to far-red results in lower concentrations of this pigment (Tripathy and Brown, 1995; Miyashita et al., 1997; Tanaka et al., 1998; Huq et al., 2004; Kim et al., 2004a; Moon et al., 2006). Positive effects of blue light irradiation on carotenoid accumulation have been demonstrated in spinach (Ohashi-Kaneko et al., 2007; Li et al., 2009), while red light improved their concentration in cucumber and pea seedlings (Wu et al., 2007; Cui et al., 2009). In kale, red and blue light promoted the accumulation of lutein and b-carotene, respectively (Lefsrud et al., 2008). In tomato, UV-B exposure before harvest increases b-carotene accumulation, while it reduces the content of this pigment and lutein in post-harvest ripe fruits (Becatti et al., 2009). Anthocyanin synthesis is typically associated with blue light through cryptochromes, as illustrated by the enhanced accumulation of anthocyanin in tomato fruits and eggplant peels exposed to blue light. In the Brassica oleracea cultivar "Red Rookie", anthocyanin content was increased by irradiation with red light at 50 mmol·m^{-2}·s^{-1} (Cashmore, 1999; Giliberto et al., 2005; Mizuno et al., 2009; Li et al., 2015). UV-A radiation also improves anthocyanin accumulation, showing plants grown under UV transparent spectral filters higher pigment content than those cultivated under UV-blocking films (Paul, 2006; Park et al., 2007; Tsormpatsidis et al., 2008; Li and Kubota, 2009) In contrast, far-red light has been proven to negatively affect anthocyanin, chlorophyll, xanthophyll and b-carotene content. In lettuce irradiated with different wavelengths of supplemental LED light, anthocyanin concentration was in the order 505>455>470>535 nm, demonstrating the reverting effect of green light on blue light-induced anthocyanin accumulation. Similarly, green light reduced total pigment, phenolic and antioxidant concentration in beet (Yanovski et al., 1998; Samuoliene et al., 2012; Zhang and Folta, 2012; Stagnari et al., 2014). In the Caryophyllales order, excluding the Caryophyllaceae and

Molluginaceae families, betacyanins replace anthocyanins. Unlike these, betacyanins do not increase their concentration upon irradiation with blue and UV-A light, but with red light, in a process thought to be controlled by phytochromes (Elliot, 1979; Sakuta, 2014). UV-A radiation shows the highest efficiency in optimizing the content of flavonoids, partly because these compounds are synthesized to protect plants from photodamage. In eggplant peels, the content in flavonoids and phenolic compounds is however increased under red light (Solovchenko and Schmitz-Eiberger, 2003; Schreiner et al., 2012; Li et al., 2015).

The flavor and aroma of crops can also be altered by manipulating the light quality. UV-B exposure increases oil and volatile content in numerous herb species such as sage, basil, lemon balm, lemon catmint, sweet flag (Acorus calamus L), Japanese mint (Mentha arvensis L. var piperascens) and Cynbopogon citratus (Kumari et al., 2009, 2010; Hikosaka et al., 2010; Bertoli et al., 2013; Manukyan, 2013). In basil, blue light was found to increase the leaf content of essential oils. Other radiations had little effect on them, but green light promoted crop biomass. However, the levels of rosmaric acid rose from 3 to 6 mg/l when 100% blue light treatment was replaced by 100% red, which could be due to an enhanced production of secondary metabolites as a result of the increased photosynthetic rate. In turn, aroma and phenolic compounds were mostly produced under green and yellow light. Combinations of blue/red/yellow or blue/red/green light favored the production of monoterpenoid volatiles, while blue/red/far-red treatments increased the accumulation of sesquiterpenoids (Loughrin and Kasperbauer, 2001, 2003; Shiga et al., 2009; Amaki et al., 2011; Manukyan, 2013; Carvalho et al., 2016). In petunia, red and far-red post-harvest treatments increase several volatile compounds involved in flower scent (Colquhoun et al., 2013).

Finally, light quality also alters the accumulation of substances that can be harmful for human consumption. Nitrate is not toxic per se, but it can be converted into nitrite in the human body, which causes metahaemoglobinaemia, production of nitrosamines and certain forms of cancer (Gangolli et al., 1994; Cassens, 1997; Chan, 2011). Red light is more efficient than blue at reducing nitrate content in plants by stimulating the activity of nitrate reductase. However, further studies in spinach and lettuce showed high efficiency of both radiations in preventing nitrate buildup (Lillo and Appenroth, 2001; Maevskaya and Bukhov, 2005;

Ohashi-Kaneko et al., 2006, 2007; Qi et al., 2007). This phenomenon could be explained by the enhanced plant growth and carbohydrate levels provided by these wavelengths, which would stimulate nitrogen metabolism (Champigny, 1995; Yorio et al., 2001; Matsuda et al., 2004). Oxalic acid salts are widely distributed among crop plants, especially in spinach. Soluble oxalates are of great concern to human health owing to their capacity to form calcium oxalate urinary stones and their negative effect on nutritional iron and calcium uptake (Santamaria et al., 1999; Radek and Savage, 2008). In plants the main precursor of oxalic acid is ascorbate. Given its role in photoprotection, its concentration increases upon UV irradiation. Blue light, as well as a combination of red and blue, demonstrated to increase ascorbate content in lettuce, spinach and komatsuna (Grace and Logan, 1996; Logan et al., 1996; Ohashi-Kaneko et al., 2007; Chen et al., 2011; Preswood et al., 2012). Additionally, a study by Qi and collaborators (2007) demonstrated that white and yellow light promoted the highest accumulation of oxalate in spinach leaf blades, while red and blue wavelengths caused the lowest content.

REFERENCES

1. Ahmad, M., Cashmore, A.R. (1993) HY4 gene of A. thaliana encodes a protein with characteristics of a blue-light photoreceptor. Nature. 366, 162- 166
2. Amaki, W., Yamazaki, N., Ichimura, M., Watanabe, H. (2011) Effects of light quality on the growth and essential oil content in sweet basil. Acta Horticulturae (ISHS). 907, 91-94
3. Anders, K., von Stetten, D., Mailliet, J., Kiontke, S., Sineshchekov, V.A., Hildebrandt, P., Hughes, J., Essen, L.O. (2011) Spectroscopic and photochemical characterization of the red-light sensitive photosensory module of Cph2 from Synechocystis PCC 6803. Photochem Photobiol. 87, 160-173
4. Atwell, B., Kriedemann, P., Turnbull, C., (1999) Plants in Action – Adaptation in Nature, Performance in Cultivation. Macmillan Education Australia Pty Ltd
5. Aukerman, M.J., Hirschfeld, M., Wester, L., Weaver, M., Clack, T., Amasino, R.M., Sharrock, R.A. (1997) A deletion in the PHYD

gene of the Arabidopsis Wassilewskija ecotype defines a role for phytochrome D in red/far-red light sensing. Plant Cell. 9, 1317-1326

6. Bae, G., Choi, G. (2008) Decoding of light signals by plant phytochromes and their interacting proteins. Annu Rev Plant Biol. 59, 281-311
7. Ballaré, C.L., Barnes, P.W., Flint, S.D. (1995) Inhibition of hypocotyl elongation by ultraviolet-B radiation in de-etiolating tomato seedlings. 1. The photoreceptor. Physiol. Plant. 93, 584-592
8. Banerjee, R., Schleicher, E., Meier, S., Viana, R.M., Pokorny, R., Ahmad, M. Bittl, R., Batschauer, A. (2007) The signaling state of Arabidopsis cryptochrome 2 contains flavin semiquinone. J. Biol. Chem. 282, 14916-14922
9. Barnes, C., Bugbee, B. (1991) Morphological responses of wheat to changes in phytochrome photoequilibnum. Plant Physiol. 97, 359-65
10. Barnes, C., Tibbitts, T., Sager, J., Deitzer, G., Bubenheim, D., Koerner, G., Bugbee, B. (1993) Accuracy of quantum sensors measuring yield photon flux and photosynthetic photon flux. Hort-Science. 28(12), 1197-1200
11. Barta, D.J., Tibbitts, T.W., Bula, R.J., Morrow, R.C. (1992) Evaluation of light emitting diode characteristics for a space-based plant irradiation source. Space. 12, 141-149
12. Batschauer, A. (1999) Light perception in higher plants. Cell Mol Life Sci. 55(2), 153-166
13. Baudry, A., Ito, S., Song, Y.H., Strait, A.A., Kiba, T., Lu, S., Henriques, R., Pruneda-Paz, J.L., Chua, N.H., Tobin, E.M., Kay, S.A., Imaizumi, T. (2010) F-box proteins FKF1 and LKP2 act in concert with ZEITLUPE to control Arabidopsis clock progression. Plant Cell. 22, 606-622
14. Becatti, E., Petroni, K., Giuntini, D., Castagna, A., Calvenzani, V., Serra, G., Mensuali-Sodi, A., Tonelli, C., Ranieri, A. (2009) Solar UV-B radiation influences carotenoid accumulation of tomato fruit through both ethylene-dependent and-independent mechanisms. J Agric Food Chem. 57, 10979-10989
15. Beggs, C., Wellman, E. (1985). Analysis of light-controlled anthocyanin formation in coleoptiles of Zea mays L.: The role of

UV-B, blue, red and far-red light. Photochemistry and Photobiology. 41, 481-486

16. Bertoli, A., Lucchesini, M., Menuali-Sodi, A., Leonardi, M., Doveri, S., Magnabosco, A., Pitelli, L. (2013) Aroma characterization and UV elicitation of purple basil from different plant tissue cultures. Food Chemistry. 141, 776-787
17. Bian, Z.H., Yang, Q.C., Liu, W.K. (2015) Effects of light quality on the accumulation of phytochemicals in vegetables produced in controlled environments: a review. J Sci Food Agric. 95, 869-877
18. Blaauw, G., Blaauw-Jansen, O.H. (1966) Effect of red light on irreversible and reversible expansion of Avena coleoptile sections. Planta. 71, 291-304
19. Boccalandro, H.E., Mazza, C.A., Mazzella, M.A., Casal, J.J., Ballaré, C.L. (2001). Ultraviolet B radiation enhances a phytochrome B-mediated photomorphogenic response in Arabidopsis. Plant Physiol. 126, 780-788
20. Boccalandro, H.E., Rugnone, M.L., Moreno, J.E., Ploschuk, E.L., Serna, L., Yanovsky, M.J., Casal, J.J. (2009) Phytochrome B enhances photosynthesis at the expense of water-use-efficiency in Arabidopsis. Plant Physiol. 150, 1083-1092
21. Botto, J.F., Sánchez, R.A., Whitelam, G.C., Casal, J.J. (1996) Phytochrome A mediates the promotion of seed germination by very low fluences of light and canopy shade-light in Arabidopsis. Plant Physiol. 110, 439-444
22. Brautigam, C.A., Smith, B.S., Ma, Z., Palnitkar, M., Tomchick, D.R., Machius, M., Deisenhofer, J. (2004) Structure of the photolyase-like domain of cryptochrome 1 from Arabidopsis thaliana. Proc. Natl Acad. Sci. USA. 101, 12142-12147
23. Brazaitytė, A., Duchovskis, P., Urbonavičiūtė, A., Samuolienė, G., Jankauskienė, J., Sakalauskaite, J., Šabajeviene, G., Sirtautas, R., Novičkovas, A. (2010) The effect of light-emitting diodes lighting on the growth of tomato transplants. Zemdirbyste-Agriculture. 97, 89-97
24. Brown, B.A., Cloix, C., Jiang, G.H., Kaiserli, E., Herzyk, P., Kliebenstein, D.J., Jenkins, G.I. (2005) A UV-B-specific signaling

component orchestrates plant UV protection. Proc. Natl Acad Sci USA. 102, 18225-18230

25. Brown, B.A., Jenkins, G.I. (2008) UV-B signaling pathways with different fluence-rate response profiles are distinguished in mature Arabidopsis leaf tissue by requirement for UVR8, HY5, and HYH. Plant Physiol. 146, 576-588
26. Brown, C.S., Schuerger, A.C., Sager, J.C. (1995) Growth and photomorphogenesis of pepper plants under red light-emitting diodes with supplemental blue or far-red lighting. J Am Soc Hortic Sci. 120, 808-813
27. Butler, W.L., Norris, K.H., Siegelman, H.W., Hendricks, S.B. (1959) Detection, assay, and preliminary purification of the pigment controlling photoresponsive development in plants. Proc Natl Acad Sci USA. 45, 1703-1708
28. Bula, R.J., Morrow, R.C., Tibbitts, T.W., Barta, D.J., Ignatius, R.W., Martin, T.S. (1991) Light-emitting diodes as a Radiation Source for Plants. Hort Science. 26, 203-205
29. Carvalho, S.D., Schwieterman, M.L., Abrahan, C.E., Colquhoun, T.A., Folta, K.M. (2016) Light Quality Dependent Changes in Morphology, Antioxidant Capacity, and Volatile Production in Sweet Basil (Ocimum basilicum). Front Plant Sci. 7, 1328
30. Casal, J.J., Sánchez, R.A., Yanovsky, M. J. (1997) The function of phytochrome A. Plant Cell Environ. 20, 813-819
31. Cashmore, A.R. (1999) Cryptochromes: Blue light receptors for plants and animals. Science. 284, 760-765
32. Cashmore, A.R. (2003) Cryptochromes: enabling plants and animals to determine circadian time. Cell. 114, 537-543
33. Casierra-Posada, F., Peña-Olmos, J.E., Ulrichs, C. (2012) Basic growth analysis in strawberry plants (Fragaria sp.) exposed to different radiation environments. Agronomía Colombiana. 30, 25-33
34. Cassens, R.G. (1997) Residual nitrite in cured meat. Food Technol. 51, 53-55
35. Chalker-Scott, L. (1999) Environmental significance of anthocyanins in plant stress responses. Photochemistry and Photobiology. 70, 1-9

36. Champigny, M.L. (1995) Integration of photosynthetic carbon and nitrogen metabolism in higher plants. Photosynth Res. 46, 117-127
37. Chan, T.Y. (2011) Vegetable-borne nitrate and nitrite and the risk of methaemoglobinaemia. Toxicol Lett. 200, 107-108
38. Chang, T.T., Liu, X.Y., Xu, Z.G., Yang, Y. (2010) Effects of light spectral energy distribution on growth and development of tomato seedlings. Sci Agric Sin. 8, 27-35
39. Chang, H., Wang, Y.P., Wang, D., Zhang, F. (2009) Effects of light quality on microtuber induction of Solanum tuberosum L. Ying Yong Sheng Tai Xue Bao. 20(8), 1891-1895
40. Chaves, I., Pokorny, R., Byrdin, M., Hoang, N., Ritz, T., Brettel, K., Essen, L.O., van der Horst, G.T., Batschauer, A., Ahmad, M. (2011) The cryptochromes: blue light photoreceptors in plants and animals. Annu Rev Plant Biol. 62, 335-364
41. Chen, W.H., Xu, Z.G., Liu, X.Y., Yang, Y.,Wang, Z.H.M., Song, F.F. (2011) Effect of LED light source on the growth and quality of different lettuce varieties. Acta Bot Boreal-Occident Sin. 31, 1434-1440
42. Cherry, J.R., Hondred, D., Walker, J.M., Vierstra, R.D. (1992) Phytochrome requires the 6-kDa N-terminal domain for full biological activity. Proc Natl Acad Sci USA. 89, 5039-5043
43. Cho, H.Y., Tseng, T.S., Kaiserli, E., Sullivan, S., Christie, J.M., Briggs, W.R. (2007) Physiological roles of the light, oxygen, or voltage domains of phototropin 1 and phototropin 2 in Arabidopsis. Plant Physiol. 143, 517-529
44. Christie, J.M., Blackwood, L., Petersen, J., Sullivan, S. (2015) Plant flavoprotein photoreceptors. Plant Cell Physiol. 56, 401-4113
45. Christie, J.M., Gawthorne, J., Young, G., Fraser, N.J., Roe, A.J. (2012a) LOV to BLUF: flavoprotein contributions to the optogenetic toolkit. Mol. Plant. 5, 533-544
46. Christie, J.M., Arvai, A.S., Baxter, K.J., Heilmann, M., Pratt, A.J., O'Hara, A., Kelly, S.M., Hothorn, M., Smith, B.O., Hitomi, K., Jenkins, G.I., Getzoff, E.D. (2012b) Plant UVR8 photoreceptor senses UV-B by tryptophan-mediated disruption of cross-dimer salt bridges. Science. 335, 1492-1496

47. Christie, J.M., Salomon, M., Nozue, K., Wada, M., Briggs, W.R. (1999) LOV (light, oxygen, or voltage) domains of the blue-light photoreceptor phototropin (nph1): binding sites for the chromophore flavin mononucleotide. Proc. Natl Acad. Sci. USA. 96, 8779-8783

48. Christie, J.M., Swartz, T.E., Bogomolni, R.A., Briggs, W.R. (2002) Phototropin LOV domains exhibit distinct roles in regulating photoreceptor function. Plant J. 32, 205-219

49. Clack, T., Shokry, A., Moffet, M., Liu, P., Faul, M., Sharrock, R.A. (2009) Obligate heterodimerization of Arabidopsis phytochromes C and E and interaction with the PIF3 basic helix-loop-helix transcription factor. Plant Cell. 21, 786-799

50. Cloix, C., Kaiserli, K., Heilmann, M., Baxter, K.J., Brown, B.A., O'Hara, A., Smith, B.O., Christie, J.M., Jenkins, G.I. (2012) The C-terminal region of the UV-B photoreceptor UVR8 initiates signaling through interaction with COP1. Proc Natl Acad Sci USA. 109, 16366-16370

51. Clough, R.C., Vierstra, R.D. (1997) Phytochrome degradation. Plant, Cell and Environment. 20, 713-721

52. Conte, M., de Simone, S., Simmons, S.J., Ballaré, C.L., Stapleton, A.E. (2010) Chromosomal loci important for cotyledon opening under UV-B in Arabidopsis thaliana. BMC Plant Biol. 10, 112

53. Cosgrove, D. J., Green, P.B. (1981): Rapid suppression of growth by blue light. Biophysical mechanism of action. Plant Physiol. 68, 1447-1453

54. Colquhoun, T.A., Schwieterman, M.L., Gilbert, J.L., Jaworski, E.A., Langer, K.M., Jones, C.R., Rushing, G.V., Hunter, T.M., Olmstead, J., Clark, D.G., Folta, K.M. (2007) Light modulation of volatile organic compounds from petunia flowers and select fruits. Postharvest Biology and Technology. 86, 37-44

55. Cui, J., Ma, Z.H., Xu, Z.G., Zgang, H., Chang, T.T., Liu, H.J. (2009) Effects of supplemental lighting with different light qualities on growth and physiological characteristics of cucumber, pepper and tomato seedlings. Acta Hortic Sin. 5, 663-670

56. Davey, M.P., Susanti, N.I., Wargent, J.J., Findlay, J.E., Paul Quick, W., Paul, N.D., Jenkins, G.I. (2012). The UV-B photoreceptor UVR8 promotes photosynthetic efficiency in Arabidopsis thaliana

exposed to elevated levels of UV-B. Photosynth. Res. 114, 121-131

57. Demarsy, E., Schepens, I., Okajima, K., Hersch, M., Bergmann, S., Christie, J., Shimazaki, K., Tokutomi, S., Fankhauser, C. (2012) Phytochrome Kinase Substrate 4 is phosphorylated by the phototropin 1 photoreceptor. EMBO J. 31, 3457-3467
58. Devlin, P.F., Halliday, K.J., Harberd, N.P., Whitelam, G.C. (1996) The rosette habit of Arabidopsis thaliana is dependent upon phytochrome action: novel phytochromes control internode elongation and flowering time. Plant J. 10, 1127-1134
59. Devlin, P.F., Kay, S.A. (2000) Cryptochromes are required for phytochrome signaling to the circadian clock but not for rhythmicity. Plant Cell. 12, 2499-2509
60. Devlin, P.F., Patel, S.R., Whitelam, G.C. (1998) Phytochrome E influences internode elongation and flowering time in Arabidopsis. Plant Cell. 10, 1479-1487
61. Elliott, D.C. (1979) Temperature-sensitive responses of red light-dependent Betacyanin Synthesis. Plant Physiol. 64, 521-524
62. Eltbaakh, Y.A., Ruslan, M.H., Alghoul, M.A., Othman, M.Y., Sopian, K., Fadhel, M.I. (2011) Measurement of total and spectral solar irradiance: Overview of existing research. Renewable and Sustainable Energy Reviews. 15, 1403-1426
63. Eskins, K., (1992) Light-quality effects on Arabidopsis development. Red, blue and far-red regulation of flowering and morphology. Physiologia Plantarum. 86, 439-444
64. Essen, L.O., Mailliet, J., Hughes, J. (2008). The structure of a complete phytochrome sensory module in the Pr ground state. Proc Natl Acad Sci USA. 105, 14709-14714
65. Favory, J.J., Stec, A., Gruber, H., Rizzini, L., Oravecz, A., Funk, M., Albert, A., Cloix, C., Jenkins, G.I., Oakeley, E.J., Seidlitz, H.K., Nagy, F., Ulm, R. (2009). Interaction of COP1 and UVR8 regulates UV-B-induced photomorphogenesis and stress acclimation in Arabidopsis. EMBO J. 28, 591-601
66. Folta, K.M., Kaufman, L.S. (2003) Phototropin 1 is required for highfluence blue-light-mediated mRNA destabilization. Plant Mol. Biol. 51, 609-618

67. Folta, K.M., Lieg, E.J., Durham, T., Spalding, E.P. (2003) Primary inhibition of hypocotyl growth and phototropism depend differently on phototropin-mediated increases in cytoplasmic calcium induced by blue light. Plant Physiol. 133, 1464-1470

68. Fornara, F., Panigrahi, K.C., Gissot, L., Sauerbrunn, N., Ruhl, M., Jarillo, J.A., Coupland, G. (2009) Arabidopsis DOF transcription factors act redundantly to reduce CONSTANS expression and are essential for a photoperiodic flowering response. Dev. Cell. 17, 75-86

69. Franklin, K.A., Davis, S.J., Stoddart, W.M., Vierstra, R.D., Whitelam, G.C. (2003a) Mutant analyses define multiple roles for phytochrome C in Arabidopsis thaliana photomorphogenesis. Plant Cell. 15, 1981-1989

70. Franklin, K.A., Praekelt, U., Stoddart, W.M., Billingham, O.E., Halliday, K.J., Whitelam, G.C. (2003b) Phytochromes B, D and E act redundantly to control multiple physiological responses in Arabidopsis. Plant Physiol. 131, 1340-1346

71. Franklin, K.A., Whitelam, G.C. (2007) Light-quality regulation of freezing tolerance in Arabidopsis thaliana. Nature Genetics. 39, 1410-1413

72. Fujiwara, S., Wang, L., Han, L., Suh, S.S., Salome, P.A., McClung, C.R., Somers, D.E. (2008) Post-translational regulation of the Arabidopsis circadian clock through selective proteolysis and phosphorylation of pseudo-response regulator proteins. J. Biol. Chem. 283, 23073-23083

73. Galvão, V.C., Fankhauser, C. (2015) Sensing the light environment in plants: photoreceptors and early signaling steps. Curr Opin Nuerobiol. 34, 46-53

74. Gangolli, S.D., Van den Brandt, P.A., Feron, V.J., Janzowsky, C., Koeman, J.H., Speijers, G.J., Spiegelhalder, B., Walker, R., Wisnok, J.S. (1994) Nitrate, nitrite and N-nitroso compounds. Eur J Pharmacol. 292, 1-38

75. Gardner, G., Lin, C., Tobin, E.M., Loehrer, H., Brinkman, D. (2009) Photobiological properties of the inhibition of etiolated Arabidopsis seedling growth by ultraviolet-B irradiation. Plant, Cell and Environment. 32, 1573-1583

76. Giliberto, L., Perrotta, G., Pallara, P.,Weller, J.L., Fraser, P.D., Bramley, P.M., Fiore, A., Tavazza, M., Giuliano, G. (2005) Manipulation of the blue light photoreceptor cryptochrome 2 in tomato affects vegetative development, flowering time, and fruit antioxidant content. Plant Physiol. 137, 199-208

77. Gitz, D.C., Liu-Gitz, L., Britz, S.J., Sullivan, J.H. (2005) Ultraviolet-B effects on stomatal density, water-use efficiency, and stable carbon isotope discrimination in four glasshouse-grown soybean (Glyicine max) cultivars. Environmental and Experimental Botany. 53, 343-355

78. Godish, T. (2004) Air quality. CRC Press LLC

79. Goins, G.D., Yorio, N.C. Sanwo-Lewandowski, M.M., Brown, C.S. (1998) Life Cycle experiments with Arabidopsis grown under red light-emitting diodes (LEDs). Life support & biosphere science: International journal of earth science. 52, 143-149

80. Goins, G.D., Yorio, N.C., Sanwo, M.M., Brown, C.S. (1997) Photomorphogenesis, photosynthesis, and seed yield of wheat plants grown under red light-emitting diodes (LEDs) with and without supplemental blue lighting. J Exp Bot. 48, 1470-1413

81. Goosey, L., Palecanda, L., Sharrock, R.A. (1997) Differential patterns of expression of the Arabidopsis PHYB, PHYD and PHYE phytochrome genes. Plant Physiol. 115, 959-969

82. Grace, S.C., Logan, B.A. (1996) Acclimation of foliar antioxidant systems to growth irradiance in three broad-leaved evergreen species. Plant Physiol. 112, 1631-1640

83. Grüber, H., Heijde, M., Heller, W., Albert, A., Seidlitz, H.K., Ulm, R. (2010) Negative feedback regulation of UV-B-induced photomorphogenesis and stress acclimation in Arabidopsis. Proc Natl Acad Sci USA. 107, 20132-20137

84. Halliday, K.J., Koornneef, M., Whitelam, G.C. (1994) Phytochrome B, and at least one other phytochrome, mediate the accelerated £owering response of Arabidopsis thaliana L. to low red:far-red ratio. Plant Physiol. 104, 1311-1315

85. Halliday, K.J., Salter, M.G., Thingnaes, E., Whitelam, G.C. (2003) Phytochrome control of flowering is temperature sensitive and

correlates with expression of the floral integrator FT. Plant J. 33, 875-885

86. Halliday, K.J., Whitelam, G.C. (2003) Changes in photoperiod or temperature reveal roles for phyD and phyE. Plant Physiol. 131, 1913-1920
87. Han, L., Mason, M., Risseeuw, E.P, Crosby, W.L., Somers, D.E. (2004) Formation of an SCFZTL complex is required for proper regulation of circadian timing. Plant J. 40, 291-301
88. Heijde, M., Binkert, M., Yin, R., Ares-Orpel, F., Rizzini, L., Van De Slijke, E., Persiau, G., Nolf, J., Gevaert, K., De Jaeger, G., Ulm, R. (2013) Constitutively active UVR8 photoreceptor variant in Arabidopsis. Proc Natl Acad Sci USA. 110, 20326-20331
89. Heijde, M., Ulm, R. (2013) Reversion of the Arabidopsis UV-B photoreceptor UVR8 to the homodimeric ground state. Proc Natl Acad Sci USA. 110, 1113-1118
90. Hennig, L., Stoddart, W.M., Dieterle, M., Whitelam, G.C., Schäfer, E. (2002) Phytochrome E controls light-induced germination of Arabidopsis. Plant Physiol. 128, 194-200
91. Heschel, M.S., Selby, J., Butler, C., Whitelam, G.C., Sharrock, R.A., Donohue, K. (2007) A new role for phytochromes in temperaturedependent germination. New Phytologist. 174, 735-741
92. Higa, T., Suetsugu, N., Wada, M. (2014) Plant nuclear photorelocation movement. J. Exp. Bot. 65, 2873-2881
93. Hikosaka, S., Ito, K., Goto, E. (2010) Effects of Ultraviolet Light on Growth, Essential Oil Concentration, and Total Antioxidant Capacity of Japanese Mint. Environmental Control in Biology. 48, 185-190
94. Hoang, N., Bouly, J.P., Ahmad, M. (2008) Evidence of a light-sensing role for folate in Arabidopsis cryptochrome blue-light receptors. Mol. Plant. 1, 68-74
95. Hogewoning, S.W., Trouwborst, G.H.M., Poorter, H., van Ieperen, W., Harbinson, J. (2010). Blue light dose–responses of leaf photosynthesis, morphology, and chemical composition of Cucumis sativus grown under different combinations of red and blue light. J Exp Bot. 61, 3107-3117

96. Hogewoning, S.W., Wientjes, E., Douwstra, P., Trouwborst, G., van Ieperen, W., Croce R., Harbinson, J. (2012) Photosynthetic quantum yield dynamics: From photosystems to leaves. Plant Cell. 24(5), 1921-1935

97. Huang, X., Ouyang, X., Yang, P., Lau, O.S., Chen, L., Wei, N., Deng, X.W. (2013) Conversion from CUL4-based COP1–SPA E3 apparatus to UVR8–COP1–SPA complexes underlies a distinct biochemical function of COP1 under UV-B. Proc Natl Acad Sci USA. 110, 16669-16674

98. Huang, X., Ouyang, X., Yang, P., Lau, O.S., Li, G., Li, J., Chen, H., Deng, X.W. (2012) Arabidopsis FHY3 and HY5 positively mediate induction of COP1 transcription in response to photomorphogenic UV-B light. Plant Cell. 24, 4590-4606

99. Huang, X., Yang, P., Ouyang, X., Chen, L., Deng, X.W. (2014) Photoactivated UVR8–COP1 module determines photomorphogenic UV-B signaling output in Arabidopsis. PLoS Genet. 10, e1004218

100. Hughes, J. (2010) Phytochrome three-dimensional structures and functions. Biochem Soc Trans. 38, 710-716

101. Hulstron, R.L. (1989) Solar resources. Massachusetts Institute of Technology

102. Huq, E., Al-Sady, B., Hudson, M.E., Kim, C., Apel, K., Quail, P.H. (2004) PHYTOCHROME-INTERACTING FACTOR 1 is a critical bHLH regulator of chlorophyll biosynthesis. Science. 305, 1937-1941

103. van Ieperen, W., Savvides, A., Fanourakis, D. (2012) Red and blue light effects during growth on hydraulic and stomatal conductance in leaves of young cucumber plants. Acta Horticulturae. 956, 223-230

104. Imaizumi, T., Tran, H.G., Swartz, T.E., Briggs, W.R., Kay, S.A. (2003) FKF1 is essential for photoperiodic-specific light signalling in Arabidopsis. Nature. 426, 302-306

105. Imaizumi, T., Schultz, T.F., Harmon, F.G., Ho, L.A., Kay, S.A. (2005) FKF1 F-box protein mediates cyclic degradation of a repressor of CONSTANS in Arabidopsis. Science 309, 293-297

106. Indorf, M., Cordero, J., Neuhaus, G., Rodríguez-Franco, M. (2007)

Salt tolerance (STO), a stress-related protein, has a major role in light signalling. Plant J. 51, 563-574

107. Inoue, S., Kinoshita, T., Matsumoto, M., Nakayama, K.I., Doi, M. Shimazaki, K. (2008) Blue light-induced autophosphorylation of phototropin is a primary step for signaling. Proc. Natl Acad. Sci. USA. 105, 5626-5631
108. Ito, S., Song, Y.H., Imaizumi, T. (2012) LOV domain-containing F-box proteins: light-dependent protein degradation modules in Arabidopsis. Mol Plant. 5, 573-582
109. Jeong, R.D., Chandra-Shekara, A.C., Barman, S.R., Navarre, D., Klessig, D.F., Kachroo, A., Kachroo, P. (2010) Cryptochrome 2 and phototropin 2 regulate resistance protein-mediated viral defense by negatively regulating an E3 ubiquitin ligase. Proc. Natl Acad. Sci. USA. 107, 13538-13543
110. Jeong, S.W., Park, S., Jin, J.S., Seo, O.N., Kim, G.-S., Kim, Y.-H., Bae, H., Lee, G., Kim, S.T., Lee, W.S., Shin, S.C. (2012) Influences of four different light-emitting diode lights on flowering and polyphenol variations in the leaves of Chrysanthemum (Chrysanthemum morifolium). Journal of Agricultural and Food Chemistry. 60, 9793-9800
111. Jiang, L., Wang, Y., Li, Q.F., Bjorn, L.O., He, J.X., Li, S.S. (2012) Arabidopsis STO/BBX24 negatively regulates UV-B signaling by interacting with COP1 and repressing HY5 transcriptional activity. Cell Res. 22, 1046-1057
112. Jiang, L., Wang, Y., Bjorn, L.O., Li, S. (2009) Arabidopsis RADICAL-INDUCED CELL DEATH1 is involved in UV-B signaling. Photochem Photobiol Sci. 8, 838-846
113. Johnson, E., Bradley, J.M., Harberd, N.P., Whitelam, G.C. (1994) Photoresponses of light-grown phyA mutants of Arabidopsis: phytochrome A is required for the perception of daylength extensions. Plant Physiol. 105, 141-149
114. Johnson, C.F., Brown, C.S., Wheeler, R.M., Sager, J.C., Chapman, D.K., Deitzer. G.F. (1996) Infrared light-emitting diode radiation causes gravitropic and morphological effects in dark-grown oat seedlings. Photochem. Photobiol. 63, 238-242
115. Jones, M.A., Feeney, K.A., Kelly, S.M., Christie, J.M. (2007)

Mutational analysis of phototropin 1 provides insights into the mechanism underlying LOV2 signal transmission. J. Biol. Chem. 282, 6405-6414

116. Jose, A.M. (1977) Photoreception and photoresponses in the radish hypocotyl. Planta. 136, 125-129

117. Kagawa, T., Sakai, T., Suetsugu, N., Oikawa, K., Ishiguro, S., Kato, T., Tabata, S., Okada, K., Wada, M. (2001) Arabidopsis NPL1: a phototropin homolog controlling the chloroplast high-light avoidance response. Science. 291, 2138-2141

118. Kami, C., Allenbach, L., Zourelidou, M., Ljung, K., Schutz, F., Isono, E., Watahiki, M.K., Yamamoto, K.T., Schwechheimer, C., Fankhauser, C. (2014) Reduced phototropism in pks mutants may be due to altered auxin-regulated gene expression or reduced lateral auxin transport. Plant J. 77, 393-403

119. Kami, C., Lorrain, S., Hornitschek, P., Fankhauser, C. (2010) Light-regulated plant growth and development. Curr Top Dev Biol. 91, 29-66

120. Kennedy, M.J., Hughes, R.M., Peteya, L.A., Schwartz, J.W., Ehlers, M.D., Tucker, C.L. (2010) Rapid blue-light-mediated induction of protein interactions in living cells. Nat. Methods 7, 973-975

121. Kevei, É., Gyula, P., Hall, A., Kozma-Bognár, L., Kim, W.Y., Eriksson, M.E., Tóth, R., Hanano, S., Fehér, B., Southern, M.M., Bastow, R.M., Viczián, A., Hibberd, V., Davis, S.J., Somers, D.E., Nagy, F., Millar, A.J. (2006) Forward genetic analysis of the circadian clock separates the multiple functions of ZEITLUPE. Plant Physiol. 140, 933-945

122. Kiba, T., Henriques, R., Sakakibara, H., Chua, N.H. (2007) Targeted degradation of PSEUDO-RESPONSE REGULATOR5 by an SCFZTL complex regulates clock function and photomorphogenesis in Arabidopsis thaliana. Plant Cell. 19, 2516-2530

123. Kim, J., Geng, R., Gallenstein, R.A., Somers, D.E. (2013) The F-box protein ZEITLUPE controls stability and nucleocytoplasmic partitioning of GIGANTEA. Development 140, 4060-4069

124. Kim, H.H., Goins, G.D., Wheeler, R.M., Sager, J.C. (2004a) Green light supplementation for enhanced lettuce growth under red and blue light emitting diodes. HortScience. 39, 1617-1622

125. Kim, H.H., Goins, G.D., Wheeler, R.M., Sager, J.C. (2004b) Stomatal conductance of lettuce grown under or exposed to different light qualities. Annals of Botany. 94, 691-697

126. Kim, W.Y., Fujiwara, S., Suh, S.S., Kim, J., Kim, Y., Han, L., David, K., Putterill, J., Nam, H.G., Somers, D.E. (2007) ZEITLUPE is a circadian photoreceptor stabilized by GIGANTEA in blue light. Nature. 449, 356-360

127. Kim, B.C., Tennessen, D.J., Last, R.L. (1998). UV-B-induced photomorphogenesis in Arabidopsis thaliana. Plant J. 15, 667-674

128. Kinoshita, T., Doi, M., Suetsugu, N., Kagawa, T., Wada, M., Shimazaki, K. (2001) Phot1 and phot2 mediate blue light regulation of stomatal opening. Nature. 414, 656-660

129. Kinoshita, T., Shimazaki, K. (1999) Blue light activates the plasma membrane H(+)-ATPase by phosphorylation of the C-terminus in stomatal guard cells. EMBO J. 18, 5548-5558

130. Kiyosue, T., Wada, M. (2000) LKP1 (LOV kelch protein 1): a factor involved in the regulation of flowering time in Arabidopsis. Plant J. 23, 807-815

131. Klein, R.M. (1978) Plants and near-ultraviolet radiation. Bot. Rev. 44, 1-127

132. Kleine, T., Lockhart, P. Batschauer, A. (2003) An Arabidopsis protein closely related to Synechocystis cryptochrome is targeted to organelles. Plant J. 35, 93-103

133. Kleiner, O., Kircher, S., Harter, K., Batschauer, A. (1999) Nuclear localization of the Arabidopsis blue light receptor cryptochrome 2. Plant J. 19, 289-296

134. Kliebenstein, D.J., Lim, J.E., Landry, L.G., Last, R.L. (2002) Arabidopsis UVR8 regulates ultraviolet-B signal transduction and tolerance and contains sequence similarity to human regulator of chromatin condensation 1. Plant Physiol. 130, 234-243

135. Kong, S.G., Suzuki, T., Tamura, K., Mochizuki, N., Hara-Nishimura, I., Nagatani, A. (2006) Blue light-induced association of phototropin 2 with the Golgi apparatus. Plant J. 45, 994-1005

136. Koornneef, M., Rolff, E., Spruit, C.J.P. (1980) Genetic control of light-inhibited hypocotyl elongation in Arabidopsis thaliana L.

Heynh. Z. Pflanzenphysiol.. 100, 147-160

137. Kostman, T.A., Tarlyn, N.M., Loewus, F.A., Franceschi, V.R. (2001) Biosynthesis of L-ascorbic acid and conversion of carbons 1 and 2 of L-ascorbic acid to oxalic acid occurs within individual calcium oxalate crystal idioblasts. Plant Physiol. 125, 634-640
138. Kubota, C., Chia, P., Yang, Z., Li, Q. (2012) Applications of far-red light emitting diodes in plant production under controlled environments. Acta Horticulturae (ISHS) 952, 59-66
139. Kumari, R., Agrawal, S.B. (2010) Supplemental UV-B induced changes in leaf morphology, physiology and secondary metabolites of an Indian aromatic plant Cymbopogon citratus (D.C.) Staph under natural field conditions. International Journal of Environmental Studies 67, 655-675
140. Kumari, R., Agrawal, S.B., Singh, S., Dubey, N.K. (2009) Supplemental ultraviolet-B induced changes in essential oil composition and total phenolics of Acorus calamus L. (sweet flag). Ecotoxicology and Environmental Safety. 72, 2013-2019
141. Langenbacher, T., Immeln, D., Dick, B., Kottke, T. (2009) Microsecond light-induced proton transfer to flavin in the blue light sensor plant cryptochrome. J. Amer. Chem. Soc. 131, 14274-14280
142. Lau, O.S., Deng, X.W. (2012) The photomorphogenic repressors COP1 and DET1: 20 years later. Trends Plant Sci. 17, 584-593
143. Lee, S.-H., Tewari, R.K., Hahn, E.-J., Paek, K.-Y. (2007) Photon flux density and light quality induces changes in growth, stomatal development, photosynthesis and transpiration of Withania somnifera (L.) Dunal. plantlets. Plant Cell, Tissue and Organ Culture. 90, 141
144. Lefsrud, M.G., Kopsell, D.A., Sams, C.E. (2008) Irradiance from distinct wavelength light-emitting diodes affect secondary metabolites in Kale. Hortscience. 43, 2243-2244
145. Li, Y.H., Chen, L., Gao, R.G., Yang, F.J., Wang, X.F., Wei, M., Shi, Q.H., Mi, Q.H. (2015) Effects of LED qualities on quality and antioxidation capacity of eggplant fruits. Ying Yong Sheng Tai Xue Bao. 26, 2728-2734
146. Li, J., Hikosaka, S., Goto, E. (2009) Effects of light quality and photosynthetic photon flux on growth and carotenoid pigments

in spinach (Spinacia oleracea L.), in VI International Symposium on Light in Horticulture 907, ed. by Goto E, Hikosaka S. ISHS, Belgium, pp. 105-110

147. Li, Q., Kubota, C. (2009) Effects of supplemental light quality on growth and phytochemicals of baby leaf lettuce. Environmental Experimental Botany. 67, 59-64
148. Li, S.S., Pan, R.C. (1994) Effect of blue light on the metabolism of carbohydrate andprotein in rice (Oryza sativa L.) seedlings. Acta Phytophysiol Sin. 21, 22–28
149. Li, H., Xu, Z., Tang, C. (2010) Effect of light-emitting diodes on growth and morphogenesis of upland cotton (Gossypium hirsutum L.) plantlets in vitro. Plant Cell Tissue Organ Culture 103, 155-163
150. Lian, H.L., He, S.B., Zhang, Y.C., Zhu, D.M., Zhang, J.Y., Jia, K.P., Sun, S.X., Li, L., Yang, H.Q. (2011) Blue-light-dependent interaction of cryptochrome 1 with SPA1 defines a dynamic signaling mechanism. Genes Dev. 25, 1023-1028
151. Lillo, C., Appenroth, K.J. (2001) Light regulation of nitrate reductase in higher plants: which photoreceptors are involved? Plant Biol. 3, 455-465
152. Lin, K.H., Huang, M.Y., Huang, W.D., Hsu, M.H., Yang, Z.W., Yang, C.M. (2013) The effects of red, blue, and white light-emitting diodes on the growth, development, and edible quality of hydroponically grown lettuce (Lactuca sativa L. var. capitata). Sci Hort. 150, 86-91
153. Lin, C., Robertson, D.E., Ahmad, M., Raibekas, A.A., Jorns, M.S., Dutton, P.L., Cashmore, A.R. (1995) Association of flavin adenine dinucleotide with the Arabidopsis blue light receptor CRY1. Science. 269, 968-970
154. Lin, C., Shalitin, D. (2003) Cryptochrome structure and signal transduction. Annu Rev Plant Biol 54, 469-496
155. Liu, Y., Li, X., Li, K., Liu, H., Lin, C. (2013) Multiple bHLH proteins form heterodimers to mediate CRY2-dependent regulation of flowering-time in Arabidopsis. PLoS Genet. 9, e1003861
156. Liu, Q., Lian, H.F., Liu, S.Q., Sun, Y.L., Yu, X.H., Guo, H.P. (2015) Effects of different LED light qualities on photosynthetic characteristics, fruit production and quality of strawberry. Ying Yong Sheng Tai Xue Bao. 26, 1743-1750

157. Liu, H., Liu, B., Zhao, C., Pepper, M., Lin, C. (2011) The action mechanisms of plant cryptochromes. Trends Plant Sci. 16, 684-691

158. Liu, H., Yu, X., Li, K., Klejnot, J., Yang, H., Lisiero, D., Lin, C. (2008) Photoexcited CRY2 interacts with CIB1 to regulate transcription and floral initiation in Arabidopsis. Science. 322, 1535-1539

159. Logan, B., Barker, D., Demmig-Adams, B., Adams, W. (1996) Acclimation of leaf carotenoid composition and ascorbate levels to gradients in the light environment within an Australian rainforest. Plant Cell Environ. 19, 1083-1090

160. Loughrin, J.H., Kasperbauer, M.J. (2001) Light reflected from colored mulches affects aroma and phenol content of sweet basil (Ocimum basilicum L.) leaves. J Agric Food Chem. 49, 1331-1335

161. Loughrin, J.H., Kasperbauer, M.J. (2003) Aroma content of fresh basil (Ocimum basilicum L.) leaves is affected by light reflected from colored mulches. J Agric Food Chem. 51, 2272-2276

162. Lu, N., Maruo, T., Johkan, M., Hohjo, M., Tsukagoshi, S., Ito, Y., Ichimura, T., Shinohara, Y. (2012) Effects of supplemental lighting with light-emitting diodes (LEDs) on tomato yield and quality of single-truss tomato plants grown at high planting density. Environmental Control in Biology. 50, 63-74

163. Ma, L., Li, J., Qu, L., Hager, J., Chen, Z., Zhao, H., Deng, X.W. (2001) Light control of Arabidopsis development entails coordinated regulation of genome expression and cellular pathways. Plant Cell. 13, 2589-2607

164. Maevskaya, S., Bukhov, N. (2005) Effect of light quality on nitrogen metabolism of radish plants. Russ J Plant Physiol. 52, 304-310

165. Makde, R.D., England, J.R., Yennawar, H.P., Tan, S. (2010) Structure of RCC1 chromatin factor bound to the nucleosome core particle. Nature. 467, 562-566

166. Malhotra, K., Kim, S.T., Batschauer, A., Dawut, L., Sancar, A. (1995) Putative blue-light photoreceptors from Arabidopsis thaliana and Sinapis alba with a high degree of sequence homology to DNA photolyase contain the two photolyase cofactors but lack DNA repair activity. Biochemistry. 34, 6892-6899

167. Manukyan, A. (2013) Effects of PAR and UV-B Radiation on Herbal Yield, Bioactive Compounds and Their Antioxidant Capacity of

Some Medicinal Plants Under Controlled Environmental Conditions. Photochemistry and Photobiology. 89, 406-414

168. Más, P., Kim, W.Y., Somers, D.E., Kay, S.A. (2003) Targeted degradation of TOC1 by ZTL modulates circadian function in Arabidopsis thaliana. Nature. 426, 567-570
169. Matsuda, R., Ohashi-Kaneko, K., Fujiwara, K., Goto, E., Kurata, K. (2004) Photosynthetic characteristics of rice leaves grown under red light with or without supplemental blue light. Plant Cell Physiology. 45, 1870-1874
170. McCree, K.J. (1971) The action spectrum, absorptance and quantum yield of photosynthesis in crop plants. Agricultural Meteorology. 9, 191-216
171. McCree, K.J. (1972a) The action spectrum, absorptance and quantum yield of photosynthesis in crop plants. Agricultural Meteorology. 9, 191-216.
172. McCree, K.J. (1972b) Test of current definitions of photosynthetically active radiation against leaf photosynthesis data. Agricultural Meteorology 10, 443-453
173. McCree, K.J. (1981) Photosynthetically active radiation. In: Lange, O.L., Nobel, P.S., Osmond, C.B., Ziegler, H., eds. Physiological plant ecology I. Berlin, Germany: Springer-Verlag. 41-55
174. Miyashita, Y.T., Kinura, Y., Kitaya, C., Kubota, C., Kozai, T. (1997) Effects of red light on the growth and morphology of potato plantlets in vitro using light emitting diodes (LEDs) as a light source for micropropagation. Acta Horticulturae (ISHS). 418, 169-173
175. Mizuno, T., Amaki, W., Watanabe, H. (2009) Effects of monochromatic light irradiation by LED on the growth and anthocyanin contents in leaves of cabbage seedlings, in VI International Symposium on Light in Horticulture 907, ed. by Goto E, Hikosaka S. ISHS, Belgium, pp. 179–184
176. Moni, A., Lee, A.Y., Briggs, W.R., Han, I.S. (2015) The blue light receptor Phototropin 1 suppresses lateral root growth by controlling cell elongation. Plant Biol. 17, 34-40
177. Monte, E., Alonso, J.M., Ecker, J.R., Zhang, Y., Li, X., Young, J., Austin-Phillips, S., Quail, P.H. (2003) Isolation and characterization of phyC mutants in Arabidopsis reveals complex cross-talk between

phytochrome signalling pathways. Plant Cell. 15, 1962-1980

178. Moon, H.K., Park, S.-Y., Kim, Y.W., Kim, C.S. (2006) Growth of Tsuru-rindo (Tripterospermum japonicum) cultured in vitro under various source of light-emitting diode (LED) irradiation. Journal of Plant Biology. 49, 174-179
179. Mora, S.J., Demers, S., Vernet, M. (2000) The effects of UV radiation in the marine environment. Cambridge University Press
180. Nagashima, A., Uehara, Y., Sakai, T. (2008) The ABC subfamily B auxin transporter AtABCB19 is involved in the inhibitory effects of N-1-naphthyphthalamic acid on the phototropic and gravitropic responses of Arabidopsis hypocotyls. Plant Cell Physiol. 49, 1250-1255
181. Nagatani, A. (2010) Phytochrome: structural basis for its functions. Curr Opin Plant Biol. 13, 565-570
182. Nakasako, M., Zikihara, K., Matsuoka, D., Katsura, H., Tokutomi, S. (2008) Structural basis of the LOV1 dimerization of Arabidopsis phototropins 1 and 2. J. Mol. Biol. 381, 718-733
183. Nakata, P.A. (2003) Advances in our understanding of calcium oxalate crystal formation and function in plants. Plant Sci. 164, 901-909
184. Nanya, K., Ishigami, Y., Hikosaka, S., Goto, E. (2012) Effects of blue and red light on stem elongation and flowering of tomato seedlings. Acta Horticulturae. 956, 264-266
185. Nelson, D.C., Lasswell, J., Rogg, L.E., Cohen, M.A., Bartel, B. (2000) FKF1, a clock-controlled gene that regulates the transition to flowering in Arabidopsis. Cell. 101, 331-340
186. Nhut, D.T., Hong, L.T.A., Watanabe, H., Goi, M., Tanaka, M. (2002) Growth of banana plantlets cultured in vitro under red and blue light-emitting diode (LED) irradiation source. Acta Horticulturae (ISHS). 575, 117-124
187. Nhut, D.T., Takamura, T., Watanabe, H., Okamoto, K., Tanaka, M. (2003) Responses of strawberry plantlets cultured in vitro under superbright red and blue light-emitting diodes (LEDs). Plant Cell, Tissue and Organ Culture. 73, 43-52
188. Nito, K., Wong, C.C., Yates III, J.R., Chory, J. (2013). Tyrosine

phosphorylation regulates the activity of phytochrome photoreceptors. Cell Reports. 3, 1970-1979

189. Noh, B., Bandyopadhyay, A., Peer, W.A., Spalding, E.P., Murphy, A.S. (2003) Enhanced gravi- and phototropism in plant mdr mutants mislocalizing the auxin efflux protein PIN1. Nature. 423, 999-1002
190. Oguchi, R., Douwstra, P., Fujita, T., Chow, W.S., Terashima, I. (2011) Intra-leaf gradients of photoinhibition induced by different color lights: implications for the dual mechanisms of photoinhibition and for the application of conventional chlorophyll fluorometers. New Phytol. 191, 146-159
191. O'Hara, A., Jenkins, G.I. (2012) In vivo function of tryptophans in the Arabidopsis UV-B photoreceptor UVR8. Plant Cell. 24, 3755-3766
192. Ohashi-Kaneko, K., Matsuda, R., Goto, E., Fujiwara, K., Kurata, K. (2006) Growth of rice plants under red light with or without supplemental blue light. Soil Sci Plant Nutr. 52, 444-452
193. Ohashi-Kaneko, K., Takase, M., Kon, N., Fujiwara, K., Kurata, K. (2007) Effect of light quality on growth and vegetable quality in leaf lettuce, spinach and komatsuna. Environ Control Biol. 45, 189-198
194. Ohgishi, M., Saji, K., Okada, K., Sakai, T. (2004) Functional analysis of each blue light receptor, cry1, cry2, phot1, and phot2, by using combinatorial multiple mutants in Arabidopsis. Proc. Natl Acad. Sci. USA. 101, 2223-2228
195. Okajima, K., Aihara, Y., Takayama, Y., Nakajima, M., Kashojiya, S., Hikima, T., Oroguchi, T., Kobayashi, A., Sekiguchi, Y., Yamamoto, M., Suzuki, T., Nagatani, A., Nakasako, M., Tokutomi, S. (2014) Light-induced conformational changes of LOV1 (light oxygen voltage-sensing domain 1) and LOV2 relative to the kinase domain and regulation of kinase activity in Chlamydomonas phototropin. J. Biol. Chem. 289, 413-422
196. Oravecz, A., Baumann, A., Máté, Z., Brzezinska, A., Molinier, J., Oakeley, E.J., Ádám, É., Schäfer, E., Nagy, F., Ulm, R. (2006) CONSTITUTIVELY PHOTOMORPHOGENIC1 is required for the UV-B response in Arabidopsis. Plant Cell. 18, 1975-1990
197. Osterlund, M.T., Hardtke, C.S., Wei, N., Deng, X.W. (2000) Targeted destabilization of HY5 during light-regulated development

of Arabidopsis. Nature. 405, 462-466

198. Ouzounis, T., Fretté, X., Rosenqvist, E., Ottosen, C.O. (2014) Spectral effects of supplementary lighting on the secondary metabolites in roses, chrysanthemums, and campanulas. Journal of Plant Physiology. 171, 1491-1499
199. Pan, R.C., Chen, F.G. (1991) Retardation of senescence in detached leaves of mung bean seedling by blue light. J Trop Subtrop Bot. 1, 67-72
200. Paradiso, R., Meinen, E., Snel, J.F.H., De Visser, P., Van Leperen, W., Hogewoning, S.W., Marcelis, L.F.M. (2011) Spectral dependence of photosynthesis and light absorptance in single leaves and canopy in rose. Scientia Horticulturae. 127, 548-554
201. Park, J.-S., Choung, M.-G., Kim, J.-B., Bae, S.C., Roh, K.H., Kim, Y.H., Cheon, C.I., Sung, M.K., Cho, K.J. (2007) Gene up-regulated during red colouration in UV-B irradiated lettuce leaves. Plant Cell Reports. 26, 507-516
202. Paul, N.D. (2006) The use of supplementary lighting in protected ornamental and edible crops: beyond the maximisation of biomass. HDC Project PC237 Final report
203. Paul N.D., Gwynn-Jones, D. (2003) Ecological roles of solar UV radiation: towards an integrated approach. Trends in Ecology and Evolution. 18, 48-55
204. Paul, N.D., Moore, J.M., Huey, M. (2006) The potential benefits of three modified plastic crop covers in Hardy Ornamental Nursery Stock production: initial investigations on a grower holding (Garden Centre Plants, Preston). HDC Project PC19a Final report
205. Pinedo, J.L., Mireles, F., Ríos, C., Quirino, L.L., Dávila, J.I. (206) Spectral signature of ultraviolet solar irradiance in Zacatecas. Geofísica Internacional. 45(4), 263-269
206. Pokorny, R., Klar, T., Hennecke, U., Carell, T., Batschauer, A., Essen, L.O. (2008) Recognition and repair of UV lesions in loop structures of duplex DNA by DASH-type cryptochrome. Proc. Natl Acad. Sci. USA. 105, 21023-21027
207. Poppe, C., Schäfer, E. (1997) Seed germination of Arabidopsis thaliana phyA/phyB double mutants is under phytochrome control. Plant Phys. 114, 1487-1492

208. Poudel, P.R., Kataoko, I., Mochioka, R. (2008) Effect of red- and blue-light-emitting diode on growth and morphogenesis of grapes. Plant Cell Tissue Organ Culture. 92, 147-153

209. Presswood, H.A., Hofmann, R., Savage, G.P. (2012) Effects of UV-B radiation on oxalate concentration of silver beet leaves. J Food Res. 1, 1-6

210. Pudasaini, A., Zoltowski, B.D. (2013) Zeitlupe senses blue-light fluence to mediate circadian timing in Arabidopsis thaliana. Biochemistry. 52, 7150-7158

211. Qi, L.D., Liu, S.H.Q., Xu, L., Yu, W.Y., Lang, Q.L., Hao, S.H.Q. (2007) Effects of light qualities on accumulation of oxalate, tannin and nitrate in spinach. Trans Chin Soc Agric Eng. 4, 201-205

212. Quail, P. H. (1994) Phytochrome genes and their expression. In Photomorphogenesis in plants, 2nd edn (ed. R.E. Kendrick & G.H.M. Kronenberg), pp. 71-104. Dordrecht, The Netherlands: Kluwer

213. Radek, M., Savage, G. (2008) Oxalates in some Indian green leafy vegetables. Int J Food Sci Nutr. 59, 246-260

214. Reed, J.W., Nagpal, P., Poole, D.S., Furuya, M., Chory, J. (1993) Mutations in the gene for red/far-red light receptor phytochrome B alter cell elongation and physiological responses throughout Arabidopsis development. Plant Cell. 5, 147-157

215. Reed, J.W., Nagatani, A., Elich, T.D., Fagan, M., Chory, J. (1994) Phytochrome A and phytochrome B have overlapping but distinct functions in Arabidopsis development. Plant Physiol. 104, 1139-1149

216. Renault, L., Nassar, N., Vetter, I., Becker, J., Klebe, C., Roth, M., Wittinghofer, A. (1998) The 1.7 Å crystal structure of the regulator of chromosome condensation (RCC1) reveals a seven-bladed propeller. Nature. 392, 97-101

217. Rizzini, L., Favory, J.-J., Cloix, C., Faggionato, D., O'Hara, A., Kaiserli, E., Baumeister, R., Schäfer, E., Nagy, F., Jenkins, G.I., Ulm, R. (2011) Perception of UV-B by the Arabidopsis UVR8 protein. Science. 332, 103-106

218. Rockwell, N.C., Su, Y.S., Lagarias, J.C. (2006). Phytochrome structure and signaling mechanisms. Annu Rev Plant Biol. 57, 837-

858

219. Rosenfeldt, G., Viana, R.M., Mootz, H.D., von Arnim, A.G., Batschauer, A. (2008) Chemically induced and light-independent cryptochrome photoreceptor activation. Mol Plant. 1, 4-14
220. Runkle, E.S., Heins, R.D. (2001) Specific functions of red, far red, and blue light in flowering and stem extension of long-day plants. Journal of the American Society of Horticulture. 126, 275-282
221. Ryan, K.G., Swinny, E.E., Winefield, C., Markham, K.R. (2001) Flavonoids and UV photoprotection in Arabidopsis mutants. Z. Naturforsch., C, J. Biosci. 56, 745-754
222. Safrany, J., Haasz, V., Mate, Z., Ciolfi, A., Feher, B., Oravecz, A., Stec, A., Dallmann, G., Morelli, G., Ulm, R., Nagy, F. (2008) Identification of a novel cis-regulatory element for UV-B-induced transcription in Arabidopsis. Plant J. 54, 402-414
223. Sager, J.C., J.C. McFarlane. (1997) Radiation, p. 1–29. In: Langhans, R.W. and T.W. Tibbitts (eds.). Plant growth chamber handbook. Iowa State Univ. Press: North Central Region Research Publication No. 340, Iowa Agriculture and Home Economics Experiment Station Special Report no. 99, Ames, IA
224. Sakai, T., Kagawa, T., Kasahara, M., Swartz, T.E., Christie, J.M., Briggs, W.R., Wada, M., Okada, K. (2001) Arabidopsis nph1 and npl1: blue light receptors that mediate both phototropism and chloroplast relocation. Proc. Natl Acad. Sci. USA. 98, 6969-6974
225. Sakamoto, K., Briggs, W.R. (2002) Cellular and subcellular localization of phototropin 1. Plant Cell. 14, 1723-1735
226. Sakuta, M. (2014) Diversity in plant red pigments: anthocyanins and betacyanins. Plant Biotechnology Reports. 8, 37-48
227. Salomon, M., Christie, J.M., Knieb, E., Lempert, U., Briggs, W.R. (2000) Photochemical and mutational analysis of the FMN-binding domains of the plant blue light receptor, phototropin. Biochemistry. 39, 9401-9410
228. Samuoliene, G., Sirtautas, R., Brazaityte, A., Duchovskis, P. (2012) LED lighting and seasonality effects antioxidant properties of baby leaf lettuce. Food Chemistry. 143, 1494-1499
229. Samuoliene, G., Viršile, A., Brazaityte, A., Jankauskiene,

J., Duchovskis, P., Novickovas, A., et al., (2003) Effect of supplementary pre-harvest LED lighting on the antioxidant and nutritional properties of green vegetables, in XXVIII International Horticultural Congress on Science and Horticulture for People (IHC2010): International Symposium on 939, ed. by Desjardins Y. ISHS, Belgium, pp. 85–91

230. Sancar, A. (2003) Structure and function of DNA photolyase and cryptochrome blue-light photoreceptors. Chem Rev. 103, 2203-2237

231. Sang, Y., Li, Q.H., Rubio, V., Zhang, Y.C., Mao, J., Deng, X.W., Yang, H.Q. (2005) N-terminal domain-mediated homodimerization is required for photoreceptor activity of Arabidopsis CRYPTOCHROME 1. Plant Cell. 17, 1569-1584

232. Santamaria, P., Elia, A., Serio, F., Todaro, E. (1999) A survey of nitrate and oxalate concentrationin fresh vegetables. J Sci Food Agric. 79, 1882-1888

233. Savvides, A., Fanourakis, D., van Ieperen, W. (2012) Co-ordination of hydraulic and stomatal conductance across light qualities in cucumber leaves. J Exp Bot. 63, 1135–1143

234. Sawa, M., Kay, S.A. (2011) GIGANTEA directly activates Flowering Locus T in Arabidopsis thaliana. Proc. Natl Acad. Sci. USA. 108, 11698-11703

235. Sawa, M., Nusinow, D.A., Kay, S.A., Imaizumi, T. (2007) FKF1 and GIGANTEA complex formation is required for day-length measurement in Arabidopsis. Science. 318, 261-265

236. Schreiner, M., Mewis, I., Huyskens-Keil, S., Jansen, M., Zrenner, R.,Winkler, J., et al. (2012) UV-B-induced secondary plantmetabolites-potential benefits for plant and human health. Crit Rev Plant Sci. 31, 229-240

237. Schultz, T.F., Kiyosue, T., Yanovsky, M., Wada, M., Kay, S.A. (2001) A role for LKP2 in the circadian clock of Arabidopsis. Plant Cell. 13, 2659-2670

238. Schwartz, A., Zeiger E. (1984): Metabolic energy for stomatal opening. Roles of photophosphorylation and oxidative phosphorylation. Planta. 161, 129-136

239. Selby, C.P., Sancar, A. (2006) A cryptochrome/photolyase class of

enzymes with single-stranded DNA-specific photolyase activity. Proc. Natl Acad. Sci. USA. 103, 17696-17700

240. Shimazaki, K., Doi, M., Assmann, S.M., Kinoshita, T. (2007) Light regulation of stomatal movement. Annu. Rev. Plant Biol. 58, 219-247

241. Shinomura, T., Nagatani, A., Chory, J., Furuya, M. (1994) The induction of seed-germination in Arabidopsis thaliana is regulated principally by phytochrome B and secondarily by phytochrome A. Plant Physiol. 104, 363-371

242. Sharrock, R.A., Clack, T. (2002) Patterns of expression and normalized levels of the five Arabidopsis phytochromes. Plant Physiol. 130, 442-456

243. Shiga, T., Shoji, K., Shimada, H., Hashida, S., Goto, F., Yoshihara, T. (2009) Effect of light quality on rosmarinic acid content and antioxidant activity of sweet basil, Ocimum basilicum L. Plant Biotechnology. 26, 255-259

244. Shinomura, T., Nagatani, A., Hanzawa, H., Kubota, M., Watanabe, M., Furuya, M. (1996) Action spectra for phytochrome A- and B-specific photoinduction of seed germination in Arabidopsis thaliana. Proc Natl Acad Sci USA. 93, 8129-8133

245. Smith, H. (1995) Physiological and ecological function within the phytochrome family. Annu Rev Plant Physiol Plant Molec Biol. 46, 289-315

246. Smith, H., Whitelam, G.C. (1990) Phytochrome, a family of photoreceptors with multiple physiological roles. Plant Cell Environ. 13, 695-707

247. Smith, H., Whitelam, G.C. (1997) The shade avoidance syndrome: multiple responses mediated by multiple phytochromes. Plant Cell Environ. 20, 840-844

248. Solovchenko, A., Schmitz-Eiberger, M. (2003) Significance of skin flavonoids for UV-B protection in apple fruits. J Exp Bot. 54, 1977-1984

249. Somers, D.E., Devlin, P.F., Kay, S.A. (1998) Phytochromes and cryptochromes in the entrainment of the Arabidopsis circadian clock. Science. 282, 1488-1490

250. Somers, D.E., Kim, W.Y., Geng, R. (2004) The F-box protein ZEITLUPE confers dosage-dependent control on the circadian clock, photomorphogenesis, and flowering time. Plant Cell. 16, 769-782

251. Song, Y.H., Smith, R.W., To, B.J., Millar, A.J., Imaizumi, T. (2012) FKF1 conveys timing information for CONSTANS stabilization in photoperiodic flowering. Science. 336, 1045-1049

252. Spalding, E.P. (2013) Diverting the downhill flow of auxin to steer growth during tropisms. Amer. J. Bot. 100, 203-214

253. Stagnari, F., Galieni, A., Cafiero, G., Pisante, M. (2014) Application of photo-selective films to manipulate wavelength of transmitted radiation and photosynthate composition in red beet (Beta vulgaris var. conditiva Alef.). J Sci Food Agric. 94, 713-720

254. Stutte, G.W., Edney, S., Skerritt, T. (2009) Photoregulation of bioprotectant content of red leaf lettuce with light-emitting diodes. Hort Science. 44, 79-82

255. Suesslin, C., Frohnmeyer, H. (2003). An Arabidopsis mutant defective in UV-B light-mediated responses. Plant J. 33, 591-601

256. Suetsugu, N., Wada, M. (2013) Evolution of three LOV blue light receptor families in green plants and photosynthetic stramenopiles: phototropin, ZTL/FKF1/LKP2 and aureochrome. Plant Cell Physiol. 54, 8-23

257. Sullivan, S., Thomson, C.E., Lamont, D.J., Jones, M.A., Christie, J.M. (2008) In vivo phosphorylation site mapping and functional characterization of Arabidopsis phototropin 1. Mol. Plant. 1, 178-194

258. Sun, J., Nishio, J.N., Vogelmann, T.C. (1998) Green Light Drives CO_2 Fixation Deep within Leaves. Plant Cell Physiology. 39, 1020-1026

259. Swartz, T.E., Corchnoy, S.B., Christie, J.M., Lewis, J.W., Szundi, I., Briggs, W.R., Bogomolni, R.A. (2001) The photocycle of a flavin-binding domain of the blue light photoreceptor phototropin. J. Biol. Chem. 276, 36493-36500

260. Takala, H., Björling, A., Berntsson, O., Lehtivuori, H., Niebling, S., Hoernke, M., Kosheleva, I., Henning, R., Menzel, A., Ihalainen, J.A., Westenhoff, S. (2014) Signal amplification and transduction in

phytochrome photosensors. Nature. 509, 245-248

261. Takase, T., Nishiyama, Y., Tanihigashi, H., Ogura, Y., Miyazaki, Y., Yamada, Y., Kiyosue, T. (2011) LOV KELCH PROTEIN2 and ZEITLUPE repress Arabidopsis photoperiodic flowering under non-inductive conditions, dependent on FLAVIN-BINDING KELCH REPEAT F-BOX1. Plant J. 67, 608-621
262. Takemiya, A., Inoue, S., Doi, M., Kinoshita, T., Shimazaki, K. (2005) Phototropins promote plant growth in response to blue light in low light environments. Plant Cell. 17, 1120-1127
263. Takemiya, A., Sugiyama, N., Fujimoto, H., Tsutsumi, T., Yamauchi, S., Hiyama, A., Tada, Y., Christie, J.M., Shimazaki, K. (2013) Phosphorylation of BLUS1 kinase by phototropins is a primary step in stomatal opening. Nat. Commun. 4, 2094
264. Tanaka, M., Takamura, T., Watanabe, H., Endo, M., Yanagi, T., Okamoto, K. (1998) In vitro growth of Cymbidium plantlets cultured under super bright and blue light-emitting diodes (LEDs). Journal of Horticultural Science and Biotechnology. 73, 39-44
265. Terashima, I., Fujita, T., Inoue, T., Chow, W.S., Oguchi, R. (2009) Green light drives leaf photosynthesis more efficiently than red light in strong white light: Revisiting the enigmatic question of why leaves are green. Plant Cell Physiology. 50, 684-697
266. Tevini, M., Iwanzik, W.M., Thoma, U. (1981) Some effects of enhanced UV-B irradiation on the growth and composition of plants. Planta. 153, 388-394
267. Tong, H., Leasure, C.D., Hou, X., Yuen, G., Briggs, W., He, Z.H. (2008) Role of root UV-B sensing in Arabidopsis early seedling development. Proc Natl Acad Sci USA. 105, 21039-21044
268. Tripathy, B.C., Brown, C.S. (1995) Root-shoot interaction in the greening of wheat seedlings grown under red light. Plant Physiol. 107, 407-411
269. Tsormpatsidis, E., Henbest, R., Davis, F.J., Battey, N., Hadley, P., Wagstaffe, A. (2008) UV irradiance as a major influence on growth, development and secondary products of commercial importance in Lollo Rosso lettuce 'Revolution'grown under polyethylene films. Environ Exp Bot. 63, 232-239
270. Wagner, J.R., Brunzelle, J.S., Forest, K.T., Vierstra, R.D. (2005) A

light-sensing knot revealed by the structure of the chromophore-binding domain of phytochrome. Nature. 438, 325-331

271. Wang, G., Gu, M., Cui, J., Shi, K., Zhou, Y., Yu, J. (2009) Effects of light quality on CO2 assimilation, chlorophyll-fluorescence quenching, expression of Calvin cycle genes and carbohydrate accumulation in Cucumis sativus. Journal of Photochemistry and Photobiology B: Biology. 96, 30-37
272. Wargent, J.J., Moore, J.P., Ennos, A.R., Paul, N.D. (2009) Ultraviolet radiation as a limiting factor in leaf expansion and development. Photochemistry and Photobiology. 85, 279-286
273. Webster, J.G. (1999) The measurement, instrumentation, and sensors handbook. CRC Press LLC
274. Weller, J.L., Murfet, I.C., Reid, J.B. (1997) Pea mutants with reduced sensitivity to far-red light de¢ne an important role for phytochrome A in day-length detection. Plant Physiol. 114, 1225-1236
275. Wellmann, E. (1976) Specific ultraviolet effects in plant morphogenesis. Photochem. Photobiol. 24, 659-660
276. Wellmann, E. (1983) UV radiation in photomorphogenesis. In Encyclopedia of Plant Physiology New Series, Vol. 16B, W. Shropshire Jr. and H. Mohr, eds (Berlin: Springer), pp. 745-756
277. Whitelam, G.C., Devlin, P.F. (1996) Novel phytochromes control germination and end-of-day far-red light responses of Arabidopsis thaliana. In Current topics in plant physiology. 17. Regulation of plant growth and development by light (ed. W.R. Briggs, R.L. Heath & E. Tobin), pp. 171-179. ASPP Press
278. Winkel-Shirley, B. (2002) Biosynthesis of flavonoids and effects of stress. Curr. Opin. Plant Biol. 5, 218-223
279. Wu, D., Hu, Q., Yan, Z., Chen, W., Yan, C., Huang, X., Zhang, J., Yang, P., Deng, H., Wang, J., Deng, X., Shi, Y. (2012) Structural basis of ultraviolet-B perception by UVR8. Nature. 484, 214-219
280. Wu, M., Hou, C., Jiang, C.,Wang, Y., Wang, C., Chen, H., Chang, H. (2007) A novel approach of LED light radiation improves the antioxidant activity of pea seedlings. Food Chem. 101, 1753-1758
281. Wu, H.-C. (2006) Improving in vitro propagation of Protea cynaroides L. (King Protea) and the roles of starch and phenolic

compounds in the rooting of cuttings. PhD thesis. University of Pretoria, Pretoria.

282. Wu, H.-C., Lin, C.-C. (2012) Red light-emitting diode light irradiation improves root and leaf formation in difficult-to-propagate Protea cynaroides L. plantlets in vitro. Hortscience. 47, 1490-1494

283. Wu,H.-C.,duToit,E.S.,Reinhardt,C.F.,Rimando,A.M.,vanderKooy, F., Meyer, J.J.M. (2007) The phenolic 3,4-dihydroxybenzoicacid, is an endogenous regulator of rooting in Protea cynaroides. Plant Growth Regulation. 52, 207-215

284. Yanagi T1, Okamoto K, Takita S. (1996) Effects of blue, red, and blue/red lights of two different PPF levels on growth and morphogenesis of lettuce plants. Acta Hortic. 440, 117-122

285. Yang, X., Kuk, J., Moffat, K. (2008) Crystal structure of Pseudomonas aeruginosa bacteriophytochrome: photoconversion and signal transduction. Proc Natl Acad Sci USA. 105, 14715-14720

286. Yang, H.Q., Wu, Y.J., Tang, R.H., Liu, D., Liu, Y., Cashmore, A.R. (2000) The C termini of Arabidopsis cryptochromes mediate a constitutive light response. Cell. 103, 815-827

287. Yanovsky, M.J., Alconada-Magliano, T.M., Mazzella, M.A., Gatz, C., Thomas, B., Casal, J.J. (1998) Phytochrome A affects stem growth, anthocyanin synthesis, sucrose–phosphate–synthase activity and neighbor detection in sunlight-grown potato. Planta. 205, 235-241

288. Yasuhara, M., Mitsui, S., Hirano, H., Takanabe, R., Tokioka, Y., Ihara, Komatsu, A., Seki, M., Shinozaki, K., Kiyosue, T. (2004) Identification of ASK and clock-associated proteins as molecular partners of LKP2 (LOV kelch protein 2) in Arabidopsis. J. Exp. Bot. 55, 2015-2027

289. Yorio, N.C., Goins, G.D., Kagie, H.R., Wheeler, R.M., Sager, J.C. (2001) Improving spinach, radish, and lettuce growth under red light emitting diodes (LEDs) with blue light supplementation. Hort Science. 36, 380-383

290. Yu, J.W., Rubio, V., Lee, N.Y., Bai, S., Lee, S.Y., Kim, S.S., Liu, L., Zhang, Y., Irigoyen, M.L., Sullivan, J.A., Zhang, Y., Lee, I., Xie, Q., Paek, N.C., Deng, X.W. (2008) COP1 and ELF3 control circadian

function and photoperiodic flowering by regulating GI stability. Mol Cell. 32, 617-630

291. Yu, X., Sayegh, R., Maymon, M., Warpeha, K., Klejnot, J., Yang, H. Huang, J., Lee, J., Kaufman, L., Lin, C. (2009) Formation of nuclear bodies of Arabidopsis CRY2 in response to blue light is associated with its blue light-dependent degradation. Plant Cell. 21, 118-130
292. Zhang, T., Folta, K.M. (2012) Green light signaling and adaptive response. Plant Signal Behav. 7, 75-78
293. Zhang, L.W., Liu, S.Q., Zhang, Z.K., Yang, R., Yang, X.J. (2010a) Effects of light qualities on the nutritive quality of radish sprouts. Acta Nutr Sin. 4, 26
294. Zhang, L., Liu, S., Zhang, Z., Yang, R., Yang, X. (2010b) Dynamic effects of different light qualities on pea sprouts quality. North Hort. 8, 4-7
295. Zhang, H., Xu, Z.G., Cui, J., Guo,Y.S., Gu, A.S. (2009) Effects of different spectra on growth and nutritious quality of radish sprouting seedlings. China Veg. 10, 28-32
296. Zhu, D., Maier, A., Lee, J.-H., Laubinger, S., Saijo, Y., Wang, H., Qu, L.-J., Hoecker, U., Deng, X.W. (2008) Biochemical characterization of Arabidopsis complexes containing CONSTITUTIVELY PHOTOMORPHOGENIC1 and SUPPRESSOR OF PHYA proteins in light control of plant development. Plant Cell. 20, 2307-2323
297. Zuo, Z., Liu, H., Liu, B., Liu, X., Lin, C. (2011) Blue light-dependent interaction of CRY2 with SPA1 regulates COP1 activity and floral initiation in Arabidopsis. Curr. Biol. 21, 841-847
298. Zuo, Z.C., Meng, Y.Y., Yu, X.H., Zhang, Z.L., Feng, D.S., Sun, S.F., Liu, B., Lin, C.T. (2012) A study of the blue-light-dependent phosphorylation, degradation, and photobody formation of Arabidopsis CRY2. Mol. Plant. 5, 726-733

CHAPTER

2

TRANSCRIPTIONAL REGULATION OF TETRAPYRROLE BIOSYNTHESIS IN ARABIDOPSIS THALIANA

Koichi Kobayashi and Tatsuru Masuda

Graduate School of Arts and Sciences, The University of Tokyo, Tokyo, Japan

Biosynthesis of chlorophyll (Chl) involves many enzymatic reactions that share several first steps for biosynthesis of other tetrapyrroles such as heme, siroheme, and phycobilins. Chl allows photosynthetic organisms to capture light energy for photosynthesis but with simultaneous threat of photooxidative damage to cells. To prevent photodamage by Chl and its highly photoreactive intermediates, photosynthetic organisms have developed multiple levels of regulatory mechanisms to coordinate tetrapyrrole biosynthesis (TPB) with the formation of photosynthetic and photoprotective systems and to fine-tune the metabolic flow with the varying needs of Chl and other tetrapyrroles under various developmental and environmental conditions. Among a wide range of regulatory mechanisms of TPB, this review summarizes transcriptional regulation of TPB genes during plant development, with focusing on several transcription factors characterized in *Arabidopsis thaliana*. Key TPB genes are tightly coexpressed with other photosynthesis-associated nuclear genes and are induced by light, oscillate in a diurnal and circadian manner, are coordinated with developmental and nutritional status, and are strongly downregulated in response to arrested chloroplast biogenesis. LONG HYPOCOTYL 5 and PHYTOCHROME-INTERACTING FACTORs, which are positive and negative transcription factors with a wide range of light signaling, respectively, target many TPB genes for light and circadian regulation. GOLDEN2-LIKE transcription factors directly regulate key TPB genes to fine-tune the formation of the photosynthetic apparatus with chloroplast functionality. Some transcription factors such as FAR-RED ELONGATED HYPOCOTYL3, REVEILLE1, and scarecrow-like transcription factors may directly regulate some specific TPB genes, whereas other factors such as GATA transcription factors are likely to regulate TPB genes in an indirect manner. Comprehensive transcriptional analyses of TPB genes and detailed characterization of key transcriptional regulators help us obtain a whole picture of transcriptional control of TPB in response to environmental and endogenous cues.

INTRODUCTION

The development of photosynthetic machinery in chloroplasts is strictly regulated in response to various developmental and environmental cues to achieve efficient photosynthesis while avoiding photodamage (Jarvis and López-Juez, 2013). Particularly, the entire process of chlorophyll (Chl) biosynthesis should be strictly organized during chloroplast biogenesis because most Chl intermediates readily generate singlet oxygen and toxic radicals under light and consequently damage cells (Triantaphylidès and Havaux, 2009). In addition, Chl biosynthesis shares the common biosynthetic pathway with other tetrapyrroles such as heme, siroheme, and phycobilins (Figure 1), and metabolic flux to other tetrapyrroles affects Chl biosynthesis in plant cells. To produce Chl efficiently and safely on demand, plants utilize multiple levels of regulation including fine-tuning of enzymatic activities by cofactors, redox, and feedback systems, control of protein stability and suborganelle localization, and modulation of protein complex formation with regulatory proteins (Brzezowski et al., 2015; Kobayashi and Masuda, 2016). In addition, transcriptional regulation is a central system used to coordinate each step of tetrapyrrole biosynthesis (TPB) with the formation and maintenance of the photosynthetic machinery in response to developmental and environmental status. This review focuses on recent findings in the transcriptional regulation of TPB in *Arabidopsis thaliana*.

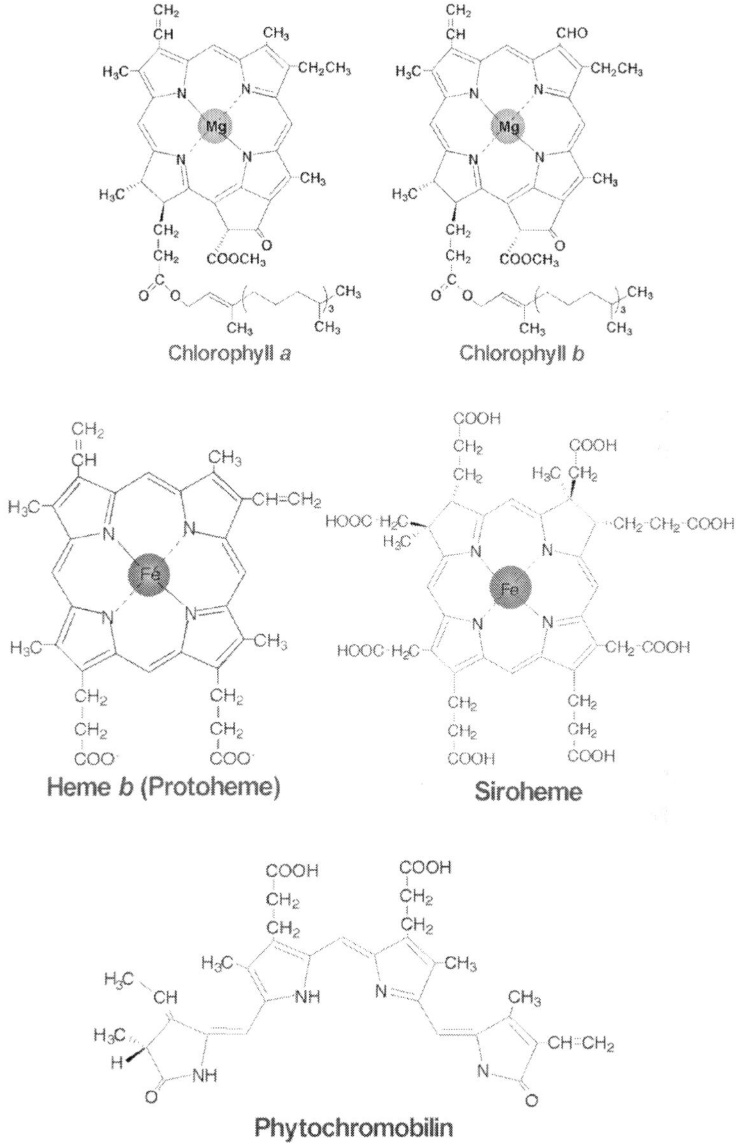

Figure 1. Structures of major tetrapyrroles in plants. Chlorophylls contain magnesium (Mg) for the central metal, whereas hemes and siroheme contain iron (Fe). Phytochromobilin is a linear tetrapyrrole functioning as a chromophore of phytochromes.

OVERVIEW OF CHL BIOSYNTHESIS IN PLANTS

In plants, all reactions of TPB take place in plastids. Figure 2 shows the TPB pathway with the involved genes. For details, readers are referred to comprehensive reviews of TPB pathways (von Wettstein et al., 1997; Beale, 1999; Heinemann et al., 2008; Tanaka et al., 2011; Brzezowski et al., 2015; Kobayashi and Masuda, 2016). Chl shares the common biosynthetic pathway with other tetrapyrroles in the first multiple steps. Plants, algae and many bacteria synthesize 5-aminolevulinic acid (ALA), the universal precursor for all tetrapyrroles, from Glu via three enzymatic steps called the C_5 pathway (Oborník and Green, 2005). Glu-$tRNA^{Glu}$ formed from Glu and $tRNA^{Glu}$ within plastids is reduced to Glu 1-semialdehyde (GSA) by Glu-tRNA reductase (GluTR). This reaction is the first committed and rate-limiting step of TPB (Papenbrock and Grimm, 2001). GSA is converted to ALA by GSA aminotransferase (GSAT), which catalyzes the intramolecular transfer of an amino group. Then two molecules of ALA are condensed asymmetrically to form the monopyrrole porphobilinogen (PBG), which is catalyzed by ALA dehydratase (ALAD; also known as PBG synthase). Four PBG molecules are then sequentially polymerized by PBG deaminase to result in the unstable linear tetrapyrrole hydroxymethylbilane (HMB). HMB is converted into the first macrocyclic tetrapyrrole, uroporphyrinogen (Urogen) III, by Urogen III synthase (UROS), then metabolized to coproporphyrinogen (Coprogen) III by Urogen III decarboxylase (UROD). A portion of Urogen III is also used for the synthesis of siroheme, the cofactor of nitrite and sulfite reductases involved in nitrogen and sulfur assimilation, respectively (Tanaka et al., 2011). Coprogen III is oxidized by Coprogen III oxidase (CPO) to form protoporphyrinogen (Protogen) IX, which is further oxidized by Protogen IX oxidase (PPO) to result in protoporphyrin (Proto) IX, the precursor for the synthesis of Chl and heme.

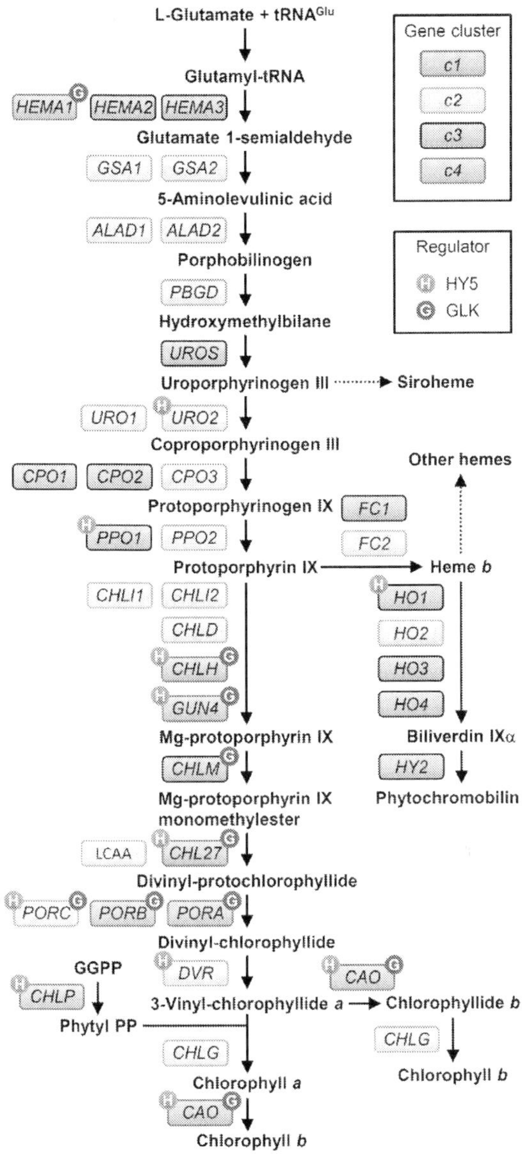

Figure 2. The tetrapyrrole biosynthesis pathway (TPB) and TPB genes in Arabidopsis. TPB genes are classified into four clusters (c1, c2, c3, and c4 in red, green, gray, and blue boxes, respectively) based on Matsumoto et al. (2004) and the ATTED-II coexpression database (Obayashi et al., 2007). Possible direct regulation of TPB genes by HY5 (Lee J. et al., 2007) and GLKs (Waters et al., 2009) is indicated by orange and purple circles, respectively, with initial letters of each regulate.

After the formation of Proto IX, the TPB pathway branches into two distinct pathways, namely the Mg- and Fe-branch for Chl and heme biosynthesis, respectively. In the Fe-branch, an Fe^{2+} ion is inserted into the Proto IX macrocycle by ferrochelatase to form protoheme (heme *b*), some of which is further converted into other hemes such as heme *a* and heme *c* or phytochromobilin, a linear tetrapyrrole functioning as a chromophore of phytochromes (Figure 1). In the Mg-branch, an Mg^{2+} ion is inserted into the Proto IX macrocycle in an ATP-dependent manner to form Mg-Proto IX. This reaction is catalyzed by Mg-chelatase consisting of the three subunits CHLD, CHIH, and CHLI in plants. GUN4, the regulatory protein of Mg-chelatase, assists this step by stabilizing the Mg-chelatase complex in membranes and mediating substrate and/or product channeling (Davison et al., 2005; Adhikari et al., 2011; Kopečná et al., 2015). Then a methyl group is added to Mg-Proto IX from S-adenosyl L-methionine by Mg-Proto IX methyltransferase, to result in Mg-Proto IX monomethyl ester (Mg-Proto ME). Mg-Proto ME is converted to 3,8-divinyl-protochlorophyllide (DV-Pchlide) by Mg-Proto ME cyclase, which forms a fifth isocyclic ring. Mg-Proto ME cyclase in plants is assumed to consist of three subunits including the membrane-bound subunits CHL27 (Rzeznicka et al., 2005) and LOW CHLOROPHYLL ACCUMULATION A (LCAA) (Albus et al., 2012; Hollingshead et al., 2012). After the reaction by Mg-Proto ME cyclase, the D ring of DV-Pchlide is reduced by Pchlide oxidoreductase (POR) to form 3,8-divinyl-chlorophyllide (DV-Chlide). Two different types of PORs, light-dependent LPOR and light-independent (dark-operative) DPOR, have been identified in photosynthetic organisms. Angiosperms possess only LPOR, which absolutely requires light for catalysis, whereas other plants, algae, and cyanobacteria possess both types of POR. Therefore, unlike other plants and cyanobacteria, angiosperms cannot synthesize Chl in darkness but instead accumulate Pchlide. DV-Chlide is converted to 3-vinyl Chlide *a* (MV-Chlide *a*) by DV-Chlide reductase (DVR). In addition, a geranylgeranyl chain is reduced by geranylgeranyl reductase encoded by *CHLP* to form a phytyl chain. Chl *a* is formed by esterification of MV-Chlide with geranylgeraniol or phytol, which is catalyzed by Chl synthase. Some of Chlide *a* or Chl *a* is converted into Chlide *b* or Chl *b* by Chlide *a* oxygenase (CAO) to diversify light-absorbing pigments in photosynthetic antenna complexes.

Genes Involved in TPB and Their Expression Profiles in *Arabidopsis*

Characterization of the photosynthetic gene cluster in *Rhodobacter* species of purple bacteria brought the first detailed understanding of genes involved in TPB (TPB genes) (Young et al., 1989; Suzuki et al., 1997). In plants, pioneering biochemical and genetic works in barley and tobacco substantially contributed to reveal TPB pathways and the involved genes (von Wettstein et al., 1997; Beale, 1999). Subsequently, almost all TPB genes have been identified and extensively characterized in *Arabidopsis* (listed in Fiable S1). In angiosperms, all TPB genes reside in the nucleus, whereas in other plant phyla, *chlL*, *chlN*, and *chlB* genes encoding DPOR subunits, which are missing in angiosperms, reside in the plastid genome (Fujita, 1996). In addition, $tRNA^{Glu}$, the starting material for all TPB, is plastid-encoded. Most TPB enzymes or subunits in *Arabidopsis* form small families encoded by paralogous genes.

Arabidopsis has two paralogous genes for GluTR isoforms, namely *HEMA1* and *HEMA2*, and two genes for ferrochelatase isoforms, *FC1* and *FC2*. Another GluTR paralog, *HEMA3*, is probably a pseudogene (Matsumoto et al., 2004). *HEMA1* and *FC2* are actively transcribed in green tissues (Ilag et al., 1994; Chow et al., 1998) and play major roles in photosynthesis (Kumar and Söll, 2000; Scharfenberg et al., 2015; Woodson et al., 2015). *HEMA2* and *FC1* are expressed in non-photosynthetic tissues and induced under several stress conditions, which reflects their roles besides photosynthesis (Nagai et al., 2007). *Arabidopsis* also features a pair of genes for GSAT, ALAD, and UROD enzymes, but functional differences between each isoform remain elusive. For the conversion from Coprogen III to Proto IX via Protogen IX, three *CPO* and two *PPO* paralogs have been identified; mutant analysis showed that *CPO1* (Ishikawa et al., 2001) and *PPO1* (Molina et al., 1999) are presumably the main isoforms for each reaction. Phylogenetic analysis revealed that *PPO2* is unique to land plants and shares homology with *PPO* in green non-sulfur bacteria (Kobayashi et al., 2014b), which suggests a unique role of this gene in land plants. The CHLI subunit of Mg-chelatase is also encoded by two isoforms: *CHLI1* plays a major role in photosynthesis (Koncz et al., 1990; Rissler et al., 2002), whereas *CHLI2* has an auxiliary role in the assembly of the Mg-chelatase complex

(Kobayashi et al., 2008). In addition, *Arabidopsis* LPOR isoforms are encoded by *PORA*, *PORB*, and *PORC*, whose expression is differentially regulated in response to light, described later. For the synthesis of a linear tetrapyrrole biliverdin IX from heme *b*, four heme oxygenase (HO) genes have been identified, with *HO1* playing a major role in the reaction (Davis et al., 1999; Muramoto et al., 1999).

To monitor the expression of TPB genes during seedling development in *Arabidopsis*, Matsumoto et al. (2004) conducted a small-scale transcriptome analysis covering most TPB genes. The authors found that TPB genes can be classified into four clusters (c1–c4) based on their expression profiles in response to light and the circadian clock. The c1 cluster contains *HEMA1*, *CHLH*, *CHL27*, and *CAO*, whose expression is repressed in dark-grown seedlings but rapidly induced by light together with *LHCB6*, which encodes a subunit of light-harvesting complex II. During photoperiodic seedling growth, these genes show well-synchronized oscillation together with *LHCB6* under diurnal and circadian rhythms. Similarly, in tobacco, a rhythmic coexpression of *HEMA1* and *CHLH* with an *LHC* gene is linked with diurnal and circadian control of tetrapyrrole metabolism (Papenbrock et al., 1999). Genes of the cl cluster form a tight coexpression network with each other and additional TPB genes, *GUN4* and *CHLP* (Figure 3A) (Masuda and Fujita, 2008) in the ATTED-II coexpression database (Aoki et al., 2016). Thus, the c1 genes including *GUN4* and *CLHP* likely have a central role in regulating Chl biosynthesis as key TPB genes. Moreover, the key TPB genes are involved in a large coexpression network consisting of photosynthesis-associated nuclear genes (PhANGs) (Kobayashi et al., 2012b). Specifically, LIGHT-HARVESTING-LIKE 3 proteins (LIL3:1 and LIL3:2), which anchor the geranylgeranyl reductase encoded by *CHLP* to the membrane via direct interaction and facilitate its enzymatic activity (Tanaka et al., 2010; Takahashi et al., 2014), are involved in the coexpression network of TPB genes (Figure 3A). Thus, these genes may share a global transcriptional regulatory system to orchestrate Chl biosynthesis with the synthesis of cognate proteins and assembly of the photosynthetic apparatus during chloroplast biogenesis and maintenance.

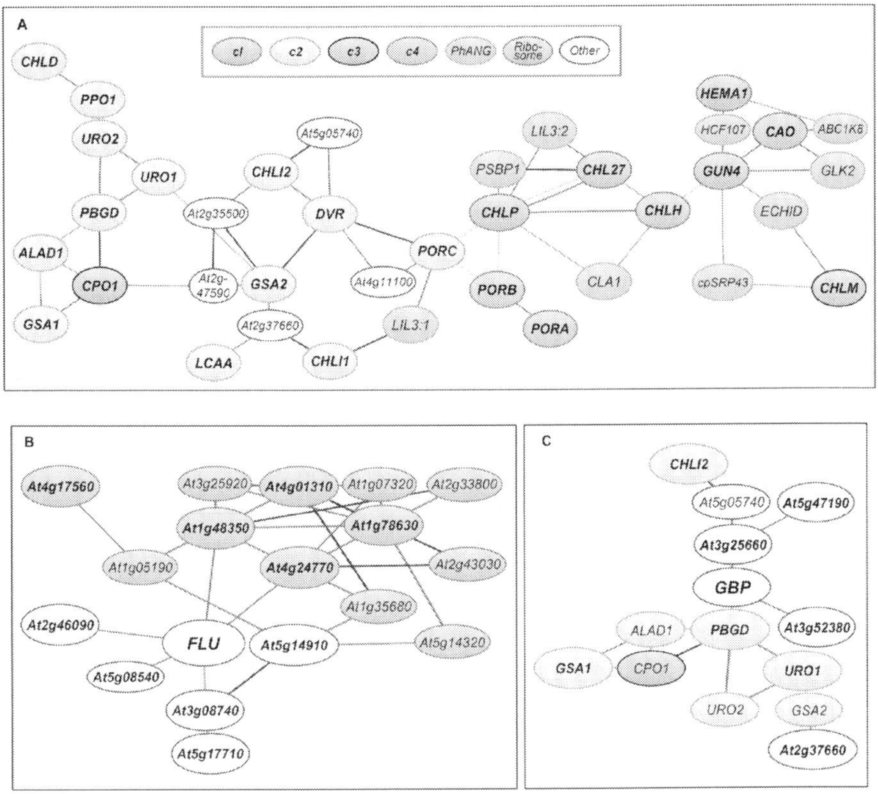

Figure 3. Coexpression networks of *Arabidopsis* genes involved in TPB. Coexpression networks of **(A)** TPB genes in cl (red), c2 (green), c3 (gray), and c4 (blue) clusters, **(B)**FLUORESCENT IN BLUE LIGHT (FLU, At3g14110), and **(C)** GluTR-binding protein (GBP, At3g21200) formed with photosynthesis-associated nuclear genes (PhANGs, orange), plastid ribosome-related genes (purple) and other nuclear genes (white). The coexpression networks were drawn by using the NetworkDrawer of the ATTED-II database v8.0 (Aoki et al., 2016; http://atted.jp/) with "add a few genes" and "Draw PPIs" options. Query genes are **(A)** all TPable S1, **(B)** *FLU* and its top 10 coexpressed e S2) and **(C)** *GBP* and its top 10 coexpresseble S3). Orange lines indicate highly reliable coexpression that is repeatedly detected in multiple coexpression data sets. Red dotted lines show links at protein–protein interaction levels. Query genes are in bold. Locus codes of PhANGs: *ABC1K8*, At5g64940; *CLA1*, At4g15560; *cpSRP43*, At2g47450; *ECHID*, At1g60550; *GLK2*, At5g44190; *HCF107*, At3g17040; *LIL3:1*, At4g17600; *LIL3:2*, At5g47110; *PSBP1*, At1g06680.

The c2 cluster is the largest cluster and widely includes genes for reactions from the earlier to later steps of TPB (Figure 2). The expression of c2 genes is induced by light and oscillated with diurnal rhythm but with smaller amplitude than that of c1 genes. Unlike c1 genes, c2 genes are not under circadian control, so light may be the primary determinant for the c2 gene cluster during a photoperiodic cycle (Matsumoto et al., 2004). The c2 genes form a coexpression network within the cluster, which connects in part with the network of the c1 cluster (Figure 3A) (Masuda and Fujita, 2008). Thus, although c2 genes may not be the main regulators of TPB, their expression is likely coordinated with the c1 genes to avoid excess accumulation of photoreactive tetrapyrrole intermediates. The *LCAA* gene in *Arabidopsis*, which encodes a membrane subunit of Mg-Proto ME cyclase, is involved in a coexpression network with c2 genes (Figure 3A). Thus, this gene may belong to the c2 cluster, whereas another membrane subunit of Mg-Proto ME cyclase, *CHL27*, belongs to the c1 cluster.

Meanwhile, genes in the c3 cluster, many involved in heme metabolism, are not responsive to light and circadian rhythms, which suggests a minor contribution of these genes to the transcriptional regulation of TPB in such conditions (Matsumoto et al., 2004). As an exception, *CHLM* in the c3 cluster connects with the c1 genes in the coexpression network (Figure 3A), so *CHLM* may be coregulated with cl genes under some conditions.

The c4 cluster comprises only two genes, *PORA* and *PORB*, whose transcripts accumulate in dark-grown seedlings and rapidly decrease with illumination. This profile is consistent with the substantial accumulation of the Pchlide–LPOR complex within etioplasts in dark-grown cotyledons and rapid breakdown of the complex after illumination (Fujita, 1996). This profile is unique to angiosperm LPOR genes. Other plants and algae essentially have DPOR genes in addition to LPOR and therefore they rarely accumulate the Pchlide–LPOR complex in the dark (Fujita, 1996). In fact, *LPOR*s in gymnosperms and mosses are upregulated by light with other photosynthesis-associated genes (Takio et al., 1998; Skinner and Timko, 1999). In *Arabidopsis*, both *PORA* and *PORB* show circadian rhythmic expression, with an oscillation peak slightly delayed as compared to c1 genes. The expression of *PORA* is much lower than that of *PORB* under the light–dark cycle (Matsumoto et al., 2004) and almost undetectable in light-adapted mature seedlings (Armstrong et al., 1995; Oosawa et al., 2000; Su et al., 2001). Meanwhile, another *LPOR*in *Arabidopsis*, *PORC*, whose expression profile greatly differs from that of *PORA* and *PORB* (Oosawa et al., 2000; Su et al., 2001), is classified in the c2 cluster with its light-inducible and diurnal cycle-dependent characteristics. Thus, circadian-regulated *PORB* and diurnal-regulated *PORC* may fine-tune Chl biosynthesis together with other TPB genes after greening, as suggested by the strong growth defects of *porB porC* double mutant during photoperiodic growth (Frick et al., 2003). Because *PORA* can complement the *porB porC* double mutant with its ectopic overexpression (Paddock et al., 2010), the different physiological roles among the POR isoforms are mainly due to their specific expression patterns. This notion is consistent with the fact that some angiosperms possess only a single *POR* gene in their genome (Spano et al., 1992; Fusada et al., 2000).

In addition to the enzymes involved in TPB, posttranslational regulatory proteins also strongly affect TPB pathways. FLUORESCENT IN BLUE LIGHT (FLU), which is a negative regulator of ALA synthesis, interacts with *HEMA1*-encoded GluTR1 to prevent excess accumulation of Chl precursors particularly Pchlide (Meskauskiene et al., 2001; Meskauskiene and Klaus, 2002; Goslings et al., 2004). The mRNA level of *FLU* is increased by light, whereas its protein level is constant during seedling growth in the dark and after transfer to light (Goslings et al., 2004; Apitz et al., 2014). Because it is a negative regulator of TPB, *FLU* is not transcriptionally linked with TPB genes in the coexpression network (FTable S2). Considering the substantial increase in *HEMA1* expression in response to light (Matsumoto et al., 2004), the increased ratio of GluTR1 to FLU during a dark-to-light transition would contribute to an increased capacity of the TPB pathway. Of note, FLU is coexpressed with many plastid ribosome-related genes (FiTable S2), so FLU may be somehow associated with protein synthesis in plastids.

A membrane bound GluTR-binding protein (GBP) is also involved in posttranslational regulation of GluTR activity. GBP interacts with GluTRs and recruits them to the thylakoid membrane (Czarnecki et al., 2011). Binding of GBP to the N-terminal domain of GluTR1 inhibits degradation of GluTR1 by the caseinolytic protease and allows for ongoing ALA synthesis for continuous heme formation (Apitz et al., 2016). Unlike

HEMA1 in the c1 cluster, *GBP* expression is almost unresponsive to light and the circadian clock and is similar to expression profiles of the c3 TPB genes (Czarnecki et al., 2011). Meanwhile, the expression profile of *GBP* in different tissues and developmental stages is similar to that of *HEMA1*. In the ATTED-II database (Aoki et al., 2016), *GBP* forms a coexpression network with many c2 genes involved in the common TPB pathway (F Table S3). Because GBP assists heme formation by maintaining GluTR activity in the membrane, the coexpression of *GBP* with the c2 TPB genes may render the TPB pathway more efficient under varying developmental and environmental conditions.

Transcriptional Factors Involved in TPB

Long Hypocotyl 5 (HY5)

Because most TPB genes are light-inducible, the light signaling pathway plays a central role in the transcriptional regulation of Chl biosynthesis. HY5 is one of the pivotal transcription factors regulating many light-associated events including photomorphogenesis and chloroplast development downstream of photoreceptors (Bae and Choi, 2008). In the dark, the HY5 protein is ubiquitinated and destabilized through CONSTITUTIVE PHOTOMORPHOGENIC1 (COP1), a ubiquitin E3 ligase regulating the abundance of various light-signaling components (Lau and Deng, 2012). Upon light perception by photoreceptors, COP1 is inactivated; HY5 is accumulated in the nucleus and mediates multiple light responses.

A genome-wide chromatin immunoprecipitation-chip (ChIP-chip) analysis suggests that many TPB genes are direct targets of HY5 together with various PhANGs (Table S1) (Lee J. et al., 2007). Only *HEMA1* in the cl cluster was not identified as the direct HY5 target in the ChIP-chip analysis. However, *HEMA1* expression in response to blue, red and far-red light was substantially decreased in a *hy5* mutant (McCormac and Terry, 2002), so *HEMA1* is also regulated by HY5. Loss of function of HY5 partially inhibits the *CHLH* expression and Chl accumulation in cotyledons during photomorphogenesis (Holm et al., 2002; Lee J. et al., 2007; Kobayashi et al., 2014a; Toledo-Ortiz et al., 2014); thus, this factor is important but not essential for Chl biosynthesis in leaf chloroplasts. Meanwhile, *hy5* roots exhibit an albino phenotype with very low expression of key TPB genes (Kobayashi et al., 2012a), which indicates that HY5 is particularly important for the expression of TPB genes in roots. In the *hy5* mutant, other factor(s) may compensate for the HY5 function in Chl biosynthesis in leaves but not roots.

HY5 specifically binds to the promoters of light-responsive genes through the G-box *cis*-element (CACGTG) (Chattopadhyay et al., 1998). ChIP-chip analysis showed significant enrichment of the G-box element in the promoter regions of HY5-targeted genes (Lee J. et al., 2007), which include key TPB genes and their coexpressed PhANGs (Kobayashi et al., 2012b). These data imply an importance of the G-box element for HY5-mediated transcriptional regulation of key TPB genes and associated genes. In fact, a deletion of the G-box sequence from the *CHLH* promoter region resulted in a decreased transcriptional activity particularly in the root (Kobayashi et al., 2012b). Moreover, Toledo-Ortiz et al. (2014) revealed that HY5 binds to G-box–containing promoter regions of *CHLH, PORC*, and other light-responsive genes. Thus, the interaction of HY5 with its target gene promoters on the G-box element would be associated with coordinated transcriptional regulation of key TPB genes with other PhANGs.

Phytochrome-Interacting Factors (PIFs)

Most TPB genes are negatively regulated in the dark to avoid excess accumulation of Chl intermediates that cause photooxidation upon illumination. PIFs, the basic helix-loop-helix transcription factors involved in broad cellular processes as a signaling hub integrating light, hormone, and other multiple developmental signals (Leivar and Quail, 2011), play an essential role in repressing Chl biosynthesis in the dark. PIFs (PIF1, PIF3, PIF4, and PIF5) accumulate in the nucleus in the dark and repress photomorphogenic responses including Chl biosynthesis. Particularly, PIF1 acts as a cofactor of COP1 to synergistically degrade HY5 in the dark (Zhu et al., 2015). However, upon illumination, PIFs are phosphorylated and degraded via the ubiquitin–proteasome system in a phytochrome-dependent manner (Leivar and Quail, 2011).

Among six well-characterized PIF proteins, PIF1, PIF3, PIF4, and PIF5 mainly function redundantly to downregulate TPB genes during seedling growth in the dark. Loss of function of PIF1 and PIF3 caused increased expression of key TPB genes such as *HEMA1*, *CHLH*, and

GUN4 and excessive accumulation of Pchlide in the dark, which led to photobleaching of etiolated seedlings upon illumination (Huq et al., 2004; Stephenson et al., 2009). Etiolated seedlings of a *pif5* mutant also showed increased *CHLH*expression and photobleaching upon illumination (Shin et al., 2009). Furthermore, a quadruple mutant (*pifq*) lacking PIF1, PIF3, PIF4, and PIF5 showed global upregulation of TPB genes along with many PhANGs (Leivar et al., 2009; Shin et al., 2009). These data indicate that PIFs are required for the coordinated repression of TPB genes and other PhANGs in the dark.

PIF3 binds to G-box regions of *CHLH* and other PhANGs (Liu et al., 2013). The interaction of PIF1 with G-box regions of *CHLH, PORC* and *CAO* was also reported (Cheminant et al., 2011; Toledo-Ortiz et al., 2014). PIFs (PIF1 and PIF4) and HY5 antagonistically regulate these target genes presumably on the same G-box–containing region (Toledo-Ortiz et al., 2014). Liu et al. (2013) revealed that histone deacetylation of target gene promoters is involved in the regulation of PhANGs by PIF3. In the dark, HISTONE DEACETYLASE15 (HDA15) interacts with PIF3 and binds to the G-box regions of target gene promoters. HDA15 recruited to the target promoters decreases their transcriptional activities via histone deacetylation. However, in the light, PIF3 is degraded by phytochrome signaling and HDA15 is dissociated from its target promoters. At the same time, histone acetyltransferases involved in light-responsive gene expression such as TAF1/HAF2 and GCN5 may acetylate histones to upregulate PhANGs (Benhamed et al., 2006). HY5 is genetically associated with these histone acetyltransferases, so HY5 may be involved in histone acetylation of target genes by recruiting these acetyltransferases in response to light.

In addition, Feng et al. (2014) reported that PIFs are involved in light-dependent gene repositioning within the *Arabidopsis* nucleus. The authors found that the *CAB* locus consisting of *LHCB1.1*, *LHCB1.2*, and *LHCB1.3* located at the nuclear interior in the dark but translocated rapidly to the nuclear periphery upon illumination before full transcriptional activation. However, in the *pifq* mutant, the *CAB* locus was positioned at the nuclear periphery even in the dark. A similar phenomenon was observed in *cop1* and *de-etiolated 1* (*det1*) mutants. Thus, PIFs may play a role to retain the *CAB* locus in the nucleus interior in the dark with COP1 and DET1. In addition, other representative light-responsive gene loci including *CHLH*showed light-dependent reposition within the nucleus as observed in the *CAB* locus (Feng et al., 2014). Because these gene loci are separately distributed throughout the *Arabidopsis* genome, the light-dependent gene locus reposition is likely a common mechanism to induce PhANGs including key TPB genes in response to light.

Golden2-Like (GLK) Transcription Factors

The GLK gene family is involved in the transcriptional regulation of chloroplast biogenesis in diverse groups of land plants. Loss of function of GLKs perturbed chloroplast biogenesis in leaves (Hall et al., 1998; Fitter et al., 2002; Wang et al., 2013), whereas their gain of function in various plants enhances chloroplast development in non-photosynthetic organs such as roots and fruits in addition to leaves (Nakamura et al., 2009; Kobayashi et al., 2012a; Powell et al., 2012).

Many plant species have GLK genes in pairs (*GLK1* and *GLK2*) (Wang et al., 2013). In *Arabidopsis*, *GLK1*and *GLK2* are functionally equivalent, and only the double knockout mutant (*glk1 glk2*) showed perturbed chloroplast development with reduced expression of PhANGs in leaves (Fitter et al., 2002). Meanwhile, transient induction of exogenous *GLK*s in the *glk1 glk2* double mutant caused immediate upregulation of the PhANGs, especially those associated with Chl biosynthesis and light harvesting, which suggests that these primary inducible genes are direct targets of GLKs (Waters et al., 2009). Moreover, a ChIP analysis confirmed the direct binding of GLK1 to the promoter regions of its target genes including key TPB genes (Waters et al., 2009). With the exception of *PORA* and *PORB*, these TPB genes form a tight coexpression network with other GLK-targeted PhANGs (Masuda and Fujita, 2008; Kobayashi et al., 2012b), which implies an involvement of GLKs in the coexpression network of key TPB genes and PhANGs as central regulators. Of note, *GLK2* itself is also involved in the coexpression network with linking to the cl genes (Figure 3), which suggests a tight transcriptional coordination between *GLK2* and the target TPB genes.

Promoter analysis of GLK-targeted genes revealed a putative GLK-recognition *cis*-element (CCAATC) (Waters et al., 2009). The GLK-recognition element is significantly co-enriched with G-box in the promoter regions of key TPB genes and coexpressed PhANGs (Kobayashi et al., 2012b). *Arabidopsis* GLK1 and GLK2 can interact with G-box binding factors in yeast (Tamai et al., 2002). Moreover, enhanced Chl accumulation in roots by GLK overexpression required a G-box binding factor HY5 (Kobayashi et al., 2012a). Thus, GLKs may function in transcriptional regulation of Chl biosynthesis in collaboration with G-box binding factors such as HY5.

Other Transcription Factors

GATA factors are transcriptional regulators that recognize a G-A-T-A core sequence of gene promoter regions. Of approximately 30 members of the GATA factor family, class B GATA factors (B-GATAs) are implicated in the regulation of Chl metabolism in addition to various developmental processes related to light and several hormonal responses (Behringer and Schwechheimer, 2015). Two B-GATAs, GNC (GATA, NITRATE-INDUCIBLE, CARBON METABOLISM INVOLVED) and GNL/CGA1 (GNC-LIKE/CYTOKININ-RESPONSIVE GATA TRANSCRIPTION FACTOR1), play a role in regulating chloroplast differentiation. Double-knockout mutations of GNC and GNL (*gnc gnl*) decreased Chl content in leaves, whereas overexpression of either gene enhanced ectopic Chl accumulation in leaf epidermis, roots and hypocotyls (Mara and Irish, 2008; Richter et al., 2010; Hudson et al., 2011, 2013; Chiang et al., 2012). The expression of *HEMA1*, *GUN4*, *PORB* and *PORC* was downregulated in the *gnc gnl* mutant but upregulated in the overexpressors of these factors (Hudson et al., 2011). However, in contrast to the direct regulation of TPB genes by HY5, PIFs, and GLKs, GNC and GNL appear to regulate these TPB genes in an indirect manner (Hudson et al., 2011). Four other B-GATAs (GATA15, GATA16, GATA17, and GATA17L) in *Arabidopsis*also function redundantly in the control of greening with GNC and GNL (Ranftl et al., 2016), and further studies are required for addressing the regulatory pathways of TPB gene expression by B-GATAs.

The transposase-derived transcription factors FAR-RED ELONGATED HYPOCOTYL3 (FHY3) and FAR-RED IMPAIREDRESPONSE1 (FAR1) are positive regulators of PHYA signaling and function in diverse developmental processes including Chl biosynthesis (Wang and Wang, 2015). Mutant analysis in *Arabidopsis* demonstrates that upregulation of *HEMA1* in response to red and far-red light stimuli requires FHY3 (McCormac and Terry, 2002). FHY3 is also important for the expression of *GUN4* and *CHLH* under far-red light (Stephenson and Terry, 2008). However, *HEMA1* but not *GUN4* and *CHLH* was identified as a putative direct target of FHY3 by ChIP-sequence analysis (Ouyang et al., 2011), which suggests that regulatory pathways by FHY3 are different between *HEMA1* and other genes (*GUN4* and *CHLH*). In addition, FHY3 and FAR1 induce the expression of *ALAD1/HEMB1* encoding an ALAD by binding to its promoter region through the FHY3/FAR1 binding site (CACGCGC) (Tang et al., 2012). The activity of FHY3 in the dark is partially repressed by PIF1 that physically interacts with the DNA binding domain of FHY3. FHY3 and FAR1 also physically interact with HY5 through their respective DNA binding domains (Li et al., 2010). The direct interaction between FHY3/FAR1 and HY5 affects PHYA signaling and the circadian clock (Wang and Wang, 2015), but its role in TPB gene regulation has not yet been elucidated.

The microRNAs miR171s (miR171a to c) and their targets scarecrow-like (SCL) transcription factors SCL6, SCL22 and SCL27 play an important role in transcriptional regulation of TPB genes, particularly those of the Chl pathway (Wang et al., 2010; Ma et al., 2014). Overexpression of miR171c enhanced degradation of the target *SCL* transcripts, which led to upregulation of *PORs* and *CAO* with increased Chl accumulation in leaves. A similar result was observed in the triple mutant of SCL6, SCL22, and SCL27. By contrast, overexpression of miR171-resistant SCL27 strongly decreased the expression of these TPB genes and Chl content in leaves. The data suggest that the balance between miR171s and their target SCLs strongly affect the expression of TPB genes involved in the Chl pathway in leaves. *In vivo* and *in vitro* analysis revealed that SCL27 can bind to the promoter of *PORC* through GT *cis*-element repeats and inhibit its expression (Ma et al., 2014). However, DELLA proteins, negative regulators of gibberellin signaling, physically interact with SCL27 and inhibit its DNA binding to the *PORC* promoter, which suggests an involvement of gibberellin signaling in TPB gene regulation via the DELLA–SCL27 interaction.

Xu et al. (2015) reported that a transcription factor REVEILLE1 (RVE1) binds to the promoter of *PORA*through a *cis*-element termed evening element (AAAATATCT) and upregulates *PORA* expression in the dark. Overexpression and loss of function of RVE1 increased and decreased the *PORA* expression in the dark, respectively. Moreover, RVE1 overexpression increased the greening rate of dark-grown seedlings with reduced reactive oxygen species (ROS) production and cell death, so RVE1 may be a crucial mediator of Chl biosynthesis and chloroplast development (Xu et al., 2015).

Key TPB genes of the c1 cluster are under transcriptional control by the circadian clock machinery (Matsumoto et al., 2004), in which TIMING OF CAB EXPRESSION1 (TOC1) plays a pivotal role as one of the core components. TOC1 functions as a transcriptional repressor by directly binding to the promoter of its target genes (Gendron et al., 2012), and *CHLH* is one of its regulatory targets. The transcript levels of *CHLH* increase in the day and decrease at night (Papenbrock et al., 1999; Matsumoto et al., 2004). However, silencing of TOC1 by RNAi induced *CHLH* expression in the night, and its overexpression suppressed *CHLH* expression in the day (Legnaioli et al., 2009). Moreover, TOC1 binding to the *CHLH*promoter was antiphasic to *CHLH* expression, which shows the direct suppression of *CHLH* expression by TOC1 to exert circadian oscillation.

Koussevitzky et al. (2007) reported that ABSCISIC ACID INSENSITIVE-4 (ABI4), an Apetala 2-type transcription factor, represses PhANG expression by binding to a specific *cis*-sequence (CCAC) in response to chloroplast dysfunction. Sun et al. (2011) proposed that the nuclear-localized N-terminal fragment of PTM protein, the chloroplast envelope-bound plant homeodomain transcription factor, is required for *ABI4* expression as a mediator of plastid-to-nucleus signaling. However, loss of ABI4 did not affect the *RBCS* promoter activity even when chloroplasts were disrupted by the photobleaching herbicide norflurazon (Acevedo-Hernandez et al., 2005). Moreover, the *abi4* mutant showed strong downregulation of *LHCB1.1* by norflurazon or by a plastidic translation inhibitor lincomycin as did the wild-type (Kerchev et al., 2011). Transcriptome analyses of the *abi4* mutant (Kerchev et al., 2011) and the ABI4 overexpression lines (Reeves et al., 2011) in *Arabidopsis* imply no direct involvement of ABI4 in TPB gene expression. Meanwhile, ABI4 is positively and negatively involved in the expression of a large number of genes associated with various cellular processes such as hormone and sugar signaling, redox homeostasis, and defense response, in addition to PhANGs (León et al., 2013). Specifically, ABI4 is required for downregulation of PhANGs under high glucose conditions (Arenas-Huertero et al., 2000; Acevedo-Hernandez et al., 2005). Moreover, ABI4 upregulates *PORA* presumably by inducing *COP1* expression in the dark (Xu et al., 2016). Therefore, multiple factors may work with ABI4 in a highly complex fashion under various conditions.

Regulation of TPB Genes during Photomorphogenesis

Angiosperms germinated in darkness undergo a dark-adapted developmental program termed skotomorphogenesis, which is characterized by a prolonged hypocotyl and etiolated cotyledons containing dark-specific plastid etioplasts. In etioplasts, a substantial amount of LPOR is accumulated along with Pchlide *a* and NADPH to form prolamellar bodies (Sundqvist and Dahlin, 1997). In *Arabidopsis* seedlings, *PORA* and *PORB* are actively expressed during skotomorphogenesis (Armstrong et al., 1995; Matsumoto et al., 2004). Mutant analysis showed that the PORA and PROB proteins accumulated in etioplasts contribute to the formation of prolamellar bodies in an additive manner (Franck et al., 2000; Paddock et al., 2010). COP1 plays an essential role in this process (Figure 4), as suggested by a remarkable decrease in *PORA* and *PORB* transcripts with loss of prolamellar bodies in dark-grown *cop1* seedlings (Sperling et al., 1998). ABI4 is required for *PORA* expression in the dark, presumably with its activity to induce *COP1*expression (Xu et al., 2016). Moreover, the regulation of the *PORA* and *PORB* expression by COP1 involves ethylene signaling mediated by an ethylene-inducible transcription factor EIN3/EIL1. Ethylene and COP1 signaling stabilize the EIN3/EIL1 protein, which then upregulates *PORA* and *PORB* by binding to their promoter regions (Zhong et al., 2009). In addition, gibberellin signaling is involved in the regulation of *PORA* and *PORB* via DELLA proteins (Cheminant et al., 2011). DELLAs can upregulate *PORA* and *PORB*expression in the dark. Although DELLAs have a function to inhibit PIF activities by direct protein–protein interactions (de Lucas et al., 2008; Feng et al., 2008), they act on *PORA* and *PORB* expression independently of PIFs (Cheminant et al., 2011). In addition, RVE1 upregulates *PORA* but not *PORB* by directly binding to the *PORA* promoter (Xu et al., 2015). The relationship between RVE1 and other regulatory factors for *PORA* regulation in darkness remains unknown. Meanwhile, *PORC* expression is almost undetectable in etiolated seedlings (Oosawa et al., 2000; Su et al., 2001) and has no important role in the formation of prolamellar bodies (Frick et al., 2003; Masuda et al., 2003; Paddock et al.**Figure 4**

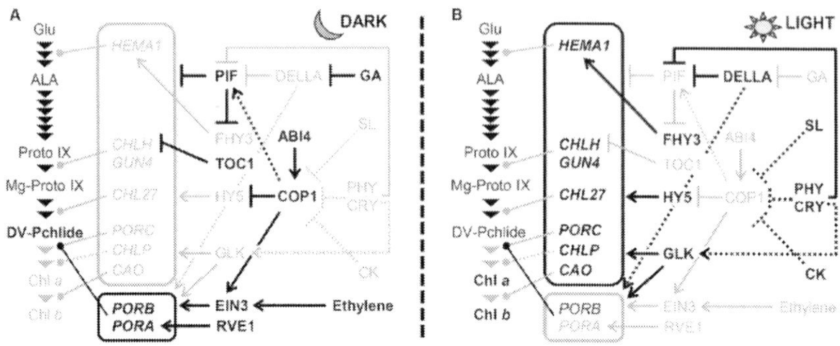

Figure 4. A model for regulating TPB genes (A) in the dark and (B) under light. The pathway for Chl biosynthesis is shown with important intermediates and enzymatic steps indicated by arrowheads. Key genes for the pathway shown in boxes are connected to each step that they involved. Arrows and bars represent positive and negative regulation, respectively. Dotted lines indicate indirect effects. Black and light blue colors represent activated and inactivated status, respectively. CK, cytokinin; CRY, cryptochrome; GA, gibberellic acid; PHY; phytochrome; SL, strigolactone.

With the exception of *POR*s actively transcribed in the dark, most TPB genes are repressed in dark-grown seedlings to prevent photodamage upon illumination (Matsumoto et al., 2004). As described earlier, PIFs play a central role in the negative regulation of TPB genes (Figure 4). Gibberellins accumulated in the dark contribute to the activation of PIFs by degrading DELLA proteins and releasing PIFs from inhibition by DELLAs (Cheminant et al., 2011). Meanwhile, major positive regulators involved in TPB gene expression are repressed in the dark. The HY5 protein is degraded in the dark via COP1/DET1-mediated ubiquitination (Osterlund et al., 2000). Moreover, the expression of *GLKs* and *B-GATAs* is low in the dark because they are light-inducible (Fitter et al., 2002; Manfield et al., 2007; Ranftl et al., 2016). Thus, activation of negative regulators and inactivation of positive regulators combined could keep the expression of key TPB genes low in the dark.

Upon illumination, Pchlide accumulated in the dark is immediately converted to Chlide by LPORs and then to Chl by Chl synthase. During this process, the expression of *PORA* and *PORB* is rapidly decreased in response to light, with possible involvement of degradation of EIN3 via inactivation of COP1 and ethylene signaling (Zhong et al., 2009). At the same time, *de novo* Chl biosynthesis is activated to develop functional chloroplasts in cotyledons. Both the phytochrome and cryptochrome photoreceptor families play a crucial role in upregulating key TPB genes such as *HEMA1*, *CHLH*, *GUN4*, and *CAO* (McCormac and Terry, 2002; Tepperman et al., 2006; Stephenson and Terry, 2008). Under light, phytochromes induce phosphorylation, ubiquitination, and subsequent degradation of PIFs (Leivar and Quail, 2011), which results in derepression of TPB genes. DELLAs also contribute to derepression of TPB genes by inhibiting PIF activities (Cheminant et al., 2011). Meanwhile, light mediates the accumulation of HY5 proteins by preventing COP1 activity and induces the expression of *GLKs* and *B-GATAs*, which together upregulate key TPB genes (Figure 4). In addition, cytokinin and strigolactone inhibit COP1 activity and increase HY5 protein level (Vandenbussche et al., 2007; Tsuchiya et al., 2010). In fact, upregulation of *HEMA1*, *CHLH* and *CHL27* in response to light was weakened in *Arabidopsis* cytokinin receptor mutants, whereas cytokinin treatment to dark-grown wild-type seedlings increased expression of these genes (Hedtke et al., 2012; Kobayashi et al., 2014a). Moreover, Cortleven et al. (2016) recently reported that type-B ARABIDOPSIS RESPONSE REGULATORs (ARRs) function in the greening process downstream of cytokinin signaling. In this process, ARR10 and ARR12 directly bind to the promoter regions of *HEMA1* in addition to *LHCB6* (Cortleven et al., 2016). These data suggest an importance of cytokinin signaling in upregulation of key TPB genes during photomorphogenesis. Melo et al. (2016) reported that in tomato, production of nitric oxide (NO) in response to light via phytochrome signaling promotes Chl accumulation presumably in part by downregulating negative regulators of photomorphogenesis such as *COP1* and *DET1* and upregulating *GLK1*. The authors proposed that negative and positive feedback loops between ethylene-NO and auxin-NO, respectively, are involved in regulating the greening of tomato seedlings (Melo et al., 2016).

Circadian Regulation of TPB Genes

After establishing photoautotrophic growth with developed chloroplasts, plants maintain homeostatic Chl biosynthesis to support continued growth with high photosynthetic activity. Chl biosynthesis is coordinated with development of photosynthetic machinery by light and the endogenous clock, with concerted circadian regulation of key TPB genes with Chl-binding apoproteins during the day–night cycle playing a crucial role (Papenbrock et al., 1999; Matsumoto et al., 2004; Stephenson et al., 2009).

As described earlier, the core evening phase factor TOC1 decreases *CHLH* transcripts at night by directly binding to the *CHLH* promoter (Legnaioli et al., 2009). Abscisic acid treatment changes the circadian rhythm of *CHLH* transcripts by upregulating *TOC1* expression, although its relevance to regulation of Chl biosynthesis is unknown. A genome-wide ChIP-sequence analysis identified only *CHLH* as the TOC1 target among TPB genes (Gendron et al., 2012). Thus, regulatory pathways in response to the circadian clock differ between *CHLH* and other cl genes, although they all show very similar circadian oscillation (Matsumoto et al., 2004).

Together with TOC1, two morning-phased transcription factors, CIRCADIAN CLOCK-ASSOCIATED1 (CCA1) and LATE ELONGATED HYPOCOTYL (LHY), form a core transcriptional feedback loop to maintain circadian oscillation. CCA1 and LHY in *Arabidopsis* colocalize in the nucleus via physical interaction and function synergistically in regulating circadian oscillation (Lu et al., 2009). CCA1 and LHY bind to the same region (CCA1-binding site) of the promoter of *LHCB1.3* to activate it in response to light (Wang et al., 1997; Lu et al., 2009). The *LHCB1.3* promoter has a G-box element to bind HY5, in close proximity to the CCA1 binding site. CCA1 affects the binding of HY5 to *LHCB1.1* and *LHCB1.3* promoters via physical interaction. HY5 is required for regulating the circadian oscillation of the *LHCB1.1* transcripts (Andronis et al., 2008). Moreover, loss of HY5 decreased the peak expression of PhANGs including *CHLH*and *PORC* in daytime (Toledo-Ortiz et al., 2014). Thus, HY5 interacting with CCA1 plays an important role in the circadian regulation of at least some of the PhANGs including key TPB genes. Besides functioning in the output of light signaling, HY5 is involved in gating light signaling to the circadian machinery in collaboration with FHY3 and FAR1 (Li et al., 2011), which reflects a diverse function of HY5 in circadian regulation.

In contrast to HY5, PIFs function to downregulate key TPB genes during circadian oscillation. Stephenson et al. (2009) showed that the periodic expression of *HEMA1*, *CHLH* and *GUN4* observed during seedling growth in darkness was strongly disordered in *pif1* and *pif3* mutants with an increasing trend. The *pif1* and *pif3* mutations did not affect the circadian expression of *TOC1*, *CCA1* and *LHY* in dark-grown seedlings, so PIF1 and PIF3 function in the output from the circadian clock under such conditions (Stephenson et al., 2009). Moreover, the expression of *CHLH* and *PORC* was increased throughout circadian oscillation in the *pifq* mutant (Toledo-Ortiz et al., 2014). Thus, PIFs would function in circadian signaling to downregulate Chl biosynthesis with the development of photosynthetic machinery at night. TOC1 can interact with PIFs and represses their transcriptional activation activity (Soy et al., 2016). However, the effect of TOC1 on the function of PIFs as a transcriptional repressor of PhANGs needs to be addressed.

The expression of *GLK2* is regulated in a circadian-dependent manner, as is the expression of key TPB genes (Fitter et al., 2002). GLK2 is a strong inducer of key TPB genes, so oscillation of *GLK2* transcript levels could be associated with the circadian regulation of key TPB genes. *GNC* and *GNL* are also under circadian regulation with expression peak at predawn (Manfield et al., 2007) and may affect the circadian oscillation of key TPB genes indirectly.

Coordinated Regulation of TPB Genes with Chloroplast Functionality

Plants need to coordinate Chl biosynthesis with the formation of the photosynthetic machinery to meet the variable demands for Chl during development and to prevent photooxidative damage by free Chl and its intermediates. Light induction of most TPB genes during chloroplast biogenesis is important to provide Chl for the newly synthesized photosynthetic apparatus. By contrast, the expression of most TPB genes is gradually decreased during plant maturation (Matsumoto et al., 2004) and strongly repressed during leaf senescence (Figure 5) (Lin and Wu, 2004). Such transcriptional regulation of Chl biosynthesis in coordination with chloroplast development is required to avoid photooxidative damage from free Chl and tetrapyrrole intermediates. In fact, deregulation of the TPB gene expression in *pif* mutants resulted in photobleaching of dark-grown seedlings with illumination (Huq et al., 2004; Leivar et al., 2009; Shin et al., 2009; Stephenson et al., 2009). Forced overexpression of *HEMA1* (Schmied et al., 2011) or *CHLH*(Shin et al., 2009) caused strong photooxidative damage to leaves in *Arabidopsis*. Moreover, a transient decrease in *CHL27* expression by dexamethasone-induced RNAi caused an imbalance in the metabolic flow of the Chl biosynthesis pathway and accumulation of Mg-Proto and Mg-Proto ME, which generates harmful ROS in the light (Schlicke et al., 2014). These data clearly show the importance of regulation of Chl biosynthesis at translevels.

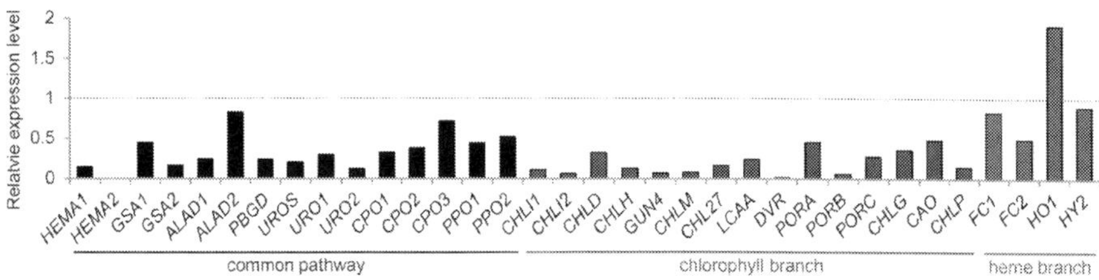

Figure 5. Expression of TPB genes during leaf senescence. Gene expression data in *Arabidopsis* were obtained from the AtGenExpress Visualization Tool (http://jsp.weigelworld.org/expviz/expviz.jsp) in the public transcriptome database. The expression of each gene in senesced leaves of 35-day-old plants was normalized to that in the sixth true leaves of 17-day-old seedlings (dotted line).

The expression of key TPB genes is strongly associated with the biogenesis of the thylakoid membrane, where Chl forms complexes with photosynthetic proteins and other cofactors. The coordinated expression of TPB genes and other PhANGs with thylakoid membrane biogenesis was revealed in mutants defective in thylakoid lipid biosynthesis (Kobayashi, 2016). Impaired biosynthesis of galactolipids, the major lipid constituents of the thylakoid membrane, caused strong downregulation of TPB genes and other PhANGs concomitant with disrupted thylakoid lipid-bilayer formation (Kobayashi et al., 2013; Fujii et al., 2014). Similar results were observed in a mutant deficient in another major thylakoid lipid, phosphatidylglycerol (Kobayashi et al., 2015). However, in these lipid mutants, activation of thylakoid biogenesis via alternative lipid biosynthetic pathways by P starvation restored the decreased TPB gene expression (Kobayashi et al., 2013, 2015). By contrast, photosynthetic activity was still dysfunctional in these mutants, presumably because of lack of essential lipid molecules in photosynthetic complexes. These data suggest that TPB gene expression is coupled with the formation of the thylakoid lipid bilayer independent of photosynthetic activities. Moreover, the expression of TPB genes and galactolipid biosynthesis genes is coordinated with each other by light and cytokinin signaling (Kobayashi et al., 2014a), which may also contribute to the orchestration of Chl biosynthesis and thylakoid biogenesis.

Land plants are at constant risk of photooxidative damage in chloroplasts by fluctuating strong light and other biotic and abiotic stresses. Sudden exposure to high light and low temperature stresses, which cause over-reduction of the photosynthetic electron transport chain, quickly downregulates many TPB genes as well as PhANGs, presumably to minimize the potential of chloroplasts to generate ROS (Bode et al., 2016). Although the redox state of photosynthetic electron components has been implicated as a source of signals to regulate PhANGs (Oswald et al., 2001; Pfannschmidt, 2003), the mode of action remains unclear (Piippo et al., 2006; Bode et al., 2016). Although modification of Mg porphyrin levels by induced silencing of *CHLH*, *CHLM*, and *CHL27* did not immediately affect the expression of TPB genes and other PhANGs, the prolonged silencing of *CHL27* decreased the expression of some TPB genes along with PhANGs, presumably because of the generation of singlet oxygen (Schlicke et al., 2014). Moreover, a knockdown mutation of *Arabidopsis CHL27* by T-DNA insertion in its promoter region downregulated many TPB genes and PhANGs (Bang et al., 2008). These data suggest that modification of Mg porphyrin metabolism secondarily affects the transcriptional activity of TPB genes via ROS-mediated signaling.

When the chloroplast function is severely impaired, the expression of PhANGs, including most TPB genes, is shut off via the plastid-to-nucleus retrograde signaling. Several *Arabidopsis* mutants, with disrupted retrograde plastid signaling, have been identified and are named *genomes-uncoupled* (*gun*) mutants (Susek et al., 1993). In *gun* mutants, chloroplast dysfunction by herbicide or antibiotic treatments could not strongly decrease the expression of key TPB genes and other PhANGs (Strand et al., 2003; Moulin et al., 2008), which suggests that key TPB genes are under strict control of plastid signaling similar to other PhANGs. A plastidic pentatricopeptide repeat protein, GUN1, plays a central role in plastid signaling by integrating multiple plastid signals to shut off the expression of PhANGs in response to chloroplast dysfunction (Chi et al., 2013). Genetic evidence suggests that GUN1 signaling activates ABI4 to downregulate PhANGs in response to chloroplast dysfunction (Koussevitzky et al., 2007), although the significance of ABI4 in this process is controversial (Acevedo-Hernandez et al., 2005; Kakizaki et al., 2009; Kerchev et al., 2011). In addition, *GLK1* and *GLK2* are under transcriptional control by plastid signaling (Waters et al., 2009; Leister and Kleine, 2016). Furthermore, overexpressors of *GLK1* and *GLK2*showed a *gun* phenotype after norflurazon or lincomycin treatment (Leister and Kleine, 2016). GUN1 somehow downregulates *GLK1* expression in response to chloroplast dysfunction, which results in the downregulation of GLK targets including key TPB genes independent of ABI4 signaling (Kakizaki et al., 2009; Waters et al., 2009). Transcriptional regulation of GLKs in response to chloroplast functionality would help optimize Chl synthesis under varying environmental and developmental conditions.

Although HY5 is one of the central transcriptional activators of light-responsive genes, including many PhANGs, during photomorphogenesis as described earlier, Ruckle et al. (2007) showed that this factor represses its target genes via cryptochrome 1 (CRY1) when chloroplasts are dysfunctional. In the authors' model, plastid signals convert HY5 from a positive to a negative regulator in a GUN1-independent manner and downregulate HY5-targeted genes. In fact, loss-of-function mutations in CRY1 or HY5 caused strong photodamage under high light conditions because of deregulation of PhANGs, and the additional *gun1*mutation to these mutants further enhanced the light-dependent growth defect (Ruckle et al., 2007). The data suggest that the dual transcriptional regulation of PhANGs by CRY1–HY5-mediated light signaling and GUN1 signaling is crucial for controlling the development of photosynthetic machinery according to light conditions and chloroplast functionality. We should note that *LHCB* genes are particularly subject to negative regulation by light signaling after lincomycin-induced plastid dysfunction and many other PhANGs including TPB genes showed only reduced light response with lincomycin treatment (Ruckle et al., 2012). Meanwhile, Kindgren et al. (2011) propose that HY5 is involved in the plastid signaling mediated by Mg-Proto IX. Feeding of Mg-Proto IX via roots strongly repressed *LHCB1.1* and *RBCS*expression in green tissues of wild-type *Arabidopsis* but not in the *hy5* mutant (Kindgren et al., 2011). Nonetheless, several biochemical and genetic studies have not supported the direct relationship between the expression of PhANGs and *in vivo* Mg-Proto IX levels (Mochizuki et al., 2008; Moulin et al., 2008; Terry and Smith, 2013; Schlicke et al., 2014), and the role of tetrapyrrole signaling in plastid signaling is still under debate.

Regulation of TPB Genes in Response to Biotic and Abiotic Stresses

In addition to producing Chl, the TPB pathway is indispensable for providing heme as a protein cofactor or a regulatory molecule for various biological processes. *Arabidopsis HEMA1* and *FC2* play major roles in photosynthesis by being abundantly expressed in green tissues as described earlier, the expression of their counterparts *HEMA2* and *FC1* is low in these tissues but is strongly upregulated in response to ROS-generating treatments such as wounding, ozone, and herbicides (Nagai et al., 2007). Of note, the expression of *HEMA2* and *FC1* is increased in the *flu* mutant exposed to light after dark incubation (op den Camp et al., 2003). The *flu* mutant lacks the ability to regulate GluTR1 activity and therefore accumulates excess Pchlide in the dark (Meskauskiene et al., 2001). With light exposure, Pchlide accumulated in the mutant generates singlet oxygen, which subsequently induces a programmed cell death response mediated by plastid-localized EXECUTER proteins (Wagner et al., 2004; Lee K.P. et al., 2007; Zhang et al., 2014). These data suggest that *HEMA2* and *FC1* are also responsive to singlet oxygen signaling. Moreover, biotic stresses such as treatment with a bacterial elicitor flagellin 22 induce strong *FC1* expression and weak *HEMA2* expression (Scharfenberg et al., 2015; Espinas et al., 2016). Although oxylipin signaling mediated by jasmonic acid (JA) and 12-oxo phytodienoic acid (OPDA), produced via peroxidation reactions of membrane lipids, is deeply involved in wounding and biotic stress responses (Reinbothe et al., 2009), *FC1*and *HEMA2* are unresponsive to these oxylipins (Taki et al., 2005; Nagai et al., 2007), which is consistent with a report that JA and OPDA do not act as second messengers during singlet oxygen signaling (Przybyla et al., 2008). Because genes encoding the endoplasmic reticulum-localized cytochrome P450 family and cytosolic ascorbate peroxidase are co-upregulated with *HEMA2* and *FC1* in response to wounding and flagellin 22 treatment, the HEMA2-FC1 pathway may supply heme for extraplastidic factors associated with defensive functions (Nagai et al., 2007; Espinas et al., 2016). In fact, mutant analysis showed that FC1-producing heme is mainly allocated to extraplastidic organelles, which contrasts with the major role of FC2-producing heme in the chloroplast (Espinas et al., 2016). Differential regulation of paralogous genes in the same reaction step as observed between *HEMA1* and *HEMA2* as well as *FC2* and *FC1* would allow plants to coordinate the TPB pathway flexibly with various kinds of developmental conditions and biotic and abiotic stresses.

PERSPECTIVES

Large-scale transcriptome analysis, targeted gene expression analysis and characterization of transcription regulators involved in TPB pathways have demonstrated a deep involvement of transcriptional regulation in TPB control during various processes of plant growth. The expression of key TPB genes is tightly coordinated with each other and with many PhANGs, with HY5 and GLKs possibly playing a central regulatory role. However, not all key TPB genes and coexpressed PhANGs are direct targets of HY5 and/or GLKs. Moreover, the classification of TPB genes into the four clusters based on light and circadian regulation is not strictly associated with the assignment of specific transcription factors to genes of each single cluster. The data suggest a complex regulation of these coexpressed genes by several transcription factors besides HY5 and GLKs under various conditions other than light and the circadian clock. Comprehensive promoter analysis of TPB genes and coexpressed PhANGs would provide further insight into the mechanism of coordinated regulation of these genes. Moreover, new molecular techniques have advanced our understanding of the involvement of chromatin remodeling and gene loci repositioning in the nucleus in transcriptional control. Thus, we anticipate an unraveling of the local regulation of TPB gene loci by specific transcription factors and also the global control of nuclear dynamics in response to endogenous and environmental stimuli.

In addition, relationships between transcriptional regulation and other potent regulatory mechanisms such as post-translational control await elucidation.

As described here, most of the evidence for the TPB gene regulation in higher plants has come from *Arabidopsis* studies. However, besides common mechanisms conserved widely among plants, different types of plants could possess different regulatory systems specific to their life styles for regulating TPB genes. An increasing amount of large-scale transcriptome data for various types of plants will help reveal a picture of transcriptional regulation of TPB that is common and specific to plant species. Future molecular and biochemical studies in diverse plant species will also advance our understanding about the regulatory mechanisms of TPB genes in response to diverse developmental and environmental conditions.

AUTHOR CONTRIBUTIONS

KK conceived the review, collected data, and wrote the manuscript. TM helped to write the manuscript.

FUNDING

This work was supported by JSPS KAKENHI Grant Numbers JP24570042, JP16K07393, and JP26711016.

CONFLICT OF INTEREST STATEMENT

The authors declare that the research was conducted in the absence of any commercial or financial relationships that could be construed as a potentiull#supplementary-material

REFERENCES

1. Acevedo-Hernandez, G. J., Leo, P., and Herrera-Estrella, L. R. (2005). Sugar and ABA responsiveness of a minimal RBCS light-responsive unit is mediated by direct binding of ABI4. Plant J. 43, 506–519. doi: 10.1f Full Text | Google Scholar
2. Adhikari, N. D., Froehlich, J. E., Strand, D. D., Buck, S. M., Kramer, D. M., and Larkin, R. M. (2011). GUN4-porphyrin complexes bind

the ChlH/GUN5 subunit of Mg-Chelatase and promote chlorophyll biosynthesis in Arabidopsis. Plant Cell 23, 1449–1467ef Full Text | Google Scholar

3. Albus, C. A., Salinas, A., Czarnecki, O., Kahlau, S., Rothbart, M., Thiele, W., et al. (2012). LCAA, a novel factor required for magnesium protoporphyrin monomethylester cyclase accumulation and feedback control of aminolevulinic acid biosynthesis in tobacco. Plant Physiol. 160, 1923–19Ref Full Text | Google Scholar
4. Andronis, C., Barak, S., Knowles, S. M., Sugano, S., and Tobin, E. M. (2008). The clock protein CCA1 and the bZIP transcription factor HY5 physically interact to regulate gene expression in Arabidopsis. Mol. Plant 1sRef Full Text | Google Scholar
5. Aoki, Y., Okamura, Y., Tadaka, S., Kinoshita, K., and Obayashi, T. (2016). ATTED-II in 2016: a plant coexpression database towards lineage-specific coexpression. Plant Cell PhysiolssRef Full Text | Google Scholar
6. Apitz, J., Nishimura, K., Schmied, J., Wolf, A., Hedtke, B., van Wijk, K. J., et al. (2016). Posttranslational control of ALA synthesis includes GluTR degradation by Clp protease and stabilization by GluTR-binding protein. Plant Physiol. 170, 20ossRef Full Text | Google Scholar
7. Apitz, J., Schmied, J., Lehmann, M. J., Hedtke, B., and Grimm, B. (2014). GluTR2 complements a hema1 mutant lacking glutamyl-tRNA reductase 1, but is differently regulated at the post-translational level. Plant Cell Physiol. 55rossRef Full Text | Google Scholar
8. Arenas-Huertero, F., Arroyo, A., Zhou, L., Sheen, J., and Leon, P. (2000). Analysis of Arabidopsis glucose insensitive mutants, gin5 and gin6, reveals a central role of the plant hormone ABA in the regulation of plant vegetative development by sugar. Genes Dev. 14, 208CrossRef Full Text | Google Scholar
9. Armstrong, G. A., Runge, S., Frick, G., Sperling, U., and Apel, K. (1995). Identification of NADPH:protochlorophyllide oxidoreductases A and B: a branched pathway for light-dependent chlorophyll biosynthesis in Arabidopsis thaliana. Plant Physiol. 108, 1 CrossRef Full Text | Google Scholar
10. Bae, G., and Choi, G. (2008). Decoding of light signals by plant

phytochromes and their interacting proteins. Annu. Rev. Plant Biol. 59, 281–311. doi: 10.| CrossRef Full Text | Google Scholar

11. Bang, W. Y., Jeong, I. S., Kim, D. W., Im, C. H., Ji, C., Hwang, S. M., et al. (2008). Role of Arabidopsis CHL27 protein for photosynthesis, chloroplast development and gene expression profiling. Plant Cell Physiol. | CrossRef Full Text | Google Scholar
12. Beale, S. I. (1999). Enzymes of chlorophyll biosynthesis. Photosynth. Res. 456
13. CrossRef Full Text | Google Scholar
14. Behringer, C., and Schwechheimer, C. (2015). B-GATA transcription factors–insights into their structure, regulation, and role in plant development. Front. Plantct | CrossRef Full Text | Google Scholar
15. Benhamed, M., Bertrand, C., Servet, C., and Zhou, D.-X. (2006). Arabidopsis GCN5, HD1, and TAF1/HAF2 interact to regulate histone acetylation required for light-responsive gene expression. Plant Cell 1act | CrossRef Full Text | Google Scholar
16. Bode, R., Ivanov, A. G., and Hüner, N. P. A. (2016). Global transcriptome analyses provide evidence that chloroplast redox state contributes to intracellular as well as long-distance signalling in response to stress and acclimation in Arabidopsis. Photosynth. Res. 12ract | CrossRef Full Text | Google Scholar
17. Brzezowski, P., Richter, A. S., and Grimm, B. (2015). Regulation and function of tetrapyrrole biosynthesis in plants and algae. Biochim. Biophys. Acta 1847,tract | CrossRef Full Text | Google Scholar
18. Chattopadhyay, S., Puente, P., Deng, X. W., and Wei, N. (1998). Combinatorial interaction of light-responsive elements plays a critical role in determining the response characteristics of light-regulated promoters in Arabidopsis. Plant J. 15stract | CrossRef Full Text | Google Scholar
19. Cheminant, S., Wild, M., Bouvier, F., Pelletier, S., Renou, J.-P., Erhardt, M., et al. (2011). DELLAs regulate chlorophyll and carotenoid biosynthesis to prevent photooxidative damage during seedling deetiolation in Arabidopsis. Plant Cebstract | CrossRef Full Text | Google Scholar
20. Chi, W., Sun, X., and Zhang, L. (2013). Intracellular signaling from plastid to nucleus. Annu. Rev. Plant Biol. 64, 559–5Abstract |

CrossRef Full Text | Google Scholar

21. Chiang, Y.-H., Zubo, Y. O., Tapken, W., Kim, H. J., Lavanway, A. M., Howard, L., et al. (2012). Functional characterization of the GATA transcription factors GNC and CGA1 reveals their key role in chloroplast development, growth, and division in Arabidopsis. Plant Ph Abstract | CrossRef Full Text | Google Scholar
22. Chow, K. S., Singh, D. P., Walker, A. R., and Smith, A. G. (1998). Two different genes encode ferrochelatase in Arabidopsis: mapping, expression and subcellular targeting of the precursor proteins. Plant J. 15d Abstract | CrossRef Full Text | Google Scholar
23. Cortleven, A., Marg, I., Yamburenko, M. V., Schlicke, H., Hill, K., Grimm, B., et al. (2016). Cytokinin regulates etioplast-chloroplast transition through activation of chloroplast-related genes. Planed Abstract | CrossRef Full Text | Google Scholar
24. Czarnecki, O., Hedtke, B., Melzer, M., Rothbart, M., Richter, A., Schröter, Y., et al. (2011). An Arabidopsis GluTR binding protein mediates spatial separation of 5-aminolevulinic acid synthesis in chloroplasts. PlaMed Abstract | CrossRef Full Text | Google Scholar
25. Davis, S. J., Kurepa, J., and Vierstra, R. D. (1999). The Arabidopsis thaliana HY1 locus, required for phytochrome-chromophore biosynthesis, encodes a protein related to heme oxygenases. Proc. Natl. Acad. Sci.bMed Abstract | CrossRef Full Text | Google Scholar
26. Davison, P. A., Schubert, H. L., Reid, J. D., Iorg, C. D., Heroux, A., Hill, C. P., et al. (2005). Structural and biochemical characterization of Gun4 suggests a mechanism for its role in chlorophyll biosynthesisubMed Abstract | CrossRef Full Text | Google Scholar
27. de Lucas, M., Davière, J.-M., Rodríguez-Falcón, M., Pontin, M., Iglesias-Pedraz, J. M., Lorrain, S., et al. (2008). A molecular framework for light and gibberellin control of cell elonPubMed Abstract | CrossRef Full Text | Google Scholar
28. Espinas, N. A., Kobayashi, K., Sato, Y., Mochizuki, N., Takahashi, K., Tanaka, R., et al. (2016). Allocation of heme is differentially regulated by ferrochelatase isoforms in Arabidopsis cells.
29. PubMed Abstract | CrossRef Full Text | Google Scholar
30. Feng, C.-M., Qiu, Y., Van Buskirk, E. K., Yang, E. J., and Chen, M. (2014). Light-regulated gene repositioning in Ara7

31. PubMed Abstract | CrossRef Full Text | Google Scholar
32. Feng, S., Martinez, C., Gusmaroli, G., Wang, Y., Zhou, J., Wang, F., et al. (2008). Coordinated regulation of Arabidopsis thalianadevelopment by light and gib48
33. PubMed Abstract | CrossRef Full Text | Google Scholar
34. Fitter, D. W., Martin, D. J., Copley, M. J., Scotland, R. W., and Langdale, J. A. (2002). GLK gene pairs regulate chloroplast development in diverse plant species. Pl0.x
35. PubMed Abstract | CrossRef Full Text | Google Scholar
36. Franck, F., Sperling, U., Frick, G., Pochert, B., van Cleve, B., Apel, K., et al. (2000). Regulation of etioplast pigment-protein complexes, inner membrane architecture, and protochlorophyllide a chemical heterogeneity by light-dependent NADPH:protochlorophyllide oxidoreductases A and B.1678
37. PubMed Abstract | CrossRef Full Text | Google Scholar
38. Frick, G., Su, Q., Apel, K., and Armstrong, G. A. (2003). An Arabidopsis porB porC double mutant lacking light-dependent NADPH:protochlorophyllide oxidoreductases B and C is highly chlorophyll-deficient and developmentally arrested. 798.x
39. PubMed Abstract | CrossRef Full Text | Google Scholar
40. Fujii, S., Kobayashi, K., Nakamura, Y., and Wada, H. (2014). Inducible knockdown of MONOGALACTOSYLDIACYLGLYCEROL SYNTHASE1 reveals roles of galactolipids in organelle differentiation in Arabidopsis cotyledon250050
41. PubMed Abstract | CrossRef Full Text | Google Scholar
42. Fujita, Y. (1996). Protochlorophyllide reduction: a key step in the greening of plants. Plant Cell Physiol. 37, 411–421. doi: 10.1093/oxfordjournals.pcp.a028962
43. CrossRef Full Text | Google Scholar
44. Fusada, N., Masuda, T., Kuroda, H., Shiraishi, T., Shimada, H., Ohta, H., et al. (2000). Nadph-protochlorophyllide oxidoreductase in cucumber is encoded by a single gene and its expression is transcriptionally enhanced by illuminati18608647
45. PubMed Abstract | CrossRef Full Text | Google Scholar
46. Gendron, J. M., Pruneda-Paz, J. L., Doherty, C. J., Gross, A. M.,

Kang, S. E., and Kay, S. A. (2012). Arabidopsis circadian clock protein, TOC1, is a DNA-binding transcription factor. Proc. Natl.200355109

47. PubMed Abstract | CrossRef Full Text | Google Scholar
48. Goslings, D., Meskauskiene, R., Kim, C., Lee, K. P., Nater, M., and Apel, K. (2004). Concurrent interactions of heme and FLU with Glu tRNA reductase (HEMA1), the target of metabolic feedback inhibition of tetrapyrrole biosynthesis, in dark- and light-grown Arabidopsispla04.02262.x
49. PubMed Abstract | CrossRef Full Text | Google Scholar
50. Hall, L. N., Rossini, L., Cribb, L., and Langdale, J. A. (1998). GOLDEN 2: a novel transcriptional regulator of cellular differentiation in pc.10.6.925
51. PubMed Abstract | CrossRef Full Text | Google Scholar
52. Hedtke, B., Alawady, A., Albacete, A., Kobayashi, K., Melzer, M., Roitsch, T., et al. (2012). Deficiency in riboflavin biosynthesis affects tetrapyrrole biosynthesis in etiolated Arabidopsis 3-011-9846-1
53. PubMed Abstract | CrossRef Full Text | Google Scholar
54. Heinemann, I. U., Jahn, M., and Jahn, D. (2008). The biochemistry of heme biosynthesis. Ab.2008.02.015
55. PubMed Abstract | CrossRef Full Text | Google Scholar
56. Hollingshead, S., Kopecná, J., Jackson, P. J., Canniffe, D. P., Davison, P. A., Dickman, M. J., et al. (2012). Conserved chloroplast open-reading frame ycf54 is required for activity of the magnesium protoporphyrin monomethylester oxidative cyclase in Synechocystis PCCbc.M112.352526
57. PubMed Abstract | CrossRef Full Text | Google Scholar
58. Holm, M., Ma, L., Qu, L., and Deng, X. (2002). Two interacting bZIP proteins are direct targets of COP1-mediated control of light-dependent gene expression /gad.969702.ing
59. PubMed Abstract | CrossRef Full Text | Google Scholar
60. Hudson, D., Guevara, D., Yaish, M. W., Hannam, C., Long, N., Clarke, J. D., et al. (2011). GNC and CGA1 modulate chlorophyll

biosynthesis and glutamate synthase (GLU1/Fd-GOGAT) expressional.pone.0026765

61. PubMed Abstract | CrossRef Full Text | Google Scholar
62. Hudson, D., Guevara, D. R., Hand, A. J., Xu, Z., Hao, L., Chen, X., et al. (2013). Rice cytokinin GATA transcription Factor1 regulates chloroplast development and plan104/pp.113.217265
63. PubMed Abstract | CrossRef Full Text | Google Scholar
64. Huq, E., Al-Sady, B., Hudson, M., Kim, C., Apel, K., and Quail, P. H. (2004). Phytochrome-interacting factor 1 is a critical bHLH regulator of chloro26/science.1099728
65. PubMed Abstract | CrossRef Full Text | Google Scholar
66. Ilag, L. L., Kumar, A. M., and Söll, D. (1994). Light regulation of chlorophyll biosynthesis at the level of 5-aminolevulinate 10.1105/tpc.6.2.265
67. PubMed Abstract | CrossRef Full Text | Google Scholar
68. Ishikawa, A., Okamoto, H., Iwasaki, Y., and Asahi, T. (2001). A deficiency of coproporphyrinogen III oxidase causes lesion formation doi: 10.1046/j.1365-313x.2001.01058.x
69. CrossRef Full Text | Google Scholar
70. Jarvis, P., and López-Juez, E. (2013). Biogenesis and homeostasis of chloroplasts and doi: 10.1038/nrm3702
71. PubMed Abstract | CrossRef Full Text | Google Scholar
72. Kakizaki, T., Matsumura, H., Nakayama, K., Che, F.-S., Terauchi, R., and Inaba, T. (2009). Coordination of plastid protein import and nuclear gene expression by plastid-to-nucleus ret 10.1104/pp.109.145987
73. PubMed Abstract | CrossRef Full Text | Google Scholar
74. Kerchev, P. I., Pellny, T. K., Vivancos, P. D., Kiddle, G., Hedden, P., Driscoll, S., et al. (2011). The transcription factor ABI4 is required for the ascorbic acid–dependent regulation of growth and regulation of jasmonate-dependent defense signaling p 10.1105/tpc.111.090100
75. PubMed Abstract | CrossRef Full Text | Google Scholar
76. Kindgren, P., Noren, L., Dios Barajas Lopez, J., de Shaikhali, J., and Strand, A. (2011). Interplay between HEAT SHOCK PROTEIN 90

and HY5 Controls PhANG Expression in Resp. doi: 10.1093/mp/ssr112

77. PubMed Abstract | CrossRef Full Text | Google Scholar
78. Kobayashi, K. (2016). Role of membrane glycerolipids in photosynthesis, thylakoid biogenesis and chl10.1007/s10265-016-0827-y
79. PubMed Abstract | CrossRef Full Text | Google Scholar
80. Kobayashi, K., Baba, S., Obayashi, T., Sato, M., Toyooka, K., Keränen, M., et al. (2012a). Regulation of root greening by light and auxin/cytokininoi: 10.1105/tpc.111.092254
81. PubMed Abstract | CrossRef Full Text | Google Scholar
82. Kobayashi, K., Fujii, S., Sasaki, D., Baba, S., Ohta, H., Masuda, T., et al. (2014a). Transcriptional regulation of thylakoid galactolipid biosynthesis coordinated with chlorophyll biosynthesis during the development of coi: 10.3389/fpls.2014.00272
83. PubMed Abstract | CrossRef Full Text | Google Scholar
84. Kobayashi, K., Fujii, S., Sato, M., Toyooka, K., and Wada, H. (2015). Specific role of phosphatidylglycerol and functional overlaps with other thylakoid lipids in Arabidopsis ci: 10.1007/s00299-014-1719-z
85. PubMed Abstract | CrossRef Full Text | Google Scholar
86. Kobayashi, K., and Masuda, T. (2016). "Regulation of chlorophyll metabolism in plants," in Handbook of Photosynthekli (Boca Raton, FL: CRC press), 173–192. doi: 10.1201/b19498-12
87. CrossRef Full Text
88. Kobayashi, K., Masuda, T., Tajima, N., Wada, H., and Sato, N. (2014b). Molecular phylogeny and intricate evolutionary history of the three isofunctional enzymes involved in the oxidatio–2155. doi: 10.1093/gbe/evu170
89. PubMed Abstract | CrossRef Full Text | Google Scholar
90. Kobayashi, K., Mochizuki, N., Yoshimura, N., Motohashi, K., Hisabori, T., and Masuda, T. (2008). Functional analysis of Arabidopsis thaliana isoforms of the Mg-c188–1195. doi: 10.1039/b802604c
91. PubMed Abstract | CrossRef Full Text | Google Scholar

92. Kobayashi, K., Narise, T., Sonoike, K., Hashimoto, H., Sato, N., Kondo, M., et al. (2013). Role of galactolipid biosynthesis in coordinated development of photosynthetic complexes and thylakoid membranes duri 250–261. doi: 10.1111/tpj.12028
93. PubMed Abstract | CrossRef Full Text | Google Scholar
94. Kobayashi, K., Obayashi, T., and Masuda, T. (2012b). Role of the G-box element in regulation of chlorophyll bi, 922–926. doi: 10.4161/psb.20760
95. PubMed Abstract | CrossRef Full Text | Google Scholar
96. Koncz, C., Mayerhofer, R., Koncz-Kalman, Z., Nawrath, C., Reiss, B., Redei, G. P., et al. (1990). Isolation of a gene encoding a n tagging in Arabidopsis thaliana. EMBO J. 9, 1337–1346.
97. PubMed Abstract | Google Scholar
98. Kopečná, J., De Vaca, I. C., Adams, N. B. P., Davison, P. A., Brindley, A. A., Hunter, C. N., et al. (2015). Porphyrin binding to Gun4 protein, facilitated by a flexible loop, controls metabolite flow through the chloro28488. doi: 10.1074/jbc.M115.664987
99. PubMed Abstract | CrossRef Full Text | Google Scholar
100. Koussevitzky, S., Nott, A., Mockler, T. C., Hong, F., Sachetto-Martins, G., Surpin, M., et al. (2007). Signals from chloroplasts convergn. Science 316, 715–719. doi: 10.1126/science. 1140516
101. CrossRef Full Text | Google Scholar
102. Kumar, A. M., and Söll, D. (2000). Antisense HEMA1 RNA expression inhibits heme and c 122, 49–56. doi: 10.1104/pp.122.1.49
103. PubMed Abstract | CrossRef Full Text | Google Scholar
104. Lau, O. S., and Deng, X. W. (2012). The photomorphogenic repressors COP193. doi: 10.1016/j.tplants.2012.05.004
105. PubMed Abstract | CrossRef Full Text | Google Scholar
106. Lee, J., He, K., Stolc, V., Lee, H., Figueroa, P., Gao, Y., et al. (2007). Analysis of transcription factor HY5 genomic binding sites revealed its hierarchical9, 731–749. doi: 10.1105/tpc.106.047688
107. PubMed Abstract | CrossRef Full Text | Google Scholar
108. Lee, K.P., Kim, C., Landgraf, F., and Apel, K. (2007). EXECUTER1- and EXECUTER2-dependent transfer of stress-related signals

from the plastid to the nucleus of Arab0270–10275. doi: 10.1073/pnas.0702061104

109. PubMed Abstract | CrossRef Full Text | Google Scholar
110. Legnaioli, T., Cuevas, J., and Mas, P. (2009). TOC1 functions as a molecular switch connecting the circ8, 3745–3757. doi: 10.1038/emboj.2009.297
111. PubMed Abstract | CrossRef Full Text | Google Scholar
112. Leister, D., and Kleine, T. (2016). Definition of a core module for the nuclear retrograde response to altered organellar gene expression iPlant 157, 297–309. doi: 10.1111/ppl.12431
113. PubMed Abstract | CrossRef Full Text | Google Scholar
114. Leivar, P., and Quail, P. H. (2011). PIFs: pivotal compon, 19–28. doi: 10.1016/j.tplants.2010.08.003
115. PubMed Abstract | CrossRef Full Text | Google Scholar
116. Leivar, P., Tepperman, J. M., Monte, E., Calderon, R. H., Liu, T. L., and Quail, P. H. (2009). Definition of early transcriptional circuitry involved in light-induced reversal of PIF-imposed repression of photoml 21, 3535–3553. doi: 10.1105/tpc.109.070672
117. PubMed Abstract | CrossRef Full Text | Google Scholar
118. León, P., Gregorio, J., and Cordoba, E. (2013). ABI4 and its lant Sci. 3:304. doi: 10.3389/fpls.2012.00304
119. PubMed Abstract | CrossRef Full Text | Google Scholar
120. Li, G., Siddiqui, H., Teng, Y., Lin, R., Wan, X., Li, J., et al. (2011). Coordinated transcriptional regulat. Cell Biol. 13, 616–622. doi: 10.1038/ncb2219
121. PubMed Abstract | CrossRef Full Text | Google Scholar
122. Li, J., Li, G., Gao, S., Martinez, C., He, G., Zhou, Z., et al. (2010). Arabidopsis transcription factor ELONGATED HYPOCOTYL5 plays a role in the Cell 22, 3634–3649. doi: 10.1105/tpc.110.075788
123. PubMed Abstract | CrossRef Full Text | Google Scholar
124. Lin, J.-F., and Wu, S.-H. (2004). Molecul, 612–628. doi: 10.1111/j.1365-313X.2004.02160.x
125. PubMed Abstract | CrossRef Full Text | Google Scholar
126. Liu, X., Chen, C.-Y., Wang, K.-C., Luo, M., Tai, R., Yuan, L., et al.

(2013). PHYTOCHROME INTERACTING FACTOR3 associates with the histone deacetylase HDA15 in repression of chlorophyll biosynthesis and nt Cell25, 1258–1273. doi: 10.1105/tpc.113.109710

127. PubMed Abstract | CrossRef Full Text | Google Scholar
128. Lu, S. X., Knowles, S. M., Andronis, C., Ong, M. S., and Tobin, E. M. (2009). CIRCADIAN CLOCK ASSOCIATED1 and LATE ELONGATED HYPOCOTYL function synerg Physiol. 150, 834–843. doi: 10.1104/pp.108.133272
129. PubMed Abstract | CrossRef Full Text | Google Scholar
130. Ma, Z., Hu, X., Cai, W., Huang, W., Zhou, X., Luo, Q., et al. (2014). Arabidopsis miR171-Targeted Scarecrow-Like proteins bind to GT cis-elements and mediate gibberellin- regulated chnet. 10:e1004519. doi: 10.1371/journal.pgen.1004519
131. PubMed Abstract | CrossRef Full Text | Google Scholar
132. Manfield, I. W., Devlin, P. F., Jen, C.-H., Westhead, D. R., and Gilmartin, P. M. (2007). Conservation, convergence, and divergence of light-responsive, circadian-regulated, and tissue-specific expression patterns durnt Physiol. 143, 941–958. doi: 10.1104/pp.106.090761
133. PubMed Abstract | CrossRef Full Text | Google Scholar
134. Mara, C. D., and Irish, V. F. (2008). Two GATA transcription factors are downstream effecant Physiol. 147, 707–718. doi: 10.1104/pp.107.115634
135. PubMed Abstract | CrossRef Full Text | Google Scholar
136. Masuda, T., and Fujita, Y. (2008). Regulam. Photobiol. Sci. 7, 1131–1149. doi: 10.1039/b807210h
137. PubMed Abstract | CrossRef Full Text | Google Scholar
138. Masuda, T., Fusada, N., Oosawa, N., Takamatsu, K., Yamamoto, Y. Y., Ohto, M., et al. (2003). Functional analysis of isoforms of NADPH: protochlorophyllide oxidoredulant Cell Physiol. 44, 963–974. doi: 10.1093/pcp/pcg128
139. PubMed Abstract | CrossRef Full Text | Google Scholar
140. Matsumoto, F., Obayashi, T., Sasaki-sekimoto, Y., Ohta, H., and Takamiya, K. (2004). Gene expression profiling of the tetrapyrrole metabPlant Physiol 135, 2379–2391. doi: 10.1104/pp.104.042408

141. PubMed Abstract | CrossRef Full Text | Google Scholar
142. McCormac, A. C., and Terry, M. J. (2002). Light-signalling pathways leading to the co-ordinated expression of HEMA1 and Lhcb duriant J. 32, 549–559. doi: 10.1046/j.1365-313X.2002.01443.x
143. PubMed Abstract | CrossRef Full Text | Google Scholar
144. Melo, N. K. G., Bianchetti, R. E., Lira, B. S., Oliveira, P. M. R., Zuccarelli, R., Dias, D. L. O., et al. (2016). Nitric oxide, ethylene, and auxin cross talk mediates greening as. Plant Physiol. 170, 2278–2294. doi: 10.1104/pp.16.00023
145. PubMed Abstract | CrossRef Full Text | Google Scholar
146. Meskauskiene, R., and Klaus, A. (2002). Interaction of FLU, a negative regulator of tetrapyrrole biosynthesis, with the glutamyl-tRNA reducta. FEBS Lett. 532, 27–30. doi: 10.1016/S0014-5793(02)03617-7
147. PubMed Abstract | CrossRef Full Text | Google Scholar
148. Meskauskiene, R., Nater, M., Goslings, D., Kessler, F., op den Camp, R., and Apel, K. (2001). FLU: a negative regulator of chlorophycad. Sci. U.S.A 98, 12826–12831. doi: 10.1073/pnas.221252798
149. PubMed Abstract | CrossRef Full Text | Google Scholar
150. Mochizuki, N., Tanaka, R., Tanaka, A., Masuda, T., and Nagatani, A. (2008). The steady-state level of Mg-protoporphyrin IX is not a determinant of plastid. Sci. U.S.A. 105, 15184–15189. doi: 10.1073/pnas.0803245105
151. PubMed Abstract | CrossRef Full Text | Google Scholar
152. Molina, A., Volrath, S., Guyer, D., Maleck, K., Ryals, J., and Ward, E. (1999). Inhibition of protoporphyrinogen oxidase expression in Arabidopsis causes a lesion-mie. Plant J. 17, 667–678. doi: 10.1046/j.1365-313X.1999.00420.x
153. PubMed Abstract | CrossRef Full Text | Google Scholar
154. Moulin, M., McCormac, A. C., Terry, M. J., and Smith, A. G. (2008). Tetrapyrrole profiling in Arabidopsis seedlings reveals that retrograde plastid nuclear signaling is not dcad. Sci. U.S.A. 105, 15178–15183. doi: 10.1073/pnas.0803054105
155. PubMed Abstract | CrossRef Full Text | Google Scholar
156. Muramoto, T., Kohchi, T., Yokota, A., Hwang, I., and Goodman,

H. M. (1999). The Arabidopsis photomorphogenic mutant hy1 is deficient in phytochrome chromophoreeme oxygenase. Plant Cell 11, 335–348. doi: 10.1105/tpc.11.3.335

157. PubMed Abstract | CrossRef Full Text | Google Scholar
158. Nagai, S., Koide, M., Takahashi, S., Kikuta, A., Aono, M., Sasaki-Sekimoto, Y., et al. (2007). Induction of isoforms of tetrapyrrole biosynthetic enzymes, AtHEMA2 and AtFC1, under stressdopsis. Plant Physiol. 144, 1039–1051. doi: 10.1104/pp.107.100065
159. PubMed Abstract | CrossRef Full Text | Google Scholar
160. Nakamura, H., Muramatsu, M., Hakata, M., Ueno, O., Nagamura, Y., Hirochika, H., et al. (2009). Ectopic overexpression of the transcription factor Os cells. Plant Cell Physiol. 50, 1933–1949. doi: 10.1093/pcp/pcp138
161. PubMed Abstract | CrossRef Full Text | Google Scholar
162. Obayashi, T., Kinoshita, K., Nakai, K., Shibaoka, M., Hayashi, S., Saeki, M., et al. (2007). ATTED-II: a database of co-expressed genes and cis elbidopsis. Nucleic Acids Res. 35, D863–D869. doi: 10.1093/nar/gkl783
163. PubMed Abstract | CrossRef Full Text | Google Scholar
164. Oborník, M., and Green, B. R. (2005). Mosaic originkaryotes. Mol. Biol. Evol. 22, 2343–2353. doi: 10.1093/molbev/msi230
165. PubMed Abstract | CrossRef Full Text | Google Scholar
166. Oosawa, N., Masuda, T., Awai, K., Fusada, N., Shimada, H., Ohta, H., et al. (2000). Identification and light-induced expression of a novel gene of NADPH-protthaliana. FEBS Lett. 474, 133–136. doi: 10.1016/S0014-5793(00)01568-4
167. PubMed Abstract | CrossRef Full Text | Google Scholar
168. op den Camp, R. G. L., Przybyla, D., Ochsenbein, C., Laloi, C., Kim, C., Danon, A., et al. (2003). Rapid induction of disygen in Arabidopsis. Plant Cell 15, 2320–2332. doi: 10.1105/tpc.014662
169. PubMed Abstract | CrossRef Full Text | Google Scholar
170. Osterlund, M. T., Hardtke, C. S., Wei, N., and Deng, X. W. (2000). development of Arabidopsis. Nature 405, 462–466. doi: 10.1038/35013076
171. PubMed Abstract | CrossRef Full Text | Google Scholar

172. Oswald, O., Martin, T., Dominy, P. J., and Graham, I. A. (2001). Plastid redox state and sugars: interactive regulato Proc. Natl. Acad. Sci. U.S.A. 98, 2047–2052. doi: 10.1073/pnas.021449998

173. PubMed Abstract | CrossRef Full Text | Google Scholar

174. Ouyang, X., Li, J., Li, G., Li, B., Chen, B., Shen, H., et al. (2011). Genome-wide binding site analysis of FAR-RED Eidopsis development. Plant Cell 23, 2514–2535. doi: 10.1105/tpc.111.085126

175. PubMed Abstract | CrossRef Full Text | Google Scholar

176. Paddock, T. N., Mason, M. E., Lima, D. F., and Armstrong, G. A. (2010). Arabidopsis protochlorophyllide oxidoreductase A (PORA) restores bulk chlorodouble mutant. Plant Mol. Biol. 72, 445–457. doi: 10.1007/s11103-009-9582-y

177. PubMed Abstract | CrossRef Full Text | Google Scholar

178. Papenbrock, J., and Grimm, B. (2001). Regulatory network of tetrapyrrole biosynthesis–studies of inttabolic and developmental control of plastids. Planta 213, 667–681. doi: 10.1007/s004250100593

179. CrossRef Full Text | Google Scholar

180. Papenbrock, J., Mock, H., Kruse, E., and Grimm, B. (1999). Expression studies in tetrapyrrole biosynthesis: inverse mrochelatase activity during cyclic photoperiods. Planta 49, 264–273. doi: 10.1007/s004250050558

181. CrossRef Full Text | Google Scholar

182. Pfannschmidt, T. (2003). its own genes. Trends Plant Sci. 8, 33–41. doi: 10.1016/S1360-1385(02)00005-5

183. PubMed Abstract | CrossRef Full Text | Google Scholar

184. Piippo, M., Allahverdiyeva, Y., Paakkarinen, V., Suoranta, U., Battchikova, N., Aro, E., et al. (2006). Chloroplast-mediated regulation of nucleat stress. Physiol. Genomics 2, 142–152. doi: 10.1152/physiolgenomics.00256.2005

185. PubMed Abstract | CrossRef Full Text | Google Scholar

186. Powell, A. L. T., Nguyen, C. V., Hill, T., Cheng, K. L., Figueroa-Balderas, R., Aktas, H., et al. (2012). Uniform ripening encodes a Goit chloroplast development. Science 336, 1711–1715. doi: 10.1126/science.1222218

187. PubMed Abstract | CrossRef Full Text | Google Scholar

188. Przybyla, D., Gobel, C., Imboden, A., Hamberg, M., Feussner, I., and Apel, K. (2008). Enzymatic, but not non-enzymatic, 1O2-mediated peroxidation of polyunsaturated fatty acids forms part of the EXECUTER1-dArabidopsis thaliana. Plant J. 54, 236–248. doi: 10.1111/j.1365-313X.2008.03409.x

189. PubMed Abstract | CrossRef Full Text | Google Scholar

190. Ranftl, Q. L., Bastakis, E., Klermund, C., and Schwechheimer, C. (2016). LLM-domain containing B-GATA factors contrt in Arabidopsis thaliana. Plant Physiol. 170, 2295–2311. doi: 10.1104/pp.15.01556

191. PubMed Abstract | CrossRef Full Text | Google Scholar

192. Reeves, W. M., Lynch, T. J., Mobin, R., and Finkelstein, R. R. (2011). Direct targets of the transcription factors ABA-Insensitive (ABI) 4 and A ABA response factors. Plant Mol. Biol. 75, 347–363. doi: 10.1007/s11103-011-9733-9

193. PubMed Abstract | CrossRef Full Text | Google Scholar

194. Reinbothe, C., Springer, A., Samol, I., and Reinbothe, S. (2009). Plant oxylipins: rolee and leaf senescence. FEBS J. 276, 4666–4681. doi: 10.1111/j.1742-4658.2009.07193.x

195. PubMed Abstract | CrossRef Full Text | Google Scholar

196. Richter, R., Behringer, C., Müller, I. K., and Schwechheimer, C. (2010). The GATA-type transcription factors GNC and GNL/CGA1 repress nd PHYTOCHROME-INTERACTING FACTORS. Genes Dev. 24, 2093–2104. doi: 10.1101/gad.594910

197. PubMed Abstract | CrossRef Full Text | Google Scholar

198. Rissler, H. M., Collakova, E., Dellapenna, D., Whelan, J., and Pogson, B. J. (2002). Chlorophyll biosynthesis. expression of a second Chl I genlimited chlorophyll synthesis. Plant Physiol. 128, 770–779. doi: 10.1104/pp.010625.770

199. PubMed Abstract | CrossRef Full Text | Google Scholar

200. Ruckle, M. E., Burgoon, L. D., Lawrence, L. A., Sinkler, C. A., and Llight signaling in Arabidopsis. Plant Physiol. 159, 366–390. doi: 10.1104/pp.112.193599

201. PubMed Abstract | CrossRef Full Text | Google Scholar

202. Ruckle, M. E., DeMarco, S. M., and Larkin, R. M. (2007). Plastid signals remodel light sroplast biogenesis in Arabidopsis. Plant Cell 19, 3944–3960. doi: 10.1105/tpc.107.054312

203. PubMed Abstract | CrossRef Full Text | Google Scholar

204. Rzeznicka, K., Walker, C. J., Westergren, T., Kannangara, C. G., von Wettstein, D., Merchant, S., et al. (2005). Xantha-l encodes a membrane subunit of the aerobic Mg-protoporphyr biosynthesis. Proc. Natl. Acad. Sci. U.S.A. 102, 5886–5891. doi: 10.1073/pnas.0501784102

205. PubMed Abstract | CrossRef Full Text | Google Scholar

206. Scharfenberg, M., Mittermayr, L., Von Roepenack-Lahaye, E., Schlicke, H., Grimm, B., Leister, D., et al.helatases in Arabidopsis thaliana. Plant Cell Environ. 38, 280–298. doi: 10.1111/pce.12248

207. PubMed Abstract | CrossRef Full Text | Google Scholar

208. Schlicke, H., Salinas, A., Firtzlaff, V., Richter, A. S., Glässer, C., Maier, K., et al. (2014). Induced deactivation of genes enetrapyrrole-mediated retrograde signaling. Mol. Plant 7, 1211–1227. doi: 10.1093/mp/ssu034

209. PubMed Abstract | CrossRef Full Text | Google Scholar

210. Schmied, J., Hedtke, B., glutamyl-tRNA reductase. J. Plant Physiol. 168, 1372–1379. doi: 10.1016/j.jplph.2010.12.010

211. PubMed Abstract | CrossRef Full Text | Google Scholar

212. Shin, J., Kim, K., Kang, H., Zulfugarov, I. S., Bae, G., Lee, C.-H., et al. (2009). Phytochromes promote seedling light responsteracting factors. Proc. Natl. Acad. Sci. U.S.A. 106, 7660–7665. doi: 10.1073/pnas.0812219106

213. PubMed Abstract | CrossRef Full Text | Google Scholar

214. Skinner, J. S., and Timko, M. P. (1999). Differential expression of genes encoding the light-dependent and light-iuring development in loblolly pine. Plant Mol. Biol. 39, 577–592. doi: 10.1023/A:1006144630071

215. PubMed Abstract | CrossRef Full Text | Google Scholar

216. Soy, J., Leivar, P., González-Schain, N., Martín, G., Diaz, C., Sentandreu, M., et al. (2016). Molecular convergence of clock and

photosensory paof target promoters. Proc. Natl. Acad. Sci. U.S.A. 113, 4870–4875. doi: 10.1073/pnas.1603745113

217. PubMed Abstract | CrossRef Full Text | Google Scholar
218. Spano, A. J., He, Z., Michel, H., Hunt, D. F., and Timko, M. P. (1992). Molecular cloning, nuclear gene structure, and oxidoreductase in pea (Pisum sativum L.). Plant Mol. Biol. 18, 967–972. doi: 10.1007/BF00019210
219. PubMed Abstract | CrossRef Full Text | Google Scholar
220. Sperling, U., Franck, F., van Cleve, B., Frick, G., Apel, K., and Armstrong, G. A. (1998). Etioplast differentiation in arabidopsis: both PORA and PORB restollide-F655 to the cop1 photomorphogenic mutant. Plant Cell 10, 283–296. doi: 10.1105/tpc.10.2.283
221. PubMed Abstract | CrossRef Full Text | Google Scholar
222. Stephenson, P. G., Fankhauserchloroplast development. Proc. Natl. Acad. Sci. U.S.A.106, 7654–7659. doi: 10.1073/pnas.0811684106
223. PubMed Abstract | CrossRef Full Text | Google Scholar
224. Stephenson, P. G., and Terry, M. J. (2008). Light signalling pathways regulating the Mg-ch-etiolation in Arabidopsis thaliana. Photochem. Photobiol. Sci. 7, 1243–1252. doi: 10.1039/b802596g
225. PubMed Abstract | CrossRef Full Text | Google Scholar
226. Strand, A., Asami, T., Alonso, J., Ecker, J. R., tion triggered by accumulation of Mg-protoporphyrinIX. Nature 421, 79–83. doi: 10.1038/nature01250.1
227. PubMed Abstract | CrossRef Full Text | Google Scholar
228. Su, Q., Frick, G., Armstrong, G., and Apel, K. (2001). POR C of Arabidopsis thaliana: a third light- athat is differentially regulated by light. Plant Mol. Biol. 47, 805–813. doi: 10.1023/A:1013699721301
229. PubMed Abstract | CrossRef Full Text | Google Scholar
230. Sun, X., Feng, P., Xu, X., Guo, H., Ma, J., Chi, W., eiption factor mediates chloroplast signals to the nucleus. Nat. Commun. 2:477. doi: 10.1038/ncomms1486
231. PubMed Abstract | CrossRef Full Text | Google Scholar
232. Sundqvist, C., and Dahlin, C. ies to light-harvesting

complexes. Physiol. Plant.100, 748–759. doi: 10.1034/j.1399-3054.1997.1000402.x

233. PubMed Abstract | CrossRef Full Text | Google Scholar
234. Susek, R. E., Ausubel, F. M., and Chory, J. (1993). Signal transd RBCS gene expression from chloroplast development. Cell 74, 787–799. doi: 10.1016/0092-8674(93)90459-4
235. PubMed Abstract | CrossRef Full Text | Google Scholar
236. Suzuki,C. E. (1997). Genetic analysis of chlorophyll biosynthesis. Annu. Rev. Genet. 31, 61–89. doi: 10.1146/annurev.genet.31.1.61
237. CrossRef Full Text | Google Scholar
238. Takahashi, K., Takabayashi, A., Tanaka, A., and Tanaka, R. (2014). Functional analysis of ligesting chlorophyll-binding motif in Arabidopsis. J. Biol. Chem. 289, 987–999. doi: 10.1074/jbc.M113.525428
239. PubMed Abstract | CrossRef Full Text | Google Scholar
240. Taki, N., Sasaki-sekimoto, Y., Obayashi, T., Kikuta, A., Kobayashi, K., Ainai, T., et al. (2005). 12-oxo-phytodienoic acid triggers expreound-induced gene expression in Arabidopsis. Plant Physiol. 139, 1268–1283. doi: 10.1104/pp.105.067058.1268
241. PubMed Abstract | CrossRef Full Text | Google Scholar
242. Takio, S., Nakao, N., Suzuki, T., Tanaka, K., Yamamoto, I., and Satoh, T. (1998). Light-dependent expression of pre in the liverwort, Marchantia paleacea var. diptera. Plant Cell Physiol. 39, 665–669. doi: 10.1093/oxfordjournals.pcp.a029420
243. CrossRef Full Text | Google Scholar
244. Tamai, H., Iwabuchi, M., and Meshi, T. (2002). Arabidopsis GARP trtivation domain shared by G-box-binding bZIP factors. Plant Cell Physiol. 43, 99–107. doi: 10.1093/pcp/pcf011
245. PubMed Abstract | CrossRef Full Text |, T. (2011). Tetrapyrrole metabolism in Arabidopsis thaliana. Arabidopsis Book 9, e0145. doi: 10.1199/tab.0145
246. PubMed Abstract | CrossRef Full Text | Google Scholar
247. Tanaka, R., Rothbart, M., Oka, S., Takabayashi, A., Takahashi, K., Shibata, M., et al. (2010). LIL3, a light-harvphyll and tocopherol biosynthesis. Proc. Natl. Acad. Sci. U.S.A. 107, 16721–16725. doi:

10.1073/pnas.1004699107

248. PubMed Abstract | CrossRef Full Text | Google Scholar
249. Tang, W., Wang, W., Chen, D., Ji, Q., Jing, Y., Wang, H., et al. (2012). Transposase-derived proteins FHY3/FAR1 interact with PHYTOCHROME-is by modulating HEMB1 during deetiolation in Arabidopsis. Plant Cell 24, 1984–2000. doi: 10.1105/tpc.112.097022
250. PubMed Abstract | CrossRef Full Text | Google Scholar
251. Tepperman, J. M., Hwang, Y. S., and Quail, P. H. (2006). phyA dominates in transductiothe initiation of Arabidopsis seedling de-etiolation. Plant J. 48, 728–742. doi: 10.1111/j.1365-313X.2006.02914.x
252. PubMed Abstract | CrossRef Full Text | Google Scholar
253. Terry, M. J., and Smith, A. G. (2013). A model r plastid-to-nucleus signaling during chloroplast biogenesis. Front. Plant Sci. 4:14. doi: 10.3389/fpls.2013.00014
254. PubMed Abstract | CrossRef Full Text | Google Scholar
255. Toledo-Ortiz, G., Johansson, H., Lee, K. P., Bou-Torrent, J., Stewart, K., Steel, G., et al. (2014).emperature control of photosynthetic gene transcription. PLoS Genet. 10:e1004416. doi: 10.1371/journal.pgen.1004416
256. PubMed Abstract | CrossRef Full Text | Google Scholar
257. Triantn plants: production, detoxification and signaling. Trends Plant Sci. 14, 219–228. doi: 10.1016/j.tplants.2009.01.008
258. PubMed Abstract | CrossRef Full Text | Google Scholar
259. Tsuchiya, Y., Vidaurre, D., Toh, S., Hanada, A., Nambara,n identifies new functions for the plant hormone strigolactone. Nat. Chem. Biol. 6, 741–749. doi: 10.1038/nchembio.435
260. PubMed Abstract | CrossRef Full Text | Google Scholar
261. Vandenbussche, F., Habricot, Y., Condiff, A. S., Maldiney, R., Van der Straeten, D., and Ahmad, M. (200 and cytokinin signalling pathways in Arabidopsis thaliana. Plant J. 49, 428–441. doi: 10.1111/j.1365-313X.2006.02973.x
262. PubMed Abstract |., Gough, S., and Kannangara, C. G. (1997). Chlorophyll biosynthesis. Plant Cell 7, 1039–1057. doi: 10.1105/tpc.7.7.1039

263. PubMed Abstract | CrossRef Full Text | Google Scholar
264. Wagner, D., Przybyla, D., Camp, R., Den Kim, C., Land of singlet oxygen–induced stress responses of Arabidopsis thaliana. Science 306, 1183–1186. doi: 10.1126/science.1103178
265. PubMed Abstract | CrossRef Full Text | Gooed roles of FHY3 and FAR1 in light signaling and beyond. Trends Plant Sci. 20, 453–461. doi: 10.1016/j.tplants.2015.04.003
266. PubMed Abstract | CrossRef Full Text | Google Scholar
267. Wang, L., Mai, Y., Zhang, Y.SCL6-II, SCL6-III, and SCL6-IV genes regulate shoot branching in Arabidopsis. Mol. Plant 3, 794–806. doi: 10.1093/mp/ssq042
268. PubMed Abstract | CrossRef Full Text | Google Scholar
269. Wang, P., Fouracral. (2013). Evolution of GOLDEN2-LIKE gene function in C3 and C4 plants. Planta 237, 481–495. doi: 10.1007/s00425-012-1754-3
270. PubMed Abstract | CrossRef Full Text | Google Scholar
271. Wang, Z., Kenigsbuch, D., Sun, L., Harel, E., Ong, M. S.,factor is involved in the phytochrome regulation of an Arabidopsis Lhcb gene. Plant Cell 9, 491–507. doi: 10.1105/tpc.9.4.491
272. PubMed Abstract | CrossRef Full Text | Google Scholar
273. Waters, M. T., Wang, P., Korkaric, M., Capper, R. G., Saunders, N. Jrs coordinate expression of the photosynthetic apparatus in Arabidopsis. Plant Cell 21, 1109–1128. doi: 10.1105/tpc.108.065250
274. PubMed Abstract | CrossRef Full Text | Google Scholar
275. Woodson, J. D., Joens, M. S., Sinson, A. B., Gilkersontin facilitates a quality-control pathway that removes damaged chloroplasts. Science 350, 450–454. doi: 10.1126/science.aac7444
276. PubMed Abstract | CrossRef Full Text | Google Scholar
277. Xu, G., Guo, H., Zhang, D., Chen, D., Jiang, Z., and Lin, R. oxidoreductase A expression and seedling greening in Arabidopsis. Photosynth. Res. 126, 331–340. doi: 10.1007/s11120-015-0146-5
278. PubMed Abstract | CrossRef Full Text | Google Scholar
279. Xu, X., Chi, W., Sun, X., of light and chloroplast signals for de-etiolation through ABI4–HY5 and COP1. Nat. Plants 2, 16066. doi: 10.1038/nplants.2016.66

280. PubMed Abstract | CrossRef Full Text | Google Scholar

281. Young, D. A., Bauer, C. E., Williams, J. C., and Marrs, B. L. (1989). Gentic evidsynthesis pigments and pigment-binding proteins in Rhodobacter capsulatus. Mol. Genet. Genomics 218, 1–12. doi: 10.1007/BF00330558

282. PubMed Abstract | CrossRef Full Text | Google Scholar

283. Zhang, S., Apel, K., and Kim, C. (2014). Singlet oxygen-mediatedArabidopsis thaliana exposed to light stress. Philos. Trans. R. Soc. Lond. B. Biol. Sci. 369, 20130227. doi: 10.1098/rstb.2013.0227

284. PubMed Abstract | CrossRef Full Text | Google Scholar

285. Zhong, S., Zhao, M., Shi, T., Shi, H., An, F., Zhao, Q., et al. (2009ation and to promote greening of Arabidopsis seedlings. Proc. Natl. Acad. Sci. U.S.A. 106, 21431–21436. doi: 10.1073/pnas.0907670106

286. PubMed Abstract | CrossRef Full Text | Google Scholar

287. Zhu, L., Bu, Q., Xu, X., Paik, I., Huang, X., Hoecker, U., et al. (2015). CUL4 forms an E3 ligase with COP1 and SPA to promote light-induced degradation of PIF1. Nat. Commun. 6, 8245. doi: 10.1038/ncomms8245

CHAPTER

3

TETRAPYRROLE SIGNALING IN PLANTS

Robert M. Larkin*

Key Laboratory of Horticultural Plant Biology (Ministry of Education), College of Horticulture and Forestry Sciences, Huazhong Agricultural University, Wuhan, China

Tetrapyrroles make critical contributions to a number of important processes in diverse organisms. In plants, tetrapyrroles are essential for light signaling, the detoxification of reactive oxygen species, the assimilation of nitrate and sulfate, respiration, photosynthesis, and programed cell death. The misregulation of tetrapyrrole metabolism can produce toxic reactive oxygen species. Thus, it is not surprising that tetrapyrrole metabolism is strictly regulated and that tetrapyrrole metabolism affects signaling mechanisms that regulate gene expression. In plants and algae, tetrapyrroles are synthesized in plastids and were some of the first plastid signals demonstrated to regulate nuclear gene expression. In plants, the mechanism of tetrapyrrole-dependent plastid-to-nucleus signaling remains poorly understood. Additionally, some of experiments that tested ideas for possible signaling mechanisms appeared to produce conflicting data. In some instances, these conflicts are potentially explained by different experimental conditions. Although the biological function of tetrapyrrole signaling is poorly understood, there is compelling evidence that this signaling is significant. Specifically, this signaling appears to affect the accumulation of starch and may promote abiotic stress tolerance. Tetrapyrrole-dependent plastid-to-nucleus signaling interacts with a distinct plastid-to-nucleus signaling mechanism that depends on GENOMES UNCUOPLED1 (GUN1). GUN1 contributes to a variety of processes, such as chloroplast biogenesis, the circadian rhythm, abiotic stress tolerance, and development. Thus, the contribution of tetrapyrrole signaling to plant function is potentially broader than we currently appreciate. In this review, I discuss these aspects of tetrapyrrole signaling.

INTRODUCTION

Tetrapyrroles contribute a number of functions to processes that are critical for the viability of diverse organisms including essential functions to the light reactions photosynthesis, which underpins life on earth. Tetrapyrroles contain four pyrroles— aromatic rings containing four carbon atoms and one nitrogen atom—often linked together with methine groups, in linear (e.g., bilins) or cyclic arrangements (e.g., porphyrins). Porphyrins chelate metal ions, such as cobalt, iron, or magnesium ions. These structural features along with a variety of ring substituents contribute to the diverse functions of tetrapyrroles (Tanaka and Tanaka, 2007; Tanaka et al., 2011; Brzezowski et al., 2015; Figure 1).

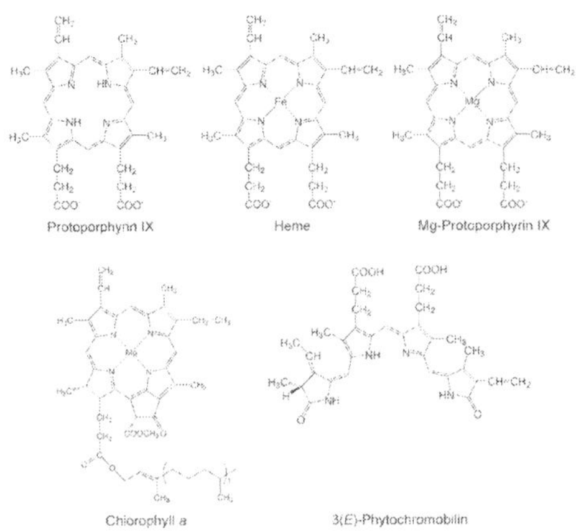

Figure 1. Selected tetrapyrroles from plants. The structures were adapted from Tanaka and Tanaka (2007).

In plants and algae, tetrapyrroles are synthesized ultimately from glutamate by a branched biosynthetic pathway localized entirely within plastids. The end products of this branched pathway are siroheme, heme, phytochromobilin, chlorophyll a, and chlorophyll b (Figure 2). These tetrapyrroles provide critical functions to important processes, such as light signaling, the detoxification of reactive oxygen species, the assimilation of nitrate and sulfate, respiration, programmed cell death, and the light-harvesting reactions of photosynthesis. In vivo, many tetrapyrroles, such as porphyrins, are sequestered to prevent them from generating singlet oxygen ($^1O_2$), a toxic reactive oxygen species that causes stress and cell death. Electronically excited versions of porphyrins produce $^1O_2$ when they transfer energy to molecular oxygen (Tanaka and Tanaka, 2007; Tanaka et al., 2011; Brzezowski et al., 2015).

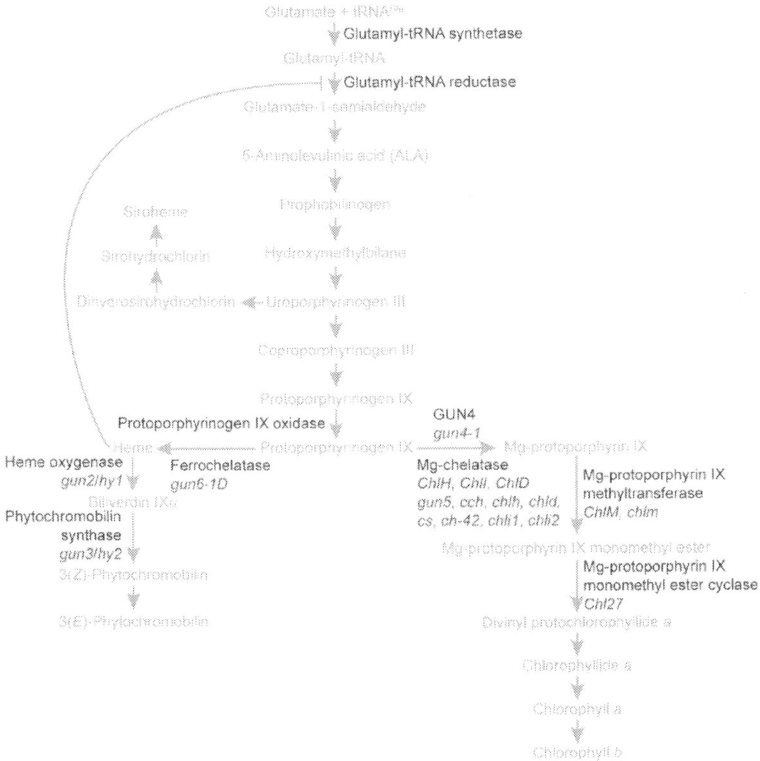

Figure2. The branched tetrapyrrole biosynthetic pathway of plants**.** The names of the precursors and end products of the branched tetrapyrrole biosynthetic

pathway are indicated with gray. Enzyme catalyzed reactions are indicated with gray arrows. The feedback inhibition of glutamyl tRNA reductase by heme is indicated with a gray T bar. The names of proteins, such as regulatory proteins and enzymes, are indicated with black. The tetrapyrrole biosynthetic pathway was adapted from Tanaka and Tanaka (2007) and Tanaka et al. (2011). Genes encoding enzymes or enzyme subunits are indicated with blue. Mutants discussed in the text are indicated with red. gun5, cch, and chlh are mutant alleles of ChlH/GUN5. chld is a mutant allele of ChlD. cs, ch-42, and chli1 are mutant alleles of ChlI1. chli2 is a mutant allele of ChlI2.

Considering the large number of critical functions performed by tetrapyrroles and the potentially lethal consequences of misregulating tetrapyrrole biosynthesis, it is not surprising that the levels of tetrapyrroles are strictly controlled (Tanaka and Tanaka, 2007; Tanaka et al., 2011; Brzezowski et al., 2015) and that tetrapyrroles affect signaling mechanisms that regulate gene expression in diverse organisms (Terry and Smith, 2013; Brzezowski et al., 2015; Zhang et al., 2015). Indeed, tetrapyrroles were some of the first molecules suggested to serve as plastid signals that regulate nuclear gene expression in both plants and algae (Susek and Chory, 1992). The idea that in addition to serving as the metabolic centers of plant cells, plastids serve as sensors that emit signals that regulate gene expression in the nucleus was first suggested almost forty years ago (Bradbeer et al., 1979). Since the early days of plastid-to-nucleus signaling research, researchers have studied the plastid regulation of photosynthesis-associated nuclear genes (PhANGs), especially the genes encoding the type I proteins of the major light harvesting complex of photosystem II (Lhcb1) (Susek and Chory, 1992; Gray et al., 2003). The Lhcb1 proteins accumulate in the thylakoid membranes where they bind chlorophylls, carotenoids, and serve as antennae for photosystem II (PSII) (Jansson et al., 1992). The expression of Lhcb genes is well studied. Additionally, when chloroplasts experience dysfunction, the expression of Lhcb genes is severely downregulated relative to other PhANGs (Susek and Chory, 1992).

Since the early days of plastid-to-nucleus signaling research, mutant alleles, and inhibitors that specifically block chloroplast biogenesis by a variety of distinct mechanisms were used to induce the plastid-to-nucleus signaling mechanisms that downregulate the expression of Lhcb genes and other PhANGs (Susek and Chory, 1992; Gray et al., 2003). Chloroplast biogenesis refers to the development of chloroplasts from non-photosynthetic proplastids. Chloroplast biogenesis occurs during germination and leaf development (Pogson et al., 2015). Subsequently, chloroplast division and a mechanism that establishes the size of the chloroplast compartment drive the proliferation of chloroplasts (Osteryoung and Pyke, 2014; Larkin et al., 2016). Experiments with inhibitors and mutant alleles that block chloroplast biogenesis indicated that the expression of PhANGs depends on chloroplast biogenesis and that the plastid-to-nucleus signaling that is activated by blocking chloroplast biogenesis typically downregulates the transcription of PhANGs (Susek and Chory, 1992; Gray et al., 2003). However, there is evidence for plastid signals using a posttranscriptional mechanism to regulate the expression of a PhANG. The plastid regulated expression of PetE from pea requires an mRNA with an intact 5′ end and depends on translation in transgenic tobacco and transgenic Arabidopsis (Sullivan and Gray, 2002; Brown et al., 2005). During the early days of plastid-to-nucleus signaling research, a porphyrin was considered an attractive candidate for a plastid signal that regulates nuclear gene expression because at the time, heme was known to regulate nuclear gene expression in Saccharomyces cerevisiae (Forsburg and Guarente, 1989) and an inverse correlation was reported between the levels of the chlorophyll precursor Mg-protoporphyrin IX monomethyl ester (Mg-ProtoME) and the levels of the mRNAs that encode the Lhcb proteins and the small subunit of RuBisCO in Chlamydomonas reinhardtii (Johanningmeier and Howell, 1984).

Now, there is a larger body of evidence favoring the idea that porphyrins and bilins serve as plastid signals that regulate nuclear gene expression in both plants and algae (Terry and Smith, 2013; Brzezowski et al., 2015; Zhang et al., 2015). Additionally, we now know that there are a number of non-porphyrin plastid signals that activate distinct plastid-to-nucleus signaling mechanisms. We also know that plastid-to-nucleus signaling is not only a feedback mechanism that optimizes chloroplast function but that plastid-to-nucleus signaling affects extraplastidic processes (Larkin, 2014; Chan et al., 2016). In addition to affecting entire cells by regulating nuclear gene expression, plastid signals affect individual chloroplasts by promoting the degradation of individual chloroplasts that suffer from photooxidative stress, apparently by stimulating the ubiquitination of chloroplast proteins that are located on the cytosolic surface of chloroplasts (Woodson et al., 2015). Indeed, plastid-to-nucleus signaling mechanisms broadly contribute to plant function by serving as major regulators of plant cells.

Plastid-to-nucleus signaling research is frequently and extensively reviewed. Here, I review the data indicating that porphyrins regulate nuclear gene expression and the evidence for particular mechanisms of porphyrin-regulated gene expression in plants. I also review possible biological functions for porphyrin-dependent plastid-to-nucleus signaling and our understanding of the interactions between a porphyrin-dependent signaling mechanism with a signaling mechanism that does not appear to require porphyrins for its activation.

EARLY EXPERIMENTS RELEVANT TO TETRAPYRROLE SIGNALING IN PLANTS

During the early days of plastid-to-nucleus signaling research, a number experiments were performed that tested whether porphyrins might contribute to the plastid regulation of nuclear gene expression in plants, but there were major technical obstacles. First, plants were thought to not take up porphyrins, such as the chlorophyll precursor Mg-ProtoME that was implicated as a plastid signal in C. reinhardtii(Johanningmeier and Howell, 1984), or similar porphyrins, such as Mg-protoporphyrin IX (Mg-Proto), and hemin. Second, the porphyrin precursors of chlorophyll, such as Mg-Proto and Mg-ProtoME, are potent photosensitizers. Third, the model genes for studying plastid-to-nucleus signaling were PhANGs. The expression of PhANGs is strongly induced by qualities and quantities of light that would induce severe photooxidative stress if porphyrins were to accumulate and diffuse freely throughout plant cells.

Researchers overcame these technical barriers by feeding 5-aminolevulinic acid (ALA) to plants. ALA is the 5-carbon universal precursor of tetrapyrroles. ALA is hydrophilic and is readily taken up by plants. Exogenous ALA was demonstrated to flood the tetrapyrrole biosynthetic pathway with intermediates, induce increases in the levels of chlorophyll precursors (Granick, 1959; Gough, 1972; Mascia, 1978), and downregulate the expression of Lhcb genes in Lepidium sativum and Arabidopsis thaliana seedlings in non-photosensitizing conditions, such as after a non-photosensitizing pulse of red light, continuous illumination with non-photosensitizing far-red light, or the dark (Kittsteiner et al., 1991; Vinti et al., 2000). These data provide evidence that the accumulation of chlorophyll precursors, such as Mg-Proto or Mg-ProtoME, or that the accumulation of other plastid-derived tetrapyrroles regulate the expression of PhANGs in plants. Similar results were obtained after treating plants with inhibitors that induce increases in the levels of Mg-ProtoMe and Mg-Proto in non-photosensitizing conditions (Oster et al., 1996; La Rocca et al., 2001). More recently, ALA was suggested to mediate the effects of Mg-porphyrin metabolism on nuclear gene expression (Czarnecki et al., 2012)

THE GUN MUTANT SCREEN

Developing mutant screens that specifically interrogate particular mechanisms is an objective approach for identifying completely unknown mechanisms. Thus, developing a specific mutant screen was an attractive approach for studying the plastid-to-nucleus signaling that is activated when chloroplast biogenesis is blocked. Developing specific mutant screens remains an attractive approach for studying diverse plastid-to-nucleus signaling mechanisms because our current knowledge of plastid-to-nucleus signaling mechanisms indicates that many plastid-to-nucleus signaling mechanisms are not intuitively obvious (Chan et al., 2016). The first specific mutant screen developed that specifically interrogates plastid-to-nucleus signaling mechanisms was the genomes uncoupled (gun) mutant screen (Susek and Chory, 1992; Susek et al., 1993). gun mutant screens use reporter genes to screen for Arabidopsis mutants that transcribe elevated levels of PhANGs, such as Lhcb genes, when chloroplast biogenesis is blocked with an herbicide treatment (Susek et al., 1993; Koussevitzky et al., 2007; Ruckle et al., 2007; Cottage et al., 2008; Woodson et al., 2011). gun alleles do not affect the expression of PhANGs in green seedlings containing well-functioning chloroplasts. gun alleles may attenuate the activity of negative regulators of PhANG expression, or gun alleles may induce the activity of positive regulators of PhANG expression (Susek et al., 1993; Mochizuki et al., 2001; Koussevitzky et al., 2007; Ruckle et al., 2007; Woodson et al., 2011). In addition to regulating the expression of PhANGs when chloroplast biogenesis is blocked, the plastid-to-nucleus signaling that depends on the GUN genes was shown to regulate the expression of PhANGs when chloroplast and etioplast function is somewhat attenuated (Koussevitzky et al., 2007; Sun et al., 2011; Woodson et al., 2013; Tadini et al., 2016).

gun mutant screens yielded a large number of mutant alleles of genes that encode a chloroplastic pentatricopeptide repeat protein named GUN1 (Susek et al., 1993; Koussevitzky et al., 2007; Cottage et al., 2008), a photoreceptor that is activated by blue light named cryptochrome1 (cry1) (Ruckle et al., 2007), and genes that contribute to tetrapyrrole metabolism (Mochizuki et al., 2001; Larkin et al., 2003; Adhikari et al., 2011). These data indicate that gun mutant screens specifically disrupt a few distinct mechanisms that downregulate the expression of Lhcb genes when chloroplast biogenesis is blocked and that when chloroplast biogenesis is blocked, specific mechanisms downregulate the expression of PhANGs. The specificity of the gun mutant screen was first demonstrated more than 20 years ago (Susek et al., 1993). The data from Susek et al. (1993) and the data from subsequent analyses of gun mutant screens (Mochizuki et al., 2001; Larkin et al., 2003; Koussevitzky et al., 2007; Ruckle et al., 2007; Cottage et al., 2008; Adhikari et al., 2011) conflict with a commonly promoted alternative interpretation that unnatural and complex metabolism downregulates the expression of PhANGs when chloroplast biogenesis is blocked (Pfannschmidt, 2010; Voigt et al., 2010; Barajas-López et al., 2012; Kindgren et al., 2012; Terry and Smith, 2013).

GUN2, GUN3, GUN4, GUN5, AND GUN6

The gun screen yielded several mutant alleles of genes that contribute to tetrapyrrole metabolism named GUN2, GUN3, GUN4, GUN5, and GUN6 (Figure 2). GUN2 encodes heme oxygenase. GUN3 encodes phytochromobilin synthase (Mochizuki et al., 2001). These enzymes contribute to the biosynthesis of phytochromobilin, the chromophore of phytochromes (Tanaka and Tanaka, 2007). gun2 and gun3 are allelic to hy1 and hy2, which disrupt photomorphogenesis (Koornneef et al., 1980). A hy1 mutant was independently demonstrated to exhibit a gun phenotype (Vinti et al., 2000). GUN5 encodes the 140-kDa subunit of Mg-chelatase (Mochizuki et al., 2001). GUN4 encodes an activator of Mg-chelatase that binds the 140-kDa subunit of Mg-chelatase and the porphyrin substrate and product of Mg-chelatase (Larkin et al., 2003; Davison et al., 2005; Verdecia et al., 2005; Adhikari et al., 2009). GUN4 is only found in organisms that perform oxygenic photosynthesis

(Larkin et al., 2003). GUN4 was demonstrated to shield porphyrins from collisions with molecular oxygen that produce $^1O_2$ (Brzezowski et al., 2014), as was suggested previously (Larkin et al., 2003; Verdecia et al., 2005). However, the subsequent finding that porphyrin-binding pocket of GUN4 is partially opened (Chen et al., 2015) and that in vitro, porphyrins produce more $^1O_2$ when they are bound to GUN4 (Tarahi Tabrizi et al., 2016) indicates that GUN4 requires additional factors to shield porphyrins from collisions with molecular oxygen. Thus, GUN4 and GUN5 divert protoporphyrin IX from heme biosynthesis to chlorophyll biosynthesis by inserting a magnesium ion into protoporphyrin IX, yielding Mg-Proto. GUN6 encodes ferrochelatase1 (FC1), which synthesizes heme from protoporphyrin IX (Woodson et al., 2011). The cloning of so many mutant alleles that disrupt tetrapyrrole metabolism indicates that gun mutant screens specifically disrupt a plastid-to-nucleus signaling mechanism that is regulated by tetrapyrrole metabolism. Consistent with these data, porphyrins serve as plastid signals in diverse organisms. Null alleles of GUN4 affected nuclear gene expression in C. reinhardtii—even in dark-grown cultures (Formighieri et al., 2012), both Mg-Proto and heme were reported to serve as plastid signals in algae (Beck, 2005; von Gromoff et al., 2008; Voß et al., 2011; Tanaka and Hanaoka, 2013), and heme was demonstrated to regulate various signaling mechanisms in diverse organisms, such as bacteria, red algae, yeast, and animals (Tanaka and Hanaoka, 2013; Terry and Smith, 2013; Kobayashi et al., 2016).

In the gun mutant screen, norflurazon is typically used to block chloroplast biogenesis. This herbicide specifically inhibits phytoene desaturase, which plays an essential role in carotenoid biosynthesis by catalyzing two sequential dehydrogenation reactions on phytoene, yielding ζ-carotene. Thus, norflurazon treatments inhibit the accumulation of carotenoids. Carotenoids are essential for quenching triplet chlorophyll and $^1O_2$ in green tissues (Susek and Chory, 1992; Gray et al., 2003; Nisar et al., 2015). Thus, when green leaves are treated with norflurazon, $^1O_2$ accumulates and leaves suffer from severe photooxidative stress (Kim and Apel, 2013). However, in the gun mutant screen, Arabidopsis seeds are germinated on a growth medium containing norflurazon and subsequently, seedlings are grown for several days on the same medium until their gene expression phenotypes are scored. In these conditions, chlorophyll does not accumulate, the thylakoid membranes do not develop, chloroplast biogenesis appears arrested at an early stage, and plastids resemble proplastids (Susek et al., 1993). Norflurazon was proposed to block chloroplast biogenesis by causing photooxidative stress during the conversion of proplastids to chloroplasts (Susek and Chory, 1992; Gray et al., 2003). This model assumes that when seedlings are germinated on a medium containing norflurazon—during the biogenesis of chloroplasts from non-photosynthetic proplastids—free chlorophyll or free chlorophyll precursors accumulate without the concomitant accumulation of carotenoids. This model assumes that these chlorophylls or chlorophyll precursors drive the accumulation of $^1O_2$ and that the resulting photooxidative stress blocks chloroplast biogenesis (Susek and Chory, 1992; Gray et al., 2003).

A variety of independent experiments—including experiments with $^1O_2$-detecting dyes and $^1O_2$-regulated genes—unambiguously indicated that $^1O_2$ accumulates in green Arabidopsis leaves treated with norflurazon (Kim and Apel, 2013). In stark contrast, the same methods for detecting $^1O_2$ indicated that $^1O_2$does not appear to accumulate when Arabidopsis seeds are germinated on a medium containing norflurazon (Kim and Apel, 2013). One interpretation of these data is that norflurazon might use a different mechanism to photobleach green leaves and to block chloroplast biogenesis during germination. For instance, the coordinated downregulation of both carotenoid and chlorophyll biosynthesis was suggested to block chloroplast biogenesis by preventing the biogenesis of the thylakoid membranes, without inducing photooxidative stress (Kim and Apel, 2013). Indeed, neither chlorophyll nor its precursors accumulate when Arabidopsis seedlings are germinated on a medium containing norflurazon (Mochizuki et al., 2008; Moulin et al., 2008).

An alternative interpretation is that a transient production of ROS blocks chloroplast biogenesis when Arabidopsis seeds are germinated on a medium containing norflurazon and because of the transient nature of this increase in the levels of ROS, these ROS are impractical or impossible to detect (Mochizuki et al., 2010; Kim and Apel, 2013). Consistent with this interpretation, when seeds were germinated on a medium containing norflurazon, the expression of one $^1O_2$-inducible gene was higher after 3 days than after 5 days of growth (Kim and Apel, 2013). However, changes in the expression of more than one $^1O_2$-inducible gene are required to provide evidence for elevated levels of $^1O_2$ because genes are typically regulated by more than one signal. Nonetheless, light influences the ability of norflurazon to downregulate the expression of nuclear genes that are regulated by plastid signals, which provides more evidence for a mechanism that depends on photooxidative stress (Susek and Chory, 1992). For this reason, norflurazon-treated seedlings are typically grown in bright white light (e.g., at least 100 μmol m^{-2} s^{-1}) to test for gunphenotypes. However, norflurazon appeared to block chloroplast biogenesis when Arabidopsis seedlings were grown in 10 μmol m^{-2} s^{-1} white light (Voigt et al., 2010). The impact of chloroplast function on light signaling (Lepistö and Rintamäki, 2012; Larkin, 2014) may at least partially explain the effect of the fluence rate on norflurazon treatments. Light signaling generally induces the expression of PhANGs in green seedlings but when chloroplast biogenesis is blocked, light signaling becomes a negative regulator of Lhcb expression (Ruckle et al., 2007 and Figure 1 from Ruckle et al., 2012) and severely attenuates the expression of many other PhANGs (Ruckle et al., 2012). This conversion of light signaling from a positive to a negative regulator of Lhcb genes while only attenuating the light-induced expression of other PhANGs may explain why blocking chloroplast biogenesis with norflurazon treatments downregulates the expression of Lhcb genes more than other PhANGs (Susek and Chory, 1992).

A number of authors have promoted the idea that the gun mutants with deficiencies in tetrapyrrole metabolism are partially resistant to norflurazon without discussing the data that conflict with this interpretation. Their interpretation is that norflurazon is less effective in mutants that are partially chlorophyll deficient because lower levels of chlorophyll biosynthesis reduce the amount of ROS produced during chloroplast biogenesis in norflurazon-treated seedlings. Therefore, in these gun mutants, they propose that the norflurazon-induced inhibition of chloroplast biogenesis is not as severe as in wild-type seedlings. Consequentially, they propose that PhANGs are expressed at higher levels in these gun mutants than in wild type because PhANG expression is correlated with chloroplast biogenesis (Mochizuki et al., 2008; Moulin et al., 2008; Kleine et al., 2009; Galvez-Valdivieso and Mullineaux, 2010; Mochizuki et al., 2010; Armbruster et al., 2011; Jarvis and López-Juez, 2013; Kleine and Leister, 2013; Terry and Smith, 2013; Bobik and Burch-Smith, 2015).

The first indication that the chlorophyll-deficient gun mutants are not resistant to norflurazon was the lack of correlation between chlorophyll-deficient phenotypes and the gun phenotypes, which was apparent after the characterization of a number of tetrapyrrole biosynthesis mutants from Arabidopsis. For instance, the 40-kDa subunit of Mg-chelatase is encoded by two genes named ChlI1 and ChlI2. A ChlI1mutant named ch42-1 was reported to accumulate 11% of the chlorophyll found in wild type or to not accumulate chlorophyll (Mochizuki et al., 2001; Rissler et al., 2002). When ChlI2 was knocked out, chlorophyll levels were significantly less but similar to wild type (Kobayashi et al., 2008). Thus, CHLI1encodes the predominant isoform of this Mg-chelatase subunit. cs is a leaky loss-of-function allele of ChlI1(Koncz et al., 1990). cs and gun4-1 accumulated less than 40% of the chlorophyll found in wild type. cchand gun5 are GUN5 mutants that accumulated approximately 35 and 70% of the chlorophyll found in wild type (Mochizuki et al., 2001). In contrast to cch, gun4-1, and gun5, neither cs nor ch42-1 expressed higher levels of Lhcb1 than wild type when chloroplast biogenesis was blocked with a norflurazon treatment (i.e., cch, gun4-1, and gun5 are gun mutants. cs and ch42-1 are not gun mutants) (Mochizuki et al., 2001). Additionally, the gun phenotype appeared significantly attenuated in the completely albino hy1 gun5double mutant relative to the partially chlorophyll deficient gun5 and hy1 (i.e., gun2) mutants (Vinti et al., 2000). Thus, there is no correlation between chlorophyll deficient phenotypes and gun phenotypes.

T-DNA insertion alleles of the genes encoding the subunits of Mg-chelatase that induce severe chlorophyll deficiencies or albinism also induce gun phenotypes (Huang and Li, 2009). In contrast to previous work that used northern blots to test for gun phenotypes (Vinti et al., 2000; Mochizuki et al., 2001), Huang and Li (2009) quantified the expression of Lhcb1 genes using qRT-PCR. They found an approximately twofold increase in Lhcb1 expression in the severely chlorophyll-deficient chli1 mutants and 4.5- to 8-fold increases in the albino chld and chlh mutants, which are null alleles of the genes encoding the 80- and 140-kDa subunits of Mg-chelatase, respectively. The finding that CHLI1 mutants were either not gunmutants or were subtle gun mutants depending on whether Lhcb1 expression was quantified by northern blotting or by qRT-PCR and that CHLH mutants exhibited robust gun phenotypes regardless of the method used for quantifying Lhcb1 expression are essentially the same results.

Additional data conflict with the idea that partial resistance to norflurazon explains the gun phenotypes of the gun mutants that have defects in tetrapyrrole metabolism. (1) Attempts to show that norflurazon-treated gun4-1 and gun5 accumulate less reactive oxygen species (ROS) than norflurazon-treated wild type by quantifying ROS-responsive gene expression, staining with ROS responsive dyes, and by treating plants with ROS scavengers were not successful (Strand et al., 2003; Voigt et al., 2010; Zhang et al., 2011a; Kim and Apel, 2013). (2) No gun mutants were obtained from screens for norflurazon resistant mutants (i.e., the happy on norflurazon mutants) (Saini et al., 2011). (3) gun4-1 and gun5 were not resistant to low concentrations of norflurazon, in contrast to the happy on norflurazon mutants (Saini et al., 2011). (4) Blocking chloroplast biogenesis with a norflurazon treatment activated plastid-to-nucleus signaling mechanisms that are distinct from the $^1O_2$-dependent plastid-to-nucleus signaling mechanisms that were activated when green seedlings were treated with norflurazon and consequentially the levels of chloroplastic $^1O_2$ were increased (Kim and Apel, 2013). (5) High-fluence-rate light and a ROS-inducing methyl viologen treatment required different promoter elements than norflurazon treatments that blocked chloroplast biogenesis to downregulate the expression of the Lhcb1*2 promoter from Nicotiana plumbaginifolia (Staneloni et al., 2008). (6) Null alleles of sigma factor2 (sig2) reduced chlorophyll levels to 20% of the chlorophyll found in wild type. Thus, in contrast to norflurazon treatments, sig2 attenuates but does not block chloroplast biogenesis. sig2 mutants expressed lower levels of PhANGs than wild type. Etiolated gun5 sig2 expressed higher levels Lhcb2.2 and other PhANGs than etiolated sig2 (Woodson et al., 2013). Thus, perturbing tetrapyrrole metabolism affected nuclear gene expression in dark-grown Arabidopsis seedlings that were not treated with norflurazon. (7) Tetrapyrrole metabolism regulated starch-related nuclear gene expression in Bright Yellow-2 (BY2) cells derived from tobacco, which lack photosynthetic plastids (Enami et al., 2011). (8) In C. reinhardtii, mutant alleles of GUN4 and ChlM—the gene encoding Mg-protoporphyrin IX methyltransferase—upregulated the expression of genes that encode the LHC proteins and genes that contribute to tetrapyrrole metabolism in cells that were not treated with norflurazon

(Meinecke et al., 2010; Formighieri et al., 2012; Brzezowski et al., 2014). (9) Phytoene desaturase was inhibited in norflurazon-treated gun mutants with deficiencies in tetrapyrrole metabolism, as a class. Thus, alternative mechanisms of partial resistance do not explain the phenotypes of these gunmutants (Voigt et al., 2010).

The idea that no evidence of partial resistance to norflurazon was obtained in so many independent experiments because during chloroplast biogenesis, the norflurazon-induced increase in ROS is transient or localized (Mochizuki et al., 2010; Kim and Apel, 2013) is difficult to reconcile with the data. This interpretation is especially difficult to reconcile with the findings that there was no correlation between the chlorophyll-deficient phenotypes and the gun phenotypes (Vinti et al., 2000; Mochizuki et al., 2001), etiolated gun5 sig2 expressed higher levels of several PhANGs than etiolated sig2 (Woodson et al., 2013), and neither gun4-1 nor gun5 were resistant to low concentrations of norflurazon—in contrast to norflurazon-resistant mutants (Saini et al., 2011). If norflurazon treatments block chloroplast biogenesis in Arabidopsis seedlings by inducing photooxidative stress, (1) chloroplast biogenesis in Arabidopsisseedlings would need to exhibit such extreme sensitivity to this photooxidative stress that the perturbations in the tetrapyrrole biosynthetic pathway exhibited by gun2, gun3, gun4-1, gun5, and gun6-1D would not affect the photobleaching activity of norflurazon and (2) the perturbations in tetrapyrrole metabolism induced by gun5 would need to affect a distinct mechanism to affect the expression of PhANGs in the dark.

THE MECHANISM OF TETRAPYRROLE SIGNALING IN PLANTS: CHLOROPHYLL BIOSYNTHETIC ENZYMES AND CHLOROPHYLL PRECURSORS

The mutant alleles of GUN2/HY1, GUN3/HY2, GUN4, and GUN5 that were isolated from gun mutant screens are loss-of-function alleles (Mochizuki et al., 2001; Larkin et al., 2003; Adhikari et al., 2011). In contrast, the mutant allele of GUN6 that was isolated by Woodson et al. (2011) is a gain-of-function allele. These mutant alleles reduce the flux through the chlorophyll branch by two distinct mechanisms: (1) directly inhibiting chlorophyll biosynthesis or (2) promoting the accumulation of heme. Heme inhibits chlorophyll biosynthesis by attenuating the activity of glutamyl tRNA reductase. These mutant alleles may also induce increases in the levels of heme by diverting protoporphyrin IX from the chlorophyll branch to the heme branch, increasing ferrochelatase activity, or by attenuating the activities of heme oxygenase and phytochromobilin synthase (Vinti et al., 2000; Mochizuki et al., 2001; Larkin et al., 2003; Woodson et al., 2011).

Based on the finding that chlorophyll biosynthesis is attenuated in gun4, gun5, cch, cs, and ch42-1, and that in contrast to gun4, gun5, and cch, neither cs nor the ch42-1 are gun mutants, tetrapyrrole-dependent plastid-to-nucleus signaling was proposed to not depend on the accumulation of a particular chlorophyll precursor, such as Mg-Proto or Mg-ProtoME. Instead, GUN5 was proposed to participate in plastid-to-nucleus signaling (Mochizuki et al., 2001). The turnover of a porphyrin, such as Mg-Proto, or changes in the subchloroplastic localization of GUN4, GUN5, or a GUN4-GUN5 complex associated or not associated with a porphyrin ligand was suggested affect plastid-to-nucleus signaling (Mochizuki et al., 2001, 2008, 2010; Larkin et al., 2003; Sun et al., 2011). Consistent with these ideas, a number of proteins perform more than one function, such as distinct functions in metabolism and signaling (Boukouris et al., 2016). Indeed, the ortholog of GUN5 from Synechocystis sp. PCC 6803 serves as both a Mg-chelatase subunit and as an anti-sigma factor (Osanai et al., 2009).

The finding that porphyrin binding increases the affinity of GUN4 and GUN5 for chloroplast membranes (Adhikari et al., 2009, 2011) is also consistent with these ideas because the chloroplast envelope membrane may serve as an important subchloroplastic location for communication between chloroplasts and extrachloroplastic compartments.

Subsequently, the levels of Mg-Proto were reported to accumulate when chloroplast biogenesis was blocked with a norflurazon treatment, and Mg-Proto was reported to accumulate to lower levels in gun2and gun5 (Strand et al., 2003). Data from a number of additional experiments performed by Strand et al. (2003) supported the idea that the accumulation of Mg-Proto is a plastid signal that downregulates PhANG expression, including an experiment showing that exogenous Mg-Proto downregulated the expression of Lhcb genes in protoplasts and that exogenous hemin did not affect the expression of Lhcbgenes in protoplasts. The data from Strand et al. (2003) conflict with the data reported by Mochizuki et al. (2001) and provide evidence that the chlorophyll-deficient gun mutants affect plastid-to-nucleus signaling by reducing the levels of Mg-Proto. Subsequently, other work provided more evidence that the accumulation of a particular Mg-porphyrin might serve as a plastid signal. A barley mutant deficient in the Mg-ProtoME cyclase accumulated Mg-ProtoME and did not exhibit a gun phenotype (Gadjieva et al., 2005). An Arabidopsis mutant deficient in the Mg-ProtoME cyclase accumulated Mg-Proto and did not exhibit a gun phenotype (Ankele et al., 2007). An Arabidopsis mutant deficient in Mg-Proto methyltransferase accumulated Mg-Proto and did not exhibit a gun phenotype (Pontier et al., 2007). Based on these data, these authors suggested that the accumulation of either Mg-Proto (Ankele et al., 2007; Pontier et al., 2007) or Mg-ProtoME (Gadjieva et al., 2005) might serve as a plastid signal that downregulates the expression of PhANGs. Modulating the levels of Mg-Proto methyltransferase with sense or antisense expression in tobacco was reported to affect PhANG expression, which is also consistent with Mg-porphyrins serving as a plastid signal (Alawady and Grimm, 2005). Using confocal laser scanning spectroscopy, Mg-Proto or Mg-ProtoME were detected in the cytosol of ALA fed Arabidopsis plants that were treated or not treated with norflurazon (Ankele et al., 2007; Zhang et al., 2011a). In one instance, Mg-Proto or Mg-ProtoME were detected in the cytosol of cells from plants that were not fed with ALA (Zhang et al., 2011a). Thus, plastids are capable of exporting Mg-Proto or Mg-ProtoME. These data are consistent with plastids exporting Mg-Proto or Mg-ProtoME and with Mg-Proto or Mg-ProtoME activating a signaling mechanism in the cytosol that regulates the expression of PhANGs.

As discussed above, researchers commonly elevated the endogenous levels of porphyrins with exogenous ALA because plant roots were thought to not take up porphyrins, such as Mg-Proto. More recently, Arabidopsis plants were reported to take up Mg-Proto through their roots when they were simply watered with solutions containing Mg-Proto (Zhang et al., 2011a; Kindgren et al., 2012; Barajas-López et al., 2013). An increase in the efficiency of Mg-Proto and hemin feeding was reported by first vacuum infiltrating seedlings prior to root feeding (Zhang et al., 2013). In one instance, feeding Mg-Proto to plants through their roots was reported to induce an increase in the areal levels of Mg-Proto (Kindgren et al., 2012). Exogenous Mg-Proto, Mg-ProtoME, or hemin doubled the cellular RNA content in Arabidopsis plants (Zhang et al., 2011b). Feeding both Mg-Proto and hemin to plants downregulated the expression of Lhcband RbcS genes (Zhang et al., 2011a, 2013; Kindgren et al., 2012; Barajas-López et al., 2013). This reduced expression of PhANGs that was induced by exogenous porphyrins was attenuated in gun1, long hypocotyl5(hy5), aba-insentive4 (abi4), and phytochrome-associated protein phosphatase 5 (papp5) (Zhang et al., 2011a; Kindgren et al., 2012; Barajas-López et al., 2013). These findings are consistent with tetrapyrrole metabolism affecting plastid-to-nucleus signaling upstream of the chloroplastic pentatricopeptide protein GUN1 and with PAPP5 and the nuclear transcription factors HY5 and ABI4 contributing to this mechanism. Exogenous porphyrins affecting GUN1-dependent plastid-to-nucleus signaling is consistent with purified chloroplasts importing porphyrins in vitro (Adhikari et al., 2009, 2011). Feeding Mg-Proto and hemin to plants was also reported to affect the expression of genes encoding transcription factors, such as CAAT binding proteins (CBP) and the C-REPEAT BINDING FACTORS (CBF) (Zhang et al., 2013; Norén et al., 2016). In apparent conflict with these data collected from intact Arabidopsis plants, Strand et al. (2003) reported that exogenous hemin did not affect the expression of Lhcb in Arabidopsis protoplasts.

In stark contrast, three different laboratories demonstrated that Mg-Proto did not accumulate in barley and Arabidopsis when chloroplast biogenesis was blocked with norflurazon treatments (Gadjieva et al., 2005; Mochizuki et al., 2008; Moulin et al., 2008). Additionally, the norflurazon-treated chlm gun4 and the chlm gun5 double mutants accumulated elevated levels of Mg-Proto relative to gun4, gun5, and wild type but exhibited the same gun phenotypes as gun4 and gun5 when chloroplast biogenesis was blocked with a norflurazon treatment (Mochizuki et al., 2008). Exogenous ALA induced an approximately 200-fold increase in the levels of Mg-Proto and Mg-ProtoME in etiolated Arabidopsis seedlings but had no impact on the subsequent far-red light-induced expression of Lhcb1 (Mochizuki et al., 2008). Elevated levels of Mg-Proto in transgenic rice plants that overexpress Myxococcus xanthus protoporphyrinogen IX oxidase did not reduced the levels of Lhcb expression relative to wild type (Phung et al., 2011). Moulin et al. (2008)

suggested that the difference in the levels of Mg-Proto in norflurazon-treated seedlings reported by Strand et al. (2003) and Moulin et al. (2008) resulted from Strand et al. (2003) misidentifying a contaminating substance as Mg-Proto. Alternatively, a transient accumulation of Mg-Proto was suggested to explain this discrepancy (Kindgren et al., 2011, 2012). Indeed, transient accumulations of Mg-Proto were reported when plants were grown in photoperiodic light (Pöpperl et al., 1998; Papenbrock et al., 1999; Norén et al., 2016) or when chloroplasts experience photooxidative stress (Aarti et al., 2006; Kindgren et al., 2011, 2012; Zhang et al., 2011a). The problem with the proposal that a transient accumulation of Mg-Proto explains this discrepancy is that the growth conditions used to detect the gunphenotype do not fluctuate (i.e., constant temperature and continuous light of a constant fluence rate) to eliminate the possibility of the circadian regulation of PhANG expression obscuring the plastid regulation of PhANG expression (Susek et al., 1993). Indeed, transient peaks of Mg-Proto were not observed in plants grown in continuous light (Pöpperl et al., 1998; Papenbrock et al., 1999; Norén et al., 2016).

An alternative to the idea that a sustained accumulation of Mg-Proto in the chloroplast regulates gene expression in the nucleus is that a transient increase in the levels of Mg-Proto might regulate gene expression in the nucleus. Indeed, transient changes in the levels of Mg-Proto and hemin affect plastid-to-nucleus signaling in algae (Kropat et al., 2000; von Gromoff et al., 2008; Kobayashi et al., 2009; Voß et al., 2011). In 3-week-old green Arabidopsis plants, a transient increase in the levels of Mg-Proto and Mg-ProtoME occurred 48 h after an oxidative-stress inducing norflurazon treatment (Zhang et al., 2011a). Similarly, oxidative stress resulting from a methyl viologen treatment induced a rise in the levels of Mg-Proto and Mg-ProtoME in cucumber and Arabidopsis (Aarti et al., 2006; Kindgren et al., 2011, 2012). Both treatments appeared to inhibit the activity of the Mg-ProtoME cyclase (Aarti et al., 2006; Zhang et al., 2011a; Kindgren et al., 2011, 2012). These herbicide treatments downregulated the expression of Lhcbgenes (Zhang et al., 2011a; Kindgren et al., 2011, 2012). Zhang et al. (2011a) provided evidence that hydrogen peroxide and superoxide might not downregulate the expression of Lhcb genes by acting downstream of GUN5. In contrast, Kindgren et al. (2011, 2012) did not account for potential ROS effects on the expression of Lhcb genes. Zhang et al. (2011a) concluded that norflurazon treatments induce a transient increase in the levels of Mg-Proto in 3-week-old plants and based on their data in 3-week-old plants inferred that in the gun mutant screen, norflurazon might induce a transient increase in the levels of Mg-Proto that leads to the long-term down-regulation of Lhcb expression. Kim and Apel (2013) performed an experiment with 5-day-old Arabidopsis seedlings that was similar to the experiment performed by Zhang et al. (2011a) with 3-week-old plants. Five-day-old seedlings were treated with norflurazon for 12 h in the dark and then returned to continuous light. This treatment induced increases in the levels of superoxide anion radical, hydrogen peroxide, $^1O_2$, and $^1O_2$-dependent plastid-to-nucleus signaling (Kim and Apel, 2013) but had no effect on the expression of Lhcb genes 24 h after the norflurazon-treated seedlings were transferred from the dark to the light. The different ages or the different methods of treating Arabidopsis plants with norflurazon may explain why Zhang et al. (2011a)observed a decrease in Lhcb expression 24 h after transferring norflurazon treated plants to bright light and Kim and Apel (2013) did not.

In contrast, Schlicke et al. (2014) induced a transient increase in Mg-porphyrin levels without first inducing photooxidative stress. They downregulated the expression of the genes encoding the 140-kDa subunit of Mg-chelatase (ChlH), Mg-Proto methyltransferase (ChlM), and Mg-ProtoME cyclase (Chl27) in 10-day-old Arabidopsis seedlings with an inducible RNAi system. Within 24 h, Mg-ProtoME accumulated—to higher levels than observed by Zhang et al. (2011a)—in the Chl27 RNAi lines. It is not clear whether the approach of Zhang et al. (2011a) or Schlicke et al. (2014) induced larger increases in the levels of Mg-Proto. In the Chl27 RNAi lines—24 h after suppression was induced—there were no macroscopic changes in the plants and no significant differences in the expression of PhANGs or ROS-regulated genes. Four days of downregulated Chl27expression induced chlorophyll deficiencies, necrosis, and downregulated PhANG expression. Plastid-to-nucleus signaling was proposed to not cause the changes in gene expression that were observed 24 h after the RNAi-induced suppression of ChlH, ChlM, and Chl27 (Schlicke et al., 2014).

When plants are grown in photoperiodic light, transient increases in the levels of Mg-Proto and Mg-ProtoME routinely occur—apparently without causing photooxidative stress. For example, at dawn in tobacco and barley, Mg-Proto accumulated 30 (barely) and 60-fold (tobacco). Mg-ProtoME accumulated 100 (tobacco) and 200-fold (barley). This accumulation occurred within 30 min (tobacco) and 60 min (barley) and declined within another hour (Pöpperl et al., 1998). An independent study of the accumulation of Mg-Proto and Mg-ProtoME in tobacco in photoperiodic light appeared to miss the first striking peaks of Mg-Proto and Mg-ProtoME detected by Pöpperl et al. (1998) but appeared to detect the subsequent broad peak of Mg-Proto and Mg-ProtoME accumulation that occurred throughout the remainder of the light period (Papenbrock et al., 1999) that was also reported by Pöpperl et al. (1998). A similar broad peak of Mg-Proto and Mg-ProtoME accumulation was reported in Arabidopsis grown in photoperiodic light (Norén et al., 2016). In tobacco, increases in the expression of Lhcb genes preceded dawn, only increased for 6 h, and began declining 8 h after dawn (Paulsen and Bogorad, 1988), several hours after the transient and striking increases in the levels of Mg-Proto and Mg-ProtoME (Pöpperl et al., 1998). Thus, in tobacco, there is no correlation between the diurnal accumulation of these Mg-porphyrins (Pöpperl et al., 1998; Papenbrock et al., 1999) and the levels of Lhcb mRNA (Paulsen and Bogorad, 1988).

An attempt to identify the cytosolic receptor for Mg-Proto in plants revealed that a large number of different cytosolic proteins bound to an affinity column constructed by linking Mg-Proto to Affi-Gel 102 (Kindgren et al., 2011). Several different types of heat shock proteins bound to this column (Kindgren et al., 2011). The affinities of these heat shock proteins for Mg-Proto were not quantified. Although the finding that the HEAT SHOCK PROTEIN 90 (HSP90) bound to a Mg-Proto-affinity column and the finding that 20 μM Mg-Proto induced a 16% reduction in the ATPase activity of HSP90 provide evidence that HSP90s bind Mg-Proto (Kindgren et al., 2011, 2012), these data do not provide evidence for the high-affinity binding expected for a receptor.

Based on precedents for heat shock proteins contributing to heme signaling in yeast and a previously demonstrated connection between HY5 and both plastid-to-nucleus signaling and Lhcb expression, Kindgren et al. (2012) provided evidence that both HSP90 and HY5 contribute to the Mg-Proto-regulated expression of Lhcb genes. RNAi lines that reduced the expression of all HSP90.1–HSP90.4 genes exhibited lower levels of PhANG expression relative to wild type, but only norflurazon- or methyl viologen-treated lines were tested (Kindgren et al., 2012). Although gun mutants exhibit defects in PhANG expression when chloroplasts experience dysfunction, gun mutants do not exhibit defects in PhANG expression when plants contain well-functioning chloroplasts. Thus, whether these effects of RNAi suppression of HSP90 are caused by specific defects in plastid-to-nucleus signaling or by broad effects on the transcriptome would appear important to test. Indeed, HSP90s broadly contribute to the transcriptome, development, and environmental responses (Sangster et al., 2007). Additionally, the idea that HY5 specifically contributes to Mg-Proto signaling conflicts with the finding that HY5 was a gun mutant when chloroplast biogenesis was blocked with a lincomycin treatment (Ruckle et al., 2007). Lincomycin treatments do not appear to activate the plastid-to-nucleus signaling that depends on tetrapyrroles (Gray et al., 2003; Koussevitzky et al., 2007). Subsequently, phytochrome-associated protein phosphatase 5 (PAPP5)—previously identified as a regulator of phytochrome signaling (Ryu et al., 2005) and as a protein that associates with a Mg-Proto Affi-Gel 102 column (Kindgren et al., 2011)—was demonstrated to serve as a negative regulator of chloroplast function by a variety of assays in green plants not treated with norflurazon. Consistent with this finding Lhcb 2.4 and other PhANGs were expressed at higher levels in papp5 than in wild type. A connection to Mg-Proto signaling was suggested because papp5 specifically increased the levels of chlorophyll in a chlorophyll-deficient Mg-ProtoME cyclase mutant and because the downregulation of Lhcb2.4 expression by exogenous Mg-Proto and exogenous ALA was attenuated in papp5 relative to wild type (Barajas-López et al., 2013). Thus, if Mg-Proto is a plastid signal, potential components of the signaling mechanism are not unique to Mg-Proto signaling. With the addition of ZEITLUPE and PSEUDO-RESPONSE REGULATOR5, this HSP90 and HY5-dependent mechanism was proposed to help plastid signals converge with the circadian clock and regulate the expression of C-CBF genes in Arabidopsis(Norén et al., 2016).

Distinct mechanistic information on porphyrin signaling was reported by Zhang et al. (2013). They provided evidence that interactions between porphyrins and GUN1 promote the binding of ABI4 to the promoters of genes that encode Lhcb and CCAAT binding factor A (CBFA). ABI4 was reported to inhibit the transcription of these genes by competitively binding to a cis-element that a distinct transcription activator binds when ABI4 is absent. Additionally, reduced levels of CBFA promoted the accumulation of the LEAFY COTYLEDON1 subunit of the heme activator protein complex (HAP), which binds to promoters of genes encoding HAP and stimulates their transcription. HAP and CBP induced transcription in general and thus, were proposed to promote the transient increases in the levels of cellular RNA that porphyrin signaling is proposed to induce (Zhang et al., 2013). Although a role for ABI4 acting downstream of GUN1 was repeatedly reported (Koussevitzky et al., 2007; Sun et al., 2011; Zhang et al., 2013; Xu et al., 2016), other work provides evidence that additional nuclear transcription factors contribute to the plastid-to-nucleus signaling that is regulated by the GUN genes, such as HY5 and GLK1 (Acevedo-Hernández et al., 2005; Ruckle et al., 2007; Kakizaki et al., 2009; Waters et al., 2009; Kerchev et al., 2011).

THE MECHANISM OF TETRAPYRROLE SIGNALING IN PLANTS: HEME

An alternative to the idea that Mg-Proto serves as plastid signal is that heme serves as a plastid signal. For example, the accumulation and export of heme from the chloroplast to the cytosol or nucleus, or changes in the chloroplastic levels or subchloroplastic location of a particular pool of heme might serve as a plastid signal that induces the expression of PhANGs. In this model, the inductive heme signal is emitted from well-functioning chloroplasts and plastids do not emit this inductive signal when chloroplast biogenesis is blocked or attenuated. The main evidence that heme serves as plastid signal is (1) that overexpressing catalytically active FC1 within the physiologically relevant range and feeding ALA to norflurazon-treated seedlings induced an increase in the expression of PhANGs (Woodson et al., 2011) and (2) that gun5, overexpressing FC1, and exogenous ALA induced the expression of PhANGs in sig2 mutants that were not treated with norflurazon (Woodson et al., 2013). SIG2 is required for the synthesis of $tRNA^{Glu}$ in plastids (Schön et al., 1986;

Kanamaru et al., 2001). The synthesis of glutamyl tRNA from glutamate and $tRNA^{Glu}$ is the first step in the tetrapyrrole biosynthesis (Tanaka and Tanaka, 2007; Tanaka et al., 2011). The levels $tRNA^{Glu}$ and glutamyl tRNA were each reduced in sig2, which links SIG2 to tetrapyrrole biosynthesis and signaling (Woodson et al., 2013). Consistent with heme serving as an inductive plastid signal, the expression of ferrochelatase from Bradyrhizobium japonicum in the cytosol of rice was suggested to induce increases in the ALA synthesizing capacity of chloroplasts by activating a signaling mechanism that regulates gene expression in the nucleus (Kim et al., 2014).

There are two isoforms of ferrochelatase in Arabidopsis. The expression of FC1 is observed in non-photosynthetic tissues and is induced by stress. FC2 is coexpressed with PhANGs (Tanaka et al., 2011). Although plants that overexpressed FC1 exhibited gun phenotypes, plants that overexpressed FC2 did not exhibit gun phenotypes (Woodson et al., 2011). These data provide more evidence that plastid-to-nucleus signaling is activated by changes in the location or concentration of a particular pool of a particular porphyrin—perhaps a transient change in concentration—(Mochizuki et al., 2008, 2010; Woodson et al., 2011) and that this porphyrin is heme (Woodson et al., 2011, 2013).

Consistent with heme serving as a plastid signal in plants, chloroplasts are known to synthesize and export heme to a variety of subcellular locations (Thomas and Weinstein, 1990; Cornah et al., 2003; Tanaka et al., 2011), and heme serves as a signal that regulates gene expression in diverse organisms (Tanaka and Hanaoka, 2013; Terry and Smith, 2013; Kobayashi et al., 2016). Nonetheless, there are data that appear to conflict with the idea that porphyrin-dependent plastid-to-nucleus signaling is explained by an increase in the level of heme serving as a plastid signal. For instance, although elevated levels of heme were reported in gun2 relative to wild type and in gun5 sig2 relative to sig2, heme was reported to not accumulate above wild-type levels in gun4, gun5, or in plants that overexpress FC1 (Woodson et al., 2011, 2013; Espinas et al., 2012). There are conflicting data on whether heme accumulates in norflurazon-treated gun2 relative to norflurazon-treated wild type and whether heme accumulates in untreated Arabidopsis seedlings relative to norflurazon-treated seedlings (Voigt et al., 2010; Woodson et al., 2011). The caveat with these conflicting data is that different methods for quantifying the levels of free heme (i.e., heme that is not associated with proteins and that may accumulate in cellular membranes) can yield different results. Indeed, an analysis of methods for extracting and quantifying heme from plants indicated that the level of free heme is not correlated with the gun phenotype. Thus, heme-dependent plastid-to-nucleus signaling does not appear to depend on the accumulation of free heme. The heme that activates plastid-to-nucleus signaling was proposed to bind to a protein that contributes to a signaling mechanism (Espinas et al., 2012).

Although the data favoring a simple heme-based signaling mechanism is compelling, ALA feeding experiments provide evidence that heme signaling is complex or that porphyrin signaling is more complex than a heme signaling mechanism. Although exogenous ALA induced increases in the expression of PhANGs in norflurazon-treated seedlings (Woodson et al., 2011) and although feeding low quantities of ALA to etiolated sig2 and sig6 seedlings induced increases in expression of PhANGs, feeding high quantities of ALA to etiolated sig2 and sig6 seedlings downregulated the expression of Lhcb. Additionally, feeding low quantities of ALA to etiolated wild-type seedlings had no inductive effect on Lhcb expression and feeding larger quantities of ALA to etiolated wild-type seedlings downregulated the expression of Lhcb(Woodson et al., 2013). Exogenous ALA either upregulated or downregulated the expression of PhANG expression depending on the age of Lepidium sativum seedlings grown in far-red light (Kittsteiner et al., 1991). Feeding ALA to both wild-type and hy1 seedlings downregulated the expression of Lhcb in the dark and in far-red light (Vinti et al., 2000; Czarnecki et al., 2012). Feeding hemin to intact plants either downregulated or had no effect on the expression of Lhcb (Zhang et al., 2011a, 2013). Some of the apparently contradictory results from ALA feeding experiments are potentially explained by light-regulated development affecting porphyrin signaling (Woodson et al., 2013). Alternatively, Zhang et al. (2013) suggested that the repressive effect of high concentrations of exogenous ALA and heme on PhANG expression and that the inductive effect of low concentrations of exogenous ALA and the overexpression of FC1 on PhANG expression provides evidence that heme can either downregulate or upregulate the expression of PhANGs depending on whether heme is present at high or low levels in the cell. Although some reports of exogenous ALA downregulating the expression of PhANGs is potentially explained by non-specific and complex effects of unnaturally high levels of porphyrins on PhANG expression, extremely high porphyrin levels would appear necessary to explain a non-specific mechanism or toxicity that is not based on ROS because 200-fold increases in the levels of Mg-Proto and Mg-ProtoME that were induced by exogenous ALA were reported to not affect the expression of Lhcb1 (Mochizuki et al., 2008).

In the conditions used for the gun mutant screen, the effects of the gun4 and gun5 alleles on nuclear gene expression are potentially explained by the influence of the defects in Mg-porphyrin biosynthesis on heme biosynthesis (Papenbrock et al., 2000; Alawady and Grimm, 2005; Woodson et al., 2011). Conversely, the data supporting the idea that heme serves as an inductive plastid signal in seedlings containing well-functioning chloroplasts (Woodson et al., 2011, 2013) is also consistent with increased heme biosynthesis affecting the chlorophyll branch of the tetrapyrrole biosynthetic pathway. Perhaps increases in heme biosynthesis might affect the subchloroplastic localizaton of GUN4, GUN5, or a GUN4-GUN5 complex bound to or not bound to a porphyrin ligand, which might serve as a repressive signal (Woodson et al., 2011).

INTERACTIONS BETWEEN TETRAPYRROLE SIGNALING AND GUN1-DEPENDENT SIGNALING

The plastid-to-nucleus signaling mechanism that is affected by tetrapyrrole metabolism appears to interact with the plastid-to-nucleus signaling that depends on GUN1. Tetrapyrrole-dependent and GUN1-dependent plastid-to-nucleus signaling mechanisms appeared to downregulate the expression of PhANGs in a partially redundant manner because a synergistic enhancement of the gun phenotype was observed in the gun1-1 gun4-1 and the gun1-1 gun5 double mutants relative to the single mutants (Mochizuki et al., 2001). However, the alleles used by Mochizuki et al. (2001) were leaky. Nonetheless, there was no evidence that the gun phenotype was enhanced in the gun4-1 gun5 double mutant relative to the single mutants (Mochizuki et al., 2001). Subsequently, a double mutant was constructed with gun5 and gun1-9, a non-sense allele of GUN1. The gun phenotype of this double mutant was indistinguishable from gun1-9. These data provide evidence that the tetrapyrrole signal acts upstream of GUN1 (Koussevitzky et al., 2007). The synergistic increase in the gun phenotype observed by Mochizuki et al. (2001) was suggested to result from leaky alleles (Koussevitzky et al., 2007). Transcriptome analyses were also consistent with significant interactions between the plastid signaling mechanisms disrupted by mutant alleles of GUN1and GUN5 (Koussevitzky et al., 2007). The finding that exogenous Mg-Proto downregulated the expression of RbcS and Lhcb in wild type but not in gun1, papp5, abi4, or hy5 (Zhang et al., 2011a; Kindgren et al., 2012; Barajas-López et al., 2013) provides more evidence that tetrapyrrole-dependent plastid signaling acts upstream of GUN1. In contrast, exogenous hemin downregulated the expression of a gene that encodes the CBP in gun1-9 in 3-week-old Arabidopsis plants (Zhang et al., 2013). Thus, heme signaling may not absolutely require GUN1 in 3-week-old Arabidopsis plants.

The finding that there were increases in the levels of ALA, heme, and chlorophyll in gun1 sig2 relative to sig2 provides more evidence for interactions between these two plastid-to-nucleus signaling mechanisms (Woodson et al., 2013). Consistent with these and other data, GUN1 was suggested to act upstream of porphyrin biosynthesis and to affect plastid-to-nucleus signaling by serving as a negative regulator of tetrapyrrole biosynthesis (Terry and Smith, 2013). In this model for plastid-to-nucleus signaling, GUN1 acts upstream of tetrapyrrole biosynthesis rather than downstream of tetrapyrrole biosynthesis as previously suggested (Koussevitzky et al., 2007). The finding that although gun1 and gun4-1 suppress the chlorophyll-deficient phenotype of slow green1, gun5 enhances the chlorophyll-deficient phenotype of slow green1 (Hu et al., 2014) provides evidence for a complex interaction between tetrapyrrole metabolism and GUN1. The finding that light-regulated development affects porphyrin-dependent plastid-to-nucleus signaling (Woodson et al., 2013) indicates additional complexity in this type of plastid-to-nucleus signaling.

In general, pentatricopeptide repeat proteins, such as GUN1, contribute to the expression of the plastidic and mitochondrial genomes (Barkan and Small, 2014). The simplest interpretation of the current data is that the influence of GUN1 on plastid-to-nucleus signaling is probably related to a role for GUN1 in the expression of the chloroplast genome and that GUN1 and tetrapyrrole metabolism interact by an indirect mechanism because the expression of the chloroplast genome and tetrapyrrole metabolism are distinct processes. However, direct interactions between GUN1 and tetrapyrrole metabolism seem possible because the biochemical function of GUN1 is poorly understood and because there is considerable genetic evidence for interactions between GUN1 and tetrapyrrole metabolism. Heme was suggested to possibly interact with GUN1-dependent plastid-to-nucleus signaling by affecting PHD type transcription factor with transmembrane domains (PTM), which acts downstream of GUN1 (Sun et al., 2011; Terry and Smith, 2013).

One approach for learning more about the biochemical function of GUN1 and the interactions between GUN1 and tetrapyrrole metabolism is to identify other molecules that bind GUN1, such as other proteins. However, purifying a multisubunit protein complex containing native GUN1 is potentially difficult because GUN1 is thought to accumulate at low levels (Liu et al., 2013). Tadini et al. (2016) provided evidence that GUN1 and the enzymes that perform tetrapyrrole biosynthesis might directly interact by performing two hybrid and BiFC experiments and by purifying a GFP-tagged version of GUN1 from plants that overexpressed GUN1-GFP after treating chloroplasts that were purified from these plants with a crosslinking agent. GUN1 is capable of interacting with four proteins that contribute to tetrapyrrole biosynthesis—the 79-kDa subunit of Mg-chelatase (also known as ChlD), porphobilinogen deaminase, uroporphyrinogen III decarboxylase, and ferrochelatase I—when these proteins are expressed at abnormally high levels in two-hybrid and BiFC experiments. Nearly 300 different proteins that contribute to diverse chloroplastic processes were crosslinked to GUN1-GFP after GUN1-GFP was overexpressed in Arabidopsis. One of these crosslinked proteins was ChlD. Thus, GUN1 has a propensity for interacting with many other proteins when it is overexpressed in yeast and plants (Tadini et al., 2016). The relevance of these numerous interactions for native GUN1 expressed at normal levels in wild-type plants is not known.

BIOLOGICAL FUNCTION OF TETRAPYRROLE SIGNALING IN PLANTS

The finding that in addition to Arabidopsis, both maize and barley mutants that are deficient in Mg-chelatase exhibited gun phenotypes (Burgess and Taylor, 1988; Gadjieva et al., 2005) indicates that selection pressure has maintained this signaling mechanism in diverse plants and that the influence of tetrapyrrole metabolism on plastid-to-nucleus signaling serves an important function in plants. Nonetheless, the biological function of tetrapyrrole-dependent plastid-to-nucleus signaling is not well established.

Although the influence of tetrapyrrole-dependent plastid-to-nucleus signaling on the expression of PhANGs is commonly studied, transcriptome analyses of C. reinhardtii are consistent with tetrapyrrole-dependent plastid-to-nucleus signaling not serving as a major regulator of PhANG expression. Indeed, the largest group of genes regulated by Mg-Proto and hemin in C. reinhardtii contribute to proteolysis and protein folding. The expression of more than half of these genes is regulated by heat shock. Based on these data, porphyrin signaling was proposed to help C. reinhardtii adapt to changing environmental conditions (Voß et al., 2011). The finding that the overexpression of stress-inducible FC1 but not the overexpression of photosynthesis-associated FC2 affects plastid-to-nucleus signaling in Arabidopsis is consistent with porphyrin-dependent plastid-to-nucleus signaling affecting non-photosynthetic and stress-related functions in Arabidopsis (Woodson et al., 2011). Nonetheless, transcriptome analyses provided evidence that porphyrin-dependent plastid-to-nucleus signaling affects PhANG expression in Arabidopsis (Strand et al., 2003). Unfortunately, a plant system similar to C. reinhardtii in the sense that tetrapyrrole-dependent plastid-to-nucleus signaling is rapidly induced without introducing complicating secondary effects, such as ROS, does not exist.

Tetrapyrrole-dependent plastid-to-nucleus signaling may broadly affect plant cells. Heme signaling appears to affect the expression of genes that contribute to starch biosynthesis and the accumulation of starch during amyloplast biogenesis in BY2 cells (Enami et al., 2011). gun5 has less basal thermotolerance than wild type (Miller et al., 2007). Porphyrin signaling in general was suggested to promote oxidative stress tolerance

(Zhang et al., 2011b). gun5 and cch exhibit reduced cold tolerance relative to wild type (Kindgren et al., 2015). Overexpressing Myxococcus xanthus protoporphyrinogen IX oxidase in Oryza sativa enhanced drought tolerance relative to wild type (Phung et al., 2011). Although Phung et al. (2011)proposed that overexpressing protoporphyrinogen IX oxidase promoted drought tolerance by inducing increases in the levels of ROS that helped acclimate chloroplasts to stress, they did not detect elevated levels of ROS in plants that overexpressed protoporphyrinogen IX oxidase. Additionally, the lines overexpressing protoporphyrinogen IX oxidase exhibited an increased capacity for porphyrin biosynthesis that resulted in elevated levels of heme. Interesting, their data show a correlation between elevated levels of ferrochelatase activity, heme, and Lhcb expression, although the levels of heme appeared to increase only transiently (Phung et al., 2011). The data from Phung et al. (2011) are consistent with tetrapyrrole signaling, such as heme signaling, promoting drought tolerance. Indeed, there is a growing body of evidence supporting a role for heme signaling in drought tolerance in plants (Nagahatenna et al., 2015). Porphyrin-derived ROS, heme signaling, or both may contribute to drought tolerance in the transgenic plants described by Phung et al. (2011). A mutant allele that specifically disrupts tetrapyrrole-dependent plastid signaling without directly affecting an enzyme that contributes to tetrapyrrole biosynthesis would help to establish the biological function of tetrapyrrole-dependent plastid-to-nucleus signaling.

The interactions between GUN1- and tetrapyrrole-dependent plastid-to-nucleus signaling are consistent with additional biological functions for tetrapyrrole-dependent plastid-to-nucleus signaling because GUN1is known to promote diverse biological functions in plants. Loss-of-function alleles of GUN1 affect chloroplast biogenesis (Mochizuki et al., 1996; Ruckle et al., 2007), plastid-to-nucleus signaling when plants are grown in diverse conditions (Susek et al., 1993; Koussevitzky et al., 2007; Ruckle et al., 2007; Kakizaki et al., 2009; Woodson et al., 2013; Tanaka et al., 2015; Tadini et al., 2016), the circadian rhythm (Hassidim et al., 2007), the accumulation of anthocyanins (Ruckle and Larkin, 2009; Cottage et al., 2010; Cheng et al., 2012), abiotic stress tolerance (Miller et al., 2007; Zhang et al., 2013), and the development of seedlings and leaves (Ruckle and Larkin, 2009; Cottage et al., 2010; Tameshige et al., 2013; Martín et al., 2016; Xu et al., 2016).

SUMMARY

Since the early days of plastid-to-nucleus signaling research, porphyrins were attractive candidates for plastid signals that regulate nuclear gene expression. A number of independent experiments with exogenous ALA, exogenous porphyrins, inhibitors, a variety of mutants with defects in tetrapyrrole metabolism, and both the overexpression and antisense suppression of enzymes that contribute to tetrapyrrole biosynthesis provide a large body of evidence favoring the idea that porphyrins regulate the expression of nuclear genes. Results from the gun mutant screen indicate that when chloroplast biogenesis is blocked, tetrapyrrole metabolism regulates the expression of PhANGs. There is a large body of evidence indicating that in the conditions used for the gun mutant screen, the effects of tetrapyrrole metabolism on nuclear gene expression are difficult or impossible to explain by changes in the levels of porphyrin-derived ROS. The regulation of nuclear gene expression by tetrapyrrole metabolism is conserved in Arabidopsis, barely, and maize. Thus, tetrapyrrole signaling makes an important contribution to plants.

The simplest interpretation of the current data is that in the conditions used for the gun mutant screen, heme serves as a plastid signal that regulates nuclear gene expression. The current data are also consistent with other mechanisms, such as a moonlighting function (Boukouris et al., 2016) for an enzyme or regulatory factor that contributes to tetrapyrrole metabolism in plastid-to-nucleus signaling, such as GUN5 (Mochizuki et al., 2001), GUN4, or a GUN4-GUN5 complex (Larkin et al., 2003) in plastid-to-nucleus signaling (Figure 3). The current data are also consistent with changes in the flux through the tetrapyrrole biosynthetic pathway activating a plastid-to-nucleus signaling mechanism (Mochizuki et al., 2008). There are precedents for each of these mechanisms in non-plant systems (Osanai et al., 2009; Terry and Smith, 2013). Interactions between tetrapyrrole-dependent plastid-to-nucleus signaling and GUN1-dependent plastid-to-nucleus signaling potentially link tetrapyrrole-dependent plastid-to-nucleus signaling to a network of plastidic and extraplastidic signaling mechanisms that is broadly significant.

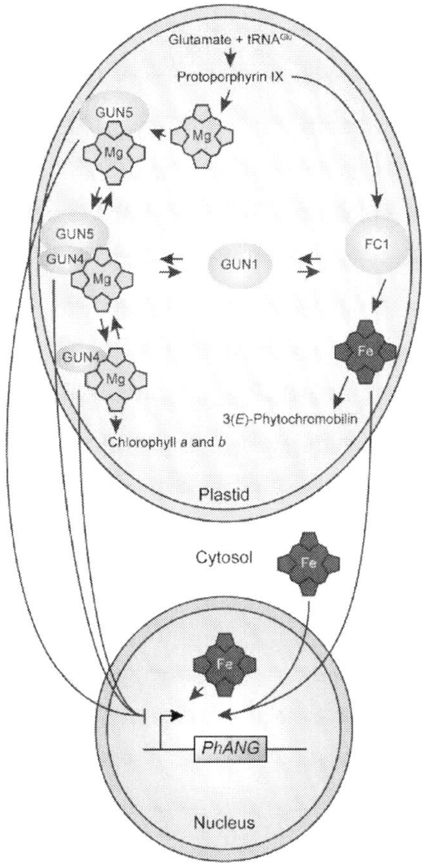

Figure 3. A model for tetrapyrrole-dependent plastid-to-nucleus signaling in plants. The simplest interpretation of the current data is that heme accumulates in well-functioning chloroplasts and that the accumulation of heme in chloroplasts or the export of heme to the cytosol or nucleus activates a mechanism that induces the expression of PhANGs. The current data do not exclude alternative mechanisms, such as a moonlighting function for an enzyme or regulatory factor that contributes to tetrapyrrole metabolism, such as GUN4, GUN5, or a GUN4-GUN5 complex that affects gene expression in the nucleus. Either the liganded or unliganded versions of GUN4, GUN5, or a GUN4-GUN5 complex might contribute to a plastid-to-nucleus signaling mechanism that regulates the expression of PhANGs. GUN1 interacts with tetrapyrrole-dependent plastid-to-nucleus signaling by an unknown mechanism.

The plastid-to-nucleus signaling mechanism that is specifically interrogated by the gun mutant screen in young seedlings does not appear to require a transient or sustained accumulation of Mg-Proto or Mg-ProtoME. However, in mature plants, the photooxidative stress that (1) induces increases in the levels of Mg-Proto and Mg-ProtoME and (2) is correlated with downregulated expression of PhANGs was observed under distinct developmental and environmental conditions relative to the conditions used to screen for gun mutants. The evidence consistent with oxidative stress inducing an increase in the levels of Mg-Proto or Mg-ProtoME that affects gene expression in the nucleus is also consistent with oxidative stress attenuating heme signaling or at least in some instances, ROS signaling

At this point, it is perhaps most important to understand the mechanism of tetrapyrrole signaling in plants. Understanding this mechanism may help us to understand why perturbations in tetrapyrrole metabolism sometimes upregulate, downregulate, or have no effect on the expression of nuclear genes. Understanding this mechanism may help us to understand the interactions between tetrapyrrole signaling and GUN1-dependent plastid-to-nucleus signaling, interactions between tetrapyrrole signaling and light-regulated development, and the biological function of tetrapyrrole signaling. Identifying a protein that specifically contributes to tetrapyrrole signaling (e.g., a protein that serves as a heme receptor in the plastid, cytosol, or nucleus) is perhaps the highest priority. Mutants with defects in such a gene are essential for learning the biological function of tetrapyrrole signaling in plants. The evidence for tetrapyrrole signaling contributing to abiotic stress tolerance and for the interactions between tetrapyrrole metabolism and GUN1-dependent plastid-to-to-nucleus signaling are consistent with an important role for tetrapyrrole signaling in plants. Thus, future research in this area is expected to advance our understanding of plant biology and possibly contribute to the development of new germplasm that will help to meet the needs of agriculture in our changing environment.

AUTHOR CONTRIBUTIONS

The author confirms being the sole contributor of this work and approved it for publication.

CONFLICT OF INTEREST STATEMENT

The author declares that the research was conducted in the absence of any commercial or financial relationships that could be construed as a potential conflict of interest.

ACKNOWLEDGMENTS

Research in the author's laboratory is supported by Huazhong Agricultural University, Wuhan, China.

REFERENCES

1. Aarti, D. P., Tanaka, R., and Tanaka, A. (2006). Effects of oxidative stress on chlorophyll biosynthesis in cucumber (Cucumis sativus) cotyledons. Physiol. Plant. 128, 186–197. doi: 10.1111/j.1399-3054.2006.00720.x
2. Acevedo-Hernández, G. J., León, P., and Herrera-Estrella, L. R. (2005). Sugar and ABA responsiveness of a minimal RBCS light-responsive unit is mediated by direct binding of ABI4. Plant J. 43, 506–519. doi: 10.1111/j.1365-313X.2005.02468.x
3. Adhikari, N. D., Froehlich, J. E., Strand, D. D., Buck, S. M., Kramer, D. M., and Larkin, R. M. (2011). GUN4-porphyrin complexes bind the ChlH/GUN5 subunit of Mg-chelatase and promote chlorophyll biosynthesis in Arabidopsis. Plant Cell 23, 1449–1467. doi: 10.1105/tpc.110.082503
4. Adhikari, N. D., Orler, R., Chory, J., Froehlich, J. E., and Larkin, R. M. (2009). Porphyrins stabilize the association of GENOMES UNCOUPLED 4 and a Mg-chelatase subunit with chloroplast membranes. J. Biol. Chem. 284, 24783–24796. doi: 10.1074/jbc.M109.025205
5. Alawady, A. E., and Grimm, B. (2005). Tobacco Mg protoporphyrin IX methyltransferase is involved in inverse activation of Mg porphyrin and protoheme synthesis. Plant J. 41, 282–290. doi: 10.1111/j.1365-313X.2004.02291.x
6. Ankele, E., Kindgren, P., Pesquet, E., and Strand, Å. (2007). In vivo visualization of Mg-protoporphyrin IX, a coordinator of

photosynthetic gene expression in the nucleus and the chloroplast. Plant Cell 19, 1964–1979. doi: 10.1105/tpc.106.048744

7. Armbruster, U., Pesaresi, P., Pribil, M., Hertle, A., and Leister, D. (2011). Update on chloroplast research: new tools, new topics, and new trends. Mol. Plant 4, 1–16. doi: 10.1093/mp/ssq060

8. Barajas-López, J., Kremnev, D., Shaikhali, J., Piñas-Fernández, A., and Strand, Å. (2013). PAPP5 is involved in the tetrapyrrole mediated plastid signalling during chloroplast development. PLoS ONE 8:e60305. doi: 10.1371/journal.pone.0060305

9. Barajas-López, J. D. D., Blanco, N. E., and Strand, Å. (2012). Plastid-to-nucleus communication, signals controlling the running of the plant cell. Biochim. Biophys. Acta 1833, 425–437. doi: 10.1016/j.bbamcr.2012.06.020

10. Barkan, A., and Small, I. (2014). Pentatricopeptide repeat proteins in plants. Annu. Rev. Plant Biol. 65, 415–442. doi: 10.1146/annurev-arplant-050213-040159

11. Beck, C. F. (2005). Signaling pathways from the chloroplast to the nucleus. Planta 222, 743–756. doi: 10.1007/s00425-005-0021-2

12. Bobik, K., and Burch-Smith, T. M. (2015). Chloroplast signaling within, between and beyond cells. Front. Plant Sci. 6:781. doi: 10.3389/fpls.2015.00781

13. Boukouris, A. E., Zervopoulos, S. D., and Michelakis, E. D. (2016). Metabolic enzymes moonlighting in the nucleus: metabolic regulation of gene transcription. Trends Biochem. Sci. doi: 10.1016/j.tibs.2016.05.013 [Epub ahead of print].

14. Bradbeer, J. W., Atkinson, Y. E., Börner, T., and Hagemann, R. (1979). Cytoplasmic synthesis of plastid polypeptides may be controlled by plastid-synthesised RNA. Nature 279, 816–817. doi: 10.1038/279816a0

15. Brown, N. J., Sullivan, J. A., and Gray, J. C. (2005). Light and plastid signals regulate the expression of the pea plastocyanin gene through a common region at the 5′ end of the coding region. Plant J. 43, 541–552. doi: 10.1111/j.1365-313X.2005.02474.x

16. Brzezowski, P., Richter, A. S., and Grimm, B. (2015). Regulation and function of tetrapyrrole biosynthesis in plants and algae. Biochim.

Biophys. Acta 1847, 968–985. doi: 10.1016/j.bbabio.2015.05.007

17. Brzezowski, P., Schlicke, H., Richter, A., Dent, R. M., Niyogi, K. K., and Grimm, B. (2014). The GUN4 protein plays a regulatory role in tetrapyrrole biosynthesis and chloroplast-to-nucleus signalling in Chlamydomonas reinhardtii. Plant J. 79, 285–298. doi: 10.1111/tpj.12560
18. Burgess, D. G., and Taylor, W. C. (1988). The chloroplast affects the transcription of a nuclear gene family. Mol. Gen. Genet. 214, 89–96. doi: 10.1007/BF00340185
19. Chan, K. X., Phua, S. Y., Crisp, P., McQuinn, R., and Pogson, B. J. (2016). Learning the languages of the chloroplast: retrograde signaling and beyond. Annu. Rev. Plant Biol. 67, 25–53. doi: 10.1146/annurev-arplant-043015-111854
20. Chen, X., Pu, H., Wang, X., Long, W., Lin, R., and Liu, L. (2015). Crystal structures of GUN4 in complex with porphyrins. Mol. Plant 8, 1125–1127. doi: 10.1016/j.molp.2015.04.013
21. Cheng, J., Yuan, S., Zhang, Z. W., Zhu, F., Tang, H., Xu, F., et al. (2012). Plastid-signalling-mediated anthocyanin accumulation in mature Arabidopsis rosettes. Plant Growth Regul. 68, 223–230. doi: 10.1007/s10725-012-9710-6
22. Cornah, J. E., Terry, M. J., and Smith, A. G. (2003). Green or red: what stops the traffic in the tetrapyrrole pathway? Trends Plant Sci. 8, 224–230. doi: 10.1016/S1360-1385(03)00064-5
23. Cottage, A., Mott, E. K., Kempster, J. A., and Gray, J. C. (2010). The Arabidopsis plastid-signalling mutant gun1 (genomes uncoupled1) shows altered sensitivity to sucrose and abscisic acid and alterations in early seedling development. J. Exp. Bot. 61, 3773–3786. doi: 10.1093/jxb/erq186
24. Cottage, A. J., Mott, E. K., Wang, J.-H., Sullivan, J. A., MacLean, D., Tran, L., et al. (2008). "GUN1 (GENOMES UNCOUPLED1) encodes a pentatricopeptide repeat (PPR) protein involved in plastid protein synthesis-responsive retrograde signaling to the nucleus," in Photosynthesis. Energy from the Sun: 14th International Congress on Photosynthesis, eds J. F. Allen, E. Gnatt, J. H. Golbeck, and B. Osmond (Berlin: Springer), 1201–1205.
25. Czarnecki, O., Gläßer, C., Chen, J. G., Mayer, K. F., and Grimm,

B. (2012). Evidence for a contribution of ALA synthesis to plastid-to-nucleus signaling. Front. Plant Sci. 3:236. doi: 10.3389/fpls.2012.00236

26. Davison, P. A., Schubert, H. L., Reid, J. D., Iorg, C. D., Heroux, A., Hill, C. P., et al. (2005). Structural and biochemical characterization of Gun4 suggests a mechanism for its role in chlorophyll biosynthesis. Biochemistry 44, 7603–7612. doi: 10.1021/bi050240x
27. Enami, K., Ozawa, T., Motohashi, N., Nakamura, M., Tanaka, K., and Hanaoka, M. (2011). Plastid-to-nucleus retrograde signals are essential for expression of nuclear starch biosynthesis genes during amyloplast differentiation in tobacco BY-2 cultured cells. Plant Physiol. 157, 518–530. doi: 10.1104/pp.111.178897
28. Espinas, N. A., Kobayashi, K., Takahashi, S., Mochizuki, N., and Masuda, T. (2012). Evaluation of unbound free heme in plant cells by differential acetone extraction. Plant Cell Physiol. 53, 1344–1354. doi: 10.1093/pcp/pcs067
29. Formighieri, C., Ceol, M., Bonente, G., Rochaix, J. D., and Bassi, R. (2012). Retrograde signaling and photoprotection in a gun4 mutant of Chlamydomonas reinhardtii. Mol. Plant 5, 1242–1262. doi: 10.1093/mp/sss051
30. Forsburg, S. L., and Guarente, L. (1989). Communication between mitochondria and the nucleus in regulation of cytochrome genes in the yeast Saccharomyces cerevisiae. Annu. Rev. Cell Biol. 5, 153–180. doi: 10.1146/annurev.cb.05.110189.001101
31. Gadjieva, R., Axelsson, E., Olsson, U., and Hansson, M. (2005). Analysis of gun phenotype in barley magnesium chelatase and Mg-protoporphyrin IX monomethyl ester cyclase mutants. Plant Physiol. Biochem. 43, 901–908. doi: 10.1016/j.plaphy.2005.08.003
32. Galvez-Valdivieso, G., and Mullineaux, P. M. (2010). The role of reactive oxygen species in signalling from chloroplasts to the nucleus. Physiol. Plant. 138, 430–439. doi: 10.1111/j.1399-3054.2009.01331.x
33. Gough, S. (1972). Defective synthesis of porphyrins in barley plastids caused by mutation in nuclear genes. Biochim. Biophys. Acta 286, 36–54. doi: 10.1016/0304-4165(72)90086-4
34. Granick, S. (1959). Magnesium porphyrins formed by barley

seedlings treated with d-aminolevulinic acid. Plant Physiol. 34:XVIII.

35. Gray, J. C., Sullivan, J. A., Wang, J. H., Jerome, C. A., and MacLean, D. (2003). Coordination of plastid and nuclear gene expression. Philos. Trans. R. Soc. Lond. B Biol. Sci. 358, 135–144. doi: 10.1098/rstb.2002.1180

36. Hassidim, M., Yakir, E., Fradkin, D., Hilman, D., Kron, I., Keren, N., et al. (2007). Mutations in CHLOROPLAST RNA BINDING provide evidence for the involvement of the chloroplast in the regulation of the circadian clock in Arabidopsis. Plant J. 51, 551–562. doi: 10.1111/j.1365-313X.2007.03160.x

37. Hu, Z., Xu, F., Guan, L., Qian, P., Liu, Y., Zhang, H., et al. (2014). The tetratricopeptide repeat-containing protein slow green1 is required for chloroplast development in Arabidopsis. J. Exp. Bot. 65, 1111–1123. doi: 10.1093/jxb/ert463

38. Huang, Y. S., and Li, H. M. (2009). Arabidopsis CHLI2 can substitute for CHLI1. Plant Physiol. 150, 636–645. doi: 10.1104/pp.109.135368

39. Jansson, S., Pichersky, E., Bassi, R., Green, B. R., Ikeuchi, M., Melis, A., et al. (1992). A nomenclature for the genes encoding the chlorophyll a/b-binding proteins of higher plants. Plant Mol. Biol. Rep. 10, 242–253. doi: 10.1007/BF02668357

40. Jarvis, P., and López-Juez, E. (2013). Biogenesis and homeostasis of chloroplasts and other plastids. Nat. Rev. Mol. Cell Biol. 14, 787–802. doi: 10.1038/nrm3702

41. Johanningmeier, U., and Howell, S. H. (1984). Regulation of light-harvesting chlorophyll-binding protein mRNA accumulation in Chlamydomonas reinhardtii. Possible involvement of chlorophyll synthesis precursors. J. Biol. Chem. 259, 13541–13549.

42. Kakizaki, T., Matsumura, H., Nakayama, K., Che, F. S., Terauchi, R., and Inaba, T. (2009). Coordination of plastid protein import and nuclear gene expression by plastid-to-nucleus retrograde signaling. Plant Physiol. 151, 1339–1353. doi: 10.1104/pp.109.145987

43. Kanamaru, K., Nagashima, A., Fujiwara, M., Shimada, H., Shirano, Y., Nakabayashi, K., et al. (2001). An Arabidopsis sigma factor (SIG2)-dependent expression of plastid-encoded tRNAs in

chloroplasts. Plant Cell Physiol. 42, 1034–1043. doi: 10.1093/pcp/pce155

44. Kerchev, P. I., Pellny, T. K., Vivancos, P. D., Kiddle, G., Hedden, P., Driscoll, S., et al. (2011). The transcription factor ABI4 is required for the ascorbic acid-dependent regulation of growth and regulation of jasmonate-dependent defense signaling pathways in Arabidopsis. Plant Cell 23, 3319–3334. doi: 10.1105/tpc.111.090100
45. Kim, C., and Apel, K. (2013). 1O2-mediated and EXECUTER dependent retrograde plastid-to-nucleus signaling in norflurazon-treated seedlings of Arabidopsis thaliana. Mol. Plant 6, 1580–1591. doi: 10.1093/mp/sst020
46. Kim, J. G., Back, K., Lee, H. Y., Lee, H. J., Phung, T. H., Grimm, B., et al. (2014). Increased expression of Fe-chelatase leads to increased metabolic flux into heme and confers protection against photodynamically induced oxidative stress. Plant Mol. Biol. 86, 271–287. doi: 10.1007/s11103-014-0228-3
47. Kindgren, P., Dubreuil, C., and Strand, Å. (2015). The recovery of plastid function is required for optimal response to low temperatures in Arabidopsis. PLoS ONE 10:e0138010. doi: 10.1371/journal.pone.0138010
48. Kindgren, P., Eriksson, M., Benedict, C., Mohapatra, A., Gough, S. P., Hansson, M., et al. (2011). A novel proteomic approach reveals a role for Mg-protoporphyrin IX in response to oxidative stress. Physiol. Plant. 141, 310–320. doi: 10.1111/j.1399-3054.2010.01440.x
49. Kindgren, P., Norén, L., López, J. D., Shaikhali, J., and Strand, Å (2012). Interplay between Heat Shock Protein 90 and HY5 controls PhANG expression in response to the GUN5 plastid signal. Mol. Plant 5, 901–913. doi: 10.1093/mp/ssr112
50. Kittsteiner, U., Brunner, H., and Rüdiger, W. (1991). The greening process in cress seedlings. II. Complexing agents and 5-aminolevulinate inhibit accumulation of cab-mRNA coding for the light-harvesting chlorophyll a/b protein. Physiol. Plant. 81, 190–196. doi: 10.1111/j.1399-3054.1991.tb01726.x
51. Kleine, T., and Leister, D. (2013). Retrograde signals galore. Front. Plant Sci. 4:45. doi: 10.3389/fpls.2013.00045
52. Kleine, T., Voigt, C., and Leister, D. (2009). Plastid signalling to

the nucleus: messengers still lost in the mists? Trends Genet. 25, 185–192. doi: 10.1016/j.tig.2009.02.004

53. Kobayashi, K., Mochizuki, N., Yoshimura, N., Motohashi, K., Hisabori, T., and Masuda, T. (2008). Functional analysis of Arabidopsis thaliana isoforms of the Mg-chelatase CHLI subunit. Photochem. Photobiol. Sci. 7, 1188–1195. doi: 10.1039/b802604c

54. Kobayashi, Y., Hiroyuki, A., Hanaoka, M., and Tanaka, K. (2016). Abscisic acid participates in the control of cell cycle initiation through heme homeostasis in the unicellular red alga Cyanidioschyzon merolae. Plant Cell Physiol. 57, 953–960. doi: 10.1093/pcp/pcw054

55. Kobayashi, Y., Kanesaki, Y., Tanaka, A., Kuroiwa, H., Kuroiwa, T., and Tanaka, K. (2009). Tetrapyrrole signal as a cell-cycle coordinator from organelle to nuclear DNA replication in plant cells. Proc. Natl. Acad. Sci. U.S.A. 106, 803–807. doi: 10.1073/pnas.0804270105

56. Koncz, C., Mayerhofer, R., Koncz-Kalman, Z., Nawrath, C., Reiss, B., Redei, G. P., et al. (1990). Isolation of a gene encoding a novel chloroplast protein by T-DNA tagging in Arabidopsis thaliana. EMBO J. 9, 1337–1346.

57. Koornneef, M., Rolff, E., and Spruit, C. J. P. (1980). Genetic control of lightinhibited hypocotyl elongation in Arabidopsis thaliana (L) Heynh. Z. Pflanzenphysiol. 100, 147–160. doi: 10.1016/S0044-328X(80)80208-X

58. Koussevitzky, S., Nott, A., Mockler, T. C., Hong, F., Sachetto-Martins, G., Surpin, M., et al. (2007). Signals from chloroplasts converge to regulate nuclear gene expression. Science 316, 715–719. doi: 10.1126/science.%201140516

59. Kropat, J., Oster, U., Rüdiger, W., and Beck, C. F. (2000). Chloroplast signalling in the light induction of nuclear HSP70 genes requires the accumulation of chlorophyll precursors and their accessibility to cytoplasm/nucleus. Plant J. 24, 523–531. doi: 10.1046/j.1365-313x.2000.00898.x

60. La Rocca, N., Rascio, N., Oster, U., and Rüdiger, W. (2001). Amitrole treatment of etiolated barley seedlings leads to deregulation of tetrapyrrole synthesis and to reduced expression of Lhc and RbcS genes. Planta 213, 101–108. doi: 10.1007/s004250000477

61. Larkin, R. M. (2014). Influence of plastid signals on light signaling

and development. Philos. Trans. R. Soc. Lond. B Biol. Sci. 369, 20130232. doi: 10.1098/rstb.2013.0232

62. Larkin, R. M., Alonso, J. M., Ecker, J. R., and Chory, J. (2003). GUN4, a regulator of chlorophyll synthesis and intracellular signaling. Science 299, 902–906. doi: 10.1126/science.1079978
63. Larkin, R. M., Stefano, G., Ruckle, M. E., Stavoe, A. K. H., Sinkler, C., Brandizzi, F., et al. (2016). REDUCED CHLOROPLAST COVERAGE genes from Arabidopsis thaliana help to establish the size of the chloroplast compartment. Proc. Natl. Acad. Sci. U.S.A. 113, E1116–E1125. doi: 10.1073/pnas.1515741113
64. Lepistö, A., and Rintamäki, E. (2012). Coordination of plastid and light signaling pathways upon development of Arabidopsis leaves under various photoperiods. Mol. Plant 5, 799–816. doi: 10.1093/mp/ssr106
65. Liu, S., Melonek, J., Boykin, L. M., Small, I., and Howell, K. A. (2013). PPR-SMRs: ancient proteins with enigmatic functions. RNA Biol.10, 1501–1510. doi: 10.4161/rna.26172
66. Martín, G., Leivar, P., Ludevid, D., Tepperman, J. M., Quail, P. H., and Monte, E. (2016). Phytochrome and retrograde signalling pathways converge to antagonistically regulate a light-induced transcriptional network. Nat. Commun. 7, 11431. doi: 10.1038/ncomms11431
67. Mascia, P. (1978). An analysis of precursors accumulated by several chlorophyll biosynthetic mutants of maize. Mol. Gen. Genet. 167, 237–244. doi: 10.1007/BF00330996
68. Meinecke, L., Alawady, A., Schroda, M., Willows, R., Kobayashi, M. C., Niyogi, K. K., et al. (2010). Chlorophyll-deficient mutants of Chlamydomonas reinhardtii that accumulate magnesium protoporphyrin IX. Plant Mol. Biol. 72, 643–658. doi: 10.1007/s11103-010-9604-9
69. Miller, G., Suzuki, N., Rizhsky, L., Hegie, A., Koussevitzky, S., and Mittler, R. (2007). Double mutants deficient in cytosolic and thylakoid ascorbate peroxidase reveal a complex mode of interaction between reactive oxygen species, plant development, and response to abiotic stresses. Plant Physiol. 144, 1777–1785. doi: 10.1104/pp.107.101436

70. Mochizuki, N., Brusslan, J. A., Larkin, R., Nagatani, A., and Chory, J. (2001). Arabidopsis genomes uncoupled 5 (GUN5) mutant reveals the involvement of Mg-chelatase H subunit in plastid-to-nucleus signal transduction. Proc. Natl. Acad. Sci. U.S.A. 98, 2053–2058. doi: 10.1073/pnas.98.4.2053

71. Mochizuki, N., Susek, R., and Chory, J. (1996). An intracellular signal transduction pathway between the chloroplast and nucleus is involved in de-etiolation. Plant Physiol. 112, 1465–1469. doi: 10.1104/pp.112.4.1465

72. Mochizuki, N., Tanaka, R., Grimm, B., Masuda, T., Moulin, M., Smith, A. G., et al. (2010). The cell biology of tetrapyrroles: a life and death struggle. Trends Plant Sci. 15, 488–498. doi: 10.1016/j.tplants.2010.05.012

73. Mochizuki, N., Tanaka, R., Tanaka, A., Masuda, T., and Nagatani, A. (2008). The steady-state level of Mg-protoporphyrin IX is not a determinant of plastid-to-nucleus signaling in Arabidopsis. Proc. Natl. Acad. Sci. U.S.A. 105, 15184–15189. doi: 10.1073/pnas.0803245105

74. Moulin, M., McCormac, A. C., Terry, M. J., and Smith, A. G. (2008). Tetrapyrrole profiling in Arabidopsis seedlings reveals that retrograde plastid nuclear signaling is not due to Mg-protoporphyrin IX accumulation. Proc. Natl. Acad. Sci. U.S.A. 105, 15178–15183. doi: 10.1073/pnas.0803054105

75. Nagahatenna, D. S., Langridge, P., and Whitford, R. (2015). Tetrapyrrole-based drought stress signalling. Plant Biotechnol. J. 13, 447–459. doi: 10.1111/pbi.12356

76. Nisar, N., Li, L., Lu, S., Khin, N. C., and Pogson, B. J. (2015). Carotenoid metabolism in plants. Mol. Plant 8, 68–82. doi: 10.1016/j.molp.2014.12.007

77. Norén, L., Kindgren, P., Stachula, P., Rühl, M., Eriksson, M. E., Hurry, V., et al. (2016). Circadian and plastid signaling pathways are integrated to ensure correct expression of the CBF and COR genes during photoperiodic growth. Plant Physiol. 171, 1392–1406. doi: 10.1104/pp.16.00374

78. Osanai, T., Imashimizu, M., Seki, A., Sato, S., Tabata, S., Imamura, S., et al. (2009). ChlH, the H subunit of the Mg-chelatase, is an anti-

sigma factor for SigE in Synechocystis sp. PCC 6803. Proc. Natl. Acad. Sci. U.S.A. 106, 6860–6865. doi: 10.1073/pnas.0810040106

79. Oster, U., Brunner, H., and Rüdiger, W. (1996). The greening process in cress seedlings. V. Possible interference of chlorophyll precursors, accumulated after thujaplicin treatment, with light-regulated expression of Lhc genes. J. Photochem. Photobiol. B 36, 255–261. doi: 10.1016/S1011-1344(96)07388-5

80. Osteryoung, K. W., and Pyke, K. A. (2014). Plastid and proplastid division. Ann. Rev. Plant. Biol. 65, 443–472. doi: 10.1146/annurev-arplant-050213-035748

81. Papenbrock, J., Mock, H.-P., Kruse, E., and Grimm, B. (1999). Expression studies in tetrapyrrole biosynthesis: inverse maxima of magnesium chelatase and ferrochelatase activity during cyclic photoperiods. Planta 208, 264–273. doi: 10.1007/s004250050558

82. Papenbrock, J., Pfündel, E., Mock, H. P., and Grimm, B. (2000). Decreased and increased expression of the subunit CHL I diminishes Mg chelatase activity and reduces chlorophyll synthesis in transgenic tobacco plants. Plant J. 22, 155–164. doi: 10.1046/j.1365-313x.2000.00724.x

83. Paulsen, H., and Bogorad, L. (1988). Diurnal and circadian rhythms in the accumulation and synthesis of mRNA for the light-harvesting Chlorophyll a/b-Binding protein in tobacco. Plant Physiol. 88, 1104–1109. doi: 10.1104/pp.88.4.1104

84. Pfannschmidt, T. (2010). Plastidial retrograde signalling – a true "plastid factor" or just metabolite signatures? Trends Plant Sci. 15, 427–435. doi: 10.1016/j.tplants.2010.05.009

85. Phung, T. H., Jung, H. I., Park, J. H., Kim, J. G., Back, K., and Jung, S. (2011). Porphyrin biosynthesis control under water stress: sustained porphyrin status correlates with drought tolerance in transgenic rice. Plant Physiol. 157, 1746–1764. doi: 10.1104/pp.111.188276

86. Pogson, B. J., Ganguly, D., and Albrecht-Borth, V. (2015). Insights into chloroplast biogenesis and development. Biochim. Biophys. Acta1847, 1017–1024. doi: 10.1016/j.bbabio.2015.02.003

87. Pontier, D., Albrieux, C., Joyard, J., Lagrange, T., and Block, M. A. (2007). Knock-out of the magnesium protoporphyrin IX methyltransferase gene in Arabidopsis. Effects on chloroplast

development and on chloroplast-to-nucleus signaling. J. Biol. Chem. 282, 2297–2304. doi: 10.1074/jbc.M610286200

88. Pöpperl, G., Oster, U., and Rüdiger, W. (1998). Light-dependent increase in chlorophyll precursors during th day-night cycle in tobacco and barley seedlings. J. Plant Physiol. 153, 40–45. doi: 10.1016/S0176-1617(98)80042-4

89. Rissler, H. M., Collakova, E., DellaPenna, D., Whelan, J., and Pogson, B. J. (2002). Chlorophyll biosynthesis. Expression of a second chl I gene of magnesium chelatase in Arabidopsis supports only limited chlorophyll synthesis. Plant Physiol. 128, 770–779. doi: 10.1104/pp.010625

90. Ruckle, M. E., Burgoon, L. D., Lawrence, L. A., Sinkler, C., and Larkin, R. M. (2012). Plastids are major regulators of light signaling in Arabidopsis thaliana. Plant Physiol. 159, 366–390. doi: 10.1104/pp.112.193599

91. Ruckle, M. E., DeMarco, S. M., and Larkin, R. M. (2007). Plastid signals remodel light signaling networks and are essential for efficient chloroplast biogenesis in Arabidopsis. Plant Cell 19, 3944–3960. doi: 10.1105/tpc.107.054312

92. Ruckle, M. E., and Larkin, R. M. (2009). Plastid signals that affect photomorphogenesis in Arabidopsis thaliana are dependent on GENOMES UNCOUPLED 1 and cryptochrome 1. New Phytol. 182, 367–379. doi: 10.1111/j.1469-8137.2008.02729.x

93. Ryu, J. S., Kim, J. I., Kunkel, T., Kim, B. C., Cho, D. S., Hong, S. H., et al. (2005). Phytochrome-specific type 5 phosphatase controls light signal flux by enhancing phytochrome stability and affinity for a signal transducer. Cell 120, 395–406. doi: 10.1016/j.cell.2004.12.019

94. Saini, G., Meskauskiene, R., Pijacka, W., Roszak, P., Sjögren, L. L., Clarke, A. K., et al. (2011). 'happy on norflurazon' (hon) mutations implicate perturbance of plastid homeostasis with activating stress acclimatization and changing nuclear gene expression in norflurazon-treated seedlings. Plant J. 65, 690–702. doi: 10.1111/j.1365-313X.2010.04454.x

95. Sangster, T. A., Bahrami, A., Wilczek, A., Watanabe, E., Schellenberg, K., McLellan, C., et al. (2007). Phenotypic diversity

and altered environmental plasticity in Arabidopsis thaliana with reduced Hsp90 levels. PLoS ONE 2:e648. doi: 10.1371/journal.pone.0000648

96. Schlicke, H., Hartwig, A. S., Firtzlaff, V., Richter, A. S., Glässer, C., Maier, K., et al. (2014). Induced deactivation of genes encoding chlorophyll biosynthesis enzymes disentangles tetrapyrrole-mediated retrograde signaling. Mol. Plant 7, 1211–1227. doi: 10.1093/mp/ssu034
97. Schön, A., Krupp, G., Gough, S., Berry-Lowe, S., Kannangara, C. G., and Söll, D. (1986). The RNA required in the first step of chlorophyll biosynthesis is a chloroplast glutamate tRNA. Nature 322, 281–284. doi: 10.1038/322281a0
98. Staneloni, R. J., Rodriguez-Batiller, M. J., and Casal, J. J. (2008). Abscisic acid, high-light, and oxidative stress down-regulate a photosynthetic gene via a promoter motif not involved in phytochrome-mediated transcriptional regulation. Mol. Plant 1, 75–83. doi: 10.1093/mp/ssm007
99. Strand,Å, Asami, T., Alonso, J., Ecker, J. R., and Chory, J. (2003). Chloroplast to nucleus communication triggered by accumulation of Mg-protoporphyrinIX. Nature 421, 79–83. doi: 10.1038/nature01204
100. Sullivan, J. A., and Gray, J. C. (2002). Multiple plastid signals regulate the expression of the pea plastocyanin gene in pea and transgenic tobacco plants. Plant J. 32, 763–774. doi: 10.1046/j.1365-313X.2002.01464.x
101. Sun, X., Feng, P., Xu, X., Guo, H., Ma, J., Chi, W., et al. (2011). A chloroplast envelope-bound PHD transcription factor mediates chloroplast signals to the nucleus. Nat. Commun. 2:477. doi: 10.1038/ncomms1486
102. Susek, R., and Chory, J. (1992). A tale of two genomes: role of a chloroplast signal in coordinating nuclear and plastid gene expression. Aust. J. Plant Physiol. 19, 387–399. doi: 10.1071/PP9920387
103. Susek, R. E., Ausubel, F. M., and Chory, J. (1993). Signal transduction mutants of Arabidopsis uncouple nuclear CAB and RBCS gene

expression from chloroplast development. Cell 74, 787–799. doi: 10.1016/0092-8674(93)90459-4

104. Tadini, L., Pesaresi, P., Kleine, T., Rossi, F., Guljamow, A., Sommer, F., et al. (2016). GUN1 controls accumulation of the plastid ribosomal protein S1 at the protein level and interacts with proteins involved in plastid protein homeostasis. Plant Physiol. 170, 1817–1830. doi: 10.1104/pp.15.02033
105. Tameshige, T., Fujita, H., Watanabe, K., Toyokura, K., Kondo, M., Tatematsu, K., et al. (2013). Pattern dynamics in adaxial-abaxial specific gene expression are modulated by a plastid retrograde signal during Arabidopsis thaliana leaf development. PLoS Genet.9:e1003655. doi: 10.1371/journal.pgen.1003655
106. Tanaka, H., Maruta, T., Tamoi, M., Yabuta, Y., Yoshimura, K., Ishikawa, T., et al. (2015). Transcriptional control of vitamin C defective 2 and tocopherol cyclase genes by light and plastid-derived signals: the partial involvement of GENOMES UNCOUPLED 1. Plant Sci. 231, 20–29. doi: 10.1016/j.plantsci.2014.11.007
107. Tanaka, K., and Hanaoka, M. (2013). The early days of plastid retrograde signaling with respect to replication and transcription. Front. Plant Sci. 3:301. doi: 10.3389/fpls.2012.00301
108. Tanaka, R., Kobayashi, K., and Masuda, T. (2011). Tetrapyrrole metabolism in Arabidopsis thaliana. Arabidopsis Book 9:e0145. doi: 10.1199/tab.0145
109. Tanaka, R., and Tanaka, A. (2007). Tetrapyrrole biosynthesis in higher plants. Annu. Rev. Plant Biol. 58, 321–346. doi: 10.1146/annurev.arplant.57.032905.105448
110. Tarahi Tabrizi, S., Sawicki, A., Zhou, S., Luo, M., and Willows, R. D. (2016). GUN4-protoporphyrin IX is a singlet oxygen generator with consequences for plastid retrograde signaling. J. Biol. Chem. 291, 8978–8984. doi: 10.1074/jbc.C116.719989
111. Terry, M. J., and Smith, A. G. (2013). A model for tetrapyrrole synthesis as the primary mechanism for plastid-to-nucleus signaling during chloroplast biogenesis. Front. Plant Sci. 4:14. doi: 10.3389/fpls.2013.00014

112. Thomas, J., and Weinstein, J. D. (1990). Measurement of heme efflux and heme content in isolated developing cotyledons. Plant Physiol.94, 1414–1423. doi: 10.1104/pp.94.3.1414

113. Verdecia, M. A., Larkin, R. M., Ferrer, J. L., Riek, R., Chory, J., and Noel, J. (2005). Structure of the Mg-chelatase cofactor GUN4 reveals a novel hand-shaped fold for porphyrin binding. PLoS Biol. 3:e151. doi: 10.1371/journal.pbio.0030151

114. Vinti, G., Hills, A., Campbell, S., Bowyer, J. R., Mochizuki, N., Chory, J., et al. (2000). Interactions between hy1 and gun mutants of Arabidopsis, and their implications for plastid/nuclear signalling. Plant J. 24, 883–894. doi: 10.1046/j.1365-313x.2000.00936.x

115. Voigt, C., Oster, U., Börnke, F., Jahns, P., Dietz, K. J., Leister, D., et al. (2010). In-depth analysis of the distinctive effects of norflurazon implies that tetrapyrrole biosynthesis, organellar gene expression and ABA cooperate in the GUN-type of plastid signalling. Physiol. Plant. 138, 503–519. doi: 10.1111/j.1399-3054.2009.01343.x

116. von Gromoff, E. D., Alawady, A., Meinecke, L., Grimm, B., and Beck, C. F. (2008). Heme, a plastid-derived regulator of nuclear gene expression in Chlamydomonas. Plant Cell 20, 552–567. doi: 10.1105/tpc.107.054650

117. Voß, B., Meinecke, L., Kurz, T., Al-Babili, S., Beck, C. F., and Hess, W. R. (2011). Hemin and magnesium-protoporphyrin IX induce global changes in gene expression in Chlamydomonas reinhardtii. Plant Physiol. 155, 892–905. doi: 10.1104/pp.110.158683

118. Waters, M. T., Wang, P., Korkaric, M., Capper, R. G., Saunders, N. J., and Langdale, J. A. (2009). GLK transcription factors coordinate expression of the photosynthetic apparatus in Arabidopsis. Plant Cell 21, 1109–1128. doi: 10.1105/tpc.108.065250

119. Woodson, J. D., Joens, M. S., Sinson, A. B., Gilkerson, J., Salomé, P. A., Weigel, D., et al. (2015). Ubiquitin facilitates a quality-control pathway that removes damaged chloroplasts. Science 350, 450–454. doi: 10.1126/science.aac7444

120. Woodson, J. D., Perez-Ruiz, J. M., and Chory, J. (2011). Heme synthesis by plastid ferrochelatase I regulates nuclear gene expression in plants. Curr. Biol. 21, 897–903. doi: 10.1016/j.cub.2011.04.004

121. Woodson, J. D., Perez-Ruiz, J. M., Schmitz, R. J., Ecker, J. R., and

Chory, J. (2013). Sigma factor mediated plastid retrograde signals control nuclear gene expression. Plant J. 73, 1–13. doi: 10.1111/tpj.12011

122. Xu, X., Chi, W., Sun, X., Feng, P., Guo, H., Li, J., et al. (2016). Convergence of light and chloroplast signals for de-etiolation through ABI4-HY5 and COP1. Nat. Plants 2:16066. doi: 10.1038/nplants.2016.66

123. Zhang, Z. W., Feng, L. Y., Cheng, J., Tang, H., Xu, F., Zhu, F., et al. (2013). The roles of two transcription factors, ABI4 and CBFA, in ABA and plastid signalling and stress responses. Plant Mol. Biol. 83, 445–458. doi: 10.1007/s11103-013-0102-8

124. Zhang, Z. W., Yuan, S., Feng, H., Xu, F., Cheng, J., Shang, J., et al. (2011a). Transient accumulation of Mg-protoporphyrin IX regulates expression of PhANGs - New evidence for the signaling role of tetrapyrroles in mature Arabidopsis plants. J. Plant Physiol. 168, 714–721. doi: 10.1016/j.jplph.2010.10.016

125. Zhang, Z. W., Yuan, S., Xu, F., Yang, H., Chen, Y. E., Yuan, M., et al. (2011b). Mg-protoporphyrin, haem and sugar signals double cellular total RNA against herbicide and high-light-derived oxidative stress. Plant Cell Environ. 34, 1031–1042. doi: 10.1111/j.1365-3040.2011.02302.x

126. Zhang, Z. W., Zhang, G. C., Zhu, F., Zhang, D. W., and Yuan, S. (2015). The roles of tetrapyrroles in plastid retrograde signaling and tolerance to environmental stresses. Planta 242, 1263–1276. doi: 10.1007/s00425-015-2384-3

CHAPTER

4

BIOCHEMISTRY AND MOLECULAR BIOLOGY OF CAROTENOID BIOSYNTHESIS IN CHILI PEPPERS (CAPSICUM SPP.)

María del Rocío Gómez-García [1] and Neftalí Ochoa-Alejo [1,2,]

[1]Department of Plant Genetic Engineering, Center for Research and Advanced Studies-Irapuato Unit, National Polytechnic Institute, Km 9.6 libramiento norte carretera Irapuato-León, Irapuato, Guanajuato 36821, México

[2]Department of Biotechnology and Biochemistry, Center for Research and Advanced Studies-Irapuato Unit, National Polytechnic Institute, Km 9.6 libramiento norte carretera Irapuato-León, Irapuato, Guanajuato 36821, México

ABSTRACT

Capsicum species produce fruits that synthesize and accumulate carotenoid pigments, which are responsible for the fruits' yellow, orange and red colors. Chili peppers have been used as an experimental model for studying the biochemical and molecular aspects of carotenoid biosynthesis. Most reports refer to the characterization of carotenoids and content determination in chili pepper fruits from different species, cultivars, varieties or genotypes. The types and levels of carotenoids differ between different chili pepper fruits, and they are also influenced by environmental conditions. Yellow-orange colors of chili pepper fruits are mainly due to the accumulation of α- and β-carotene, zeaxanthin, lutein and β-cryptoxanthin. Carotenoids such as capsanthin, capsorubin and capsanthin-5,6-epoxide confer the red colors. Chromoplasts are the sites of carotenoid pigment synthesis and storage. According to the most accepted theory, the synthesis of carotenoids in chili peppers is controlled by three loci: *c1*, *c2* and *y*. Several enzymes participating in carotenoid biosynthesis in chili pepper fruits have been isolated and characterized, and the corresponding gene sequences have been reported. However, there is currently limited information on the molecular mechanisms that regulate this biosynthetic pathway. Approaches to gain more knowledge of the regulation of carotenoid biosynthesis are discussed.

Keywords: *Capsicum*; carotenoids; chili pepper; biosynthesis; regulation

INTRODUCTION

Carotenoids are widely distributed in nature and are synthesized by all photosynthetic organisms (cyanobacteria, algae and plants) as well as non-photosynthetic microorganisms, such as fungi and some bacteria [1]. Since the first independent reports elucidating the structure of β-carotene by Richard Kuhn and Paul Karrer in 1928–1930, more than 700 natural carotenoids had been described prior to 2004 [2]. It is estimated that the annual worldwide carotenoid production by nature is equivalent to approximately 100 million tons [3], and more than 20 new structures are reported each year [2].

Carotenoids are isoprenoid compounds that consist of eight isoprene (ip) units attached in a head-tail pattern where the double bond order is inverted at the center of the molecule [4]. Carotenoids comprise a wide group of lipophilic pigments that provide a series of colors, including yellow, orange and red. These pigments are responsible for the coloring of many flowers such as marigolds [5], daffodils, *Gentiana lutea* and *Sandersonia aurantiaca* [6]; fruits such as tomatoes [7], papayas [8], mandarins and oranges [9]; and roots (carrots) [10]. Carotenoids are mostly found intracellularly at the chloroplast and chromoplast membranes in plants. Traditionally, they have been structurally classified as carotenoids, including α-carotene, β-carotene and xanthophylls such as β-cryptoxanthin, lutein, zeaxanthin, violaxanthin, neoxanthin and fucoxanthin [4]. Chromophore length determines the absorption spectrum of a carotenoid molecule and thus its color to the eye [11].

The main function of carotenoids is the protection of cells and organelles against oxidative damage, which they accomplish by interacting with singlet oxygen molecules and scavenging peroxy radicals, thus preventing the accumulation of harmful oxygen species [12]. They are also involved in photosynthesis (participating in the light-harvesting process and as photoprotectors of the photosynthetic apparatus), within the xanthophyll cycle (protecting against light damage) [4], and as precursors of abscisic acid. Additionally, carotenoids have a paramount ecological function because they act as attractants for pollinators and seed dispersal agents [13]. Furthermore, oxidative cleavage of carotenoids by a family of carotenoid cleavage dioxygenases (CCDs; enzymes that cleave double bonds) leads to the production of apocarotenoids, compounds with a variety of biological important activities such as phytohormones (ABA, and strigolactones, a group of terpene lactones with hormone activity that promote germination of root parasitic plants, stimulate symbiotic interactions between plants and arbuscular mycorrhizal fungi, and inhibit shoot axillary branching), the visual and signaling molecules retinal (chromophore of various visual pigments in animals) and retinoic acid (nuclear receptor ligand that is a major signal controlling a wide range of transcriptional processes), and the aromatic volatiles β-ionone (pollinator attractant and fruit or vegetable flavor), β-cyclocitral, geranial, geranyl acetone, theaspirone, α-damascenone and β-damascenone responsible for the flavor and aroma/scent of a number of flowers and a diversity of foods [14–16].

Carotenoids have an essential function in human nutrition and health; humans are unable to *de novo* synthesize vitamin A from endogenous isoprenoid precursors, but plant carotenoids (β-carotene, α-carotene, γ-carotene and β-cryptoxanthin; provide the primary dietary source of provitamin A (meaning they can be converted into retinol) [17]. In addition to their nutritional value, carotenoids, acting as antioxidants, have been implicated in reducing the risk of cancer and cardiovascular diseases [18]. α- and β-carotene suppress tumorigenesis in the skin, lung, liver and colon [19]. Lycopene prevents cardiovascular diseases and possibly prostate cancer [20]. Likewise, it was reported that a diet rich in carotenoids is directly connected to a reduced risk of age-related macular degeneration [21]. Similarly, zeaxanthin and lutein (essential components of the macular pigment in the eye) showed the strongest association between dietary intake and reduced risk of macular degeneration [1,22]. Apocarotenoids have also showed interesting multifunctional activities, and can be useful in the prevention of cancer and other degenerative diseases [17]. Chili pepper fruits produce and accumulate apocarotenoids such as apo-14′-zeaxanthinal, apo-13-zeaxanthinone, apo-12′-capsorubinal, apo-8′-capsorubinal, 9,9′-diapo-10,9′-*retro*-carotene-9,9′-dione, apo-8′-zeaxanthinal, apo-10′-zeaxanthinal, apo-12′-zeaxanthinal, apo-15-zeaxanthinal, apo-11-zeaxanthinal and apo-9-zeaxanthinone [23].

At the commercial level, carotenoids have a broad variety of applications, such as nutrient supplements, for pharmaceutical purposes, in animal feeds [12], coloring foods, cosmetics and nutraceutical agents as well [4].

CAPSICUM FRUIT CAROTENOIDS

Chili pepper fruits synthesize and accumulate a variety of compounds, such as the characteristic capsaicinoids (hot compounds), vitamins (Vitamins A, C and B), and pigments (anthocyanins and carotenoids) (Figure 1). Studies of *Capsicum*fruit carotenoid pigments began in the nineteen-century. Henri Braconnot reported the first investigations of *Capsicum annuum* (paprika) pigments in 1817 [24]. One pigment from *Capsicum annuum* (paprika) was obtained in a crystalline form in 1927, and the name capsanthin was proposed [25]. Similarly, von Zechmeister and von Cholnoky [26] published a series of articles on the carotenoid content of paprika and found carotenoids such as β-carotene, cryptoxanthin and zeaxanthin that had not been previously described, in addition to capsanthin and capsorubin [27]. Other new carotenoids from paprika, such as lutein epoxide, antheraxanthin, violaxanthin, cryptocapsin, and mutatoxanthin, were also reported [28]. Later, it was found that capsanthin and capsorubin contained one and two cyclopentane rings, respectively, adjacent to their keto groups, which were part of the conjugated double bond system [29,30].

Figure 1. Chili pepper fruits (*Capsicum annuum*; "Bolivian") showing carotenoids and anthocyanins in the pericarp tissue.

CAROTENOID CONTENT AND COMPOSITION DURING CHILI PEPPER FRUIT RIPENING

Chili pepper fruits undergo profound morphological, physiological and metabolic transformations in terms of pigment composition and content during ripening. These changes in fruit composition are affected by the genotype, maturity and growth conditions [31,32]. Chlorophyll is responsible for the green color of chili pepper fruits, anthocyanins are violet/purple pigments, and the yellow-orange colors are provided by α- and β-carotene, zeaxanthin, lutein and β-cryptoxanthin. Carotenoids with oxygenated forms such as those bearing acylcyclopentanol end groups control the red color in chili pepper fruits. The most characteristic examples are capsanthin, capsorubin and capsanthin-5,6-epoxide [33].

Capsanthin is a red color pigment that is lipophilic and is mostly synthesized and accumulated in red *Capsicum* spp. fruits but also in *Lilium*, *Aesculus* and *Berberis* species and in fruits from *Asparagus officinalis* [34–42]. Capsanthin levels may represent up to 50% of total carotenoid fruit content during the ripening stage. The capsanthin structure contains

11 conjugated double bonds, a conjugated keto group and a cyclopentane ring. Due to these characteristics, capsanthin is a powerful antioxidant (good radical scavenger) [43]. Though β-carotene and β-cryptoxanthin possess the same number of double bonds as capsanthin, their antioxidant abilities are lower than that of capsanthin. Although capsanthin does not have provitamin A properties, it is considered a functional compound due to its antioxidant and anti-tumor (colon cancer) activities [44].

Carotenoids are not randomly dispersed or freely accessible in chromoplasts, but rather 95% of these pigments accumulate in specific substructures: the fibrils [45]. These fibrils are aligned along the longitudinal axes of chromoplasts, which explains their birefringence. Purified fibrils contain galactolipids, phospholipids and a single 32 kDa protein named fibrillin. A model for the chromoplast fibril architecture proposes that carotenoids accumulate in the center of the fibrils and are surrounded by a layer of polar lipids (galactolipids and phospholipids), which in turn are surrounded by an outer layer of fibrillin that is directly attached to the plastid stroma [45].

The content and composition of the carotenoid profiles in chili pepper fruits during ripening is determined by two metabolic processes: (1) transformation of existing photosynthetic pigments; and (2) *de novo* carotenoid biosynthesis. For example, studies of the pigment content during several ripening stages (green, color break I, color break II, red I, red II), in the chili pepper cultivars Bola and Agridulce revealed that the chlorophyllic pigments (lutein and neoxanthin) disappeared in the red I and II stages (mature fruit), whereas the concentrations of intermediates in the synthesis of red pigments (β-carotene and violaxanthin) increased, and zeaxanthin, capsanthin, capsorubin, β-cryptoxanthin and capsolutein were synthesized *de novo* [46]. However, in yellow-orange chili pepper fruits from *C. baccatum*, *C. pubescens* and *C. annuum*, violaxanthin was the major carotenoid (37% to 68% of total carotenoids), followed by *cis*-violaxanthin, antheraxanthin and lutein (5% to 14%) [47].

Rates of carotenoid accumulation in mature red chili pepper fruits (*C. annuum*) have often been found to be as follows, in decreasing order: ketoxanthophylls (capsanthin, capsorubin and capsolutein), xanthophylls (zeaxanthin, neoxanthin and violaxanthin), epoxyxanthophylls (capsanthin-5,6-epoxide and capsanthin-3,6 epoxide) and hydrocarbons (lycopene) [48].

In ripe chili pepper fruits, most carotenoids are esterified with fatty acids, making them liposoluble [49] and facilitating their accumulation in the lipophilic globule at the chromoplasts [44]. The esterification process takes place mostly in *de novo*biosynthesized pigments such as capsanthin, capsorubin, zeaxanthin and β-cryptoxanthin, although it also occurs in previously synthesized pigments such as violaxanthin [50]. *De novo* biosynthesis and esterification occur simultaneously during chili pepper fruit ripening. Red xanthophylls, such as capsanthin, are esterified by short chains of saturated fatty acids such as lauric (12:0), myristic (14:0) and palmitic acid (16:0), while yellow xanthophylls are mainly esterified by myristic (14:0), palmitic (16:0) and an unsaturated linoleic acid form ($18{:}2^{\Delta 9,12}$). For this reason, the higher numbers of double bonds in the fatty acid chains in yellow xanthophylls make them less stable compared to their red counterparts [51]. Similarly, in full ripe chili pepper fruits (*Capsicum annuum* cv. Bola), the free carotenoids and partially and totally esterified forms of these pigments account for 21.3%, 35.6% and 43.1% of the total, respectively [51]. Likewise, the balance among these three esterified fractions in fully ripe chili pepper fruits (*C. annuum*) appears to be well conserved in Numex, Mana, Belrubi, Negral and Delfin cultivars, and these might be used as a "ripeness index". Interestingly, the "Negral" chili pepper, which retains chlorophyll at the fully ripe stage, showed the same degree of esterification as other red chili pepper fruit cultivars, supporting the idea that the carotenogenesis process is independent from the chlorophyll catabolism [50].

Total carotenoid content in mature red chili pepper fruits can increase from two up to 60 times compared to the immature fruits. For instance, in black paprika fruits (*C. annuum*), the total carotenoid content during ripening ranges from 48.5 mg/100 g (d.wt.) in immature green fruits up to 3211 mg/100 g in mature red fruits, which represents a 66-fold increase (Table 1) [54]. Carotenoid composition also undergoes drastic changes during ripening. In general, the most abundant carotenoids in immature green chili pepper fruits are lutein, β-carotene and neoxanthin; their percentages in green bell peppers are: lutein (40.8%), neoxanthin (15.1%) and β-carotene (13.4%) (Table 1) [57]. However, carotenoid composition changes as fruit ripens; for example, in red bell peppers, the main carotenoids are capsanthin (35%), β-carotene (11.6%) and violaxanthin (10%) (Table 1) [27]. These modifications to carotenoid composition have been observed in other chili pepper varieties as well (Table 1). Of the total carotenoid content (9.15 mg/100 g f.wt.) in the Signal Red variety (*C. annuum*), capsanthin (predominant pigment) represented 46% [57]. Likewise, the total carotenoid content is also dependent on variety. Generally, "Signal Red" > "Signal Orange" ≈ "Signal Yellow" ≈ "Signal Green", and red pepper fruits have the highest antioxidant activity.

A detailed study of the relative carotenoid content in Red paprika (*Capsicum annuum* var. *lycopersiciforme rubrum*) at six ripening stages was described [55]. Thirty-five different known carotenoids varied with the developmental stage. Lutein was the most abundant in the green stage, then decreased with further ripening stages and was absent in the red or deep red stages. Capsanthin levels were low in the early stages and increased gradually until the red stage, at which the highest content was recorded. β-carotene was almost constant throughout the growth and ripening of fruits.

Table 1. Carotenoid contents in fruits from several *Capsicum annuum* varieties.

Chili pepper type	Ripening stage	Main carotenoids	Total carotenoids	Reference
Red bell pepper	Mature (Red)	Capsanthin (34.7%), β-carotene (11.6%) and violaxanthin (9.9%)	284 and 127 mg per kg f.wt. (as β-carotene) (two lots of peppers)	[27]
Green bell pepper	Immature (Green)	Lutein (40.8%), neoxanthin (15.1%), violaxanthin (13.8%) and β-carotene (13.4%)	10.6, 11.2 and 9.0 mg per kg f.wt. (as β-carotene) (three lots of green peppers)	[52]
Yellow pepper	Immature (Green)	Violaxanthin (34%), antheraxanthin (10.5%), lutein (9.2%) and zeaxanthin (8.5%)	13.2 mg/100 g d.wt.	[53]
	Mature (Orange)	Lutein (37.8%), β-carotene (19.8%) and neoxanthin (5.5%)	488.6 mg/100 g d.wt.	
Black paprika	Immature (Black)	Lutein (28.5%) and zeaxanthin (11.96%)	48.5 mg/100 g of d.wt.	[54]
	Mature (Red)	Capsanthin (42%), zeaxanthin (8%), capsorubin (3.2%) and β-carotene (7%)	3211 mg/100 g d.wt.	
Szentesi Kosszarvú	Immature (Green)	Lutein (31.9%) and β-carotene (11.3%)	11.5 mg/100 g d.wt.	[48]
	Mature (Red)	Capsanthin (29%) and zeaxanthin (15%)	994.7 mg/100 g d.wt.	

Chili pepper type	Ripening stage	Main carotenoids	Total carotenoids	Reference
Capsicum annuum var. *lycopersiciforme rubrum*	Immature (Green)	Lutein (31.6%) and β-carotene (13.7%)	19.6 mg/100 g d.wt.	[55]
	Mature (Red)	Capsanthin (37%), zeaxanthin (8%), and β-carotene (9%)	1297.1 mg/100 g d.wt.	
Sweet peppers	Immature (Green)	Lutein (2.3 mg/100 g f.wt.) and β-carotene (1.7 mg/100 g f.wt.)	5.1 mg/100 g f.wt.	[32]
	Green	Lutein (1.4 mg/100 g f.wt.) and β-carotene (2.1 mg/100 g f.wt.)	4.9 mg/100 g f.wt.	
	Immature (Red)	β-carotene (1.9 mg/100 g f.wt.) and zeaxanthin (2.9 mg/100 g f.wt.)	9.5 mg/100 g f.wt.	
	Red	β-carotene (4.3 mg/100 g f.wt.) and capsanthin (19.9 mg/100 g f.wt.)	45.6 mg/100 g f.wt.	
Ancho, guajillo and mulato	Ancho (Mature)	β-carotene (20.9%), and violaxanthin (14.5%)	7.5 mg/100 g d.wt.	[56]
	Guajillo (Mature)	β-carotene (17.9%), violaxanthin (13.2%)	6.8 mg/100 g d.wt.	
	Mulato (Mature)	Violaxanthin (22%) β-carotene (14.9%)	7.2 mg/100 g d.wt.	

In mature yellow pepper fruits, the carotenoid content and composition is quite different as compared to the red fruits. Violaxanthin (34%), antheraxanthin (10.5%), lutein (9.2) and zeaxanthin (8.5) were the most abundant carotenoids in immature (green) yellow pepper fruits. These fruits were devoid of red carotenoids (capsanthin, capsorubin and cryptocapsin) [54].

Chili pepper fruits from some cultivars are capable of retaining chlorophyll (delayed chlorophyll senescence) at the fully ripening stage, producing a dark-brown color due to the combination of chlorophyll and red carotenoids [58]. For example, "Negral" retained chlorophyll at the fully ripe stage (almost 14% of chlorophylls at the green stage), whereas "Bola Roja" fruits reached the fully ripe stage without any chlorophyll [59]. These differences between chili pepper varieties or cultivars are attributed to differences in the plastid ultrastructure. The persistence of a long peripheral thylakoids in the chromoplasts of "Negral" chili peppers was most likely associated with the presence of chlorophyll in fully ripe fruits of this cultivar. This study concluded that "Negral" chili peppers have undergone a minor plastid evolution compared to their "Bola Roja" counterparts [59].

Analyzing the changes in the accumulation of individual carotenoids in red fruit chili peppers in different cultivars (*Capsicum annuum* L.

"Mana", "Numex", "Belrubi", "Delfin" and "Negral" (a chlorophyll-retaining mutant at the ripe stage)) during fruit ripening has revealed an inverse relationship between total carotenoid content and the red to yellow isochromic pigment fraction ratio (R/Y) as well as the capsanthin-to-zeaxanthin ratio (Caps/Zeax) [50]. This relationship appears to be related to the carotenogenic capacity of the cultivar, and thus selection and breeding should not only seek higher total carotenoid content but also attempt to increase these ratios. For example, the Mana cultivar had the highest total carotenoid content (13,208 mg/kg d.wt.) but the lowest R/Y (1.25) and Caps/Zeax (3.38) ratios, and therefore these are the parameters to improve. Cultivar Negral had a high carotenoid content (8797 mg/kg d.wt.) and high R/Y and Caps/Zeax ratios as well, and thus it could be used to transfer these characteristics into Mana cultivar by direct crosses. Interestingly, chili pepper fruits undergo substantial structural changes during the late ripening stages [60]. In general, chloroplasts developed into chromoplasts, which are rich in carotenoids. The typical chloroplast thylakoid structure disintegrates and is replaced by non-chlorophyllous single thylakoids, which are derived in part from the inner envelope membrane. These structural changes correlate with an enhanced accumulation of keto-carotenoids (capsanthin, capsorubin and cryptocapsin). Lipid metabolism is affected as well, as galactolipid levels are diminished while phospholipids accumulate.

Carotenoid composition analysis of red paprika fruits (*C. annuum*, var. *lycopersiciforme rubrum*) during ripening showed that the major component was capsanthin (37%), but numerous minor carotenoids were also present. These included violaxanthin, antheraxanthin and capsanthin-5,6-epoxide (5,6-epoxy-end group); capsanthin-3,6-epoxide and cucurbitaxanthin B (3,6-epoxy end group); 5,6-diepikarpoxanthin (3,5,6-trihydroxy end group); capsanthone (3-oxo-κ end group); nigroxanthin (γ end group); and capsorubin, cryptocapsin and several furanoid oxides and *cis* isomers [55]. A possible biosynthetic pathway for the formation of minor carotenoids containing 3,5,6-trihydroxy-β, 3,6-epoxy-β, and 6-hydroxy-γ end groups was proposed.

Recently, a study of red, orange and yellow chili pepper fruits from 28 cultivars (*C. annuum* (24), *C. frutescens* (1) and *C. chinense* (3)) was conducted to examine the relationship between fruit color and β-carotene content, to characterize the content of six carotenoids in orange-fruited varieties and to analyze the DNA sequences of four structural biosynthetic genes to identify metabolic and genetic differences between them [61]. The most abundant carotenoids in pericarp tissues were β-carotene, capsanthin and capsorubin. A large degree of variability in terms of β-carotene, capsanthin and total carotenoids was observed in all cultivars. Orange color was mainly determined by the presence of β-carotene but in two cases was due to the accumulation of red and yellow carotenoids. Examination of phytoene synthase (*PSY*), lycopene β-cyclase (*LCY-B*), a β-carotene hydroxylase (*BCH; CRTZ-2*) and capsanthin-capsorubin synthase (*CCS*) gene sequences indicated that different alleles for specific carotenoid biosynthetic enzymes were associated with the accumulation of specific carotenoids in orange fruits that could be used as molecular markers for high β-carotene (high provitamin A) fruit content selection.

The antioxidant activities of chili pepper fruits are also affected by ripening. Fully ripe chili pepper fruits exhibit higher antioxidant activity than green fruits. This difference is easily explained by the increased amounts of carotenoids, phenolics, flavonoids and ascorbic acid that are present in fully ripe chili pepper fruits [33]. The levels of these antioxidants in chili pepper fruits may be affected by genotype, plant growth conditions and post-harvest management [62].

Finally, in non-photosynthetic organs such as seeds, flowers or red fruits, carotenoids have complex structural diversity, as these compounds undergo a series of secondary metabolic reactions such as oxidation, cleavage of polyene chains, epoxidation, and (*Z*/*E*) (*cis-trans*) isomerization, among others. Therefore, novel carotenoid structures are detected in these plant organs [44]. The improvement of analytical instruments such as NMR, MS and HPLC has made it possible to determine the structures of very minor carotenoids found in nature. For instance, the new carotenoid compounds 3′-deoxy-capsanthin and 3,4-dehydroxy-3′-deoxycapsanthin, which are minor components of mature paprika fruits (*C. annuum*), were isolated and characterized [63]. Their chemical structures were determined to be (3*R*, 5′*R*)-3-hydroxy-β,κ-caroten-6′-one and (5′*R*)-3,4-didehydro-β,κ-caroten-6′-one, respectively, using UV-vis, NMR, CD (Circular Dichroism), HRFABMS (High Resolution Fast Atom Bombardment Mass Spectrometry) and FABMS (Fast Atom Bombardment Mass Spectrometry)/MS spectra. These compounds are the first examples of carotenoids that possess a 6-oxo-κ end group.

CAROTENOID BIOSYNTHESIS PATHWAY IN PLANTS

The carotenoid biosynthesis pathway in plants is tightly linked to other biosynthetic routes such as those for gibberellins, tocopherols, chlorophylls and phylloquinones via the five-carbon (C5) compound isopentenyl pyrophosphate (IPP). Since the discovery of the mevalonate (MVA) pathway in the 1950s, it was assumed that IPP was synthesized from acetyl-CoA via mevalonate [3]. Although the MVA pathway may contribute to carotenoid biosynthesis in some cases, such as in etiolated seedlings [22], plant carotenoids are mainly produced through the methylerythritol phosphate (MEP)-derived pathway in light-grown plants [64–66].

All plant carotenoids are synthesized from units of five carbon compounds, IPP and its isomer dimethylallyl pyrophosphate (DMAPP) (Figure 2). The addition of three IPP molecules to a DMAPP molecule generates geranylgeranyl pyrophosphate (GGPP), the immediate precursor of carotenoids. This reaction is catalyzed by GGPP synthase (GGPS) [1]. This enzyme was isolated as a homodimer (74 kDa) from chili pepper chromoplasts [67], and its corresponding cDNA was identified and sequenced [68]. GGPP is the common precursor for various plastid isoprenoids, including gibberellins, chlorophylls, tocopherols, phylloquinones, plastoquinones and carotenoids [1].

The first step in plant carotenoid biosynthesis pathway is the formation of phytoene (Figure 2); this is a two-step condensation reaction with two GGPP molecules catalyzed by PSY [1]. Phytoene synthesis in the stroma of chloroplasts, etioplasts and amyloplasts from tissues of different plant species (*Capsicum annuum*, *Pisum sativum*, *Spinacia oleracea*, *Hordeum vulgare*, *Triticum aestivum* and *Zea mays*) was first reported by Dogbo *et al.* [78]. PSY biochemical characterization was carried out using chili pepper tissues [79]. The phytoene molecule is made of a 40-carbon isoprenoid polyene chain with conjugated double bonds that form the backbone of plant carotenoids and determine their physical and biological properties [1]. The second step is the formation of lycopene, which occurs after four phytoene desaturation and isomerization reactions leading from 15-*cis*-phytoene to all-*trans*-lycopene (Figure 2). These desaturations and isomerization reactions sequentially produce

9,15-*cis*-phytofluene, 9,15,9′-tri-*cis*-ζ-carotene, 9,9′-di-*cis*-ζ-carotene, 7,9,9′-tri-*cis*-neurosporene and finally all-*trans*-lycopene (Figure 2) [77] and increase the conjugated double bond series, which constitutes the chromophore, transforming non-colored phytoene into red lycopene [80]. In plants, the four desaturation reactions are catalyzed in two steps by two phylogenetically related enzymes: phytoene desaturase (PDS) and ζ-carotene desaturase (ZDS). The enzyme responsible for the desaturation of phytoene to ζ-carotene is PDS. Phytoene desaturase genes were identified and isolated from chili peppers [81]. Phytoene (15-*cis*-phytoene) suffers two sequential desaturations by PDS to produce 9,15-*cis*-phytofluene and 9,15,9′-*cis*-ζ-carotene, which can be isomerized into ζ-carotene by light (Figure 2). However, carotenogenesis in organs under dark conditions, such as roots and etiolated leaves, suggests the participation of a ζ-carotene isomerase (Z-ISO) that transforms 9,15,9′-tri-*cis*-ζ-carotene into 9,9′-di-*cis*-ζ-carotene (Figure 2) [76,77,82]. Recently, a *Z-ISO* gene from maize and *Arabidopsis* was isolated and characterized and was found to be important for both light-exposed and tissues under dark conditions [83]. ZDS catalyzes two desaturation steps of 9,9′-di-*cis*-ζ-carotene to produce 7,9,9′-tri-*cis*-neurosporene, and 7,9,7′9′-tetra-*cis*-lycopene; and finally, a carotenoid isomerase (CRTISO) converts this later compound into all-*trans*-lycopene (Figure 2) [76,77]. A cDNA encoding the gene for ZDS was cloned from chili peppers and expressed in *E. coli* [84]. The deduced amino acid sequence of this cDNA encoded a protein with a mature size of approximately 59 kDa derived from a precursor polypeptide with a NH_2-terminal extension resembling transit peptides for plastid targeting.

In *Arabidopsis* etioplasts and tomato chromoplasts, carotenoid isomerase (CRTISO) activity is necessary to transform pro-lycopene (product of PDS and ZDS activities) into lycopene [85,86]. However, it appears that photo-isomerization can substitute for CRTISO in chloroplasts. Thus, it is possible that CRTISO in plants functions in carotenoid biosynthesis in non-photosynthetic tissues [1]. More recently, two CRTISO (*Zmcrtiso1* and *Zmcrtiso2*) cDNAs were identified from corn (*Zea mays*) mapping to different chromosomes [76]. ZmCRTISO1 exhibited activity on tetra-*cis*-prolycopene and produced all-*trans*-lycopene, but was not capable of acting on the 15-*cis* double bond of 9,15,9′-tri-*cis*-ζ-carotene; ZmCRTISO2 was inactivated by a premature termination codon in the inbred line B73, but the mutation was absent in other corn cultivars and the activity was the same as ZmCRTISO1.

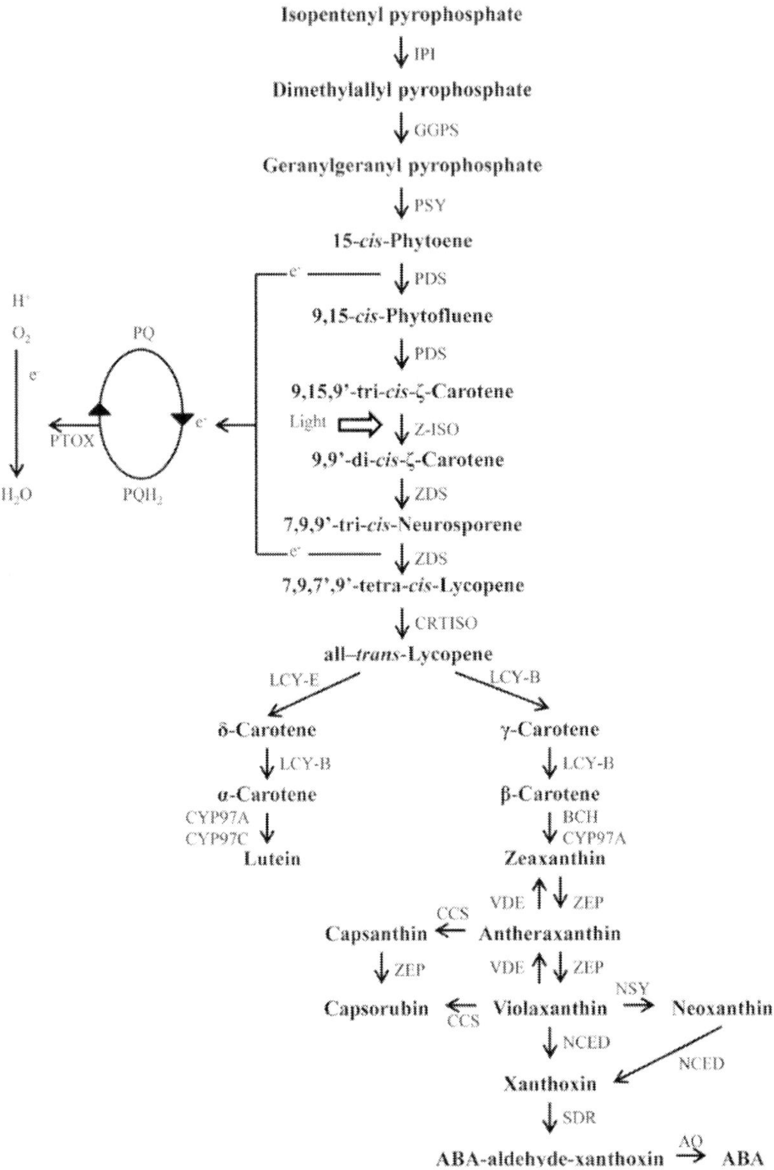

Figure 2. Carotenoid biosynthesis pathway in plants. IPI, isopentenyl pyrophosphate isomerase; GGPS, geranylgeranyl pyrophosphate synthase; PSY, phytoene synthase; PDS, phytoene desaturase; Z-ISO, ζ-carotene isomerase; ZDS, ζ-carotene desaturase; CRTISO, carotene or carotenoid

isomerase; LCY-B, lycopene-β-cyclase; LCY-E, lycopene-ε-cyclase; BCH, β-carotene hydroxylase or carotene β-hydroxylase (non-heme di-iron type); CYP97A, β-carotene hydroxylase (cytochrome 450 type); CYP97C, ε-carotene hydroxylase (cytochrome 450 type); ZEP, zeaxanthin epoxidase; CCS, capsanthin-capsorubin synthase; VDE, violaxanthin de-epoxidase; NSY, neoxanthin synthase; NCED, 9-*cis*-epoxycarotenoid dioxygenase (a carotenoid cleavage dioxygenase; CCD); SDR, short-chain dehydrogenase/reductase; AO, aldehyde oxidase; ABA, abscisic acid; PTOX, plastid terminal oxidase; PQ, oxidized plastoquinone; PQH_2, reduced plastoquinone. Adapted from [69–77].

It has been proposed that phytoene desaturation involving PDS and ZDS requires the transfer of two electrons to oxidized plastoquinones, which in turn are re-oxidized by a plastid terminal oxidase (PTOX) using O_2 as a terminal acceptor (Figure 2) [70]. Mutant plants affected in *PTOX* gene (*immutants* and *ghost*) are defective in carotenoid production and accumulate phytoene; as a result, these mutants exhibit photobleaching in high light conditions because of a lack of photoprotective carotenoids [69,70,87–90]. PTOX is a plastid-located plastoquinol:oxygen oxidoreductase linked to carotenoid biosynthesis and chlororespiration to prevent photooxidative damage [91]. PTOX is synthesized as a precursor polypeptide, which is imported into chloroplasts and inserted into the thylakoid membrane [69]. In *Arabidopsis* a *PTOX*cDNA showing a 350-amino acid ORF encoding a putative 40.5 kD protein was isolated [69]. The genomic sequence of PTOX revealed a structure composed of nine exons and 8 introns [69].

The third step in carotenoid biosynthesis is the cyclization of lycopene (Figure 2). Carotenoid cyclization is limited to the formation of a six-membered ring at one or both ends of the acyclic precursor. These end groups merely differ in the position of a double bond in the cyclohexane ring. The type of end group depends on the nature of the cyclase enzyme [3]. The mechanism of cyclization involves a proton attack at C2 and C2′ of lycopene. The resulting carbonium ion intermediate is stabilized by loss of a proton from either C1 or C4 to yield a β- or ϵ-ring, respectively [55]. The formation of β-carotene is catalyzed by a LCY-B, which adds a β ring to each final end of lycopene to produce β-carotene. In the case of α-carotene, the lycopene ϵ-cyclase (LCY-E) attaches a ε ring to lycopene and generates δ-carotene; then, LCY-B joints a β ring to the final end of δ-carotene to yield α-carotene (Figure 2) [1]. LCY-B has been isolated from a variety of red paprikas [92,93].

The fourth step in the carotenoid biosynthesis pathway is a hydroxylation step (Figure 2). Xanthophylls are oxidation products derived from α and β-carotene. Zeaxanthin and lutein are formed by hydroxylation at the 3 and 3′ carbon atoms of β,β-carotene or β,ϵ-carotene, respectively, by separate hydroxylases, each one specific for β or ϵ rings [55]. β,β and β,ε xanthophylls are produced by two different classes of carotene hydroxylases: heme-(cytochrome P450) (CYP97) and non-heme (di-iron) (BCH; also abbreviated β-CH, CHY, CHYB, CRTR-B, HYD or HYDB by different authors) (Figure 2) [94]. Hydroxylation of the two β-ionone rings in β-carotene generates zeaxanthin, whereas hydroxylation of the one β-ring and one ε-ring in α-carotene leads to lutein formation. Hydroxylation of β-rings in the carotenes is potentially mediated by either the P450-type CYP97A or di-iron β-CH enzymes, while hydroxylation of the ε-ring of α-carotene is carried out by CYP97C, another P450 enzyme. Zeaxanthin is easily converted into violaxanthin, via antheraxanthin, introducing the 5,6-epoxy group to the 3-hydroxy β-rings. This reaction is catalyzed by a zeaxanthin epoxidase (ZEP) (Figure 2). Conversely, in leaves growing under high light intensity, violaxanthin de-epoxidase (VDE) catalyzes the de-epoxidation reaction in two steps, allowing the transformation of violaxanthin into zeaxanthin (this reverse reaction is much more efficient at dissipating the excess excitation energy) (Figure 2). However, when light conditions return to normal values, zeaxanthin is transformed into violaxanthin. This reversible inter-conversion is known as the xanthophyll cycle and is of key importance for NPQ (non-photochemical quenching) and for plant adaptation to changes in environmental conditions [95].

The last step in carotenoid biosynthesis is the conversion of violaxanthin into neoxanthin by the action of a neoxanthin synthase (NSY; also abbreviated NXS by some authors) (Figure 2) [1,71]. Violaxanthin can also be converted into xanthoxin and subsequently into abscisic acid through the action of a 9-*cis*-epoxycarotenoid dioxygenase (NCED) and an aldehyde oxidase (AO) (Figure 2). It has been proposed that neoxanthin can be also converted into xanthoxin by the NCED (Figure 2) [71,72].

CAROTENOID BIOSYNTHESIS IN CHILI PEPPER FRUITS

Capsicum has been one of the most important models to study the chemistry and biosynthesis of carotenoids in plants. Chili pepper fruits accumulate yellow, orange or red carotenoids during the ripening process. Chili pepper fruits exhibit green color due to the presence of chloroplasts during the early stages, but these organelles suffer profound modifications during the late fruit ripening stages to become chromoplasts, which are carotenoid-accumulating plastids [60].

Labeling experiments demonstrated that *Capsicum* chromoplasts incubated with labeled isopentenyl pyrophosphate synthesized several labeled carotenes, among them phytoene, *cis*-phytofluene and *trans*-phytofluene, β-carotene and ζ-carotene [96]. However, α-carotene was poorly labeled. These results demonstrated that chromoplasts are the site of carotenoid biosynthesis and that those carotenoids derived from β-carotene are predominantly synthesized during fruit ripening. Moreover, phytoene is biosynthesized in the stroma, and the membrane fraction (chromoplast envelope plus achlorophyll lamellae derived in part from the inner envelope membrane) is the site where desaturation and cyclization reactions occur during the biogenesis of colored carotenoids. A protocol for the isolation and purification of *Capsicum annuum* chromoplasts was reported [97]. In summary, the lipid composition, carotenoid content and the incorporation of [1-^{14}C]isopentenyl pyrophosphate into carotenoids in purified chromoplasts has been described.

Xanthophylls are oxidation-derived products of carotenoids. In *Capsicum*, xanthophylls are responsible for the yellow, orange and red colors in fruits [57]. The keto-xanthophylls (capsanthin, capsorubin and cryptocapsin) are found in red pepper fruits synthesized in the same way as β-carotene via a rearrangement of the epoxy-cyclohexenyl groups of β,β-xanthophylls (antheraxanthin and violaxanthin) (Figure 2) [98,99]. The suggested mechanism involves a pinacolic rearrangement. Conversion of epoxy-xanthophylls (antheraxanthin and violaxanthin) into keto-xanthophylls (capsanthin and capsorubin) is carried out at the *Capsicum* chromoplast membranes (Figure 2) [100].

A metabolic correlation analysis for a number of carotenoids including β-carotene, β-cryptoxanthin, cryptocapsin, capsanthin, capsorubin, neoxanthin, mutatoxanthin and luteoxanthin was carried out by means of a Pearson matrix calculated from the observed pigment concentrations in two *Capsicum* cultivars (Bola Roja and Negral) during fruit maturation as an approach to investigate the participation of different carotenoid precursors and intermediates [59]. The conclusions of this study were: (1) a high linear correlation between β-carotene, β-cryptoxanthin and cryptocapsin was observed, suggesting that these carotenoids were involved in the same metabolic route and that they might be intermediaries in the same pathway; and (2) a high negative correlation between the levels of these three carotenoids and xanthophylls (capsanthin, capsorubin, neoxanthin, mutatoxanthin, luteoxanthin and cryptoflavin) was observed, indicating that the synthesis of these compounds during maturation was paralleled by a decrease in β-carotene, β-cryptoxanthin and cryptocapsin, which are potential precursors in *Capsicum* carotenoid metabolism. Likewise, it was suggested that β-carotene, β-cryptoxanthin and zeaxanthin were violaxanthin and antheraxanthin precursors, which in turn were the precursors for capsorubin and capsanthin, respectively. β-cryptoxanthin is the β-cryptoxanthin epoxide precursor (Figure 2). Based on these results, a carotenoid pathway in *Capsicum* was proposed (Figure 2) [101,102].

A series of articles were published describing the isolation and characterization of enzymes that are involved in the carotenoid biosynthetic pathway in chili peppers as well as genetic aspects of this pathway. The isopentenyl pyrophosphate isomerase (IPI) and GGPS enzymes were first isolated and purified to homogeneity from chili pepper chromoplast stroma using aminophenethyl pyrophosphate-affinity chromatography [67]. IPI was revealed to be a monomeric enzyme with a molecular weight of 33,500 ± 500 Da and a K_m of 6 μM for isopentenyl pyrophosphate. GGPS was found to be dimeric with a native molecular weight of 74,000 ± 2000 Da resulting from the association of two identical subunits with molecular weights of 37,000 + 1000. GGPS catalyzes the prenyl transfer reaction using isopentenyl pyrophosphate (K_m = 3 μM) and either dimethylallyl pyrophosphate (K_m = 0.95 μM), geranyl pyrophosphate (K_m = 1 μM) or farnesyl pyrophosphate (K_m = 1.2 μM). It appears that GGPS is non-covalently bound but is always associated with IPI or the PSY enzyme complex in the chromoplast stroma of *Capsicum* [67]. A corresponding cDNA was further identified, isolated using specific antibodies, and sequenced [68]. The cloned cDNA encoded a precursor of 369 amino acids with a transit peptide of approximately 60 amino acids. GGPS was exclusively immunolocalized in plastids. Expression analysis demonstrated that this gene was strongly induced during the chloroplast to chromoplast transition in ripening chili pepper fruits and correlated with an increase in the enzyme activity. The *GGPS* gene was cloned, and the nucleotide sequence indicated an open reading frame of 369 codons that was identical to the cDNA without any interruption by introns, and Southern analysis of *C. annuum* genomic DNA suggested the presence of a single copy [103]. *GGPS* gene expression was induced in chili pepper fruit tissues by wounding [104].

The formation of phytoene from the condensation of two GGPP molecules, a two-step reaction catalyzed by PSY, is the first step in plant carotenoid biosynthesis (Figure 2) [79]. The site of phytoene biosynthesis was localized exclusively to the plastid compartment [105]. This enzyme was isolated and purified to homogeneity by affinity chromatography from the chromoplast stroma of chili pepper fruits and was further characterized, exhibiting a monomeric structure with a molecular size (M_r) of 47,500 Da and a Michaelis-Menten kinetics value of 0.30 μM for geranylgeranyl pyrophosphate. The enzymatic activity required Mn^{2+} and was inhibited by inorganic pyrophosphate and hydroxyphenylglyoxal (an arginine-specific reagent) [79]. DNA sequence analysis of the *PSY* genes from seven orange fruit chili pepper varieties revealed the length of the gene to be 2849 bp with six exons, five introns, and a predicted protein consisting of 420 amino acids [61].

PDS in chili peppers carries out a two-step desaturation reaction of phytoene, which is subsequently converted into phytofluene and ζ-carotene, the first visible carotenoid (Figure 2). PDS was isolated from *Capsicum annuum* chromoplast membranes and exhibited a 56 ± 2 kDa molecular weight, an apparent optimum pH of 7.2–8, and a K_m of 7 and 4 μM for phytoene and phytofluene, respectively. Additionally, FAD bound to the protein [81]. Antibodies raised against purified PDS were used to isolate a full-length cDNA (2000 bp) that corresponded to an open reading frame of 582 codons and to a 65 kDa protein; the deduced primary structure confirmed a characteristic dinucleotide-binding site and an 81% identity with a PDS from soybean (*Glycine max*). Heterologous expression of a *PDS* cDNA in *E. coli* permitted the production of a recombinant desaturase, which exhibited the same properties as chromoplast PDS. Analysis of *PDS* mRNA levels showed expression in seedlings, young expanding leaves, senescing leaves and fruits. Low expression in green fruits and a slight increase before the detection of carotenoid synthesis was observed, and levels then remained constant through ripening. However, PDS activity and protein levels increased significantly during the transition from chloroplast to chromoplast [81].

The two-step desaturation reaction conversion of ζ-carotene into neurosporene and lycopene is catalyzed by ZDS, a chromoplast membrane-bound enzyme (Figure 2). This enzyme was purified and characterized after heterologous expression of a *ZDS* cDNA from *Capsicum annuum* in *E. coli* [84,106]. The deduced sequence of the *ZDS* cDNA encoded a mature protein of approximately 59 kDa, which is synthesized as a precursor with a transit peptide for plastid localization [84]. ZDS exhibited a monomeric structure with lipophilic quinones as cofactors (decyl-plastoquinone or decyl-ubiquinone increased its activity) [106]. Evidence of dimerization was established, but no differences in activity between the monomeric and dimeric forms were observed. Very similar K_m values were found for ζ-carotene (8.4 μM) and neurosporene (9.0 μM), indicating that both carotenes are converted at similar rates in the first and second desaturation steps [106].

As it was previously mentioned, the reactions catalyzed by PDS and ZDS have been proposed to be linked or coupled to a plastid terminal oxidase. A *PTOX* gene cDNA (1387 bp) was cloned from chili pepper fruits (*C. annuum*) [107]. Genomic analysis suggested the presence of a single gene copy, and the expression correlated with that of *PDS* and *ZDS* genes during fruit ripening.

Lycopene is the precursor of cyclic carotenoids such as β-carotene, an important component of the reaction centers and antenna of the photosynthetic apparatus. Lycopene is also a substrate for the biosynthesis of various other carotenoids (xanthophylls, zeaxanthin, antheraxanthin, violaxanthin and neoxanthin). Abscisic acid synthesis occurs via β-carotene and the derived carotenoids (Figure 2). β-carotene is the most important precursor of vitamin A for human and animal food. Carotenoid cyclization reactions are associated with chromoplast membranes [96], and LCY-B was isolated from *C. annuum*fruit chromoplast membranes [108]. The membrane-solubilized enzyme was capable of incorporating labeled lycopene into β-carotene but not into α-carotene. No NADP or FAD was required as a cofactor, and sulfhydryl reagents severely inhibited its activity, indicating that SHgroups were involved in the cyclization process. A cDNA encoding the enzyme catalyzing the cyclization of lycopene to β-carotene (LCY-B) was also isolated from *Capsicum annuum* fruits and expressed in *E. coli* [109]. The deduced amino acid sequence of the cloned *LCY-B* cDNA was 498 residues long with a calculated molecular weight of 55.6 kDa for the precursor polypeptide, including the transit peptide for chloroplast/chromoplast targeting, whereas the molecular weight of the mature protein was approximately 50 kDa. A potential dinucleotide-binding site was also observed at the NH_2 terminus. Alignment with the CCS sequence revealed an identity of 55% and a similarity of 72%. *LCY-B* was expressed in young and senescent leaves and showed constitutive expression during chili pepper fruit development. Recently, sequence analysis of *LCY-B* genes from seven orange fruit chili pepper varieties indicated a size of 1495 bp and the absence of introns in the genomic DNA for all seven varieties [61]. The predicted protein was 499 amino acids long.

With regard to the synthesis of zeaxanthin from β-carotene, BCH is the enzyme that together with a CYP97A enzyme catalyzes this two-step reaction (Figure 2). A *BCH* type gene sequence analysis (*CRTZ-2*) from the genomic DNA of seven orange fruit varieties indicated the gene to be 2150 bp in size; the coding region was 2026 bp (316 amino acids) and contained six introns and seven exons [61].

ZEP is the enzyme that catalyzes the transformation reaction of zeaxanthin into antheraxanthin and later into violaxanthin (Figure 2). ZEP also catalyzes the conversion reaction of β-cryptoxanthin to β-cryptoxanthin epoxide. A *ZEP*cDNA from chili pepper (*C. annuum*) was cloned using a *Nicotiana plumbaginifolia* cDNA probe, and the corresponding enzyme was over-expressed in *E. coli* and purified [102]. This enzyme specifically acted on the β-ring of xanthophylls such as β-cryptoxanthin, zeaxanthin and antheraxanthin in the presence of numerous carotenoid substrates possessing ϵ- (α-carotene and lutein) and β-rings (β-carotene, β-cryptoxanthin and antheraxanthin). Purified ZEP incubated with zeaxanthin in the presence of the stromal proteins ferredoxin or rubredoxin produced violaxanthin and antheraxanthin, as determined by HPLC [102]. Similarly, when ZEP was incubated with β-cryptoxanthin, the reaction product was β-cryptoxanthin-5,6-epoxide. These results indicate that ZEP catalyzes the following reactions: (1) β-cryptoxanthin to β-cryptoxanthin epoxide; (2) zeaxanthin to antheraxanthin; and (3) antheraxanthin to violaxanthin (Figure 2). The deduced peptide sequence showed a characteristic FAD binding domain, and the recombinant enzyme (65 kDa) required NADPH-reduction power to be transferred to the epoxidase via ferredoxin for the cyclohexenyl-carotenoid epoxidation. This gene was found to be upregulated in oxidatively stressed seedlings and during the chloroplast-chromoplast differentiation process in chili pepper fruits.

CCS is the enzyme that synthesizes the final carotenoid products (capsanthin and capsorubin) in red chili pepper fruits (Figure 2). This enzyme catalyzes the transformation of the 5,6-epoxycarotenoids antheraxanthin and violaxanthin into capsanthin and capsorubin, respectively. The *in vitro* conversion of labeled violaxanthin and antheraxanthin into capsorubin and capsanthin by a chromoplast-enriched fraction from chili pepper fruits was demonstrated early on [98,99]. The capsanthin-capsorubin synthase name was proposed due to its bi-functionality. CCS was isolated and purified from *Capsicum annuum* chromoplasts [101]. The purified enzyme was found to be a 50 kDa monomer. Cloning of a *CCS* cDNA and the expression pattern analysis in leaves and fruits during several ripening stages and in different tissues demonstrated that the *CCS* gene was specifically expressed during chromoplast development in mature chili pepper fruits. The *CCS* gene was exclusively expressed in fruits that accumulated keto-carotenoids but not in impaired mutants during this biosynthetic step (yellow-fruited mutants lacking the red pigments capsanthin and capsorubin) [101]. *CCS* gene expression was induced in chili pepper fruits by wounding [104]. Promoter sequence analysis (−588 bp) of the *CCS* gene revealed the presence of *ABA* (abscisic acid-responsive), *ELI* (elicitor-responsive), *HD* (homeodomain) and *WUN* (wound-responsive) *cis*-acting elements [104]. When this promoter was fused to *GUS*-encoding sequences, expression was induced by ROS (reactive oxygen species), indicating that *CCS* gene expression is controlled by ROS. CCS also showed LCY-B activity (catalyzing the conversion of lycopene into β-carotene) when a *CCS* cDNA was expressed in *E. coli* [109]. A proposed molecular model for carotenoid channeling in *C. annuum* fruits suggested that massive and specific channeling of linear carotenoids into the β-carotene pathway in red *C. annuum* fruits is due to the concomitant action of LCY-B and CCS on β-cyclic carotenoids [109]. This model takes into consideration the fact that *LCY-B* gene is not upregulated during fruit ripening and does not require negative regulation of α-carotene biosynthesis. The *LCY-B* gene and *CCS* sequences from *C. annuum* share 55% identity, which suggests that both genes originated from a common ancestral gene [109]. The *CCS* gene was found to be 1494 bp long with no introns and a predicted size of 499 amino acids when genomic DNA from six orange chili pepper fruit varieties was analyzed [61].

As it was previously mentioned, violaxanthin can be converted into capsorubin in chili peppers, but this carotenoid can also be potentially transformed into xanthoxin, a precursor of ABA, by the action of NCED, a carotenoid cleavage dioxygenase (CCD). Until now, no reports on this type of enzymes in chili peppers have been published, but just a 1776 bp sequence (Y14164) has been deposited in the NCBI.

REGULATION OF CAROTENOID BIOSYNTHESIS IN *CAPSICUM*

Plants have evolved complex regulatory mechanisms to control carotenoid biosynthesis and its accumulation. In general, the carotenoid concentration represents a steady state resulting from a combination of biosynthesis and degradation [110]. Carotenoid biosynthesis genes have been cloned from several plants, such as *Arabidopsis*, tomato, chili pepper, daffodil and marigold. These genes are highly regulated and directly connected to plant phenotype color [111]. It has been shown in tomato fruits and marigold petals that carotenoid accumulation is mainly controlled by the transcriptional regulation of carotenoid biosynthetic genes [112–114]. In the case of *Capsicum*, carotenoid-biosynthesis regulation at the gene and enzyme levels is not fully understood. It is well known that the total carotenoid content is quite diverse among *Capsicum* species (*C. baccatum*, *C. chacoense*, *C. chinense*, *C. frutescens* and *C. annuum*) [115]. *Capsicum annuum* lines very often exhibit the highest carotenoid content, ranging between 390 and 16,600 μg/g (d.wt.) [115]. These data suggest that there may be several regulation steps in carotenoid biosynthesis, depending on the genotype.

Several approaches have been used to study carotenoid biosynthesis regulation, such as genetic analysis of the fruit color trait, characterization of mutant plants defective in the production of various carotenoids, expression analysis of structural biosynthetic genes, and over-expression of structural genes.

Fine genetic control of carotenoid production is responsible for the type and quantity of carotenoid accumulation in chili pepper fruits. It has been established that mature chili pepper fruit color is determined by three loci (three independent pairs of genes): *c1*, *c2* and *y* [116]. Furthermore, using RFLP and specific-PCR to analyze DNA sequences and thus determine polymorphisms for the *CCS* gene in F_2 progeny (this filial was derived from a cross between red and yellow fruit-producing chili pepper plants) showed that *CCS* completely segregated with the red fruit color and that locus *y*(encoding for *CCS*) controlled the red character [117]. These results support the hypothesis that the yellow chili pepper fruit color phenotype might be the result of a *CCS* gene deletion. A co-segregation of the *y* locus and *CCS* in populations generated from crosses between chili pepper plants bearing red × white and red × yellow fruits indicated the correspondence of the two genes [118]. Similarly, the *CCS* gene completely co-segregated with fruit color in the F_2population, as assessed by a PCR polymorphism analysis of the *CCS* gene and the generation of a TLC separation carotenoid profile, suggesting that *CCS* determined the chili pepper fruit color by altering the carotenoid pattern.

A genetic map was established using RFLP and AFLP markers and an F_2 population derived from an interspecific cross between *Capsicum annuum* cv. TF68 and *Capsicum chinense* cv. Habanero [119]. The TF68 ripe fruit was red and its Habanero counterpart was orange. The red color was dominant over the orange in the F_1 population. To identify the gene(s) or marker(s) tightly linked to the red/orange locus, several candidate genes involved in the carotenoid biosynthetic pathway, including *GGPS*, *PSY*, *PDS*, *LCY-B* and *CCS*, were examined. The results showed that only the *PSY* gene completely co-segregated with the color in the F_2 population, suggesting that this gene corresponded to the *c2* locus responsible for the accumulation of red color [119]. Similarly, the *PSY* locus was associated with the content of individual pigments such as capsanthin, capsorubin and zeaxanthin. These results strongly suggested that carotenoid levels in fruits were determined by the composition of the *PSY* allele. In the same way, QTL and HPLC analysis of F_2 lines demonstrated that *PSY* was the locus responsible for fruit color in ripe red peppers (*Capsicum* spp.) [119]. The *PSY* locus was not linked to the other candidate *CCS* gene, which implies that *PSY* and *CCS* segregated independently. Therefore, the orange-color phenotype might be the result of reduced PSY activity rather than a lack of red pigment. Thus, this enzyme was proposed to be the rate-limiting step of carotenoid biosynthesis [119]. More recently, a splicing mutation in the *PSY* gene was demonstrated to cause orange coloration in Habanero (*C. chinense*) chili pepper fruits [120]. This mutant exhibited a recessive *c2*homozygous allele with a point mutation at the splice acceptor site of the fifth intron of the *PSY* gene, which causes both a frame-shift and a premature translational termination and is responsible for the orange fruit color.

Borovsky *et al.* [121] described an EMS-induced chili pepper mutant affected in the *CHY2* gene, which encodes a β-carotene hydroxylase. This mutant produced and accumulated orange carotenoids (β-carotene) in the fruits instead of the red ones of the red-fruited progenitor "Maor". Mutation was caused by an A^{709} to G transition in the *CHY2* gene. These authors hypothesized that *CHY2* controls the orange mutation in chili pepper. *chy2* mutant also exhibited changes in volatile compounds (higher norisoprenoid levels).

When *CCS*-defective mutants were characterized the presence of a 240 bp deletion at the 5′ end in the coding region of the *CCS* gene from yellow-fruited lines of *Capsicum* was demonstrated, indicating that this gene determines the red color in the pericarp of chili pepper fruits [118]. A deletion in the upstream region of the *CCS* gene in orange-fruit pepper plants was also found [120]. Using Southern hybridization and sequencing analysis, a 211 bp sequence of the downstream region of the gene was detected in plants with orange fruits, while no transcripts of the *CCS* gene were demonstrated by RT-PCR in mature orange fruits [122]. Therefore, the orange color of chili pepper fruits was the consequence of a *CCS* gene deletion. *C. chinense* varieties Y2 (PI 800065) and Y3 (PI 164918), which do not accumulate capsanthin in the fruits, possess the encoding and promoter regions of the *CCS* gene in their genomes [111]. Two structural mutations in Y2 and Y3 chili peppers were detected by multiple alignment analysis of the nucleotide sequences from the

CCS coding regions. One of these mutations resulted in a frame-shift, causing early termination of translation via an 8 bp insertion at position 1431 of the gene in Y2. The *CCS* mutation in Y3 produced a premature stop-codon via a single base change at position 599. The Y2 chili pepper could generate a truncated CCS protein of 495 amino acids, while the Y3 pepper could generate a truncated CCS protein of 119 amino acids. This PTC phenomenon can be explained by a mechanism called "nonsense-mediated mRNA decay" [123], in which mRNAs harboring premature translation-termination codons decay rapidly because the resulting *C*-terminally truncated proteins can function as dominant negative inhibitors of the full-length protein. Likewise, *CCS* mutations are a key example of carotenoid biosynthesis regulation in ripe yellow chili pepper fruits [111]. Similarly, it was observed that in *C. baccatum*, *C. chinense*, *C. annuum* and *C. pubescens* red pepper fruits, the induction of *PSY*, *PDS* and *BCH*, in addition to *LCY-B*, *VDE* and *CCS* genes, during ripening was required to accumulate high total carotenoid levels; therefore, the expression levels of carotenoid biosynthetic genes are the critical for the accumulation of high total carotenoid levels in chili pepper fruits [111]. Recently, a *CCS* variant gene (*ccs-3*) was described in an orange-fruited chili pepper variety (Fogo) that does not accumulate capsanthin and capsorubin [61]. This *ccs-3* variant gene contained an early termination stop codon due to the deletion of a cytosine at nucleotide position 1283, which caused a frame-shift in the coding sequence, thus giving rise to a polypeptide product of 423 amino acids. Li *et al.* [124] described a new *ccs* chili pepper (*C. annuum*) variant in the yellow fruit line CK7 with a premature stop codon derived from a C to G change (1095 bp downstream of the start codon), and also a downstream frame-shift caused by a 1 bp nucleotide deletion (1265 bp downstream of the start codon). This variant exhibited positive expression of the mutant CCS protein.

Carotenoid biosynthesis and accumulation is influenced by different factors. The role of reactive oxygen species (ROS) as regulators of *in vivo* carotenoid biosynthesis was tested in chromoplasts from mature-green stage bell pepper fruits (*C. annuum* cv. Yolo Wonder) [104]. The addition of ROS-inducing compounds such as menadione, tert-butylhydroperoxide or paraquat and pro-oxidants such as diamide or buthionine sulfoximine to green pericarp discs from chili pepper fruits rapidly and dramatically induced the simultaneous expression of multiple carotenogenic gene mRNAs. Menadione induced *GGPS* and *CCS* gene expression, and tert-butylhydroperoxide triggered *GGPS*, *PSY*, *PDS*, *ZEP* and *CCS* gene expression but not *LCY-B*. Pericarp disks treated with paraquat showed a strong induction of *CCS* mRNA accumulation, and those treated with diamide and buthionine sulfoximine exhibited over-expression of *GGPS* and *CCS*. Similarly, down-regulation of catalase by amitrole (which inhibits catalase activity) induced the expression of carotenogenic gene mRNAs (*GGPS*, *PSY*, *PDS*, *LCY-B*, *ZEP* and *CCS*), leading to the synthesis of capsanthin in excised green pericarp discs. These results suggested that ROS acted as a novel class of second messengers that mediated intense carotenoid synthesis during chromoplast differentiation. As previously mentioned [104], characterization of the *CCS* promoter revealed the presence of *ABA* (abscisic acid-responsive), *ELI* (elicitor-responsive), *HD* (homeodomain) and *WUN* (wound-responsive) *cis*-acting elements, suggesting the possibility of *CCS* expression regulation by all of these factors.

The light-dark regulation of carotenoid biosynthesis in chili pepper (*Capsicum annuum*) leaves has been investigated [110]. The contribution of carotenoid biosynthesis to total carotenoid leveles was studied by treating chili pepper plants with 0.1 mmol/l norflurazon (NFZ; an herbicide that blocks carotenoid biosynthesis by inhibiting PDS leading to an accumulation of phytoene) under light and dark conditions. Herbicide-treated leaves from green chili pepper plants exhibited a phytoene concentration of 1.53 mg/g d.wt. under a light period of 280 $\mu mol\ m^{-2}s^{-1}$ for 48 h. However, when NFZ was applied to plants grown in dark, no accumulation of phytoene was observed. The amount of cyclic carotenoids was approximately the same with and without NFZ. These results indicated that carotenoid biosynthesis was completely blocked when herbicide-treated plants were transferred to dark conditions. Additionally, it was observed that transcripts from all analyzed genes decreased under dark conditions. For instance, *ZDS*, *PTOX* (plastid terminal oxidase) and *LCY-B*transcripts in dark-grown plants were 19%, 24% and 44%, respectively, of their particular levels observed in light-grown plants. With regard to *PSY* and *PDS* transcript levels, dark-treated plants exhibited less than 10% of the levels in their light-grown counterparts. The very low *PSY* transcript levels in non-illuminated plants coincided quite well with the absence of phytoene accumulation in dark

NFZ-treated plants. Consequently, the down-regulation of *PSY* appeared to be the central switch in shutting down carotenoid biosynthesis in the dark.

In some cases, the function of a specific gene can be demonstrated by heterologous over-expression. When the *CCS* gene was over-expressed in *Nicotiana benthamiana* plants by means of a viral RNA vector (*Tobacco mosaic virus*) bearing the *CCS*cDNA under the Tobamovirus subgenomic promoter, *CCS*-infected plants developed an orange color phenotype in their leaves as a consequence of the accumulation of high capsanthin levels (up to 36% of total carotenoids) [125]. Interestingly, the capsanthin was not esterified. This carotenoid accumulation was associated with thylakoid membrane distortion and reduction of grana stacking. The high capsanthin accumulation was balanced by a decrease in xanthophyll content, suggesting that a subtle auto-regulatory system acts to control carotenoid biosynthesis and accumulation in the chloroplast.

Carotenoid biosynthesis is a multi-step (multi-gene) pathway that theoretically must be regulated in a very fine and coordinate way. In general terms, biosynthetic complex pathways are regulated by transcription factors, which coordinate the expression of the participating genes. In the case of regulation of the carotenoid biosynthetic pathway, current understanding of regulation at the transcription level is very limited. Recently, a quantitative comparative expression analysis of the *PYS, LCY-B, BCH* (*CRTZ-2*) and *CCS* genes in orange-fruited chili pepper cultivars (*C. annuum*) showed a particular carotenoid accumulation profile and also a characteristic expression pattern for each chili pepper type [126]. In one cultivar with the *ccs-3* mutant allele (Fogo), *CCS* gene transcripts were detected, but no CCS enzyme was produced, whereas in two other cultivars (Orange Grande and Oriole) with four wild-type genes, no *CCS* transcripts were recorded, and no capsanthin or capsorubin was produced. In the case of "Canary", this cultivar expressed the four wild-type genes, but no CCS enzyme was produced, and consequently no red carotenoids accumulated. These results suggest that non-structural genes (potentially transcription factors) might regulate carotenoid biosynthesis.

FUTURE RESEARCH ON CHILI PEPPER CAROTENOID BIOSYNTHESIS

Carotenoid biosynthesis and their accumulation in *Capsicum* fruits is a broad field of study. Chili pepper fruits contain a wide variety of carotenoid pigments, which comprise a large amount of structural diversity, that are almost exclusively found in *Capsicum* members. For instance, two new carotenoid pigments in *C. annuum* fruits were reported [64]. Moreover, not all carotenoid biosynthesis pathways in chili pepper fruits are completely known and understood, and more research is thus needed to elucidate these carotenoid pathways, including those that produce minor carotenoids, at both the enzymatic and genetic levels. The central role of carotenoids in plant development and adaptation is of paramount importance, as their biosynthesis is coordinated with other developmental processes such as plastid formation and flower and fruit development [3]. Unfortunately, the regulation of carotenoid biosynthesis in *Capsicum* at the gene and enzyme level is currently poorly understood. As of now, only a few studies of transcription factors or other genes that impose global regulatory functions on carotenoid metabolism in plants have been described [74]. Only two types of transcription factors (RAP2.2 and PIFs) have been identified that directly interact with the *Arabidospsis* PSY promoter [127,128]. To our knowledge, no reports of transcription factors that regulate carotenoid biosynthesis in *Capsicum* have been revealed so far. As a result, it is imperative to focus more research on this specific genetic issue to unravel carotenoid production in chili pepper fruits.

Another approach to better understand carotenoid biosynthesis regulation in chili pepper fruits is the study and characterization of promoters from structural biosynthetic genes. Until now, only the *CCS* gene promoter has been partially studied, and some *cis*-element sequences have been reported [81,124]. Characterization of promoter sequences for all of the structural biosynthetic carotenogenic genes should surely reveal possible interactions with transcription factors and also developmental or environmental regulatory elements.

PLAUSIBLE APPLICATIONS

Capsicum fruits possess great diversity in terms of color, flavor, pungency and nutritional value. These fruits contain high pigment content including carotenoids and anthocyanins, and vitamins such as C, A and E [32]. It was demonstrated that chili pepper pigments, including carotenoids, possess high antioxidant and anticarcinogenic activities [3,33,129–131]. Due to their high carotenoid pigment content, ripe chili pepper fruits are in high demand in industry, as they are used as colorants for cosmetics and foods [4]. Additionally, some chili pepper fruit carotenoids have provitamin A activity, and so their consumption is beneficial for human and animal health, as both humans and animals do not synthesize carotenoids *de novo* and rely upon the diet as the source of these essential products [3]. Therefore, it is essential to have knowledge about the carotenoid biosynthesis steps and their regulation at the genetic, molecular and enzymatic levels to understand the accumulation process in chili pepper fruits. If those chili pepper carotenoids that show maximal health benefits can be increased, doing so may be beneficial in reducing the incidence of human diseases, including age-related macular degeneration and cancer. Furthermore, carotenogenic genes from *Capsicum* species could be useful to increase the beneficial value of other important crops through genetic engineering approaches [132]. For example, a construct bearing a recombinant *PSY* gene from *Capsicum* and a carotene desaturase (*CrtI*) gene from *Pantoea* under the control of either the β-*conglycinin* gene or *CaMV-35S* promoter was used to transform *Glycine max* via *A. tumefaciens* infection, and the resulting transgenic plants showed a 62-fold increase in total carotenoids in the seeds, of which 77% corresponded to β-carotene [133].

CONCLUSIONS

Carotenoids are responsible for the yellow, orange and red colors of chili pepper fruits. The carotenoid biosynthetic pathway in *Capsicum* chili pepper fruits has been fully established. Mutants that are impaired in some carotenoid structural genes, thus resulting in a great diversity and variety of chili pepper fruit colors, have been characterized to understand carotenoid biosynthesis and regulation. Some carotenoid biosynthetic enzymes have been isolated and characterized, and the genes encoding them have been sequenced, but only in a few cases their promoters have been characterized. No regulatory proteins or their encoding genes that control carotenoid biosynthesis in chili pepper fruits have been described. Information is currently being used to select chili pepper materials with high carotenoid contents. Genetic engineering to improve or to modify the carotenoid content of chili pepper fruits is pending.

ACKNOWLEDGMENTS

The authors gratefully acknowledge financial support from Conacyt (Mexico) for the doctoral fellowship to María del Rocío Gómez-García.

CONFLICTS OF INTEREST

The authors declare no conflict of interest.

REFERENCES

1. Botella-Pavía, P.; Rodríguez-Concepción, M. Carotenoid biotechnology in plants for nutritionally improved foods. Physiol. Plant 2006, 126, 369–381.
2. Maoka, T. Recent progress in structural studies of carotenoids in animals and plants. Arch. Biochem. Biophys 2009, 483, 191–195.
3. Fraser, P.D.; Bramley, P.M. The biosynthesis and nutritional uses of carotenoids. Progr. Lipid Res 2004, 43, 228–265.
4. Delgado-Vargas, F.; Peredes-López, O. Natural Colorants for Food and Nutraceutical Uses; CRC Press: Boca Raton, FL, USA, 2003; p. 327.
5. Hadden, W.L.; Watkins, R.H.; Levy, L.W.; Regalado, E. Carotenoid composition of Marigold (Tagetes erecta) flower extract used as nutritional supplement. J. Agric. Food Chem 1999, 47, 4189–4194.
6. Jordan, B.R. The Molecular Biology and Biotechnology of Flowering, 2nd ed; CABI Publishing: Cambridge, MA, USA, 2006; p. 404.
7. Fraser, P.D.; Truesdale, M.R.; Bird, C.R.; Schuch, W.; Bramley, P. Carotenoid biosynthesis during tomato fruit development. Plant Physiol 1994, 105, 405–413.
8. Skelton, R.L.; Yu, Q.; Srinivasan, R.; Manshardt, R.; Moore, P.H.; Ray, M. Tissue differential expression of lycopene β-cyclase gene in papaya. Cell Res 2006, 16, 731–739.

9. Kato, M.; Ikoma, Y.; Matsumoto, H.; Sugiura, M.; Hyodo, H.; Yano, M. Accumulation of carotenoids and expression of carotenoid biosynthetic genes during maturation in citrus fruit. Plant Physiol 2004, 134, 824–837.
10. Bartley, G.E.; Scolnik, P.A. Plant carotenoids: Pigments for photoprotection, visual attraction, and human health. Plant Cell 1995, 7, 1027–1038.
11. Weedon, B.C.L.; Moss, G.P. Structure and Nomenclature. In Carotenoids; Britton, G., Liaaen-Jensen, S., Pfander, H.P., Eds.; Birkhäuser Verlag: Basel, Switzerland, 1995; Volume I B, Spectroscopy; pp. 27–70.
12. Sandmann, G.; Albrecht, M.; Schnurr, G.; Knörzer, O.; Böger, P. The biotechnological potential and design of novel carotenoids by gene combination in Escherichia coli. Trends Biotechnol 1999, 17, 233–236.
13. Zhu, C.; Naqvi, S.; Capell, T.; Christou, P. Metabolic engineering of ketocarotenoid biosynthesis in higher plants. Arch. Biochem. Biophys 2009, 483, 182–190.
14. Auldridge, M.E.; McCarthy, D.R.; Klee, H.J. Plant carotenoid cleavage oxygenases and their apocarotenoid products. Curr. Opin. Plant Biol 2006, 9, 315–321.
15. Walter, M.H.; Floss, D.S.; Strack, D. Apocarotenoids: Hormones, mycorrhizal metabolies and aroma volatiles. Planta2010, 232, 1–17.
16. Wang, R.K.; Lu, J.J.; Xing, G.N.; Gai, J.Y.; Zhao, T.J. Molecular evolution of two consecutive carotenoid cleavage dioxygenase genes in strigolactone biosynthesis in plants. Genet. Mol. Res 2011, 10, 3664–3673.
17. Sharoni, Y.; Linnewiel-Hermoni, K.; Khanin, M.; Salman, H.; Veprik, A.; Danilenko, M.; Levy, J. Carotenoids and apocarotenoids in cellular signaling related to cancer: A review. Mol. Nutr. Food Res 2012, 56, 259–269.
18. Li, L.; van Eck, J. Metabolic engineering of carotenoid accumulation by creating a metabolic sink. Transgenic Res 2007, 16, 581–585.
19. Nishino, H.; Tokuda, H.; Satomi, Y.; Masuda, M.; Bu, P.; Onozuka, M.; Yamaguchi, S.; Okuda, Y.; Takayasu, J.; Tsuruta, J.; et al. Cancer prevention by carotenoids. Pure Appl. Chem 1999, 71, 2273–2278.

20. Sporn, M.B.; Liby, K.T. Is lycopene an effective agent for preventing prostate cancer? Cancer Prev. Res 2013, 6, 384–386.
21. Snodderly, D.M. Evidence for protection against age-related macular degeneration by carotenoids and antioxidant vitamins. Am. J. Clin. Nutr 1995, 62, 1448S–1461S.
22. Rodríguez-Concepción, M.; Forés, O.; Martinez-Garcia, J.F.; Gonzalez, V.; Phillips, M.A.; Ferrer, A.; Boronat, A. Distinct light-mediated pathways regulate the biosynthesis and exchange of isoprenoid precursors during Arabidopsis seedling development. Plant Cell 2004, 16, 144–156.
23. Maoka, T.; Fujiwara, Y.; Hashimoto, K.; Akimoto, N. Isolation of a series of apocarotenoids from the fruits of the red paprika Capsicum annuum L. J. Agric. Food Chem 2001, 49, 1601–1606.
24. Braconnot, H. Mémoire sur le principe extractif et sur les extraits en général. J. Phys. Chim. Hist. Nat. Arts 1817, 84, 267–296.
25. Ikan, R. Natural Products: A Laboratory Guide, 2nd ed; Academic Press: San Diego, CA, USA, 1991; p. 360.
26. Von Zechmeister, L.; von Cholnoky, L. Untersuchungen über den paprika-farbstoff. XI. Isomerisierungs-Erscheinungen. Ann. Chem 1940, 543, 248–257.
27. Curl, A.L. The carotenoids of red bell peppers. J. Agric. Food Chem 1962, 10, 504–509.
28. Von Cholnoky, L.; Györgyfy, K.; Nagy, E.; Pánczél, M. Investigations of carotenoid pigments. I. Pigments of red paprika. Acta Chim. Acad. Sci. Hung 1955, 6, 143–147.
29. Barber, M.S.; Jackman, L.M.; Warren, C.K.; Weedon, B.C.L. The structures of the paprika ketones. Proc. Chem. Soc.1960, 19–20.
30. Entschel, R.; Karrer, P. Zur Konstitution des capsanthins und capsorubins. Helv. Chim. Acta 1960, 43, 89–94.
31. Howard, L.R.; Talcott, S.T.; Brenes, C.H.; Villalon, B. Changes in phytochemical and antioxidant activity of selected pepper cultivars (Capsicum species) as influenced by maturity. J. Agric. Food Chem 2000, 48, 1713–1720.
32. Marín, A.; Ferreres, F.; Tómas-Barberán, F.A.; Gill, M.I.

Characterization and quantitation of antioxidant constituents of sweet pepper (Capsicum annuum L.). J. Agric. Food Chem 2004, 52, 3861–3869.

33. Sun, T.; Xu, Z.; Wu, C.T.; Janes, M.; Prinyawiwatkul, W.; No, H.K. Antioxidant activities of different colored Sweet Bell peppers (Capsicum annuum L.). J. Food Sci 2007, 72, S98–S102.
34. Karrer, P.; Oswald, A. Carotinoide aus den staubbeuteln von Lilium tigrinum. Ein nenes carotenoid: Anthetaxanthin. Helv. Chim. Acta 1935, 13, 1303–1305.
35. Seybold, A. Untersuchungen übe den farbwechsel von blumenblättern, früchten un samenschalen. Sber. Heilb. Akad. Wiss. Math-Naturwiss Kl 1953, 4, 31–124.
36. Valadon, L.R.G.; Mummery, R.S. Carotenoids of lilies and of red pepper: Biogenesis of capsanthin and capsorubin. Z. Pflanzephysiol 1976, 82, 407–416.
37. Rüttimann, A. Synthesis and Stereochemistry of Red Pepper Carotenoids. In Carotenoid Chemistry and Biochemistry; Britton, G., Goodwin, T.W., Eds.; Pergamon Press: New York, NY, USA, 1982; pp. 71–86.
38. Märki-Fischer, E.; Eugster, C.H. Das carotinoidspektrum der antheren un petalen von Lilium tigrinum cv. “Red Night”. Helv. Chim. Acta 1985, 68, 1708–1715.
39. Partali, V.; Liaaen-Jensen, S.; Huneck, S.; Khaidav, T. Carotenoids from the flowers of Lilium pumilum. Die Pharm 1987, 42, 208.
40. Deli, J.; Matus, Z.; Tóth, G. Carotenoid composition in the fruits of Asparagus officinalis. J. Agric. Food Chem 2000, 48, 2793–2796.
41. Yamagishi, M.; Kishimoto, S.; Nakayama, M. Carotenoid composition and changes in expression of carotenoid biosynthetic genes in tepals of Asiatic hybrid lily. Plant Breed 2010, 129, 100–107.
42. Jeknic, Z.; Moré, J.T.; Jeknic, S.; Jevremovic, S.; Subotic, A.; Chen, T.H.H. Cloning and functional characterization of a gene for capsanthin-capsorubin synthase from tiger lily (Lolium lancifolium Thunb. “Splendens”). Plant Cell Physiol2012, 53, 1899–1912.
43. Matsufuji, H.; Nakamura, H.; Chino, M.; Takeda, M. Antioxidant

activity of capsanthin and the fatty acid esters in paprika (Capsicum annuum). J. Agric. Food Chem 1998, 46, 3468–3472.

44. Kim, S.; Ha, T.Y.; Hwang, I.K. Analysis, bioavailability, and potential healthy effects of capsanthin, natural red pigment from Capsicum spp. Food Rev. Int 2009, 25, 198–213.
45. Deruère, J.; Römer, S.; d'Harlingue, A.; Backhaus, R.A.; Kuntz, M.; Camara, B. Fibril assembly and carotenoid overaccumulation in chromoplasts: A model for supramolecular lipoprotein structures. Plant Cell 1994, 6, 119–133.
46. Mínguez-Mosquera, M.I.; Hornero-Méndez, D. Formation and transformation of pigments during the fruit ripening of Capsicum annuum cv. Bola and Agridulce. J. Agric. Food Chem 1994, 42, 38–44.
47. Rodríguez-Burruezo, A.; González-Mas, M.C.; Nuez, F. Carotenoid composition and vitamin A value in ají (Capsicum baccatum L.) and rocoto (C. pubescens R. & P.), 2 pepper species from the Andean region. J. Food Sci 2010, 75, S446–S453.
48. Deli, J.; Matus, Z.; Tóth, G. Carotenoid composition in the fruits of Capsicum annuum cv. Szentesi Kosszarvú during ripening. J. Agric. Food Chem 1996, 44, 711–716.
49. Mínguez-Mosquera, M.I.; Hornero-Méndez, D. Separation and quantification of the carotenoid pigments in red peppers (Capsicum annuum L.), paprika and oleoresin by reversed-phase HPLC. J. Agric. Food Chem 1993, 41, 1616–1620.
50. Hornero-Méndez, D.; Gómez-Ladrón de Guevara, R.; Mínguez-Mosquera, M.I. Carotenoid biosynthesis changes in five red pepper (Capsicum annuum L.) cultivars during ripening. Cultivar selection for breeding. J. Agric. Food Chem2000, 48, 3857–3864.
51. Mínguez-Mosquera, M.I.; Hornero-Méndez, D. Changes in carotenoid esterification during the fruit ripening of Capsicum annuum cv. Bola. J. Agric. Food Chem 1994, 42, 640–644.
52. Curl, A.L. The carotenoids of green bell peppers. J. Agric. Food Chem 1964, 12, 522–524.
53. Matus, Z.; Deli, J.; Szabolcs, J. Carotenoid composition of yellow pepper during ripening: Isolation of β-cryptoxanthin 5,6-epoxide. J. Agric. Food Chem 1991, 39, 1907–1914.

54. Deli, J.; Matus, Z.; Szabolcs, J. Carotenoid composition in the fruits of black paprika (Capsicum annuum variety longum nigrum) during ripening. J. Agric. Food Chem 1992, 40, 2072–2076.
55. Deli, J.; Molnár, P.; Matus, Z.; Tóth, G. Carotenoid composition in the fruits of red paprika (Capsicum annuum var. lycopersiciforme rubrum) during ripening; Biosynthesis of carotenoids in red paprika. J. Agric. Food Chem 2001, 49, 1517–1523.
56. Collera-Zuñiga, O.; García, J.F.; Meléndez, G.R. Comparative study of carotenoid composition in three mexican varieties of Capsicum annuum L. Food Chem 2005, 90, 109–114.
57. Matsufuji, H.; Ishikawa, K.; Nunomura, O.; Chino, M.; Takeda, M. Anti-oxidant content of different coloured sweet peppers, white, green, yellow, orange and red (Capsicum annuum L.). Int. J. Food Sci. Technol 2007, 42, 1482–1488.
58. Roca, M.; Hornero-Méndez, D.; Gandul-Rojas, B.; Mínguez-Mosquera, M.I. Stay-green phenotype slows the carotenogenic process in Capsicum annuum (L.) fruits. J. Agric. Food Chem 2006, 54, 8782–8787.
59. Almela, L.; Fernández-López, J.A.; Candela, M.E.; Egea, C.; Alcázar, M.D. Changes in pigments, chlorophyllase activity, and chloroplast ultrastructure in ripening pepper for paprika. J. Agric. Food Chem 1996, 44, 1704–1711.
60. Camara, B.; Brangeon, J. Carotenoid metabolism during chloroplast to chromoplast transformation in Capsicum annuum fruit. Planta 1981, 151, 359–364.
61. Guzman, I.; Hamby, S.; Romero, J.; Bosland, P.W.; O'Connell, M.A. Variability of carotenoid biosynthesis in orange colored Capsicum spp. Plant Sci. 2010, 179, 49–59.
62. Deepa, N.; Kaur, C.; George, B.; Singh, B.; Kapoor, H.C. Antioxidant constituents in some sweet pepper (Capsicum annuum L.) genotypes during maturity. LWT-Food Sci. Technol 2007, 40, 121–129.
63. Maoka, T.; Akimoto, N.; Fujiwara, Y.; Hashimoto, K. Structure of new carotenoids with the 6-oxo-κ end group from the fruits of paprika, Capsicum annuum. J. Nat. Prod 2004, 67, 115–117.
64. Lichtenthaler, H.K. The 1-deoxy-d-xylulose-5-phosphate pathway of isoprenoid biosynthesis in plants. Annu. Rev. Plant Physiol. Plant

Mol. Biol 1999, 50, 47–65.

65. Eisenreich, W.; Rohdich, F.; Bacher, A. Deoxyxylulose phosphate pathway to terpenoids. Trends Plant Sci 2001, 6, 78–84.
66. Rodríguez-Concepción, M.; Boronat, A. Elucidation of the methylerythritol phosphate pathway for isoprenoid biosynthesis in bacteria and plastids. A metabolic milestone achieved through genomics. Plant Physiol 2002, 130, 1079–1089.
67. Dogbo, O.; Camara, B. Purification of isopentenyl pyrophosphate isomerase and geranylgeranyl pyrophosphate synthase from Capsicum chromoplasts by affinity chromatography. Biochim. Biophys. Acta 1987, 920, 140–148.
68. Kuntz, M.; Römer, S.; Suire, C.; Hugueney, P.; Weil, J.H.; Schantz, R.; Camara, B. Identification of a cDNA for the plastid-located geranylgeranyl pyrophosphate synthase from Capsicum annuum: Correlative increase in enzyme activity and transcript level during fruit ripening. Plant J 1992, 2, 25–34.
69. Carol, P.; Stevenson, D.; Bisanz, C.; Breitenbach, J.; Sandman, G.; Mache, R.; Coupland, G.; Kuntz, M. Mutations in the Arabidopsis gene immutants cause a variegated phenotype by inactivating a chloroplast terminal oxidase associated with phytoene desaturation. Plant Cell 1999, 11, 57–68.
70. Carol, P.; Kuntz, M. A plastid terminal oxidase comes to light: Implications for carotenoid biosynthesis and chlororespiration. Trends Plant Sci 2001, 6, 31–36.
71. Seo, M.; Koshiba, T. Complex regulation of ABA biosynthesis in plants. Trends Plant Sci 2002, 7, 41–48.
72. Schwartz, S.H.; Qin, X.; Zeevaart, J.A.D. Elucidation of the indirect pathway of abscisic acid biosynthesis by mutants, genes, and enzymes. Plant Physiol 2003, 131, 1591–1601.
73. Aluru, M.R.; Rodermel, S.R. Control of chloroplast redox by IMMUTANTS terminal oxidase. Physiol. Plant 2004, 120, 4–11.
74. Lu, S.; Li, L. Carotenoid metabolism: Biosynthesis, regulation, and beyond. J. Integr. Plant Biol 2008, 50, 778–785.
75. Walter, M.H.; Strack, D. Carotenoids and their cleavage productos: Biosynthesis and functions. Nat. Prod. Rep 2011, 28, 663–692.

76. Li, Q.; Farre, G.; Naqvi, S.; Breitenbach, J.; Sanahuja, G.; Bai, C.; Sandmann, G.; Capel, T.; Christou, P.; Zhu, C. Cloning and functional charactarization of the maize carotenoid isomerase and β-carotene hydroxylase genes and their regulation during endosperm maturation. Plant Physiol. 2010, 19, 1053–1068.
77. Zhu, C.; Sanahuja, G.; Yuan, D.; Farré, G.; Arjó, G.; Berman, J.; Zorrilla-López, U.; Banakar, R.; Bai, C.; Pérez-Massot, E.; et al. Biofortification of plants with altered antioxidant content and composition: Genetic engineering strategies. Plant Biotechnol. J 2013, 11, 129–141.
78. Dogbo, O.; Bardat, F.; Laferriere, A.; Quennement, J.; Brangeon, J.; Camara, B. Metabolism of plastid terpenoids. I. Biosynthesis in plastid stroma isolated from higher plants. Plant Sci 1987, 49, 89–101.
79. Dogbo, O.; Laferriere, A.D.; d'Harlingue, A.; Camara, B. Carotenoid biosynthesis; Isolation of a bifunctional enzyme catalyzing the synthesis of phytoene. Proc. Natl. Acad. Sci. USA 1988, 85, 7054–7058.
80. Cunningham, F.X.; Gantt, E. Genes and enzymes of carotenoid biosynthesis in plants. Annu. Rev. Plant Physiol. Plant Mol. Biol 1998, 49, 557–583.
81. Hugueney, P.; Römer, S.; Kuntz, M.; Camara, B. Characterization and molecular cloning of a flavoprotein catalyzing the synthesis of phytofluene and ζ-carotene in Capsicum chromoplasts. Eur. J. Biochem 1992, 209, 399–407.
82. Li, F.; Murillo, C.; Wurtzel, E.T. Maize Y9 encodes a product essential for 15-cis zetacarotene isomerization. Plant Physiol 2007, 144, 1181–1189.
83. Chen, Y.; Li, F.; Wurtzel, E.T. Isolation and characterization of the Z-ISO gene encoding a missing component of carotenoid biosynthesis in plants. Plant Physiol 2010, 153, 66–79.
84. Albrecht, M.; Klein, A.; Hugueney, P.; Sandmann, G.; Kuntz, M. Molecular cloning and functional expression in E. coliof a novel plant enzyme mediating zeta-carotene desaturation. FEBS Lett 1995, 372, 199–202.
85. Isaacson, T.; Ronen, G.; Zamir, D.; Hirschberg, J. Cloning of

tangerine from tomato revels a carotenoid isomerase essential for the production of beta-carotene and xanthophylls in plants. Plant Cell 2002, 14, 333–342.

86. Park, H.; Kreunen, S.S.; Cuttriss, A.J.; DellaPenna, D.; Pogson, B.J. Identification of the carotenoid isomerase provides insight into carotenoid biosynthesis, prolamellar body formation, and photomorphogenesis. Plant Cell 2002, 14, 321–332.
87. Scolnik, P.A.; Hinton, P.; Greenblatt, I.M.; Giuliano, G.; Delanoy, M.R.; Spector, D.L.; Pollock, D. Somatic instability of carotenoid biosynthesis in the tomato ghost mutant and its effect on plastid development. Planta 1987, 171, 11–18.
88. Wetzel, C.M.; Jiang, C.-Z.; Meehan, L.J.; Voytas, D.F.; Ridermel, S.R. Nuclear-organelle interactions: The immutantsvariegation mutant of Arabidopsis is plastid autonomous and impaired in carotenoid biosynthesis. Plant J 1994, 6, 161–175.
89. Wu, D.; Wright, D.A.; Wetzel, C.; Voytas, D.F.; Rodermel, S. The immutants variegation locus of Arabidopsis defines a mitochondrial alternative oxidase homolog that functions during early choroplasts biogenesis. Plant Cell 1999, 11, 43–55.
90. Shahbazi, M.; Gilbert, M.; Labouré, A.-M.; Kuntz, M. Dual role of the plastid terminal oxidase in tomato. Plant Physiol2007, 145, 691–702.
91. Rumeau, D.; Peltier, G.; Cournac, L. Chlororespiration and cyclic electron flow around PSI during photosynthesis and plant stress response. Plant Cell Environ 2007, 30, 1041–1051.
92. Parker, R.S.; Swanson, J.E.; You, C.S.; Edwards, A.J.; Huang, T. Bioavailability of carotenoids in human subjects. Proc. Nutr. Soc 1999, 58, 155–162.
93. During, A.; Hussain, M.M.; Morel, D.W.; Harrison, E.H. Carotenoid uptake and secretion by CaCo-2 cells: β-carotene isomer selectivity and carotenoid interactions. J. Lip. Res 2002, 43, 1086–1095.
94. Quinlan, R.F.; Shumskaya, M.; Bradbury, L.M.T.; Beltrán, J.; Ma, C.; Kennelly, E.J.; Wurtzel, E.T. Synergistic interactions between carotene ring hydroxylases drive lutein formation in plant carotenoid biosynthesis. Plant Physiol. 2012, 160, 204–214.
95. Demming-Adams, B.; Gilmore, A.M.; Adams, W.W. In vivo

functions of carotenoids in higher plants. FASEB J 1996, 10, 403–412.

96. Camara, B.; Bardat, F.; Monéger, R. Sites of biosynthesis of carotenoids in Capsicum chromoplasts. Eur. J. Biochem 1982, 127, 255–258.
97. Camara, B.; Bardat, F.; Dogbo, O.; Brangeon, J.; Monéger, R. Terpenoid mtabolism in plastids. Isolation and biochemical characteristics of Capsicum annuum chromoplasts. Plant Physiol 1983, 73, 94–99.
98. Camara, B. Biosynthesis of keto-carotenoids in Capsicum Annuum fruits. FEBS Lett 1980, 118, 315–318.
99. Camara, B.; Monéger, R. Carotenoid biosynthesis. In vitro conversion of antheraxanthin to capsanthin by a chromoplast enriched fraction of Capsicum fruits. Biochem. Biophys. Res. Commun 1981, 99, 1117–1122.
100. Camara, B. Prenylation of chlorophyllide a in Capsicum plastids. Methods Enzymol 1985, 110, 274–281.
101. Bouvier, F.; Hugueney, P.; d'Harlingue, A.; Kuntz, M.; Camara, B. Xanthophyll biosynthesis in chromoplasts: Isolation and molecular cloning of an enzyme catalyzing the conversion of 5,6-epoxycarotenoid into ketocarotenoid. Plant J1994, 6, 45–54.
102. Bouvier, F.; d'Harlingue, A.; Hugueney, P.; Marin, E.; Marion-Poll, A.; Camara, B. Xanthophyll biosynthesis. J. Biol. Chem 1996, 271, 28861–28867.
103. Badillo, A.; Steppuhn, J.; Deruère, J.; Camara, B.; Kuntz, M. Structure of a functional geranylgeranyl pyrophosphate synthase gene from Capsicum annuum. Plant Mol. Biol 1995, 27, 425–428.
104. Bouvier, F.; Backhaus, R.A.; Camara, B. Induction and control of chromoplast-specific carotenoid genes by oxidative stress. J. Biol. Chem 1998, 273, 30651–30659.
105. Camara, B. Terpenoid metabolism in plastids. Sites of phytoene synthetase activity and synthesis in plant cells. Plant Physiol 1984, 74, 112–116.
106. Breitenbach, J.; Kuntz, M.; Takaichi, S.; Sandmann, G. Catalytic properties of an expressed and purified higher plant type ζ-carotene

desaturase from Capsicum annuum. Eur. J. Biochem. 1999, 265, 376–383.

107. Josse, E.-M.; Simkin, A.J.; Gaffé, J.; Labouré, A.-M.; Kuntz, M.; Carol, P. A plastid terminal oxidase associated with carotenoid desaturation during chromoplast differentiation. Plant Physiol 2000, 123, 1427–1436.

108. Camara, B.; Dogbo, O. Demonstration and solubilization of lycopene cyclase from Capsicum chromoplast membranes. Plant Physiol 1986, 80, 172–174.

109. Hugueney, P.; Badillo, A.; Chen, H.; Klein, A.; Hirschberg, J.; Camara, B.; Kuntz, M. Metabolism of cyclic carotenoids: A model for the alteration of this biosynthetic pathway in Capsicum annuum chromoplasts. Plant J 1995, 8, 417–424.

110. Simkin, A.J.; Zhu, C.; Kuntz, M.; Sandmann, G. Light-dark regulation of carotenoid biosynthesis in pepper (Capsicum annuum) leaves. J. Plant Physiol 2003, 160, 439–443.

111. Ha, S.H.; Kim, J.B.; Park, J.S.; Lee, S.W.; Cho, K.J. A comparison of the carotenoid accumulation in Capsicum varieties that show different ripening colours: Deletion of the capsanthin-capsorubin synthase gene is not a prerequisite for the formation of a yellow pepper. J. Exp. Bot 2007, 58, 3135–3144.

112. Ronen, G.; Cohen, M.; Zamir, D.; Hirschberg, J. Regulation of carotenoid biosynthesis during tomato fruit development: Expression of the gene for lycopene epsilon-cyclase is down-regulated during ripening and is elevated in the mutant Delta. Plant J 1999, 17, 341–351.

113. Ronen, G.; Carmel-Goren, L.; Zamir, D.; Hirschberg, J. An alternative pathway to beta-carotene formation in plant chromoplasts discovered by map-based cloning of Beta and old-gold color mutations in tomato. Proc. Natl. Acad. Sci. USA 2000, 97, 1102–1107.

114. Moehs, C.P.; Tian, L.; Osteryoung, K.W.; DellaPenna, D. Analysis of carotenoids biosynthetic gene expression during marigold petal development. Plant Mol. Biol 2001, 45, 281–293.

115. Levy, A.; Harel, S.; Palevitch, D.; Akiri, B.; Menagem, E.; Kanner, J. Carotenoid pigments and β-carotene in paprika fruits (Capsicum

spp.) with different genotypes. J. Agric. Food Chem 1995, 43, 362–366.

116. Hurtado-Hernandez, H.; Smith, P.G. Inheritance of mature fruit color in Capsicum annuum L. J. Hered 1985, 76, 211–213.

117. Lefebvre, V.; Kuntz, M.; Camara, B.; Palloix, A. The capsanthin-capsorubin synthase gene: A candidate gene for the ylocus controlling the red fruit colour in pepper. Plant Mol. Biol 1998, 36, 785–789.

118. Popovsky, S.; Paran, I. Molecular genetics of the y locus in pepper: Its relation to capsanthin-capsorubin synthase and to fruit color. Theor. Appl. Genet 2000, 101, 86–89.

119. Huh, J.H.; Kang, B.C.; Nahm, S.H.; Kim, S.; Ha, K.S.; Lee, M.H.; Kim, B.D. A candidate gene approach identifies phytoene synthase as the locus for mature fruit color in red pepper (Capsicum spp.). Theor. Appl. Genet 2001, 102, 524–530.

120. Kim, O.R.; Cho, M.-C.; Kim, B.-D.; Huh, J.H. A splicing mutation in the gene encoding phytoene synthase causes orange coloration in Habanero pepper fruits. Mol. Cells 2010, 30, 569–574.

121. Borovsky, Y.; Tadmor, Y.; Bar, E.; Meir, A.; Lewinsohn, E.; Paran, I. Induced mutation in *β-CAROTENE HYDROXYLASE* results in accumulation of β-carotene and conversion of red to orange color in pepper fruits. Theor. Appl. Genet 2013, 126, 557–565.

122. Lang, Y.Q.; Yanagawa, S.; Sasanuma, T.; Sasakuma, T. Orange fruit color in Capsicum due to deletion of capsanthin-capsorubin synthesis gene. Brccd. Sci 2004, 54, 33–39.

123. Buhler, M.; Mohn, F.; Stalder, L.; Muhlemann, O. Transcriptional silencing of nonsense codon-containing immunoglobulin minigenes. Mol. Cells 2005, 18, 307–317.

124. Li, Z.; Wang, S.; Gui, X.-L.; Chang, X.-B.; Gong, Z.-H. A further analysis of the relationship between yellow ripe-fruit color and the capsanthin-capsorubin synthase gene in pepper (Capsicum sp.) indicated a new mutant variant in C. annuum and a tandem repeat structure in promoter region. PLoS One 2013, 8, e61996.

125. Kumagai, M.H.; Keller, Y.; Bouvier, F.; Clary, D.; Camara, B. Functional integration of non-native carotenoids into chloroplasts by viral-derived expression of capsanthin-capsorubin synthase in Nicotiana benthamiana. Plant J 1998, 14, 305–315.

126. Rodríguez-Uribe, L.; Guzman, I.; Rajapakse, W.; Richins, R.D.; O'Connell, M.A. Carotenoid accumulation in orange-pigmente Capsicum fruit, regulated at multiple levels. J. Exp. Bot. 2012, 63, 517–526.

127. Welsch, R.; Maass, D.; Voegel, T.; DellaPenna, D.; Beyer, P. Transcription factor RAP2.2 and its interacting partner SINAT2: Stable elements in the carotenogenesis of Arabidopsis leaves. Plant Physiol 2007, 145, 1073–1085.

128. Toledo-Ortiz, G.; Huq, E.; Rodríguez-Concepción, M. Direct regulation of phytoene synthase gene expression and carotenoid biosynthesis by phytochrome-interacting factors. Proc. Natl. Acad. Sci. USA 2010, 107, 11626–11631.

129. Daood, H.G.; Tömöskozi-Farkas, R.; Kapitány, J. Antioxidant content of bio and conventional spice red pepper (Capsicum annuum L.) as determined by HPLC. Acta Agron. Hung 2006, 54, 133–140.

130. Guil-Guerrero, J.L.; Martínez-Guirado, C.; Rebolloso-Fuentes, M.M.; Carrique-Pérez, A. Nutrient composition and antioxidant activity of 10 pepper (Capsicum annuum) varieties. Eur. Food Res. Technol 2006, 224, 1–9.

131. Rodríguez-Burruezo, A.; Prohens, J.; Raigón, M.D.; Nuez, F. Variations for bioactive compounds in ají (Capsicum baccatum L.) and rocoto (C. pubescens R. & P.) and implications for breeding. Euphytica 2009, 170, 169–181.

132. Herbers, K. Vitamin production in transgenic plants. J. Plant Physiol 2003, 160, 821–829.

133. Kim, M.-J.; Kim, J.K.; Kim, H.J.; Pak, J.H.; Lee, J.-H.; Kim, D.-H.; Choi, H.K.; Jung, H.W.; Lee, J.-D.; Chung, Y.-S.; et al. PLos One 2012, 7, e48287.

CHAPTER

5

CHLOROPLAST GENOMES: DIVERSITY, EVOLUTION, AND APPLICATIONS IN GENETIC ENGINEERING

Henry Daniell[1], Choun-Sea Lin[2], Ming Yu[1] and Wan-Jung Chang[2]

[1]Department of Biochemistry, School of Dental Medicine, University of Pennsylvania

[2]Agricultural Biotechnology Research Center, Academia Sinica, Taipei, Taiwan.

ABSTRACT

Chloroplasts play a crucial role in sustaining life on earth. The availability of over 800 sequenced chloroplast genomes from a variety of land plants has enhanced our understanding of chloroplast biology, intracellular gene transfer, conservation, diversity, and the genetic basis by which chloroplast transgenes can be engineered to enhance plant agronomic traits or to produce high-value agricultural or biomedical products. In this review, we discuss the impact of chloroplast genome sequences on understanding the origins of economically important cultivated species and changes that have taken place during domestication. We also discuss the potential biotechnological applications of chloroplast genomes.

INTRODUCTION

Chloroplasts are active metabolic centers that sustain life on earth by converting solar energy to carbohydrates through the process of photosynthesis and oxygen release. Although photosynthesis is often recognized as the key function of plastids, they also play vital roles in other aspects of plant physiology and development, including the synthesis of amino acids, nucleotides, fatty acids, phytohormones, vitamins and a plethora of metabolites, and the assimilation of sulfur and nitrogen. Metabolites that are synthesized in chloroplasts are important for plant interactions with their environment (responses to heat, drought, salt, light, and so on) and their defense against invading pathogens. So, chloroplasts serve as metabolic centers in cellular reactions to signals and respond via retrograde signaling [1, 2]. The chloroplast genome encodes many key proteins that are involved in photosynthesis and other metabolic processes.

The advent of high-throughput sequencing technologies has facilitated rapid progress in the field of chloroplast genetics and genomics. Since the first chloroplast genome, from tobacco (*Nicotiana tabacum*), was sequenced in 1986 [3], over 800 complete chloroplast genome sequences have been made available in the National Center for Biotechnology Information (NCBI) organelle genome database, including 300 from crop and tree genomes. Insights gained from complete chloroplast genome sequences have enhanced our understanding of plant biology and diversity; chloroplast genomes have made significant contributions to phylogenetic studies of several plant families and to resolving evolutionary relationships within phylogenetic clades. In addition, chloroplast genome sequences have revealed considerable variation within and between plant species in terms of both sequence and structural variation. This information has been especially valuable for our understanding of the climatic adaptation of economically important crops, facilitating the breeding of closely related species and the identification and conservation of valuable traits [4, 5]. Improved understanding of variation among chloroplast genomes has also allowed the identification of specific examples of chloroplast gene transfer to plant nuclear or mitochondrial genomes, which has shed new light on the relationship between these three genomes in plants.

In addition to improving our understanding of plant biology and evolution, chloroplast genomics research has important translational applications, such as conferring protection against biotic or abiotic stress and the development of vaccines and biopharmaceuticals in edible crop plants. Indeed, the first commercial-scale production of a human blood protein in a Current Good Manufacturing Processes (cGMP) facility was published recently [6]. The lack of conservation of intergenic spacer regions, even among chloroplast genomes of closely related plant species, and the species specificity of regulatory sequences have facilitated the development of highly efficient transformation vectors for the integration and expression of foreign genes in chloroplasts. Because the published literature is rarely cross-referenced, this review highlights the impact of chloroplast genomes on various biotechnology applications. In addition to our enhanced understanding of chloroplast biology, we discuss in depth the roles of chloroplast genome sequences in improving our understanding of intracellular gene transfer, conservation, diversity, and the genetic basis by which chloroplast transgenes are engineered to enhance plant agronomic traits or to produce high-value agricultural or biomedical products. In addition, we discuss the impact of chloroplast genome sequences on increasing our understanding of the origins of economically important cultivated species and changes that occurred during domestication.

ADVANCES IN CHLOROPLAST GENOME SEQUENCING TECHNOLOGY

One of the important factors in the rapid advancement of the chloroplast genomics field is improvement in sequencing technologies. In studies conducted before the availability of high-throughput methods, isolated chloroplasts were used for the amplification of the entire chloroplast genome by rolling circle amplification [7, 8, 9, 10, 11, 12]. An alternative strategy is to screen bacterial artificial chromosome (BAC) or fosmid libraries using chloroplast genome sequences as probes [13, 14, 15, 16, 17, 18, 19, 20]; however, these methods are subject to many challenges, including difficulty in constructing good-quality BAC or fosmid libraries, large numbers of PCR reactions, and the possibility of contamination from other organellar DNA [21, 22, 23, 24, 25, 26, 27, 28, 29, 30, 31, 32]. The PCR approach is also difficult to apply to species that have no relatives whose chloroplast genomes have been sequenced or those with highly rearranged chloroplast genomes.

The development of next-generation sequencing (NGS) methods provided scientists with faster and cheaper methods to sequence chloroplast genomes. Moore and colleagues [33] first reported using NGS to determine chloroplast genome sequences, in *Nandina* and *Platanus*. Although multiple NGS platforms are available for chloroplast genome sequencing [34], Illumina is currently the major NGS platform used for chloroplast genomes [21, 32, 35, 36] because it allows the use of rolling circle amplification products [35, 37]. Investigators can then use bioinformatics platforms to perform de novo assembly without the need for reference genome sequences; from these assemblies it is possible to identify consensus chloroplast genome sequences [32]. A third-generation sequencer, the PacBio system which uses single molecule real-time (SMRT) sequencing, is now widely used in chloroplast genome sequencing [38, 39, 40, 41, 42, 43]. Its advantage is long read lengths [44], which facilitate de novo genome assembly, particularly in the four chloroplast junctions between the inverted repeat (IR) and single-copy regions.

The low accuracy (~85 % of the raw data) of the long reads produced by the PacBio platform [45] can be corrected by combining the latest chemistry with a hierarchical genome assembly process algorithm; accuracy rates as high as 99.999 % can be achieved after such post-error corrections [46]. Accuracy can also be increased using Illumina short reads [42]. In a study of *Potentilla micrantha*, sequencing with the Illumina platform produced seven contigs covering only 90.59 % of the chloroplast genome; by contrast, using the PacBio platform with error correction, the entire genome was successfully assembled in a single contig [39].

CHLOROPLAST GENOME STRUCTURE

The chloroplast genomes of land plants have highly conserved structures and organization of content; they comprise a single circular molecule with a quadripartite structure that includes two copies of an IR region that separate large and small single-copy (LSC and SSC) regions (Fig. 1a, b). The chloroplast genome includes 120–130 genes, primarily participating in photosynthesis, transcription, and translation. Recent studies have identified considerable diversity within non-coding intergenic spacer regions, which often include important regulatory sequences [13].

Despite the overall conservation in structure, chloroplast genome size varies between species, ranging from 107 kb (*Cathaya argyrophylla*) to 218 kb (*Pelargonium*), and is independent of nuclear genome size (Table 1). Certain lineages of land-plant chloroplast genomes also show significant structural rearrangements, with evidence of the loss of IR regions or entire gene families. Furthermore, there is also evidence for the existence of linear chloroplast genomes, as illustrated in Fig. 1b. The percentage of each form within the cell varies in different reports [47, 48].

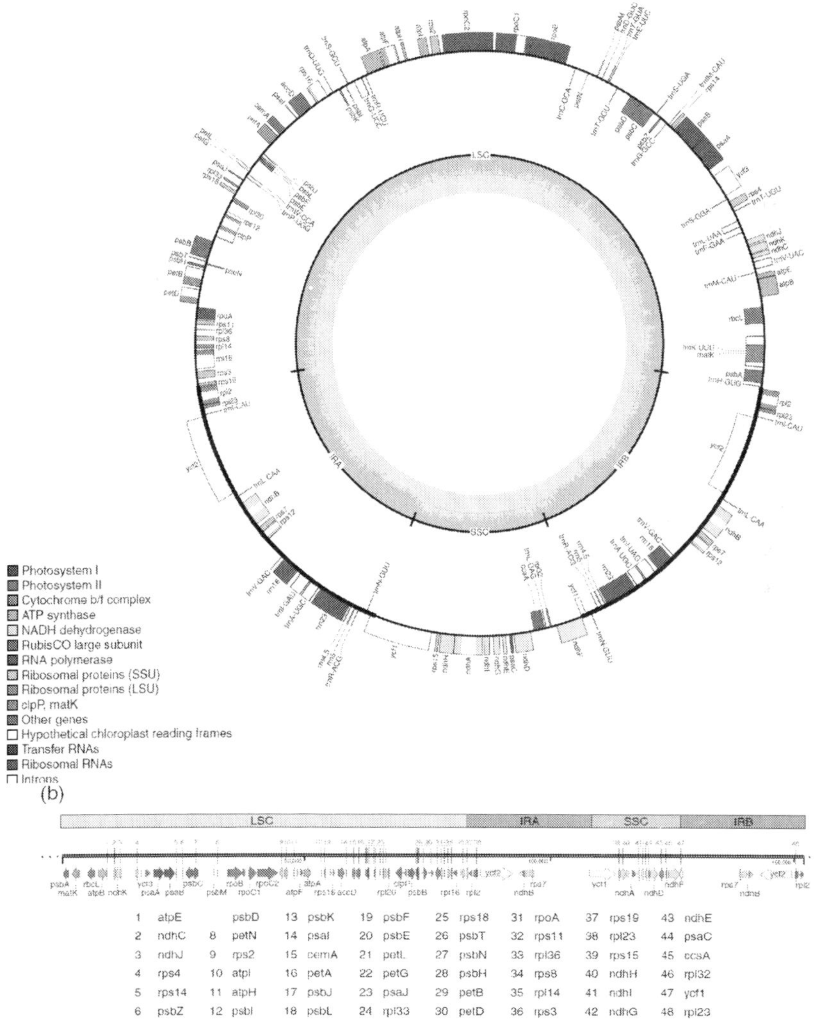

Figure 1 Map of the soybean (*Glycine max*) chloroplast genome. This genome was used to engineer biotic stress tolerance against insects and herbicides. The

quadripartite structure includes two copies of an IR region (*IRA* and *IRB*) that separate large single-copy (*LSC*) and small single-copy (*SSC*) regions [18]. **a** Circular form. The GC content graph (*gray circle* inside) marks the 50 % threshold of GC content. **b** Linear form. Different *colors* indicate genes in different functional groups. *IR* inverted repeat, *LSU* large subunit, *SSU* small subunit

Table 1. Alphabetical list of crop and tree species that have complete annotated chloroplast genome sequences

Species	Common name	Accession	Genome size (bp)	Uses	Reference(s)
Crops					
Acorus gramineus	Sweet flag	NC_026299	152849	Ornamental	[169]
Agrostis stolonifera	Creeping bent grass	NC_008591	136584	Forage	[8]
Allium cepa	Onion	NC_024813	153538	Vegetable	[170]
Ananas comosus	Pineapple	NC_026220	159636	Fruit	[171]
Anthriscus cerefolium	Chervil	NC_015113	154719	Medicinal	[172]
Artemisia frigida	Fringed sagewort	NC_020607	151076	Medicinal	[173]
Atropa belladonna	Belladonna	NC_004561	156687	Medicinal	[174]
Brassica napus (2)*	Canola	NC_016734	152860	Oil	[175]
Calanthe triplicata	Black orchid	NC_024544	158759	Flower	[176]
Cannabis sativa (2)	Marijuana	NC_027223	153854	Fiber	[177]
Capsicum annuum (2)	Pepper	NC_018552	156781	Vegetable	[178]
Carica papaya	Papaya	NC_010323	160100	Fruit	[179]
Catharanthus roseus	Madagascar periwinkle	NC_021423	154950	Flower	[180]
Cenchrus americanus	Pearl millet	NC_024171	140718	Cereals	[181]
Cicer arietinum	Chickpea	NC_011163	125319	Vegetable	[7]
Coix lacryma-jobi	Job's tears	NC_013273	140745	Cereals	[29]
Colocasia esculenta	Taro	NC_016753	162424	Vegetable	[182]

Cucumis sativus (3)	Cucumber	NC_007144	155293	Vegetable	[183]
Curcuma roscoeana	Jewel of Burma	NC_022928	159512	Medicinal	[184]
Cymbidium tortisepalum (5)	Cymbidium orchid	NC_021431	155627	Flower	[55]
Cypripedium formosanum (3)	Formosa's lady's slipper	NC_026772	178131	Flower	[32]
Daucus carota	Carrot	NC_008325	155911	Vegetable	[9]
Dendrobium catenatum	Dendrobium orchid	NC_024019	152221	Flower	[56]
Dieffenbachia seguine	Dumbcane	NC_027272	163699	Ornamental	[185]
Digitaria exilis	White fonio	NC_024176	140908	Cereals	[181]
Echinochloa oryzicola	Late barnyard grass	NC_024643	139891	Cereals	[186]
Ephedra equisetina	Ma Huang	NC_011954	109518	Medicinal	[187]
Erycina pusilla	Mini orchid	NC_018114	143164	Flower	[36]
Fagopyrum esculentum (2)	Common buckwheat	NC_010776	159599	Cereals	[188]
Festuca arundinacea (4)	Kentucky fescue	NC_011713	136048	Forage	[189]
Fragaria vesca (6)	Wild strawberry	NC_015206	155691	Fruit	[190]
Glycine max (9)	Soybean	NC_007942	152218	Oil	[18]
Glycyrrhiza glabra	Common liquorice	NC_024038	127943	Medicinal	[74]
Gossypium barbadense (22)	Sea island cotton	NC_008641	160317	Fiber	[69]
Guizotia abyssinica	Ramtilla	NC_010601	151762	Bird seed	[191]
Helianthus annuus (9)	Common sunflower	NC_007977	151104	Oil	[192]
Heliconia collinsiana	Platanillo	NC_020362	161907	Ornamental	[193]
Hordeum vulgare	Barley	NC_008590	136462	Cereals	[8]

Hyoscyamus niger	Henbane	NC_024261	155720	Medicinal	[194]
Ipomoea batatas	Sweet potato	NC_026703	161303	Vegetable	[195]
Ipomoea purpurea	Common morning glory	NC_009808	162046	Ornamental	[196]
Lactuca sativa	Lettuce	NC_007578	152765	Vegetable	[197]
Lilium superbum	Turk's-cap lily	NC_026787	152069	Flower	[198]
Lolium multiflorum (2)	Ryegrass	NC_019651	135175	Lawn	[199]
Lotus japonicus	Birdsfoot trefoil	NC_002694	150519	Forage	[200]
Manihot esculenta	Cassava	EU117376	161453	Starch crop	[20]
Masdevallia picturata (2)	Masdevallia orchid	NC_026777	157423	Flower	[32]
Musa textilis	Banana	NC_022926	161347	Fruit	[184]
Nicotiana tabacum (4)	Tobacco	Z00044	155943	Tobacco	[3]
Nuphar advena	Spatterdock	NC_008788	160866	Medicinal	[201]
Nymphaea alba (2)	White water-lily	NC_006050	159930	Flower	[24]
Oncidium hybrid	Oncidium	NC_014056	146484	Flower	[54]
Oryza sativa (6)	Rice	X15901	134525	Cereals	[202]
Panax ginseng (2)	Ginseng	NC_006290	156318	Medicinal	[203]
Panicum virgatum	Switchgrass	NC_015990	139619	Biofuel	[204]
Paphiopedilum armeniacum(2)	Slipper orchid	NC_026779	162682	Flower	[32]
Parthenium argentatum	Guayule	NC_013553	152803	Biofuel	[205]
Pelargonium (2)	Geranium	NC_008454	217942	Flower	[206]
Phalaenopsis hybrid (3)	Phalaenopsis orchid	NC_007499	148964	Flower	[51]
Phaseolus vulgaris	Kidney bean	NC_009259	150285	Bean	[78]
Pisum sativum	Pea	NC_014057	122169	Vegetable	[76]
Raphanus sativus	Radish	NC_024469	153368	Vegetable	[207]
Ravenala madagascariensis	Traveller's tree	NC_022927	166170	Ornamental	[184]

Ricinus communis	Castor bean	NC_016736	163161	Oil	[208]
Saccharum hybrid (2)	Sugarcane	NC_005878	141182	Sugar	[209]
Salvia miltiorrhiza	Redroot sage	NC_020431	151328	Medicinal	[210]
Secale cereale	Rye	NC_021761	114843	Cereals	[64]
Sesamum indicum	Sesame	NC_016433	153324	Oil	[211]
Solanum lycopersicum (11)	Tomato	NC_007898	155461	Vegetable	[13]
Solanum tuberosum	Potato	DQ231562	155312	Starch crop	[212]
Sorghum bicolor (2)	Sorghum	NC_008602	140754	Cereals	[8]
Spinacia oleracea	Spinach	NC_002202	150725	Vegetable	[213]
Trifolium grandiflorum (8)	Large-flower hop clover	NC_024034	125628	Forage	[74]
Triticum aestivum (6)	Bread wheat	NC_002762	134545	Cereals	[63]
Vanilla planifolia	Vanilla	NC_026778	148011	Fruit	[32]
Vigna radiata (3)	Mung bean	NC_013843	151271	Bean	[79]
Zea mays	Maize	NC_001666	140384	Cereals	[62]
Zingiber spectabile	True ginger	NC_020363	155890	Ornamental	[193]
Trees and perennial plants					
Abies koreana	Fir	NC_026892	121373	Wood	[214]
Actinidia chinensis (2)	Kiwifriut	NC_026690	156346	Fruit	[215]
Amentotaxus formosana	Taiwan catkin yew	NC_024945	136430	Timber	[216]
Araucaria heterophylla	Norfolk island araucaria	NC_026450	146723	Timber	[217]
Bambusa multiplex (4)	Golden goddess bamboo	NC_024668	139394	Ornamental	[91]
Bambusa oldhamii	Green bamboo	NC_012927	139350	Vegetable	[28]
Berberis bealei	Beale's mahonia	NC_022457	164792	Ornamental	[218]

Bismarckia nobilis	Bismarck palm	NC_020366	158210	Ornamental	[193]
Buxus microphylla	Japanese box	NC_009599	159010	Ornamental	[219]
Calocedrus formosana	Taiwan incense-cedar	NC_023121	127311	Timber	[220]
Calycanthus floridus	Carolina-allspice	NC_004993	153337	Medicinal	[23]
Camellia oleifera (13)	Tea oil plant	NC_023084	156971	Oil	[221]
Camellia reticulata	To-tsubaki	NC_024663	156971	Flower	[222]
Carludovica palmata	Toquilla palm	NC_026786	158545	Fiber	[198]
Castanea mollissima	Chestnut	NC_014674	160799	Fruit	[14]
Cathaya argyrophylla	Cathaya	NC_014589	107122	Timber	[223]
Cedrus deodara	Cedar	NC_014575	119299	Timber	[223]
Cephalotaxus wilsoniana (2)	Wilson plum yew	NC_016063	136196	Timber	[224]
Chrysobalanus icaco	Coco plum	NC_024061	162775	Fruit	[225]
Citrus sinensis (2)	Orange	NC_008334	160129	Fruit	[12]
Cocos nucifera	Coconut	NC_022417	154731	Oil	[226]
Coffea arabica	Coffee	NC_008535	155189	Beverage	[10]
Corymbia gummifera (4)	Red bloodwood	NC_022407	160713	Timber	[227]
Corynocarpus laevigata	Karaka nut	NC_014807	159202	Fruit	[37]
Cryptomeria japonica	Sugi	NC_010548	131810	Timber	[228]
Dendrocalamus latiflorus	Sweet giant bamboo	NC_013088	139394	Vegetable	[28]
Elaeis guineensis	African oil palm	NC_017602	156973	Oil	[229]
Eucalyptus globulus (32)	Eucalyptus	NC_008115	160286	Timber	[230]
Hevea brasiliensis	Rubber tree	NC_015308	161191	Rubber	[231]
Jasminum nudiflorum	Winter jasmine	NC_008407	165121	Ornamental	[232]

Jatropha curcas	Barbados nut	NC_012224	163856	Biofuel	[233]
Juniperus bermudiana (4)	Bermuda juniper	NC_024021	127659	Timber	[234]
Larix decidua	European larch	NC_016058	122474	Timber	[224]
Licania sprucei (3)	Licania	NC_024065	162228	Ornamental	[225]
Liquidambar formosana	Chinese sweetgum	NC_023092	160410	Timber	[30]
Liriodendron tulipifera	Tulip tree	NC_008326	159886	Timber	[235]
Metasequoia glyptostroboides	Dawn redwood	NC_027423	131887	Timber	[236]
Millettia pinnata	Indian beech	NC_016708	152968	Ornamental	[81]
Morus indica (3)	White mulberry	NC_008359	158484	White mulberry	[237]
Nageia nagi	Asian bayberry	NC_023120	133722	Timber	[220]
Nandina domestica	Heavenly bamboo	NC_008336	156599	Ornamental	[33]
Nerium oleander	Oleander	NC_025656	154903	Ornamental	[238]
Olea europaea (5)	Olive	NC_015604	155862	Oil	[239]
Phoenix dactylifera	Date palm	NC_013991	158462	Fruit	[240]
Phyllostachys edulis (4)	Moso bamboo	NC_015817	139679	Timber	[89]
Picea sitchensis (3)	Sitka spruce	NC_011152	120176	Timber	[35]
Pinus taiwanensis (12)	Taiwan red pine	NC_027415	119741	Timber	[241]
Platanus occidentalis	American sycamore	NC_008335	161791	Ornamental	[33]
Podocarpus lambertii (3)	Podocarpus	NC_023805	133734	Ornamental	[242]
Populus alba	White poplar	NC_008235	156505	Timber	[243]
Prinsepia utilis	Himalayan cherry	NC_021455	156328	Ornamental	[244]
Prunus persica (6)	Peach	NC_014697	157790	Fruit	[14]

Pseudophoenix vinifera	Florida cherry palm	NC_020364	157829	Ornamental	[193]
Pseudotsuga sinensis	Chinese douglas	NC_016064	122513	Timber	[224]
Pyrus pyrifolia (2)	Chinese pear	NC_015996	159922	Fruit	[245]
Quercus rubra (4)	Oak	NC_020152	161304	Timber	[246]
Sapindus mukorossi	Soapberries	NC_025554	160481	Medicinal	[247]
Taiwania cryptomerioides (2)	Taiwania	NC_016065	132588	Timber	[224]
Theobroma cacao	Cacao tree	HQ336404	160604	Beverage	[14]
Vaccinium macrocarpon	Large cranberry	NC_019616	176045	Fruit	[248]
Vitis vinifera	Wine grape	NC_007957	160928	Fruit	[19]
Wollemia nobilis	Wollemia	NC_027235	145630	Timber	[249]

*The number of species in the same genus as the listed species that have sequenced and annotated chloroplast genomes is shown in parentheses

Like the genes, the introns in land-plant chloroplast genomes are generally conserved, but the loss of introns within protein-coding genes has been reported in several plant species [49], including barley (*Hordeum vulgare*) [8], bamboo (*Bambusa* sp.) [28], cassava (*Manihot esculenta*) [20], and chickpea (*Cicer arietinum*) [7]. The proteins encoded by genes in which intron loss is known to occur have diverse functions; they include an ATP synthase (atpF), a Clp protease (clpP), an RNA polymerase (rpoC2), and ribosomal proteins (rpl2, rps12, and rps16) [49]. The majority of reported intron losses have been observed in specific plant groups or species, although some examples of intron loss (such as that in *clpP*) occur in diverse plant species, including monocots (Poaceae), eudicots (Onagraceae and Oleaceae) and gymnosperms (*Pinus*) [49].

DIVERSITY OF CHLOROPLAST GENOME SEQUENCES

At higher taxonomic levels (family level), protein-coding regions and conserved sequences of the chloroplast genome can be used for phylogenetic analysis and domestication studies [49]. Earlier phylogenetic analyses utilized partial chloroplast DNA sequences. The use of variable regions or multiple DNA fragments dramatically enhanced the utility of these analyses but there is insufficient information in these sequences to provide the high-resolution necessary to differentiate closely related taxa, particularly some within-species taxa whose taxonomic relationships are unclear. Complete chloroplast genome sequences are valuable for deciphering phylogenetic relationships between closely related taxa and for improving our understanding of the evolution of plant species.

In this section, we discuss several examples of comparisons of chloroplast genomes, within and between crop species, that have provided unique insight into evolutionary relationships among taxa. We also discuss the origin and geographic distribution of economically important species, as well as their adaptations to different climatic conditions and the use of genome information in their breeding and conservation.

A key application of the chloroplast genome in agriculture is the identification of commercial cultivars and the determination of their purity. DNA barcodes derived from the chloroplast genome can be used to identify varieties and in the conservation of breeding resources. Success in breeding is determined by genetic compatibility and chloroplast genomes serve as a valuable tool for identifying plants that are likely to be closely related and, therefore, genetically compatible. Understanding the genetic relationships between cultivated crops and their wild relatives informs efforts to introduce specific advantageous traits into cultivated crops. In the section below, we discuss how chloroplast genomes have been used to elucidate the evolutionary relationships and domestication history of a few major crops and how this informs breeding programs.

Breeding

The Orchidaceae is a large family that encompasses about 6–11 % of all angiosperms [50] and is important in floriculture. Many commercially important orchid species belong to the subfamily Epidendroideae and chloroplast genomes of several species from this subfamily have been sequenced [51, 52, 53, 54, 55, 56, 57, 58]. Because it is easy to perform inter-generic crossing in orchids and because the record of breeding is sometimes incomplete, it is often difficult to validate the parental origin of commercially important varieties [54]. Corrected parental information is important for breeding and variety identification. In an investigation of the Oncidiinae, a subtribe within the Epidendroideae, PCR products derived from eight conserved regions in 15 commercial varieties resolved their phylogenetic relationship at the species level [54] and helped to resolve putative errors in parental origin. Parental records had indicated that *Odontoglossum* 'Violetta von Holm', *Odontoglossum* 'Margarete Holm' and *Odontocidium* 'Golden Gate' are derived from the same female parent (*Odontoglossum bictoniense*) but phylogenetic analyses of 'Violetta von Holm' did not correlate with those of 'Golden Gate' or 'Margarete Holm' [54]. A possible reason for inconsistencies between the chloroplast DNA-based phylogenetic tree and the parental record is chloroplast capture. Chloroplast capture is the introgression of chloroplasts from one species into another after intrageneric and intergeneric hybridization [59]. Although chloroplast genomes provide useful information for phylogenetic analyses involving closely related taxa, chloroplast capture by hybridization may distort phylogenetic relationships if captured chloroplast genomes or genes included therein are used [60]. The use of both nuclear and chloroplast genomes can provide more complete phylogenies [4, 61].

Phylogenetic studies

There are several published chloroplast genomes from cereals, including those from sorghum (*Sorghum bicolor*), barley [8], maize (*Zea mays*) [62], wheat (*Triticum aestivum*) [63], rye (*Secale cereale*) [64], and rice (*Oryza sativa*) [65]. Rice is one of the world's most important crops and is the primary carbohydrate source for the global human population (http://www.ers.usda.gov/topics/crops/rice.aspx). The *Oryza* species are classified into ten genome types, including six diploids (AA, BB, CC, EE, FF, and GG) and four allotetraploids (BBCC, CCDD, HHJJ, and HHKK). Attempts to clarify the evolutionary relationships between cultivated rice and its wild relatives remain contentious and inconclusive [4]. For example, there are two wild species that have an AA genome in Australia, *Oryza meridionalis* (annual) and *Oryza rufipogon* (perennial). *Oryza sativa* was domesticated from Asian *O. rufipogon* 10,000 years ago [65]. Nevertheless, analysis of complete Australian and Asian wild rice chloroplast genomes indicated that Australian *O. rufipogon* chloroplast genomes are more similar to those of Australian *O. meridionalis* than to those of Asian *O. rufipogon* [65, 66, 67]. Using 19 chloroplast genomes of *Oryza* AA genome species, a robust phylogenetic tree was established, which will aid in improving rice crops and in conservation strategies [4, 5].

Cotton is the most important textile fiber crop and the first cotton (*Gossypium hirsutum*) chloroplast genome was published in 2006 [11]. The diploid *Gossypium* species comprise eight genome groups (A to G and K genomes). *Gossypium hirsutum* (upland cotton), the most widely planted cotton species in the world, is an allotetraploid of the ancestral A and D genome species [68]. Chloroplast genome sequences are available for 22 *Gossypium* species and these can be used to glean information about the evolution and domestication of this crop [11, 68, 69] (Table 1). Simple sequence repeat primers were used to investigate 41 species of *Gossypium*, including all eight genome groups and allotetraploid species [70]. The results indicated that two modern A-genome species, *Gossypium herbaceum* and *Gossypium arboretum*, were not cytoplasmic donors of tetraploid (AD) species; instead, the AD genome species originated from an extinct ancestor species of the modern A genome [68, 70].

Domestication

Information on chloroplast genomes is useful for understanding the domestication of several crops, particularly legumes [71]. The chloroplast genome structure of legumes is very interesting; it contains multiple rearrangements, including large inverted segments and loss of inverted repeats [72]. An example is a 51-kb inversion that was first identified in the soybean (*Glycine max*) chloroplast genome sequence [18] before being reported in most members of the subfamily Papilionoideae [7, 73, 74, 75, 76, 77]. A 78-kb reversion was subsequently confirmed in *Phaseolus* and *Vigna* chloroplast genomes [78, 79]. More recently, 36-kb [80] and 5.6-kb [81] inversions inside the 51-kb inversion were identified. There are many important genes within these inverted regions but no gene is disturbed and plant survival and performance are not affected. These unique characteristics are not only very useful in phylogenetic studies [82] but also provide important information for chloroplast transformation in legumes. Chloroplast structure is also important for the design of primers needed in the amplification of sequences for further domestication and phylogenetic analysis.

Citrus is one of the most commercially important fruit genera. In 2006, the first *Citrus* chloroplast genome, that of sweet orange (*Citrus* × *sinensis*), was published [12] and this served as a reference genome for subsequent publications [83, 84]. Phylogenetic analysis of 34 chloroplast genomes of *Citrus* (28) and *Citrus*-related genera (6) indicated that citrus fruits have the same common ancestor [84, 85]. In four genes (*matK*, *ndhF*, *ycf1*, and *ccsA*), single-nucleotide variations and insertion/deletion frequencies were clearly higher than average and showed that these genes have been positively selected. The *matK* gene encodes a maturase that is involved in splicing type II introns and the *matK* sequence is often used in phylogenetic and evolutionary studies [84]. Positive selection of *matK* is observed not only in citrus but is common in several other plant species. In fact, more than 30 plant groups have been shown to undergo positive selection of *matK* genes, indicating that the gene is subject to a number of different ecological selective pressures [86]. The *ndhF* gene encodes a subunit of the chloroplast NAD(P)H dehydrogenase (NDH) complex. Chloroplast NDH monomers are sensitive to high light stress, suggesting that the *ndh* genes may also be involved in stress acclimation [87]. These studies indicated that *matK* and *ndhF* show positive selection in Australian species, potentially contributing to their adaptation to a hot, dry climate [84, 85].

Bamboo is an economically and ecologically important forest plant in Asia [88]. Bamboo grows quickly and new culms are regenerated from the rhizome after harvesting, making it a sustainable and ecologically and environmentally friendly crop. The first two bamboo chloroplast genomes have been published [28] and many more bamboo chloroplast genomes are now available [88, 89, 90, 91, 92, 93]. Bamboo has a long juvenility and it is difficult to obtain flowers for taxonomic studies; consequently the taxonomic relationships of bamboo have proven challenging to unravel on the basis of traditional reproductive organ morphology. Furthermore, the extremely low rate of sequence divergence meant that the taxonomic and phylogenetic relationships of temperate woody bamboos at lower taxonomic levels proved difficult to resolve [88]. These relationships were eventually resolved with high-resolution phylogenetic trees using 25 bamboo chloroplast genomes [93]. In addition to woody bamboos, chloroplast genomes have also been published for herbaceous bamboo [88, 92]. An interesting phenomenon identified in herbaceous bamboo chloroplast genomes is that of gene transfer from the mitochondrial genome to the chloroplast genome. This was an unusual observation, as the chloroplast genome is thought to be nearly immune to the transfer of DNA from nuclear and mitochondrial genomes [88, 92, 94]. A possible reason for this recalcitrance to DNA transfer is the lack of an efficient DNA uptake system [94]. Prior to its observation in herbaceous bamboo, this phenomenon was only observed in two eudicot chloroplast genomes [94] and in monocots [88, 92].

TRANSFER OF CHLOROPLAST GENES TO NUCLEAR OR MITOCHONDRIAL GENOMES

There are three distinct genomes in plant cells: nuclear, mitochondrial, and plastid. Mitochondria are believed to have evolved from a single endosymbiotic event by the uptake of a proteobacterium, whereas chloroplasts evolved from endosymbiosis of a cyanobacterium, after which there was a massive transfer of genes from the chloroplast to the nucleus [95]. There are distinct translation systems in these organelles: nuclear-encoded genes are translated in the cytosol and the protein products are then transported to the locations in which they function, including chloroplasts [96], whereas chloroplast-encoded proteins are directly synthesized within the chloroplast. Multi-subunit functional protein complexes that are involved in photosynthesis or protein synthesis are also assembled within chloroplasts.

Gene content, number, and structure are conserved in the chloroplast genome sequences of most autotrophic land plants [97, 98] but some protein-encoding genes are absent in specific species [49]. The loss of genes such as *infA*, *rpl22*, and *ndh* from the chloroplast genome and their intracellular transfer to the nuclear or mitochondrial genomes provide valuable information for phylogenetic analyses and evolutionary studies. It is very easy to identify the chloroplast origin of genes in plant mitochondrial or nuclear genomes [99, 100] by intracellular gene transfer [32], but this could also lead to erroneous phylogenic relationships when short sequences are used instead of complete chloroplast genome sequences.

The chloroplast *translation initiation factor 1* (*infA*) is a homolog of the essential gene *infA* in *Escherichia coli*[101, 102]. This gene initiates translation in collaboration with two nuclear-encoded initiation factors to mediate interactions between mRNA, ribosomes, and initiator tRNA-Met [102]. Many parallel losses of chloroplast-encoded *infA* have occurred during angiosperm evolution [102] (Fig. 2). Nuclear-encoded *infA*genes have been identified in *Arabidopsis thaliana*, soybean, tomato (*Solanum lycopersicum*), and ice plant (*Mesembryanthemum crystallinum*) [102]. Protein sequences of nuclear-encoded *infA* in these four species contain chloroplast transit peptides. Studies using soybean and *A. thaliana* infA-GFP proteins have shown that nuclear-encoded *infA* genes are translated in the cytosol and transported into chloroplasts [102]. Many more chloroplast-encoded *infA* deletions have been identified recently (Fig. 2).

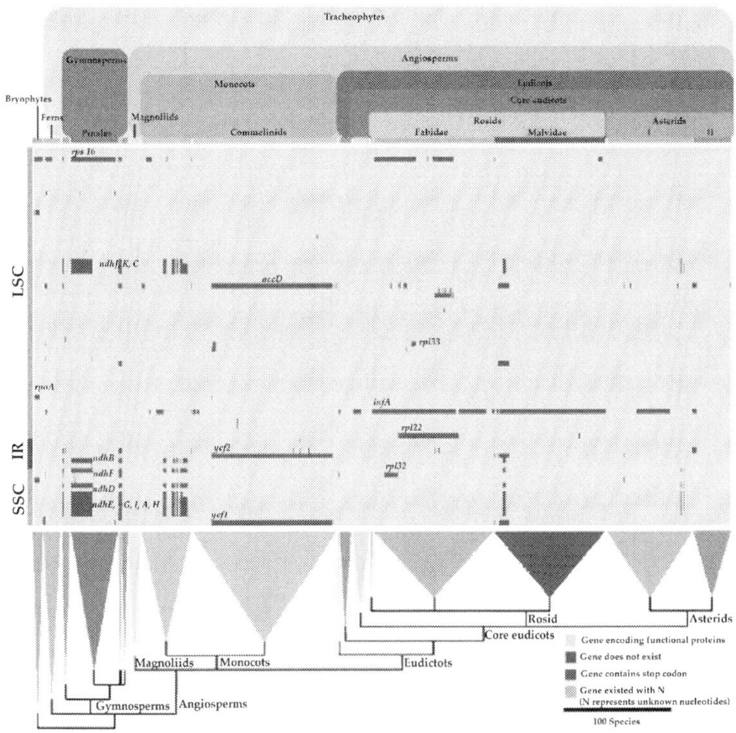

Figure 2: Chloroplast genome structure and gene expression across tracheophytes. These 658 chloroplast genomes were downloaded from NCBI Organelle Resources. The *X-axis* indicates the taxonomy of the chloroplast genome species following the Angiosperm Phylogeny Group III system and NCBI taxonomy. The *bar width* represents 100 species. The *Y-axis* shows the chloroplast genes, which were classified by different chloroplast regions. *Gray*

boxes indicate absence of genes. *Red boxes* indicate stop codons in genes. *Blue boxes* indicate unknown nucleotides (*N*) in genes. *IR* inverted repeat, *LSC* large single-copy region, *SSC* small single-copy region

There are 57 chloroplast genomes in 26 genera in which the essential gene *rpl22* is reported to have been deleted from the chloroplast and transferred to the nuclear genome (Fig. 2) [14, 103]. Nuclear-encoded *rpl22*contains a transit peptide that is predicted to deliver this protein from the cytosol to chloroplasts. These peptides are diverse, suggesting that there were two independent *rpl22* transfers in the Fabaceae and the Fagaceae [14]. Similar transfer to the nucleus has also been observed for *rpl32* deletion from chloroplast genomes [104, 105, 106].

Eleven chloroplast genes encode *ndh* subunits, which are involved in photosynthesis. The ndh proteins assemble into the photosystem I complex to mediate cyclic electron transport in chloroplasts [107, 108] and facilitate chlororespiration [109]. Some autotrophic plants lack functional *ndh* genes in their chloroplast genomes [36, 51, 54, 55, 110, 111, 112, 113, 114, 115] (Fig. 2). Unlike the single gene losses described previously, the entire family of *ndh* genes has been deleted in these plants. Seven orchid chloroplast genomes indicated at least three independent *ndh* deletions [32]. Some orchid *ndh* DNA fragments were identified in the mitochondrial genome but the complete *ndh* genes required to translate putative functional protein complexes are absent [32]. In the nuclear genome of Norway spruce, only non-functional plastid *ndh* gene fragments are present [116]. Normal photosynthesis is observed in these *ndh*-deleted species [32, 117]. Furthermore, *ndh*-deleted transformants are autotrophic and produce carbohydrates through photosynthesis [107, 118, 119, 120, 121].

Many more chloroplast-gene deletions have been observed, including deletions of *accD*, *ycf1*, *ycf2*, *ycf4*, *psaI*, *rpoA*, *rpl20*, *rpl23*, *rpl33*, and *rps16*; many unique gene deletions have been identified in only one or a few species (*psbJ*, *rps2*, *rps14*, and *rps19*) (Fig. 2). The functions of these genes, phenotypes of their knock-out mutants, and evidence for their transfer are summarized in Additional file 1. Most essential genes that have been lost from chloroplast genomes have been transferred to the nucleus to maintain the plant's photosynthetic capacity, with the exception of *ycf1* and *ycf2*.

In summary, chloroplast genome sequences are most valuable for understanding plant evolution and phylogeny. Databases of not only plant genomes but also plant transcriptomes will be useful in investigating deletion events or the transfer of chloroplast genes to other organellar genomes to complement such deletions.

ADVANCES IN CHLOROPLAST GENOME ENGINEERING

In the past century, desirable agronomic traits, including yield enhancement and resistance to pathogens or abiotic stress, were achieved by breeding cultivated crops with their wild relatives. As explained above, chloroplast genome sequences are very useful in the identification of closely related, breeding-compatible plant species. With the advent of modern biotechnology, desirable traits from unrelated species can now be readily introduced into commercial cultivars. Such genetically modified crops have revolutionized agriculture in the past two decades, dramatically reducing the use of chemical pesticides and herbicides while enhancing yield. For most commercial cultivars, herbicide- or insect-resistance genes are introduced into the nuclear genome. There are, however, a few limitations for nuclear transgenic plants, including low levels of expression (<1 % total soluble protein (TSP)) and potential escape of transgenes via pollen.

Engineering the introduction of foreign genes into the chloroplast genome addresses both of these concerns. Just two copies of transgenes are typically introduced into the nuclear genome, whereas up to 10,000 transgene copies have been engineered into the chloroplast genome of each plant cell, resulting in extremely high levels of foreign gene expression (>70 % TSP) [122]. Most importantly, chloroplast genomes are maternally inherited in most cultivated crops, minimizing or eliminating transgene escape via pollen [123].

The basic process of chloroplast engineering is explained in Fig. 3a, b. Chloroplast genome engineering is accomplished by integrating foreign genes into intergenic spacer regions without disrupting the native chloroplast genes (Fig. 3a). Two chloroplast genes are used as flanking sequences to facilitate integration of transgene cassettes. Transgene cassettes include a selectable marker gene and gene(s) of interest, both regulated by chloroplast gene promoters and untranslated regions (UTRs; Fig. 3a). Chloroplast genome sequences are essential to build transgene cassettes because they provide both flanking and regulatory sequences. Transgene cassettes that are inserted into bacterial plasmids are called chloroplast vectors and they are bombarded into plant cells using gold particles and a gene gun (Fig. 3b). Because of the presence of chloroplast DNA in the nuclear or mitochondrial genome, transgene cassettes may integrate via homologous or non-homologous recombination events; but any transgenes that are integrated within the nuclear or mitochondrial genome will not be expressed because chloroplast regulatory sequences are not functional in other genomes. If such integration occurs, the transgenes could be easily identified by evaluation of their integration site and eliminated [124].

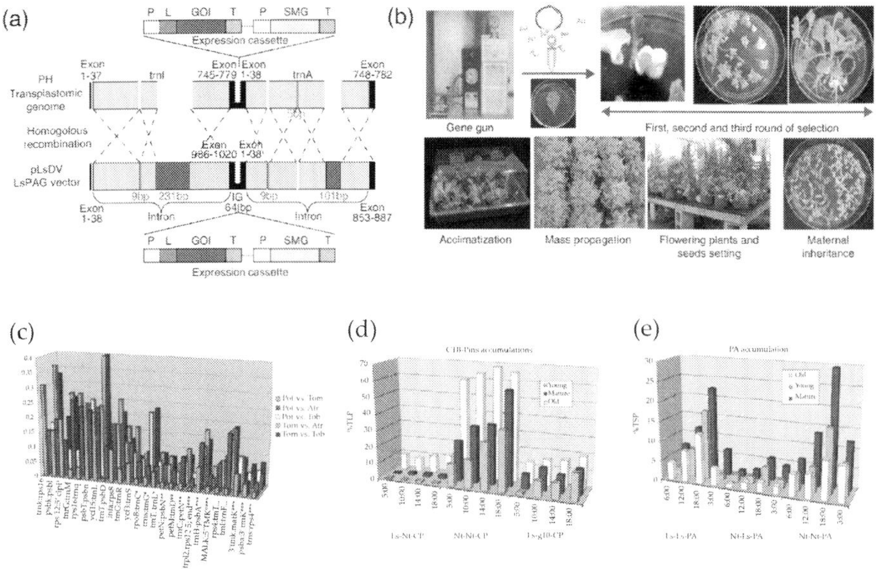

Figure 3: Basic process of chloroplast genetic engineering, diversity in intergenic spacer regions, and impact of transgene integration (endogenous versus heterologous genome sequences). **a** Complexity of heterologous sequence integration into intergenic spacer regions between lettuce and tobacco. The schematic diagram represents recombination between the tobacco transplastomic genome and the lettuce transformation vector [128]. *Purple bars* represent unique lettuce intron sequence; the *green bar* represents unique tobacco intron sequence; *black bars* are exon regions; *blue regions* are looped out sequence. The expression cassette comprises: promoters (*P*), leader sequence (*L*), gene

of interest (*GOI*), terminators (*T*), and selectable marker gene (*SMG*). *IG* intergenic spacer region. **b** Basic process of chloroplast genetic engineering. Gene delivery is performed by bombardment with gold microparticles coated with chloroplast vectors, followed by three rounds of selection to achieve homoplasmy. After confirmation of transgene integration, plants are grown in the greenhouse to increase biomass. Chloroplast transgenes are maternally inherited without Mendelian segregation of introduced traits. **c** Comparison of 21 of the most variable intergenic spacer regions among Solanaceae chloroplast genomes. *Atr Atropa*, *Pot* potato, *Tob*tobacco, *Tom* tomato. *Tier 1, **tier 2, and ***tier 3 regions reported in the paper by Shaw et al. [250]. Plotted values were converted from percentage identity to sequence divergence on a scale from 0 to 1 as shown on the *Y-axis*; these values demonstrate a wide range of sequence divergence in different regions. Nucleotide sequences were determined by a bridging shotgun method and genome annotation was performed using the Dual Organellar GenoMe Annotator [13]. **d**, **e** Decrease in the expression of transgenes regulated by heterologous psbA promoters and untranslated regions (UTRs) engineered via tobacco chloroplast genomes. When the lettuce (*La*) *psbA*regulatory region was used in tobacco (*Na*) chloroplasts or vice versa, transgene expression is dramatically reduced. **d** Accumulation of a cholera toxin B subunit (*CTB*) and proinsulin (*Pins*) fusion protein (*CP*) was quantified by densitometry and **e** anthrax protective antigen (*PA*) accumulation was estimated by enzyme-linked immunosorbent assay (ELISA). Total leaf protein (*TLP*) or total soluble protein (*TSP*) data are presented as a function of light exposure and developmental stage. The order of young, mature, and old is different in **d** and **e** because of the accumulation of more CTB-Pins in older leaves and PA in mature leaves [128]. Young (top five), mature (fully grown), and old (bottom three) leaves were fully expanded and were cut from plants grown in the greenhouse for 8–10 weeks

One of the challenges of creating chloroplast transgenic (transplastomic) plants is the elimination of all untransformed copies (>10,000 per cell) of the native chloroplast genome and replacing them with transformed genomes that contain integrated transgene cassettes. The absence of the native chloroplast genome and the presence of only the modified genomes is referred to as the homoplasmic state, which is typically achieved after two or three rounds of selection (Fig. 3b). The most effective selectable marker used is the *aadA* gene, which confers resistance to streptomycin and spectinomycin. These antibiotics bind specifically to chloroplast ribosomes and disrupt protein synthesis without interfering with any other cellular process. Efforts to transform the chloroplast genome of cereal crops have been mostly unsuccessful. This could be due to the instability of chloroplast DNA in the mature leaves of cereals [47] or to a requirement for better selectable markers [125].

Table 2 provides the first global, comprehensive summary of the power of chloroplast genetic engineering, utilizing valuable information generated by the sequencing of chloroplast genomes described in previous sections. This table includes the most complete list of chloroplast genomes that have been engineered for enhanced agronomic traits or the production of different bio-products, including biopolymers, industrial enzymes, biopharmaceuticals, and vaccines. Within Table 2, transgenes are grouped according to their functions and are organized according to their site of integration. The efficiency of transgene expression is also included in Table 2, providing important information about the regulatory sequences used to express the transgenes.

Table 2: Engineering the chloroplast genome for biotechnology applications

Site of integration	Transgenes	Regulatory sequences	Efficiency of expression	Engineered traits or products	Reference(s)
Insect or pathogen tolerance					
trnI/trnA	Bgl-1	5′psbA/3′psbA	>160-fold enzyme	Resistance against whitefly and aphid	[141]
trnI/trnA	Pta	5′psbA/3′psbA	7.1–9.2 % TSP	Broad-spectrum resistance against aphid, whitefly, Lepidopteran insects, bacterial and viral pathogens	[142]
trnI/trnA	PelB1, PelD2	5′psbA/3′psbA	~2.42 units mg^{-1} FW	Resistance against Erwinia soft rot	[150]
trnI/trnA	RC1011, PG12	5′psbA/3′	17–38 % TSP	Resistance to Erwinia soft rot and tobacco mosaic virus	[140]
trnI/trnA	cpo	Prrn/psbA/psbA	NR	Resistance to fungal pathogens in vitro (Aspergillus flavus, Fusarium verticillioides, and Verticillium dahliae) and in planta (Alternaria alternata)	[251]
trnI/trnA	Bt cry2Aa2operon	Prrn/ggagg/psbA	45.3 % TSP	100 % mortality of cotton bollworm, beet armyworm; cuboidal Bt crystals formation	[137]

trnI/trnA	Bt cry9Aa2	Prrn/ggagg/ rbcL	~10 % of TSP	Resistance to Phthorimaea operculella	[252]
trnI/trnA	msi-99	Prrn/ggagg/ psbA	21–43 % TSP	Resistance to in planta challenge of Pseudomonas syringae, Aspergillus flavus, Fusarium moniliforme, Verticillium dahlia, and Colletotrichum destructivum	[253]
trnI/trnA	sporamin1, CeCPI2, and chitinase2	Prrn/TpsbA	0.85–1 % TSP	Resistance against Spodoptera lituraand Spodoptera exigua leaf spot, as well as soft rot diseases	[254]
trnI/trnA	MSI-99	Prrn/Trps16	89.75 $\mu g\ g^{-1}$FW	Resistance against rice blast fungus	[255]
trnV/rps12/7	cry1A(c)	Prrn/rbcL/rps16	3–5 % of TSP	Resistance to larvae of Heliothis virescens, Helicoverpa zea, and Spodoptera exigua	[256]
trnV/rps12/7	cry1Ab	Prrn/T7gene10/ rbcL	NR	Resistance to caterpillar of Anticarsia gemmatalis	[145]
rbcL/accD	cry2Aa2	Prrn/ggagg/ psbA	2–3 % of TSP	Resistance to Heliothis virescens, Helicoverpa zea, and Spodoptera exigua	[257]
Abiotic stress tolerance					
trnI/trnA	tps1	Prrn/ggagg/ psbA	>169-fold transcript	Drought tolerance: growth in 6 % polyethylene glycol and rehydration after 24 days of drought	[258]
trnI/trnA	merA/merB	Prrn/ggagg/ psbA	NR	Phytoremediation: high level tolerance to the organomercurial compounds, up to 400 μM phenylmercuric acetate	[259]
trnI/trnA	badh	Prrn/T7 g10/ rps16	93–101 $\mu M\ g^{-1}$FW	Salt tolerance: carrot plants survived up to 400 mM NaCl	[135]
trnI/trnA	*-TMT*	Prrn/T7g 10/ TpsbA	>7.7 % TSP	Increased salt and heavy metal tolerance, enhanced accumulation of α-tocopherol in seeds	[153]
trnI/trnA	mt1	Prrn/T7 g10/ Trps16	NR	Phytoremediation: resistant to mercury, up to 20 μm	[260]
trnV/rps12/7	b-bar1	Prrn/TrbcL	>7 % TSP	Resistance to the herbicide phosphinothricin	[261]

trnV/rps7/12	EPSPS	Prrn/Trps16	>10 % TSP	Resistance to the herbicide glyphosate	[262]
rbcL/accD	EPSPS/aroA	Prrn/ggagg/psbA	NR	Resistance to glyphosate (>5 mM)	[129]
rbcL/accD	mALS	PpsbA/TpsbA	NR	Tolerant to pyrimidinylcarboxylate, imidazolinon, and sulfonylurea/pyrimidinylcarboxylate herbicides	[263]
rbcL/accD	Bar	Prrn/rbcL/psbA	NR	Herbicide resistance: up to 25 μg ml^{-1}glufosinate	[264]
rbcL/rbcL	Hppd	psbA/psbA/3′rbcL	5 % TSP	Resistance to herbicide	[265]
rbcL/accD	panD	Prrn/rbcL 3′	>4-fold β-alanine	Tolerance to high-temperature stress	[266]
trnfM/trnG	lycopene β-cyclase	atpI/rps16	0.28 mg g^{-1}DW	Herbicide resistance and triggers conversion of lycopene	[133]
prs14/trnG	HTP, TCY, TMT	Prrn/T7 g10/TrbcL	NR	Increase in vitamin E in fruit; cold-stress tolerance	[267]
Other agronomic traits					
trnI/trnA	phaA	Prrn/psbA/psbA	14.71β-ketothiolase mg^{-1} FW	Engineered cytoplasmic male sterility	[268]
trnI/trnA	RbcS	T7g10 or psbA	>150-fold RbcS transcript	Restoration of RuBisCO activity in rbcSmutants	[136]
rbcL/accD	TC, *-TMT*	PpsbA/Trsp16	3 nmol $h^{-1}mg^{-1}$ FW	Vitamin E accumulation in tobacco and lettuce	[269]
rbcL/accD	CrtZ, CrtW	Prrn/Trps16	NR	Accumulation of astaxanthin fatty acid esters in lettuce	[270]
trnV/orf708	BicA	psbA/psbA/psbA	~0.1 % TSP	CO_2 capture within leaf chloroplasts	[271]
trnV/3′rps12	Trx f, Trx m	prrn T7G10/rps12	NR	Starch synthesis/chloroplast redox regulation	[272]
trnfM/trnG	CV-N	Prrn/T7g10/TatpA	~0.3 % TSP	Increased mRNA stability and protein stability with the expression of CV-N in chloroplasts	[273]
trnI/trnA	Bgl-1	5′psbA/3′psbA	44.4 units Bgl1 g^{-1} FW	β-Glucosidase increased enzyme cocktail efficiently to release sugar from paper, citrus peel, and wood	[141]

trnI/trnA	ubiC	5′psbA/3′psbA	25 % DW	250-fold higher pHBA polymer accumulation than nuclear transgenic lines	[149]
trnI/trnA	man 1	5′psbA/3′psbA	25 units g^{-1}FW	Mannanase increased enzyme cocktail released sugar from paper, citrus peel, and wood	[274]
trnI/trnA	cutinase or swollenin	5′PsbA/3′PsbA	47.7 % reduction of MGDG and DGDG in cutinase and 68.5 % in swollenin	Swollenin enlarged and irreversibly unwound cotton fiber; cutinase showed esterase and lipase activity; used in enzyme cocktails	[275]
trnI/trnA	bgl1	5′psbA/3′psbA	14 units mg^{-1} FW	Enzyme cocktails produced glucose from filter paper, pine wood, or citrus peel	[150]
	swo1		NR		
	xyn2		421 units mg^{-1} FW		
	Acetyl sylan esterase		NR		
	celD		493 units mg^{-1} FW		
	celO		442 units mg^{-1} FW		
	Lipase		NR		
	Cutinase		15 units mg^{-1} FW		
trnI/trnA	PMK, MVK, MDD, AACT, HMGS, HMGRt; IPP, FPP, ADS, CYP71AV1, AACPR	Prrn/PpsbA	0.1 mg g^{-1}FW	Artemisinic acid for several isoprenoid products	[276]
trnI/trnA	Cel6A,Cel6B	Prrn/rbcL/rbcL	2–4 % TSP	Hydrolyzed crystalline cellulose	[277]
trnfM/trnG	bgl1C, cel6B, cel9A, xeg74	Prrn/T7g10/TrbcL	5– 40 % TSP	Cell wall-degrading enzyme activity	[278]
rbcL/accD	phbC, phbA, phbB	Prrn/rbcL 3′	0.16 % DW	Polyhydroxybutyrate (PHB) accumulation in leaves	[279]
rbcL/accD	crtZ, crtW	Prrn/Trps16	>0.5 % DW	Astaxanthin accumulation	[280]
trnV/rps7	EGPh	psbA/psbA/Trps16	25 % TSP	Chloroplast-derived β-1,4-endoglucanase (EGPh) was recovered from dry leaves and digested carboxymethyl cellulose (CMC) substrate	[281]

trnI/trn A	EX4	PpsbA/TpsbA	14.3 % TSP	CTB–EX4 showed increased insulin secretion similar to the commercial injectable EX4 in pancreatic β-cells and in mice fed with cells expressing EX4 in chloroplasts	[160]
trnI/trn A	MBP	PpsbA/TpsbA	2 % TSP	Amyloid loads were reduced in ex vivo studies in human Alzheimer's brain and in vivo in Alzheimer's mice fed with bio-encapsulated CTB–MBP. Abeta was also reduced in retinae and loss of retinal ganglion cells was prevented	[162]
trnI/trn A	FVIII	PpsbA/TpsbA	370 mg g^{-1}FW	Feeding of the HC/C2 antigen mixture substantially suppressed T-helper cell responses and inhibitor formation against FVIII in hemophilia A mice	[282]
trnI/trn A	HSA	PpsbA/TpsbA	26 % TSP	In vitro chaperone activity of Trx m and Trx f	[283]
trnI/trn A	EDA	PpsbA/TpsbA	2.0 % TSP	The vaccine adjuvant EDA from fibronectin retains its proinflammatory properties when expressed in tobacco chloroplasts	[284]
trnI/trn A	Proinsulin	PpsbA/TpsbA	47 % TSP in tobacco, 53 % TLP in lettuce	Oral delivery of proinsulin in plant cells lowered glucose levels comparably to injectable commercial insulin	[285]
trnI/trn A	HSA	psbA/psbA/psbA	~11 % TSP	First report of human blood protein in chloroplasts; function not evaluated	[286]
trnI/trn A	IGF	psbA/psbA/psbA	32.7 % TSP	Promoted growth of cultured HU-3 cells in a dose-dependent manner	[287]
trnI/trnA	FIX	PpsbA/TpsbA	1 mg g^{-1}DW (0.56 % TLP)	Oral delivery of CTB-FIX lettuce cells suppressed inhibitor formation against FIX in hemophilia B mice	[6]

trnI/trnA	FIX	Ppsba/TpsbA	3.8 % TSP; 0.4 mg g^{-1}FW	Tolerance induction via complex immune regulation, involving tolerogenic dendritic and T-cell subsets	[288]
trnI/trnA	GAA	Ppsba/TpsbA	5.7 mg g^{-1}DW	Reduced toxic antibody responses in enzyme replacement therapy in Pompe mice	[289]
trnI/trnA	ACE2 Ang-(1–7)	PpsbA/TpsbA	CTB–ACE2: 2.14 % TLP CTB-Ang1–7: 8.7 % TLP	Oral delivery of ACE2 and Ang (1–7) significantly improved cardiopulmonary structure and functions, decreased the elevated right ventricular systolic blood pressure and improved pulmonary blood flow in animals with induced pulmonary hypertension	[161]
trnI/trn A	BACE	Prrn/TpsbA	2.0 % TSP	Immunogenic response against the BACE antigen in mice	[290]
trnI/trn A	IFNα2b	Prrn/TpsbA	3 mg g^{-1} FW	Protected cells against VSV CPE and HIV; increased MHC I antibody on splenocytes and total number of natural killer cells and protected mice from a highly metastatic lung tumor	[291]
trnI/trn A	CTB-pins	Prrn/T7g10/TpsbA and rps16	16 % TSP in tobacco, 72 % TLP in lettuce	CTB-proinsulin-fed non-obese diabetic mice significantly decreased inflammation (insulitis); insulin-producing β cells in pancreatic islets were highly protected, increased in insulin production with lower blood or urine glucose levels; increased expression of immunosuppressive cytokines	[128, 292]
rbcL/accD	IFN-γ	PpsbA/TpsbA	6 % TSP	Protection of human lung carcinoma cells against infection by encephalomyocarditis virus	[293]
rbcL/accD	hTrx	PpsbA/Trps16	1 % TSP	Protected mouse from hydrogen peroxide	[294]

rbcL/accD	A1AT	PpsbA/TrbcL	2 % TSP	Binds to porcine pancreatic elastase	[295]
rbcL/accD	TGFβ3	Prrn/T7g10/psbC	12 % TLP	Inhibits mink lung epithelial cell proliferation	[296]
trnV/3′rps12	hCT-1	Prrn/G10L/Trps16	5 % TSP	Biologically active on human hepatocarcinoma cell line	[297]
trnV/rps7/12	hST	PpsbA or Prrn/G10L/Trps16	0.2–7.0 % TSP	Promotes growth of Nb2 cells in a dose-dependent manner	[298]
trnfM/trnG	pal, cpl-1	Prrn/T7g10/TpsbA	~30 % TSP	Bacteriolytic activity and kills Streptococcus pneumoniae, the causative agent of pneumonia	[299]
trnI/trn A	ESAT-6	5′psbA/3′psbA	~7.5 % TSP	Hemolysis of red blood cells and GM1 binding	[165]
trnI/trn A	AMA1	5′psbA/3′psbA	7.3 % TSP in tobacco, 13.2 % TSP in lettuce	Long-term immunity against cholera challenge; inhibition of malarial parasite; protection correlated with IgA and IgG1	[164]
trnI/trn A	MSP1	5′psbA/3′psbA	10.1 % TSP in tobacco, 6.1 % TSP in lettuce		
trnI/trn A	2 L21	5′psbA/3′psbA	6.0 % TSP		
trnI/trn A	Pag	5′psb/3′psbA	~29.6 % TSP	Macrophage lysis assay, systemic immune response, toxin neutralization assay, mice survived (100 %) challenge with lethal doses of anthrax toxin	[300, 301]
trnI/trn A	L1	PpsbA/TpsbA	20–26 % TSP	Induced systemic immune response and produced neutralizing antibodies in mice	[302]
trnI/trnA	RA4	PpsbA/T psbA	0.2 % TLP	Oral administration elicited both mucosal and systemic Th1/Th2 responses to reduce Toxoplasma-parasite load	[303]
trnI/trnA	rFaeG	PpsbA/TrbcL	>1 % DW	Transplastomic plants expressing the rFaeG protein could possibly be used for delivery of an oral vaccine against porcine F4+ ETEC infections	[304]

trnI/trn A	F1-V	Prrn/TpsbA	14.8 % TSP	Orally immunized mice heavily challenged with plague (Yersinia pestis) were protected better than those given IP injections	[305]
trnI/trn A	CTB-2 L21	PpsbA/TpsbA	31.1 % TSP	Immunogenic in mice following IP or oral administration	[306]
trnI/trnA	VP8*	psbA/psbA/ Trps16	600 μg g^{-1}FW	Induced strong immune response and virus neutralization	[307]
trnI/trn A	CtxB	Prrn/ggagg/ TpsbA	4.1 % TSP	Efficient GM1 ganglioside-binding	[308]
trnI/trn A	LTB	Prrn/ggagg/ TpsbA	2.5 % TSP	GM1 ganglioside-binding assay	[309]
trnI/trn A	LecA	Prrn/T7g10/ TpsbA	7 % TSP	Systemic immune response in mice	[310]
trnI/trn A	BACE	Prrn/TpsbA	2.0 % TSP	Immunogenic response against the BACE antigen in mice	[290]
rbcL/accD	OspA, OspA-T	PpsbA/TpsbA	1–10 % TSP	Systemic immune response and protection against Borrelia burgdorferi (Lyme disease)	[311]
trnN/trn R	LTB	Prrn/T7g10/ TrbcL	2.3 % TSP	GM1 ganglioside-binding assay; oral immunization partially protected mice from cholera toxin challenge	[312]
trnN/trnR	DPT	Prrn/T7g10/ TrbcL	0.8 % TSP	Immunogenic in orally inoculated mice with freeze-dried chloroplast-derived multi-epitope DPT protein	[313]
trnN/trnR	C4V3	Prrn/T7g10/ TrbcL	~15 μg mg^{-1}DW	Plant-derived C4V3 has elicited both systemic and mucosal antibody responses in mice, as well as CD4+ T cell proliferation responses	[314]
trnN/trnR	L1	Prrn/TrbcL	>2 % of TSP	Proper folding and display of conformational epitopes for L1 in the fusion protein by antigen capture ELISA	[315]
trnfM/trnG	p24	Prrn/T7g10/ TrbcL	~4 % TSP	Induced strong CD4+ and CD8+ T-cell responses in mice	[316]

trnGtrnfM	HEV E2	Prrn/psbA/ TpsbA	1.09 ng μg⁻¹TSP	Immune response in mice against hepatitis E virus	[317]
trnH/trnK	CSFV E2	Prrn/TpsbA	1–2 % TSP	Immune response in mice against swine fever	[318]
rrn16/ rps12/7	TetC	Prrn/T7 g10/ TrbcL atpB/TrbcL	10–25 % TSP	Mice developed systemic immune response and survived the tetanus toxin challenge	[319]
rrn16/trnI	E7	PpsbA/Trps	3–8 % TSP	Several therapeutic HPV-specific E7-based vaccine formulations have been tested in animal models and some have advanced into clinical trials	[320]

Abbreviations: *Ang (1–7)* Angiotensin (1–7), *BACE* human b-site APP cleaving enzyme, *Bgl* β-glucosidase, *CPE* carbapenemase-producing Enterobacteriaceae, *CTB* cholera toxin B subunit, *DGDG* digalactosyldiacylglycerol, *DPT* diphteria, pertussis, tetanus, *DW* dry weight, *EDA* extra domain A-fibronectin, *ELISA* enzyme-linked immunosorbent assay, *ETEC* enterotoxigenic *Escherichia coli*, *EX4* exendin-4, *FVIII* coagulation factor VIII, *FW* fresh weight, *HPV* human papilloma virus, *IP* intraperitoneal, *MBP* myelin basic protein, *MGDG* monogalactosyldiacylglycerol, *NR* not recorded, *RbcS* small subunit of RuBisCO, *RuBisCO*ribulose-1,5-bisphosphate carboxylase/oxygenase, *TLP* total leaf protein, *TSP* total soluble protein, *VSV* vesicular stomatitis virus

IMPACT OF SEQUENCE DIVERSITY IN THE CHLOROPLAST GENOME ON TRANSGENE INTEGRATION

Figure 3a shows examples of transplastomic genomes that have been transformed with either an endogenous or a heterologous flanking sequence. Every single nucleotide change in the heterologous sequence was subsequently edited out and corrected to achieve 100 % homology to the native sequence within the intergenic spacer region (Fig. 3a). The repetitive editing process significantly reduces the efficiency of transgene integration when using heterologous flanking sequences. This challenge is made even more difficult by inadequate conservation of intergenic spacer regions, even within the same family. Figure 3cshows comparisons of 21 of the most variable intergenic spacer regions; only four of the >150 spacer regions, including the trnl/trnA spacer region, are conserved among members of the Solanaceae. Among grass chloroplast genomes, not a single intergenic spacer region is conserved [8]. This necessitates construction of species-specific chloroplast vectors using endogenous flanking sequences and underscores the need to sequence the chloroplast genomes of economically important crop species.

IDEAL SITES IN THE CHLOROPLAST GENOME FOR TRANSGENE INTEGRATION

The selection of a suitable intergenic spacer region from among more than 100 sites found in each chloroplast genome is a major concern. Statements on the lack of positional effects in the transplastomic literature are common and are used to contrast chloroplast genetic engineering with nuclear transgene integration, which is often associated with profound differences in the expression of transgenes dependent on their site of integration. Evidence shows, however, that there are also positional effects within the chloroplast genome (Table 2). IR regions are found in duplicate in most chloroplast genomes; therefore, transgenes should be inserted within the IR region instead of the SSC or LSC regions because this should double the copy number of transgenes. Integration of a transgene cassette into one copy of the IR facilitates integration into the other copy, thereby enhancing selection pressure to achieve homoplasmy through this copy correction mechanism, a characteristic feature of the chloroplast genome [126, 127, 128]. Therefore, the site of integration plays a crucial role in transgene expression level and in enhancing homoplasmy under selection by antibiotics. Most importantly, in all sequenced chloroplast genomes within a single plant species, the DNA sequence in one copy of the IR is identical to that in the other copy, without any exception (Table 1).

An early controversy in the chloroplast genetic engineering field was the suitability of transcriptionally silent spacer regions, where native genes (for example, *rbcL*/*accD*) are located on opposite strands of the chloroplast genome, or transcriptionally active spacer regions, where native genes (for example, *trnA*/*trnI*) are located within operons on the same strand. After a herbicide resistance gene was introduced into the transcriptionally active spacer region for the first time [129], most subsequent studies preferentially used this site of integration (Table 2). The integration of transgenes into the transcriptionally active spacer region (trnl/trnA) has led to 25-fold higher expression of transgenes compared with the transcriptionally silent spacer region (rbcl/accD) [130], possibly due to the presence of multiple promoters (heterologous and endogenous) that enhance transcription. Introns present within *trnI/trnA* genes (used as flanking sequences) also provide efficient processing of native or foreign transcripts. The *trnA* gene intron includes a chloroplast origin of replication and produces more copies of the template (chloroplast vectors) for integration of the transgene cassette [131]. In fact, among 114 transgenes in different plant species in Table 2, 71 are integrated at the *trnA*/*trnI* site of the chloroplast genome, confirming the unique advantages of this site [127, 129, 130].

ROLE OF CHLOROPLAST GENOME REGULATORY SEQUENCES IN TRANSGENE EXPRESSION

In addition to the site of integration, regulatory sequences located upstream (promoter, 5′ UTR) and downstream (3′ UTR) of transgenes play a major role in determining their expression level. The *psbA*regulatory region, first used almost 25 years ago [131], still appears to be the best option for use in an expression cassette, as the *psbA* gene encodes the most highly translated protein in the chloroplast [132] and it can also mediate light-induced activation of translation [128]. Indeed, almost all highly expressed transgenes (>70 % TSP, >25 % dry weight) utilize the *psbA* regulatory region; among 114 transgenes expressed via the chloroplast genome, 84 use the *psbA* regulatory sequence (Table 2). Other endogenous regulatory sequences that are used include *rbcL* and *atpA*, which result in lower transgene expression levels than the *psbA* promoter/5′ UTR.

Using regulatory regions from photosynthetic genes has the advantage of light regulation, making them ideal for transgene expression in photosynthetic organs (leaves; Fig. 3d, e). However, when the lettuce *psbA*regulatory region was used in tobacco chloroplasts or vice versa, transgene expression was dramatically reduced (Fig. 3d, e) [128]. Nucleotide differences within the *psbA* 5′ UTR between tobacco and lettuce (*Lactuca sativa*) resulted in changes that decreased the interaction of RNA-binding proteins and produced variation in the size of the stem, bulge, and terminal loop of the UTR [128]. In addition, most regulatory proteins (including sigma factors that bind to the promoter region) are nuclear encoded and transported to chloroplasts. This underscores a caveat associated with using regulatory sequences for transgene expression: the need to make species-specific chloroplast vectors to accommodate highly specific regulatory region-binding proteins.

Heterologous regulatory sequences are necessary for transgene expression that is independent of cellular control, especially in non-photosynthetic organs such as fruits and edible roots, where chloroplast protein synthesis is poor [133]. A heterologous UTR (T7 *gene10*) was first evaluated for expression in leaves [127, 134] and was subsequently tested in non-green tissues. When the expression of *BETAINE ALDEHYDE DEHYDROGENASE* (*BADH*) was regulated by the T7 *gene10* UTR in carrot (*Daucus carota*) plants, 75 % of the expression level in leaves was observed in non-green edible roots, conferring the highest level of salt tolerance (400 mM NaCl) found in the published literature (Fig. 4i, j) [135]. Although T7 *gene10* has been successfully used to engineer salt tolerance in non-green tissues, its expression level is not as high as that of the *psbA* regulatory sequence in leaves [136]. The only other heterologous UTR that expressed transgenes at high levels is that from the *Bacillus thuringiensis* (*Bt*) operon [137]. Use of this operon produced the highest level of insecticidal toxin protein (52 % TLP) ever reported in the published literature [137]. These high levels of toxin accumulation in chloroplasts could result from the combination of high-level expression and protein stability; the Bt protein formed cuboidal crystals within chloroplasts (Fig. 4e) due to co-expression of a chaperone that facilitates folding. When fed, transplastomic leaves, cotton bollworm (*Helicoverpa* sp.) were killed with a single bite of leaf and insects that had 40,000-fold increased resistance to Bt were also killed (Fig. 4f, g). Nevertheless, expression of this transgene in tomato fruit is very poor [133, 138, 139] and further research is needed to enhance transgene expression in fruits.

Figure 4: Engineering the chloroplast genome to confer biotic/abiotic stress tolerance or expression of high-value products. **a–d** Industrial production of blood clotting factor IX (FIX) bioencapsulated in lettuce plants in a hydroponic cGMP facility. **a** Biomass production of FIX-expressing plants. **b–d** Steps in capsule preparation. After harvesting and lyophilization of fresh leaves, freeze-dried FIX-accumulating leaves were powdered and prepared as capsules [6]. **e–g** Overexpression of the Bt cry2Aa2 operon in chloroplasts leads to the formation of the Bt insecticidal crystal protein. In bioassays with the *Helicoverpa zea*, **f** eating the transplastomic leaf kills the caterpillar, while **g** the control leaf is consumed by the growing caterpillar [137]. **h** Ultrastructure of the chloroplast envelope membrane of transplastomic γ-tocopherol methyltransferase (γ-TMT) tobacco plants shows the formation of multiple layers of inner envelope membranes as the result of γ-TMT overexpression [153]. **i, j** Expression of *BETAINE ALDEHYDE DEHYDROGENASE* (*BADH*) in carrot plants. **i** Transgenic carrot plants thrived in soil irrigated with 400 mM sodium chloride, whereas untransformed carrot plants showed retarded growth in the presence of salt. **j** Carrot roots from transplastomic plants [135]. **k** Phenotypes of tomato fruits from transplastomic tomato plants expressing lycopene β-cyclase transgenes compared with wild-type plants. Fruits were harvested at different ripening stages. Orange color of ripe fruits indicates efficient conversion of red lycopene into orange β-carotene (provitamin A) [154]

ENGINEERING THE CHLOROPLAST GENOMES FOR BIOTECHNOLOGY APPLICATIONS

Conferring stress tolerance

In the past decade, chloroplast genetic engineering has focused primarily on the overexpression of target genes with the potential to enhance biotic stress tolerance, which is very important for plant protection and yield enhancement. Yield loss due to insect pests can be very serious in many countries. In addition to cotton bollworm resistance conferred by hyper-expression of Bt protein in chloroplasts [137], there are many other striking recent examples of improved biotic stress tolerance. Retrocyclin-101 and Protegrin-1 protect against *Erwinia* soft rot and tobacco mosaic virus (TMV), which result in yield loss in several cultivated crops [140]. Whitefly and aphid resistance has been accomplished by expressing β-glucosidase [141], which releases insecticidal sugar esters from hormone conjugates. Multiple resistances against aphids, whiteflies, lepidopteran insects, and bacterial and viral pathogens were achieved by expressing the *Pinellia ternata agglutinin* (*PTA*) gene in the chloroplast genome [142]. More than 40 transgenes have been stably integrated into and expressed within the chloroplast genome, conferring important agronomic traits, including insect resistance in edible crops cabbage (*Brassica oleracea*) [143], soybean [144, 145], and eggplant (*Solanum melongena*) [146].

More recently, scientists have begun to explore new strategies to downregulate specific target genes. One such approach is to express double-stranded RNAs (dsRNAs) within the chloroplast genome and to use RNA interference (RNAi) to confer the desired agronomic traits, mainly resistance to insects that cause severe yield loss. This strategy has been demonstrated by expressing long or short dsRNAs that activate RNAi and disrupt target genes in insects, providing efficient protection against insects without the need for chemical pesticides. One such example is the suppression of three essential proteins required for insect survival—lepidopteran chitin synthase (Chi), cytochrome P450 monooxygenase (P450), and V-ATPase—using dsRNAs in the tobacco chloroplast system [147]. Each dsRNA was expressed independently in chloroplasts and leaves were fed to insects. The transcription level of target genes in *Helicoverpa* insects decreased to almost undetectable levels in the midgut, resulting in a significant reduction in the net weight of larvae and in pupation rate [147]. Transplastomic potato plants producing *β-actin*-targeting long dsRNA were lethal to Colorado potato beetle (*Leptinotarsa decemlineata*) larvae, providing yet another crop protection mechanism [148].

Synthesis of enzymes and biomaterials

In addition to improved resistance against both biotic and abiotic stress, the chloroplast genome has been engineered to produce useful enzymes, biomaterials, and biofuels, or even to enhance biomass. The first report of metabolic engineering using chloroplast genomes produced the highest level of the poly(p-hydroxybenzoic acid (pHBA) polymer (25 % dry weight) in normal healthy plants despite the diversion of a major metabolic intermediate [149]. The first use of plant-derived enzyme cocktails for the production of fermentable sugars from lignocellulosic biomass was accomplished recently [150]. Unlike the single biofuel enzymes previously expressed in chloroplasts, nine different genes from bacteria or fungi were expressed in *E. coli* or tobacco chloroplasts using a new technique that enabled the insertion of fungal genes with several introns, eliminating the need to prepare cDNA libraries. Industrial fermentation systems are currently limited by high cost and low production capacity; chloroplast-derived enzyme cocktails offer several striking advantages, including significantly reduced cost, improved stability of chloroplast-derived enzymes, and no need for enzyme purification. Interestingly, expression of β-glucosidase released hormones from conjugates, resulting in elevated phytohormone levels and increased biomass [141], an unexpected outcome of enzyme expression.

Enhancing nutrition

Seed oils, such as those from soybean, rapeseed (*Brassica napus*), and maize, are the major dietary source of vitamin E. They have very low α-tocopherol content but relatively high levels of γ-tocopherol. Only a few seed oils, such as sunflower (*Helianthus annuus*) seed oil, contain high levels of α-tocopherol, an important precursor of vitamin E [151]. γ-Tocopherol is the biosynthetic precursor of α-tocopherol, suggesting that the α-tocopherol biosynthetic pathway catalyzed by γ-tocopherol methyl transferase (γ-TMT) is the rate-limiting step [152]. Engineering of the *γ-tmt* gene into the chloroplast genome resulted in the formation of multiple layers of the inner chloroplast envelope (Fig. 4h) due to γ-TMT overexpression, with around tenfold higher conversion of γ-tocopherol to α-tocopherol in seeds [153]. Likewise, introducing *lycopene β-cyclase* genes into the tomato plastid genome increased the conversion of lycopene into provitamin A (β-carotene), with obvious phenotypic changes (Fig. 4k) [154].

Biopharmaceuticals

At present, protein drugs are extremely expensive; for example, >90 % of the global population cannot afford insulin, a drug needed to treat the global diabetes epidemic. The high cost of protein drugs is due to their production in prohibitively expensive fermentation systems (which cost more than $450–700 million to build depending on their capacity [155, 156]), prohibitively expensive purification from host proteins, the need for refrigerated storage and transport, and the short shelf-life of the final product. Protein drugs made by plant chloroplasts overcome most of these challenges because they do not require expensive fermentation systems and are produced in federal drug administration (FDA)-approved hydroponic greenhouses (Fig. 4a) [157]. Lettuce leaves expressing protein drugs are lyophilized and stored indefinitely at ambient temperature without losing their efficacy (Fig. 4b–d) [6]. The plant cell wall protects protein drugs from acids and enzymes in the stomach because human enzymes do not digest plant cell wall glycans. Human gut microbes, however, have evolved to break down every glycosidic bond in the plant cell wall and therefore release the protein drug into the gut lumen, directing its delivery to the blood or immune system [158, 159].

Oral delivery of several human therapeutic proteins expressed in chloroplasts is highly efficacious in the treatment of several human diseases, including diabetes, cardiovascular disease, pulmonary hypertension, and Alzheimer's disease. Most proteins were expressed in tobacco chloroplasts for initial evaluation and were subsequently expressed in lettuce chloroplasts for advancing them to the clinic. Oral delivery of exendin-4, which modulates the secretion of insulin in a glucose-dependent manner, lowered glucose in diabetic animals by stimulating the production of insulin in a manner similar to that of the injectable drug [160]. Oral delivery of angiotensin-converting enzyme 2 (ACE2) and angiotensin (Ang) (1–7) significantly improved cardiopulmonary structure and function, decreased elevated right ventricular systolic blood pressure, and improved pulmonary blood flow in animals with induced pulmonary hypertension [161]. Oral delivery of plant cells expressing ACE2 and Ang (1–7) also reduced endotoxin-induced uveitis (EIU) and dramatically decreased cellular infiltration and retinal vasculitis, as well as damage and folding in experimental autoimmune uveoretinitis [158]. It is also possible to orally deliver protein drugs across the blood–brain barrier to the Alzheimer's brain to remove plaques [162].

The first industrial-scale production of human blood clotting factor in a cGMP facility was reported recently [6] (Fig. 4a–d). In a 1000 ft^2 hydroponic cGMP facility, it is possible to produce up to 30,000 doses for a 20-kg pediatric patient. Clotting factor made in lettuce was stable for up to 2 years when lyophilized cells were stored at ambient temperature, completely eliminating the need for the cold chain. This enables the first commercial development of an oral drug and addresses the extremely expensive purification, cold storage and transportation, and short shelf-life of current protein drugs. Oral delivery of a broad dose range was effective in the prevention of antibody formation after injection of clotting factor IX (FIX), further facilitating human clinical studies.

Vaccines against infectious diseases

The current iteration of vaccines, using attenuated bacteria or viruses, offer protection against major infectious diseases but they also present major challenges. For example, the oral polio vaccine that is used around the globe has caused severe polio resulting from mutations and recombination with other viruses [163]. In addition, all current vaccines require cold storage and transportation, making distribution in developing countries a major challenge. Many of these challenges can be overcome by using chloroplasts.

One successful chloroplast-derived vaccine conferred dual immunity against cholera and malaria in animal studies [164]. Cholera is a major disease causing high mortality, with the only licensed vaccine being not only expensive but also limited in its duration of protection. No vaccine is currently available for malaria. The cholera toxin-B subunit (CTB) of *Vibrio cholerae* was fused to the malarial vaccine antigen apical membrane antigen-1 (AMA1) and merozoite surface protein-1 (MSP1) and expressed in lettuce or tobacco chloroplasts. While no suitable models exist to test human malaria, a cholera toxin challenge using mice immunized with chloroplast-expressed CTB was highly effective and provided the longest duration of protection in the published literature [164]. These early results show that chloroplasts are ideal for producing low-cost booster vaccines against several infectious diseases [165] for which the global population has been primed previously (Table 2), but lack of an oral priming strategy is still a major limitation in this field.

MOVING FORWARD

It is amazing that the chloroplast genome can express >120 foreign genes from different organisms, including bacteria, viruses, fungi, animals, and humans. The insertion of commercially useful traits, including herbicide and insect resistance, into soybean resulted in high-level expression and superior transgene containment, with no antibiotic selectable markers; but even so, these lines were not developed commercially. Nevertheless, recurring concerns about insect resistance against biopesticides have resulted in new USDA requirements on planting *Bt* corn [122], which may eventually require utilization of the transplastomic approach to confer agronomic traits. The nuclear transgenic approach is inadequate to develop products when higher-level transgene expression is a requirement. Thus, chloroplast transformation has a unique advantage in advancing the field of molecular farming for the production of vaccines, biopharmaceuticals, or other bio-products.

Although products with high-level expression have now advanced to the clinic or are in commercial development, a better understanding of chloroplast translation is required to improve several other gene products. The availability of chloroplast genome sequences should help in the development of codon optimization programs using highly expressed chloroplast genes, but among the ~3000 cultivated crops, sequenced chloroplast genomes are available for crops from fewer than 70 genera. Major funding agencies have not supported crop chloroplast genome sequence projects because of the misconception that all chloroplast genomes are similar, as evidenced by the publication of fewer than ten crop chloroplast genome sequences between 1986 and 2004. This review illustrates the importance of sequencing more crop chloroplast genomes for various biotechnology applications. Furthermore, new selectable markers are needed to transform the chloroplast genomes of cereals, which has been elusive for the past two decades.

Chloroplast genome sequences will be valuable assets in herbal medicine. Most medicinal plants are rare species and very little information is available to confirm their identity. DNA barcodes derived from chloroplast genomes will be useful for identifying varieties and resources; this concept is also valuable in the identification of the origin of cultivated crops and their close relatives to enhance breeding or transfer of useful traits. Molecular techniques to sequence the genomes of single chloroplasts could help to eliminate chloroplast-like sequences that are present in the mitochondrial or nuclear genome. The ability to sequence chloroplast genomes using minimal leaf materials could help us to understand variations in different segments of a variegated leaf in horticultural crops. Further, determining complete chloroplast genome sequences from fossils or recently extinct plants could shed more light on chloroplast genome evolution; help us to understand these species' inadequate fitness to cope with environmental changes; and help us to build new phylogenetic trees. The technology for isolating DNA from fossils is already available [166, 167, 168]. All of these goals can be accomplished with less expensive and more accurate genome sequences, utilizing longer read sequencing technology and new bioinformatics tools.

DECLARATIONS

Acknowledgements

We acknowledge valuable contributions from editors of *Genome Biology* (Drs Dominique Morneau and Ripudaman Bains) to enhance the flow and presentation to non-specialists and reviewers for their critical comments.

Funding

Research from the Daniell Laboratory included in this review was supported by the Bill and Melinda Gates Foundation (OPP1031406), the National Institutes of Health (NIH; R01 HL107904, HL109442, GM

63879, EY 024564), Novo Nordisk, Bayer, and Department of Energy ARPA-E grants to Henry Daniell.

Authors' contributions

HD and CSL wrote this review. MY assembled Table 2 and Figs. 3 and 4 with guidance from HD. WJC assembled Table 1 and Figs. 1 and 2 with guidance from CSL. All authors read and approved the final manuscript.

REFERENCES

1. Bobik K, Burch-Smith TM. Chloroplast signaling within, between and beyond cells. Front Plant Sci. 2015;6:781.
2. Daniell H, Chan H-T, Pasoreck EK. Vaccination through chloroplast genetics: affordable protein drugs for the prevention and treatment of inherited or infectious human diseases. Annu Rev Genet. 2016. In Press.
3. Shinozaki K, Ohme M, Tanaka M, Wakasugi T, Hayashida N, Matsubayashi T, et al. The complete nucleotide sequence of the tobacco chloroplast genome: its gene organization and expression. EMBO J. 1986;5:2043–9.
4. Wambugu P, Brozynska M, Furtado A, Waters D, Henry R. Relationships of wild and domesticated rices (*Oryza* AA genome species) based upon whole chloroplast genome sequences. Sci Rep. 2015;5:13957.
5. Brozynska M, Furtado A, Henry RJ. Genomics of crop wild relatives: expanding the gene pool for crop improvement. Plant Biotechnol J. 2016;14:1070–85.
6. Su J, Zhu L, Sherman A, Wang X, Lin S, Kamesh A, et al. Low cost industrial production of coagulation factor IX bioencapsulated in lettuce cells for oral tolerance induction in hemophilia B. Biomaterials. 2015;70:84–93.
7. Jansen RK, Wojciechowski MF, Sanniyasi E, Lee SB, Daniell H. Complete plastid genome sequence of the chickpea (*Cicer arietinum*) and the phylogenetic distribution of *rps12* and *clpP* intron losses among legumes (Leguminosae). Mol Phylogenet Evol. 2008;48:1204–17.

8. Saski C, Lee SB, Fjellheim S, Guda C, Jansen RK, Luo H, et al. Complete chloroplast genome sequences of *Hordeum vulgare*, *Sorghum bicolor* and *Agrostis stolonifera*, and comparative analyses with other grass genomes. Theor Appl Genet. 2007;115:571–90.
9. Ruhlman T, Lee SB, Jansen RK, Hostetler JB, Tallon LJ, Town CD, Daniell H. Complete plastid genome sequence of *Daucus carota*: implications for biotechnology and phylogeny of angiosperms. BMC Genomics. 2006;7:222.
10. Samson N, Bausher MG, Lee SB, Jansen RK, Daniell H. The complete nucleotide sequence of the coffee (*Coffea arabica*L.) chloroplast genome: organization and implications for biotechnology and phylogenetic relationships amongst angiosperms. Plant Biotechnol J. 2007;5:339–53.
11. Lee SB, Kaittanis C, Jansen RK, Hostetler JB, Tallon LJ, Town CD, Daniell H. The complete chloroplast genome sequence of *Gossypium hirsutum*: organization and phylogenetic relationships to other angiosperms. BMC Genomics. 2006;7:61.
12. Bausher MG, Singh ND, Lee SB, Jansen RK, Daniell H. The complete chloroplast genome sequence of *Citrus sinensis* (L.) Osbeck var 'Ridge Pineapple': organization and phylogenetic relationships to other angiosperms. BMC Plant Biol. 2006;6:21.
13. Daniell H, Lee SB, Grevich J, Saski C, Quesada-Vargas T, Guda C, et al. Complete chloroplast genome sequences of *Solanum bulbocastanum*, *Solanum lycopersicum* and comparative analyses with other Solanaceae genomes. Theor Appl Genet. 2006;112:1503–18.
14. Jansen RK, Saski C, Lee SB, Hansen AK, Daniell H. Complete plastid genome sequences of three rosids (*Castanea*, *Prunus*, *Theobroma*): evidence for at least two independent transfers of *rpl22* to the nucleus. Mol Biol Evol. 2011;28:835–47.
15. Wolfe KH, Morden CW, Palmer JD. Function and evolution of a minimal plastid genome from a nonphotosynthetic parasitic plant. Proc Natl Acad Sci U S A. 1992;89:10648–52.
16. Bortiri E, Coleman-Derr D, Lazo GR, Anderson OD, Gu YQ. The complete chloroplast genome sequence of *Brachypodium distachyon*: sequence comparison and phylogenetic analysis of

eight grass plastomes. BMC Res Notes. 2008;1:61.

17. Sato S, Nakamura Y, Kaneko T, Asamizu E, Tabata S. Complete structure of the chloroplast genome of *Arabidopsis thaliana*. DNA Res. 1999;6:283–90.
18. Saski C, Lee SB, Daniell H, Wood TC, Tomkins J, Kim HG, Jansen RK. Complete chloroplast genome sequence of *Glycine max* and comparative analyses with other legume genomes. Plant Mol Biol. 2005;59:309–22.
19. Jansen RK, Kaittanis C, Saski C, Lee SB, Tomkins J, Alverson AJ, Daniell H. Phylogenetic analyses of *Vitis* (Vitaceae) based on complete chloroplast genome sequences: effects of taxon sampling and phylogenetic methods on resolving relationships among rosids. BMC Evol Biol. 2006;6:32.
20. Daniell H, Wurdack KJ, Kanagaraj A, Lee SB, Saski C, Jansen RK. The complete nucleotide sequence of the cassava (*Manihot esculenta*) chloroplast genome and the evolution of *atpF* in Malpighiales: RNA editing and multiple losses of a group II intron. Theor Appl Genet. 2008;116:723–37.
21. Wu J, Liu B, Cheng F, Ramchiary N, Choi SR, Lim YP, Wang X-W. Sequencing of chloroplast genome using whole cellular DNA and Solexa sequencing technology. Front Plant Sci. 2012;3:234.
22. Grivet D, Heinze B, Vendramin G, Petit R. Genome walking with consensus primers: application to the large single copy region of chloroplast DNA. Mol Ecol Notes. 2001;1:345–9.
23. Goremykin V, Hirsch-Ernst K, Wölfl S, Hellwig F. The chloroplast genome of the ‹basal› angiosperm *Calycanthus fertilis*–structural and phylogenetic analyses. Plant Syst Evol. 2003;242:119–35.
24. Goremykin VV, Hirsch-Ernst KI, Wolfl S, Hellwig FH. The chloroplast genome of *Nymphaea alba*: whole-genome analyses and the problem of identifying the most basal angiosperm. Mol Biol Evol. 2004;21:1445–54.
25. Goremykin VV, Hirsch-Ernst KI, Wölfl S, Hellwig FH. Analysis of the *Amborella trichopoda* chloroplast genome sequence suggests that *Amborella* is not a basal angiosperm. Mol Biol Evol. 2003;20:1499–505.
26. Dhingra A, Folta KM. ASAP: amplification, sequencing &

annotation of plastomes. BMC Genomics. 2005;6:176.

27. Heinze B. A database of PCR primers for the chloroplast genomes of higher plants. Plant Methods. 2007;3:4.
28. Wu FH, Kan DP, Lee SB, Daniell H, Lee YW, Lin CC, et al. Complete nucleotide sequence of *Dendrocalamus latiflorus* and *Bambusa oldhamii* chloroplast genomes. Tree Physiol. 2009;29:847–56.
29. Leseberg CH, Duvall MR. The complete chloroplast genome of *Coix lacryma-jobi* and a comparative molecular evolutionary analysis of plastomes in cereals. J Mol Evol. 2009;69:311–8.
30. Dong W, Xu C, Cheng T, Lin K, Zhou S. Sequencing angiosperm plastid genomes made easy: a complete set of universal primers and a case study on the phylogeny of saxifragales. Genome Biol Evol. 2013;5:989–97.
31. Mardanov AV, Ravin NV, Kuznetsov BB, Samigullin TH, Antonov AS, Kolganova TV, Skyabin KG. Complete sequence of the duckweed (*Lemna minor*) chloroplast genome: structural organization and phylogenetic relationships to other angiosperms. J Mol Evol. 2008;66:555–64.
32. Lin CS, Chen JJ, Huang YT, Chan MT, Daniell H, Chang WJ, et al. The location and translocation of *ndh* genes of chloroplast origin in the Orchidaceae family. Sci Rep. 2015;5:9040.
33. Moore MJ, Dhingra A, Soltis PS, Shaw R, Farmerie WG, Folta KM, Soltis DE. Rapid and accurate pyrosequencing of angiosperm plastid genomes. BMC Plant Biol. 2006;6:17.
34. Wang W, Messing J. High-throughput sequencing of three Lemnoideae (duckweeds) chloroplast genomes from total DNA. PLoS One. 2011;6, e24670.
35. Cronn R, Liston A, Parks M, Gernandt DS, Shen R, Mockler T. Multiplex sequencing of plant chloroplast genomes using Solexa sequencing-by-synthesis technology. Nucleic Acids Res. 2008;36, e122.
36. Pan IC, Liao DC, Wu FH, Daniell H, Singh ND, Chang C, et al. Complete chloroplast genome sequence of an orchid model plant candidate: *Erycina pusilla* apply in tropical oncidium breeding. PLoS One. 2012;7, e34738.

37. Atherton RA, McComish BJ, Shepherd LD, Berry LA, Albert NW, Lockhart PJ. Whole genome sequencing of enriched chloroplast DNA using the Illumina GAII platform. Plant Methods. 2010;6:22.

38. Jackman SD, Warren RL, Gibb EA, Vandervalk BP, Mohamadi H, Chu J, et al. Organellar genomes of white spruce (*Picea glauca*): assembly and annotation. Genome Biol Evol. 2016;8:29–41.

39. Ferrarini M, Moretto M, Ward JA, Šurbanovski N, Stevanović V, Giongo L, et al. An evaluation of the PacBio RS platform for sequencing and *de novo* assembly of a chloroplast genome. BMC Genomics. 2013;14:670.

40. Wu Z, Gui S, Quan Z, Pan L, Wang S, Ke W, et al. A precise chloroplast genome of *Nelumbo nucifera* (Nelumbonaceae) evaluated with Sanger, Illumina MiSeq, and PacBio RS II sequencing platforms: insight into the plastid evolution of basal eudicots. BMC Plant Biol. 2014;14:289.

41. Li Q, Li Y, Song J, Xu H, Xu J, Zhu Y, et al. High-accuracy de novo assembly and SNP detection of chloroplast genomes using a SMRT circular consensus sequencing strategy. New Phytol. 2014;204:1041–9.

42. Redwan R, Saidin A, Kumar S. Complete chloroplast genome sequence of MD-2 pineapple and its comparative analysis among nine other plants from the subclass Commelinidae. BMC Plant Biol. 2015;15:196.

43. Chen X, Li Q, Li Y, Qian J, Han J. Chloroplast genome of *Aconitum barbatum* var. *puberulum* (Ranunculaceae) derived from CCS reads using the PacBio RS platform. Frontiers Plant Sci. 2015;6:42.

44. Eid J, Fehr A, Gray J, Luong K, Lyle J, Otto G, et al. Real-time DNA sequencing from single polymerase molecules. Science. 2009;323:133–8.

45. English AC, Richards S, Han Y, Wang M, Vee V, Qu J, et al. Mind the gap: upgrading genomes with Pacific Biosciences RS long-read sequencing technology. PLoS One. 2012;7, e47768.

46. Chin CS, Alexander DH, Marks P, Klammer AA, Drake J, Heiner C, et al. Nonhybrid, finished microbial genome assemblies from long-read SMRT sequencing data. Nat Methods. 2013;10:563–9.

47. Oldenburg DJ, Bendich AJ. DNA maintenance in plastids and mitochondria of plants. Frontiers Plant Sci. 2015;6:883.
48. Oldenburg DJ, Bendich AJ. The linear plastid chromosomes of maize: terminal sequences, structures, and implications for DNA replication. Curr Genet. 2016;62:431–42.
49. Jansen RK, Cai Z, Raubeson LA, Daniell H, Leebens-Mack J, Müller KF, et al. Analysis of 81 genes from 64 plastid genomes resolves relationships in angiosperms and identifies genome-scale evolutionary patterns. Proc Natl Acad Sci U S A. 2007;104:19369–74.
50. Pillon Y, Chase MW. Taxonomic exaggeration and its effects on orchid conservation. Conserv Biol. 2007;21:263–5.
51. Chang CC, Lin HC, Lin IP, Chow TY, Chen HH, Chen WH, et al. The chloroplast genome of *Phalaenopsis aphrodite*(Orchidaceae): comparative analysis of evolutionary rate with that of grasses and its phylogenetic implications. Mol Biol Evol. 2006;23:279–91.
52. Jheng CF, Chen TC, Lin JY, Chen TC, Wu WL, Chang CC. The comparative chloroplast genomic analysis of photosynthetic orchids and developing DNA markers to distinguish *Phalaenopsis* orchids. Plant Sci. 2012;190:62–73.
53. Kim GB, Kwon Y, Yu HJ, Lim KB, Seo JH, Mun JH. The complete chloroplast genome of *Phalaenopsis* 'Tiny Star'. Mitochondrial DNA. 2016;27:1300–2.
54. Wu FH, Chan MT, Liao DC, Hsu CT, Lee YW, Daniell H, et al. Complete chloroplast genome of Oncidium Gower Ramsey and evaluation of molecular markers for identification and breeding in Oncidiinae. BMC Plant Biol. 2010;10:68.
55. Yang JB, Tang M, Li HT, Zhang ZR, Li DZ. Complete chloroplast genome of the genus *Cymbidium*: lights into the species identification, phylogenetic implications and population genetic analyses. BMC Evol Biol. 2013;13:84.
56. Luo J, Hou BW, Niu ZT, Liu W, Xue QY, Ding XY. Comparative chloroplast genomes of photosynthetic orchids: insights into evolution of the Orchidaceae and development of molecular markers for phylogenetic applications. PLoS One. 2014;9, e99016.
57. Li J, Chen C, Wang ZZ. The complete chloroplast genome of the

Dendrobium strongylanthum (Orchidaceae: Epidendroideae). Mitochondrial DNA. 2016;27:3048–9.PubMed

58. da Rocha Perini V, Leles B, Furtado C, Prosdocimi F. Complete chloroplast genome of the orchid *Cattleya crispata*(Orchidaceae: Laeliinae), a Neotropical rupiculous species. Mitochondrial DNA. 2015. doi:https://doi.org/10.3109/19401736.2014.1003850.
59. Tsitrone A, Kirkpatrick M, Levin DA. A model for chloroplast capture. Evolution. 2003;57:1776–82.
60. Soltis DE, Kuzoff RK. Discordance between nuclear and chloroplast phylogenies in the heuchera group (Saxifragaceae). Evolution. 1995;49:727–42.
61. Rieseberg LH, Soltis DE. Phylogenetic consequences of cytoplasmic gene flow in plants. Evol Trends Plants. 1991;5:65–84.
62. Maier RM, Neckermann K, Igloi GL, Kossel H. Complete sequence of the maize chloroplast genome: gene content, hotspots of divergence and fine tuning of genetic information by transcript editing. J Mol Biol. 1995;251:614–28.
63. Ogihara Y, Isono K, Kojima T, Endo A, Hanaoka M, Shiina T, et al. Structural features of a wheat plastome as revealed by complete sequencing of chloroplast DNA. Mol Genet Genomics. 2002;266:740–6.
64. Middleton CP, Senerchia N, Stein N, Akhunov ED, Keller B, Wicker T, Kilian B. Sequencing of chloroplast genomes from wheat, barley, rye and their relatives provides a detailed insight into the evolution of the Triticeae tribe. PLoS One. 2014;9, e85761.
65. Brozynska M, Omar ES, Furtado A, Crayn D, Simon B, Ishikawa R, Henry RJ. Chloroplast genome of novel rice germplasm identified in northern Australia. Tropical Plant Biol. 2014;7:111–20.
66. Waters DL, Nock CJ, Ishikawa R, Rice N, Henry RJ. Chloroplast genome sequence confirms distinctness of Australian and Asian wild rice. Ecol Evol. 2012;2:211–7.
67. Sotowa M, Ootsuka K, Kobayashi Y, Hao Y, Tanaka K, Ichitani K, et al. Molecular relationships between Australian annual wild rice, *Oryza meridionalis*, and two related perennial forms. Rice. 2013;6:26.

68. Xu Q, Xiong G, Li P, He F, Huang Y, Wang K, et al. Analysis of complete nucleotide sequences of 12 *Gossypium*chloroplast genomes: origin and evolution of allotetraploids. PLoS One. 2012;7, e37128.

69. Ibrahim RI, Azuma J, Sakamoto M. Complete nucleotide sequence of the cotton (*Gossypium barbadense* L.) chloroplast genome with a comparative analysis of sequences among 9 dicot plants. Genes Genet Syst. 2006;81:311–21.

70. Li P, Li Z, Liu H, Hua J. Cytoplasmic diversity of the cotton genus as revealed by chloroplast microsatellite markers. Genet Resour Crop Evol. 2014;61:107–19.

71. Pickersgill B, Debouck DG. Domestication patterns in common bean (*Phaseolus vulgaris* L.) and the origin of the Mesoamerican and Andean cultivated races. Theor Appl Genet. 2005;110:432–44.

72. Palmer JD, Osorio B, Thompson WF. Evolutionary significance of inversions in legume chloroplast DNAs. Curr Genet. 1988;14:65–74.

73. Sherman-Broyles S, Bombarely A, Grimwood J, Schmutz J, Doyle J. Complete plastome sequences from *Glycine syndetika*and six additional perennial wild relatives of soybean. G3 (Bethesda). 2014;4:2023–33.

74. Sabir J, Schwarz E, Ellison N, Zhang J, Baeshen NA, Mutwakil M, et al. Evolutionary and biotechnology implications of plastid genome variation in the inverted-repeat-lacking clade of legumes. Plant Biotechnol J. 2014;12:743–54.

75. Cai Z, Guisinger M, Kim H-G, Ruck E, Blazier JC, McMurtry V, et al. Extensive reorganization of the plastid genome of *Trifolium subterraneum* (Fabaceae) is associated with numerous repeated sequences and novel DNA insertions. J Mol Evol. 2008;67:696–704.

76. Magee AM, Aspinall S, Rice DW, Cusack BP, Semon M, Perry AS, et al. Localized hypermutation and associated gene losses in legume chloroplast genomes. Genome Res. 2010;20:1700–10.

77. Bogdanova VS, Zaytseva OO, Mglinets AV, Shatskaya NV, Kosterin OE, Vasiliev GV. Nuclear-cytoplasmic conflict in pea (*Pisum sativum* L.) is associated with nuclear and plastidic candidate genes encoding acetyl-CoA carboxylase subunits. PLoS One. 2015;10:e0119835.

78. Guo X, Castillo-Ramírez S, González V, Bustos P, Fernández-Vázquez JL, Santamaría RI, et al. Rapid evolutionary change of common bean (*Phaseolus vulgaris* L.) plastome, and the genomic diversification of legume chloroplasts. BMC Genomics. 2007;8:228.

79. Tangphatsornruang S, Sangsrakru D, Chanprasert J, Uthaipaisanwong P, Yoocha T, Jomchai N, Tragoonrung S. The chloroplast genome sequence of mungbean (*Vigna radiata*) determined by high-throughput pyrosequencing: structural organization and phylogenetic relationships. DNA Res. 2010;17:11–22.

80. Martin GE, Rousseau-Gueutin M, Cordonnier S, Lima O, Michon-Coudouel S, Naquin D, et al. The first complete chloroplast genome of the Genistoid legume *Lupinus luteus*: evidence for a novel major lineage-specific rearrangement and new insights regarding plastome evolution in the legume family. Ann Bot. 2014;113:1197–210.

81. Kazakoff SH, Imelfort M, Edwards D, Koehorst J, Biswas B, Batley J, et al. Capturing the biofuel wellhead and powerhouse: the chloroplast and mitochondrial genomes of the leguminous feedstock tree *Pongamia pinnata*. PLoS One. 2012;7, e51687.

82. Schwarz EN, Ruhlman TA, Sabir JS, Hajrah NH, Alharbi NS, Al-Malki AL, et al. Plastid genome sequences of legumes reveal parallel inversions and multiple losses of *rps16* in papilionoids. J Syst Evol. 2015;53:458–68.

83. Su HJ, Hogenhout SA, Al-Sadi AM, Kuo CH. Complete chloroplast genome sequence of Omani Lime (*Citrus aurantiifolia*) and comparative analysis within the Rosids. PLoS One. 2014;9, e113049.

84. Carbonell-Caballero J, Alonso R, Ibañez V, Terol J, Talon M, Dopazo J. A phylogenetic analysis of 34 chloroplast genomes elucidates the relationships between wild and domestic species within the genus *Citrus*. Mol Biol Evol. 2015;32:2015–35.

85. Caspermeyer J. Most comprehensive study to date reveals evolutionary history of *Citrus*. Mol Biol Evol. 2015;32:2217–8.

86. Chen SL, Xiao PG. Molecular evolution and positive Darwinian selection of the chloroplast maturase matK. J Plant Res. 2010;123:241–7.

87. Peng L, Yamamoto H, Shikanai T. Structure and biogenesis of the chloroplast NAD(P)H dehydrogenase complex. Biochim Biophys

Acta. 1807;2011:945–53.

88. Ma PF, Zhang YX, Guo ZH, Li DZ. Evidence for horizontal transfer of mitochondrial DNA to the plastid genome in a bamboo genus. Sci Rep. 2015;5:11608.
89. Zhang YJ, Ma PF, Li DZ. High-throughput sequencing of six bamboo chloroplast genomes: phylogenetic implications for temperate woody bamboos (Poaceae: Bambusoideae). PLoS One. 2011;6, e20596.
90. Burke SV, Grennan CP, Duvall MR. Plastome sequences of two New World bamboos–*Arundinaria gigantea* and *Cryptochloa strictiflora* (Poaceae)–extend phylogenomic understanding of Bambusoideae. Am J Bot. 2012;99:1951–61.
91. Gao J, Li K, Gao LZ. The complete chloroplast genome sequence of the *Bambusa multiplex* (Poaceae: Bambusoideae). Mitochondrial DNA. 2014;18:1–3.
92. Wysocki WP, Clark LG, Attigala L, Ruiz-Sanchez E, Duvall MR. Evolution of the bamboos (Bambusoideae; Poaceae): a full plastome phylogenomic analysis. BMC Evol Biol. 2015;15:50.
93. Ma PF, Zhang YX, Zeng CX, Guo ZH, Li DZ. Chloroplast phylogenomic analyses resolve deep-level relationships of an intractable bamboo tribe Arundinarieae (Poaceae). Syst Biol. 2014;63:933–50.
94. Smith DR. Mitochondrion-to-plastid DNA, transfer: it happens. New Phytol. 2014;202:736–8.
95. Timmis JN, Ayliffe MA, Huang CY, Martin W. Endosymbiotic gene transfer: organelle genomes forge eukaryotic chromosomes. Nat Rev Genet. 2004;5:123–35.
96. Li H, Chiu CC. Protein transport into chloroplasts. Annu Rev Plant Biol. 2010;61:157–80.
97. Barrett CF, Freudenstein JV, Li J, Mayfield-Jones DR, Perez L, Pires JC, Santos C. Investigating the path of plastid genome degradation in an early-transitional clade of heterotrophic orchids, and implications for heterotrophic angiosperms. Mol Biol Evol. 2014;31:3095–112.
98. Tiller N, Bock R. The translational apparatus of plastids and its role in plant development. Mol Plant. 2014;7:1105–20.

99. Hazkani-Covo E, Zeller RM, Martin W. Molecular poltergeists: mitochondrial DNA copies (numts) in sequenced nuclear genomes. PLoS Genet. 2010;6, e1000834.

100. Smith DR, Crosby K, Lee RW. Correlation between nuclear plastid DNA abundance and plastid number supports the limited transfer window hypothesis. Genome Biol Evol. 2011;3:365–71.

101. Cummings HS, Hershey JWB. Translation initiation-factor *IF1* is essential for cell viability in *Escherichia coli.* J Bacteriol. 1994;176:198–205.

102. Millen RS, Olmstead RG, Adams KL, Palmer JD, Lao NT, Heggie L, et al. Many parallel losses of infA from chloroplast DNA during angiosperm evolution with multiple independent transfers to the nucleus. Plant Cell. 2001;13:645–58.

103. Gantt JS, Baldauf SL, Calie PJ, Weeden NF, Palmer JD. Transfer of *rpl22* to the nucleus greatly preceded its loss from the chloroplast and involved the gain of an intron. EMBO J. 1991;10:3073–8.

104. Park S, Jansen RK, Park S. Complete plastome sequence of *Thalictrum coreanum* (Ranunculaceae) and transfer of the *rpl32* gene to the nucleus in the ancestor of the subfamily Thalictroideae. BMC Plant Biol. 2015;15:40.

105. Ueda M, Fujimoto M, Arimura S, Murata J, Tsutsumi N, Kadowaki K. Loss of the *rpl32* gene from the chloroplast genome and subsequent acquisition of a preexisting transit peptide within the nuclear gene in *Populus*. Gene. 2007;402:51–6.

106. Cusack BP, Wolfe KH. When gene marriages don't work out: divorce by subfunctionalization. Trends Genet. 2007;23:270–2.

107. Ueda M, Kuniyoshi T, Yamamoto H, Sugimoto K, Ishizaki K, Kohchi T, et al. Composition and physiological function of the chloroplast NADH dehydrogenase-like complex in *Marchantia polymorpha*. Plant J. 2012;72:683–93.

108. Munekage Y, Hashimoto M, Miyake C, Tomizawa K-I, Endo T, Tasaka M, Shikanai T. Cyclic electron flow around photosystem I is essential for photosynthesis. Nature. 2004;429:579–82.

109. Peltier G, Cournac L. Chlororespiration. Annu Rev Plant Biol. 2002;53:523–50.

110. Wakasugi T, Tsudzuki J, Ito S, Nakashima K, Tsudzuki T, Sugiura M. Loss of all *ndh* genes as determined by sequencing the entire chloroplast genome of the black pine *Pinus thunbergii*. Proc Natl Acad Sci U S A. 1994;91:9794–8.

111. Sanderson MJ, Copetti D, Búrquez A, Bustamante E, Charboneau JL, Eguiarte LE, et al. Exceptional reduction of the plastid genome of saguaro cactus (*Carnegiea gigantea*): loss of the *ndh* gene suite and inverted repeat. Am J Bot. 2015;102:1115–27.

112. McCoy SR, Kuehl JV, Boore JL, Raubeson LA. The complete plastid genome sequence of *Welwitschia mirabilis*: an unusually compact plastome with accelerated divergence rates. BMC Evol Biol. 2008;8:130.

113. Braukmann TWA, Kuzmina M, Stefanović S. Loss of all plastid *ndh* genes in Gnetales and conifers: extent and evolutionary significance for the seed plant phylogeny. Curr Genet. 2009;55:323–37.

114. Blazier J, Guisinger MM, Jansen RK. Recent loss of plastid-encoded *ndh* genes within *Erodium* (Geraniaceae). Plant Mol Biol. 2011;76:263–72.

115. Weng M-L, Blazier JC, Govindu M, Jansen RK. Reconstruction of the ancestral plastid genome in Geraniaceae reveals a correlation between genome rearrangements, repeats and nucleotide substitution rates. Mol Biol Evol. 2014;31:645–59.

116. Ranade SS, García-Gil MR, Rosselló JA. Non-functional plastid *ndh* gene fragments are present in the nuclear genome of Norway spruce (*Picea abies* L. Karsch): insights from in silico analysis of nuclear and organellar genomes. Mol Genet Genomics. 2016;291:935–41.

117. Ruhlman TA, Chang WJ, Chen JJ, Huang YT, Chan MT, Zhang J, et al. NDH expression marks major transitions in plant evolution and reveals coordinate intracellular gene loss. BMC Plant Biol. 2015;15:100.

118. Takabayashi A, Endo T, Shikanai T, Sato F. Post-illumination reduction of the plastoquinone pool in chloroplast transformants in which chloroplastic NAD(P)H dehydrogenase was inactivated. Biosci Biotechnol Biochem. 2002;66:2107–11.

119. Burrows PA, Sazanov LA, Svab Z, Maliga P, Nixon PJ. Identification of a functional respiratory complex in chloroplasts through analysis

of tobacco mutants containing disrupted plastid *ndh* genes. EMBO J. 1998;17:868–76.

120. Kofer W, Koop H-U, Wanner G, Steinmüller K. Mutagenesis of the genes encoding subunits A, C, H, I, J and K of the plastid NAD(P) H-plastoquinone-oxidoreductase in tobacco by polyethylene glycol-mediated plastome transformation. Mol Gen Genet. 1998;258:166–73.
121. Shikanai T, Endo T, Hashimoto T, Yamada Y, Asada K, Yokota A. Directed disruption of the tobacco *ndhB* gene impairs cyclic electron flow around photosystem I. Proc Natl Acad Sci U S A. 1998;95:9705–9.
122. Jin S, Daniell H. The engineered chloroplast genome just got smarter. Trends Plant Sci. 2015;20:622–40.
123. Daniell H. Transgene containment by maternal inheritance: effective or elusive? Proc Natl Acad Sci U S A. 2007;104:6879–80.
124. Dheeraj V, Daniell H. Chloroplast vector systems for biotechnology applications. Plant Physiol. 2007;145:1129–43.
125. Dunne A, Maple-Grødem J, Gargano D, Haslam RP, Napier JA, Chua NH, et al. Modifying fatty acid profiles through a new cytokinin-based plastid transformation system. Plant J. 2014;80:1131–8.
126. Kavanagh TA, Thanh ND, Lao NT, McGrath N, Peter SO, Horváth EM, et al. Homeologous plastid DNA transformation in tobacco is mediated by multiple recombination events. Genetics. 1999;152:1111–22.
127. Guda C, Lee SB, Daniell H. Stable expression of a biodegradable protein-based polymer in tobacco chloroplasts. Plant Cell Rep. 2000;19:257–62.
128. Ruhlman T, Verma D, Samson N, Daniell H. The role of heterologous chloroplast sequence elements in transgene integration and expression. Plant Physiol. 2010;152:2088–104.
129. Daniell H, Datta R, Varma S, Gray S, Lee SB. Containment of herbicide resistance through genetic engineering of the chloroplast genome. Nat Biotechnol. 1998;16:345–8.
130. Krichevsky A, Meyers B, Vainstein A, Maliga P, Citovsky V. Autoluminescent plants. PLoS One. 2010;5, e15461.

131. Daniell H, Vivekananda J, Nielsen B, Ye G, Tewari K, Sanford J. Transient foreign gene expression in chloroplasts of cultured tobacco cells after biolistic delivery of chloroplast vectors. Proc Natl Acad Sci U S A. 1990;87:88–92.

132. Klein RR, Mullet JE. Control of gene expression during higher plant chloroplast biogenesis. Protein synthesis and transcript levels of *psbA*, *psaA-psaB*, and *rbcL* in dark-grown and illuminated barley seedlings. J Biol Chem. 1987;262:4341–8.PubMed

133. Wurbs D, Ruf S, Bock R. Contained metabolic engineering in tomatoes by expression of carotenoid biosynthesis genes from the plastid genome. Plant J. 2007;49:276–88.

134. McBride KE, Schaaf DJ, Daley M, Stalker DM. Controlled expression of plastid transgenes in plants based on a nuclear DNA-encoded and plastid-targeted T7 RNA polymerase. Proc Natl Acad Sci U S A. 1994;91:7301–5.

135. Kumar S, Dhingra A, Daniell H. Plastid-expressed betaine aldehyde dehydrogenase gene in carrot cultured cells, roots, and leaves confers enhanced salt tolerance. Plant Physiol. 2004;136:2843–54.

136. Dhingra A, Portis Jr AR, Daniell H. Enhanced translation of a chloroplast-expressed *RbcS* gene restores small subunit levels and photosynthesis in nuclear *RbcS* antisense plants. Proc Natl Acad Sci U S A. 2004;101:6315–20.

137. De Cosa B, Moar W, Lee SB, Miller M, Daniell H. Overexpression of the Bt cry2Aa2 operon in chloroplasts leads to formation of insecticidal crystals. Nat Biotechnol. 2001;19:71–4.

138. Ruf S, Hermann M, Berger IJ, Carrer H, Bock R. Stable genetic transformation of tomato plastids and expression of a foreign protein in fruit. Nat Biotechnol. 2001;19:870–5.

139. Zhou F, Badillo-Corona JA, Karcher D, Gonzalez-Rabade N, Piepenburg K, Borchers AM, et al. High-level expression of human immunodeficiency virus antigens from the tobacco and tomato plastid genomes. Plant Biotechnol J. 2008;6:897–913.

140. Lee SB, Li B, Jin S, Daniell H. Expression and characterization of antimicrobial peptides Retrocyclin-101 and Protegrin-1 in chloroplasts to control viral and bacterial infections. Plant Biotechnol J. 2011;9:100–15.

141. Jin S, Kanagaraj A, Verma D, Lange T, Daniell H. Release of hormones from conjugates: chloroplast expression of beta-glucosidase results in elevated phytohormone levels associated with significant increase in biomass and protection from aphids or whiteflies conferred by sucrose esters. Plant Physiol. 2011;155:222–35.

142. Jin S, Zhang X, Daniell H. *Pinellia ternata* agglutinin expression in chloroplasts confers broad spectrum resistance against aphid, whitefly, Lepidopteran insects, bacterial and viral pathogens. Plant Biotechnol J. 2012;10:313–27.

143. Liu CW, Lin CC, Chen JJ, Tseng MJ. Stable chloroplast transformation in cabbage (*Brassica oleracea* L. var. *capitata* L.) by particle bombardment. Plant Cell Rep. 2007;26:1733–44.

144. Dufourmantel N, Pelissier B, Garcon F, Peltier G, Ferullo JM, Tissot G. Generation of fertile transplastomic soybean. Plant Mol Biol. 2004;55:479–89.

145. Dufourmantel N, Tissot G, Goutorbe F, Garcon F, Muhr C, Jansens S, et al. Generation and analysis of soybean plastid transformants expressing *Bacillus thuringiensis* Cry1Ab protoxin. Plant Mol Biol. 2005;58:659–68.

146. Singh AK, Verma SS, Bansal KC. Plastid transformation in eggplant (*Solanum melongena* L.). Transgenic Res. 2010;19:113–9.

147. Jin S, Singh ND, Li L, Zhang X, Daniell H. Engineered chloroplast dsRNA silences cytochrome p450 monooxygenase, V-ATPase and chitin synthase genes in the insect gut and disrupts *Helicoverpa armigera* larval development and pupation. Plant Biotechnol J. 2015;13:435–46.

148. Zhang J, Khan SA, Hasse C, Ruf S, Heckel DG, Bock R. Pest control. Full crop protection from an insect pest by expression of long double-stranded RNAs in plastids. Science. 2015;347:991–4.

149. Viitanen PV, Devine AL, Khan MS, Deuel DL, Van Dyk DE, Daniell H. Metabolic engineering of the chloroplast genome using the *Echerichia coli ubiC* gene reveals that chorismate is a readily abundant plant precursor for p-hydroxybenzoic acid biosynthesis. Plant Physiol. 2004;136:4048–60.

150. Verma D, Kanagaraj A, Jin SX, Singh ND, Kolattukudy PE, Daniell H. Chloroplast-derived enzyme cocktails hydrolyse lignocellulosic

biomass and release fermentable sugars. Plant Biotechnol J. 2010;8:332–50.

151. Schneider C. Chemistry and biology of vitamin E. Mol Nutr Food Res. 2005;49:7–30.

152. Shintani D, DellaPenna D. Elevating the vitamin E content of plants through metabolic engineering. Science. 1998;282:2098–100.

153. Jin S, Daniell H. Expression of gamma-tocopherol methyltransferase in chloroplasts results in massive proliferation of the inner envelope membrane and decreases susceptibility to salt and metal-induced oxidative stresses by reducing reactive oxygen species. Plant Biotechnol J. 2014;12:1274–85.

154. Apel W, Bock R. Enhancement of carotenoid biosynthesis in transplastomic tomatoes by induced lycopene-to-provitamin A conversion. Plant Physiol. 2009;151:59–66.

155. Grabowski H, Cockburn I, Long G. The market for follow-on biologics: how will it evolve? Health Aff. 2006;25:1291–301.

156. Spök A, Karner S, Stein AJ, Rodríguez-Cerezo E. Plant molecular farming. Opportunities and challenges. JRC Scientific and Technical Reports 2008. http://ftp.jrc.es/EURdoc/JRC43873.pdf. Accessed 17 May 2016.

157. Holtz BR, Berquist BR, Bennett LD, Kommineni VJ, Munigunti RK, White EL, et al. Commercial-scale biotherapeutics manufacturing facility for plant-made pharmaceuticals. Plant Biotechnol J. 2015;13:1180–90.

158. Kwon KC, Daniell H. Low-cost oral delivery of protein drugs bioencapsulated in plant cells. Plant Biotechnol J. 2015;13:1017–22.

159. El Kaoutari A, Armougom F, Gordon JI, Raoult D, Henrissat B. The abundance and variety of carbohydrate-active enzymes in the human gut microbiota. Nat Rev Microbiol. 2013;11:497–504.

160. Kwon KC, Nityanandam R, New JS, Daniell H. Oral delivery of bioencapsulated exendin-4 expressed in chloroplasts lowers blood glucose level in mice and stimulates insulin secretion in beta-TC6 cells. Plant Biotechnol J. 2013;11:77–86.

161. Shenoy V, Kwon KC, Rathinasabapathy A, Lin SN, Jin GY,

Song CJ, et al. Oral delivery of angiotensin-converting enzyme 2 and angiotensin-(1–7) bioencapsulated in plant cells attenuates pulmonary hypertension. Hypertension. 2014;64:1248–59.

162. Kohli N, Westerveld DR, Ayache AC, Verma A, Shil P, Prasad T, et al. Oral delivery of bioencapsulated proteins across blood–brain and blood-retinal barriers. Mol Ther. 2014;22:535–46.

163. Chan HT, Daniell H. Plant-made oral vaccines against human infectious diseases—are we there yet? Plant Biotechnol J. 2015;13:1056–70.

164. Davoodi-Semiromi A, Schreiber M, Nalapalli S, Verma D, Singh ND, Banks RK, et al. Chloroplast-derived vaccine antigens confer dual immunity against cholera and malaria by oral or injectable delivery. Plant Biotechnol J. 2010;8:223–42.

165. Lakshmi PS, Verma D, Yang X, Lloyd B, Daniell H. Low cost tuberculosis vaccine antigens in capsules: expression in chloroplasts, bio-encapsulation, stability and functional evaluation in vitro. PLoS One. 2013;8, e54708.

166. Hagelberg E, Hofreiter M, Keyser C. Ancient DNA: the first three decades. Philos Trans R Soc Lond B Biol Sci. 2015;370:20130371.

167. Golenberg EM, Giannasi DE, Clegg MT, Smiley CJ, Durbin M, Henderson D, et al. Chloroplast DNA sequence from a Miocene *Magnolia* species. Nature. 1990;344:656–8.

168. Rosselló JA. The never-ending story of geologically ancient DNA: was the model plant *Arabidopsis* the source of Miocene Dominican amber? Biol J Linn Soc. 2014;111:234–40.

169. Zhu A, Guo W, Gupta S, Fan W, Mower JP. Evolutionary dynamics of the plastid inverted repeat: the effects of expansion, contraction, and loss on substitution rates. New Phytol. 2016;209:1747–56.

170. von Kohn C, Kielkowska A, Havey MJ. Sequencing and annotation of the chloroplast DNAs and identification of polymorphisms distinguishing normal male-fertile and male-sterile cytoplasms of onion. Genome. 2013;56:737–42.

171. Nashima K, Terakami S, Nishitani C, Kunihisa M, Shoda M, Takeuchi M, et al. Complete chloroplast genome sequence of pineapple (*Ananas comosus*). Tree Genet Genomes. 2015;11:60.

172. Downie SR, Jansen RK. A comparative analysis of whole plastid genomes from the Apiales: expansion and contraction of the inverted repeat, mitochondrial to plastid transfer of DNA, and identification of highly divergent noncoding regions. Syst Botany. 2015;40:336–51.

173. Liu Y, Huo N, Dong L, Wang Y, Zhang S, Young HA, et al. Complete chloroplast genome sequences of Mongolia medicine *Artemisia frigida* and phylogenetic relationships with other plants. PLoS One. 2013;8, e57533.

174. Schmitz-Linneweber C, Regel R, Du TG, Hupfer H, Herrmann RG, Maier RM. The plastid chromosome of *Atropa belladonna* and its comparison with that of *Nicotiana tabacum*: the role of RNA editing in generating divergence in the process of plant speciation. Mol Biol Evol. 2002;19:1602–12.

175. Hu Z-Y, Hua W, Huang S-M, Wang H-Z. Complete chloroplast genome sequence of rapeseed (*Brassica napus* L.) and its evolutionary implications. Genet Resour Crop Evol. 2011;58:875–87.

176. Yang JB, Li DZ, Li HT. Highly effective sequencing whole chloroplast genomes of angiosperms by nine novel universal primer pairs. Mol Ecol Resour. 2014;14:1024–31.

177. Oh H, Seo B, Lee S, Ahn D-H, Jo E, Park J-K, Min G-S. Two complete chloroplast genome sequences of *Cannabis sativa*varieties. Mitochondrial DNA. 2016;27:2835–7.

178. Jo YD, Park J, Kim J, Song W, Hur CG, Lee YH, Kang BC. Complete sequencing and comparative analyses of the pepper (*Capsicum annuum* L.) plastome revealed high frequency of tandem repeats and large insertion/deletions on pepper plastome. Plant Cell Rep. 2011;30:217–29.

179. Ming R, Hou S, Feng Y, Yu Q, Dionne-Laporte A, Saw JH, et al. The draft genome of the transgenic tropical fruit tree papaya (*Carica papaya* Linnaeus). Nature. 2008;452:991–6.

180. Ku C, Chung WC, Chen LL, Kuo CH. The complete plastid genome sequence of Madagascar periwinkle *Catharanthus roseus* (L.) G. Don: plastid genome evolution, molecular marker identification, and phylogenetic implications in asterids. PLoS One. 2013;8, e68518.

181. Mariac C, Scarcelli N, Pouzadou J, Barnaud A, Billot C, Faye A, et al. Cost-effective enrichment hybridization capture of chloroplast genomes at deep multiplexing levels for population genetics and phylogeography studies. Mol Ecol Resour. 2014;14:1103–13.

182. Ahmed I, Biggs PJ, Matthews PJ, Collins LJ, Hendy MD, Lockhart PJ. Mutational dynamics of aroid chloroplast genomes. Genome Biol Evol. 2012;4:1316–23.

183. Plader W, Yukawa Y, Sugiura M, Malepszy S. The complete structure of the cucumber (*Cucumis sativus* L.) chloroplast genome: its composition and comparative analysis. Cell Mol Biol Lett. 2007;12:584–94.

184. Barrett CF, Specht CD, Leebens-Mack J, Stevenson DW, Zomlefer WB, Davis JI. Resolving ancient radiations: can complete plastid gene sets elucidate deep relationships among the tropical gingers (Zingiberales)? Ann Bot. 2014;113:119–33.

185. Wang B, Han L, Chen C, Wang Z. The complete chloroplast genome sequence of *Dieffenbachia seguine* (Araceae). Mitochondrial DNA. 2016;27:2913–4.PubMed

186. Ye CY, Lin Z, Li G, Wang YY, Qiu J, Fu F, Zhang H, et al. *Echinochloa* chloroplast genomes: insights into the evolution and taxonomic identification of two weedy species. PLoS One. 2014;9, e113657.

187. Wu CS, Lai YT, Lin CP, Wang YN, Chaw SM. Evolution of reduced and compact chloroplast genomes (cpDNAs) in gnetophytes: selection toward a lower-cost strategy. Mol Phylogenet Evol. 2009;52:115–24.

188. Logacheva MD, Samigullin TH, Dhingra A, Penin AA. Comparative chloroplast genomics and phylogenetics of Fagopyrum esculentum ssp. ancestrale--a wild ancestor of cultivated buckwheat. BMC Plant Biol. 2008;8:59.

189. Cahoon AB, Sharpe RM, Mysayphonh C, Thompson EJ, Ward AD, Lin A. The complete chloroplast genome of tall fescue (*Lolium arundinaceum*; Poaceae) and comparison of whole plastomes from the family Poaceae. Am J Bot. 2010;97:49–58.

190. Shulaev V, Sargent DJ, Crowhurst RN, Mockler TC, Folkerts O, Delcher AL, et al. The genome of woodland strawberry (*Fragaria*

vesca). Nat Genet. 2011;43:109–16.

191. Dempewolf H, Kane NC, Ostevik KL, Geleta M, Barker MS, Lai Z, et al. Establishing genomic tools and resources for *Guizotia abyssinica* (L.f.) Cass.–the development of a library of expressed sequence tags, microsatellite loci, and the sequencing of its chloroplast genome. Mol Ecol Resour. 2010;10:1048–58.
192. Timme RE, Kuehl JV, Boore JL, Jansen RK. A comparative analysis of the *Lactuca* and *Helianthus* (Asteraceae) plastid genomes: identification of divergent regions and categorization of shared repeats. Am J Bot. 2007;94:302–12.
193. Barrett CF, Davis JI, Leebens-Mack J, Conran JG, Stevenson DW. Plastid genomes and deep relationships among the commelinid monocot angiosperms. Cladistics. 2013;29:65–87.
194. Sanchez-Puerta MV, Abbona CC. The chloroplast genome of *Hyoscyamus niger* and a phylogenetic study of the tribe Hyoscyameae (Solanaceae). PLoS One. 2014;9, e98353.
195. Yan L, Lai X, Li X, Wei C, Tan X, Zhang Y. Analyses of the complete genome and gene expression of chloroplast of sweet potato (*Ipomoea batata*). PLoS One. 2015;10, e0124083.
196. McNeal JR, Kuehl JV, Boore JL, de Pamphilis CW. Complete plastid genome sequences suggest strong selection for retention of photosynthetic genes in the parasitic plant genus *Cuscuta*. BMC Plant Biol. 2007;7:57.
197. Kanamoto H, Yamashita A, Asao H, Okumura S, Takase H, Hattori M, et al. Efficient and stable transformation of *Lactuca sativa* L. cv. Cisco (lettuce) plastids. Transgenic Res. 2006;15:205–17.
198. Mennes CB, Lam VK, Rudall PJ, Lyon SP, Graham SW, Smets EF, Merckx VS. Ancient Gondwana break-up explains the distribution of the mycoheterotrophic family Corsiaceae (Liliales). J Biogeogr. 2015;42:1123–36.
199. Hand ML, Spangenberg GC, Forster JW, Cogan NO. Plastome sequence determination and comparative analysis for members of the *Lolium-Festuca* grass species complex. G3 (Bethsada). 2013;3:607–16.
200. Kato T, Kaneko T, Sato S, Nakamura Y, Tabata S. Complete structure of the chloroplast genome of a legume, *Lotus japonicus*. DNA Res.

2000;7:323–30.

201. Raubeson LA, Peery R, Chumley TW, Dziubek C, Fourcade HM, Boore JL, Jansen RK. Comparative chloroplast genomics: analyses including new sequences from the angiosperms *Nuphar advena* and *Ranunculus macranthus*. BMC Genomics. 2007;8:174.
202. Hiratsuka J, Shimada H, Whittier R, Ishibashi T, Sakamoto M, Mori M, et al. The complete sequence of the rice (*Oryza sativa*) chloroplast genome: intermolecular recombination between distinct tRNA genes accounts for a major plastid DNA inversion during the evolution of the cereals. Mol Gen Genet. 1989;217:185–94.
203. Kim KJ, Lee HL. Complete chloroplast genome sequences from Korean ginseng (*Panax schinseng* Nees) and comparative analysis of sequence evolution among 17 vascular plants. DNA Res. 2004;11:247–61.
204. Young HA, Lanzatella CL, Sarath G, Tobias CM. Chloroplast genome variation in upland and lowland switchgrass. PLoS One. 2011;6, e23980.
205. Kumar S, Hahn FM, McMahan CM, Cornish K, Whalen MC. Comparative analysis of the complete sequence of the plastid genome of *Parthenium argentatum* and identification of DNA barcodes to differentiate *Parthenium* species and lines. BMC Plant Biol. 2009;9:131.
206. Chumley TW, Palmer JD, Mower JP, Fourcade HM, Calie PJ, Boore JL, Jansen RK. The complete chloroplast genome sequence of *Pelargonium x hortorum*: organization and evolution of the largest and most highly rearranged chloroplast genome of land plants. Mol Biol Evol. 2006;23:2175–90.
207. Jeong YM, Chung WH, Mun JH, Kim N, Yu HJ. De novo assembly and characterization of the complete chloroplast genome of radish (*Raphanus sativus* L.). Gene. 2014;551:39–48.
208. Rivarola M, Foster JT, Chan AP, Williams AL, Rice DW, Liu X, et al. Castor bean organelle genome sequencing and worldwide genetic diversity analysis. PLoS One. 2011;6, e21743.
209. Calsa Junior T, Carraro DM, Benatti MR, Barbosa AC, Kitajima JP, Carrer H. Structural features and transcript-editing analysis of sugarcane (*Saccharum officinarum* L.) chloroplast genome. Curr

Genet. 2004;46:366–73.

210. Qian J, Song J, Gao H, Zhu Y, Xu J, Pang X, et al. The complete chloroplast genome sequence of the medicinal plant *Salvia miltiorrhiza*. PLoS One. 2013;8, e57607.

211. Yi DK, Kim KJ. Complete chloroplast genome sequences of important oilseed crop *Sesamum indicum* L. PLoS One. 2012;7, e35872.

212. Chung HJ, Jung JD, Park HW, Kim JH, Cha HW, Min SR, et al. The complete chloroplast genome sequences of *Solanum tuberosum* and comparative analysis with Solanaceae species identified the presence of a 241-bp deletion in cultivated potato chloroplast DNA sequence. Plant Cell Rep. 2006;25:1369–79.

213. Schmitz-Linneweber C, Maier RM, Alcaraz JP, Cottet A, Herrmann RG, Mache R. The plastid chromosome of spinach (*Spinacia oleracea*): complete nucleotide sequence and gene organization. Plant Mol Biol. 2001;45:307–15.

214. Yi DK, Yang JC, So S, Joo M, Kim DK, Shin CH, et al. The complete plastid genome sequence of *Abies koreana* (Pinaceae: Abietoideae). Mitochondrial DNA. 2016;27:2351–3.PubMed

215. Yao X, Tang P, Li Z, Li D, Liu Y, Huang H. The first complete chloroplast genome sequences in Actinidiaceae: genome structure and comparative analysis. PLoS One. 2015;10, e0129347.

216. Hsu CY, Wu CS, Chaw SM. Ancient nuclear plastid DNA in the yew family (Taxaceae). Genome Biol Evol. 2014;6:2111–21.

217. Ruhsam M, Rai HS, Mathews S, Ross TG, Graham SW, Raubeson LA, et al. Does complete plastid genome sequencing improve species discrimination and phylogenetic resolution in *Araucaria*? Mol Ecol Resour. 2015;15:1067–78.

218. Ma J, Yang B, Zhu W, Sun L, Tian J, Wang X. The complete chloroplast genome sequence of *Mahonia bealei*(Berberidaceae) reveals a significant expansion of the inverted repeat and phylogenetic relationship with other angiosperms. Gene. 2013;528:120–31.

219. Hansen DR, Dastidar SG, Cai Z, Penaflor C, Kuehl JV, Boore JL, Jansen RK. Phylogenetic and evolutionary implications of complete chloroplast genome sequences of four early-diverging angiosperms: *Buxus* (Buxaceae), *Chloranthus*(Chloranthaceae), *Dioscorea*

(Dioscoreaceae), and *Illicium* (Schisandraceae). Mol Phylogenet Evol. 2007;45:547–63.

220. Wu CS, Chaw SM. Highly rearranged and size-variable chloroplast genomes in conifers II clade (cupressophytes): evolution towards shorter intergenic spacers. Plant Biotechnol J. 2014;12:344–53.

221. Shi C, Liu Y, Huang H, Xia EH, Zhang HB, Gao LZ. Contradiction between plastid gene transcription and function due to complex posttranscriptional splicing: an exemplary study of *ycf15* function and evolution in angiosperms. PLoS One. 2013;8, e59620.

222. Huang H, Shi C, Liu Y, Mao SY, Gao LZ. Thirteen *Camellia* chloroplast genome sequences determined by high-throughput sequencing: genome structure and phylogenetic relationships. BMC Evol Biol. 2014;14:151.

223. Lin CP, Huang JP, Wu CS, Hsu CY, Chaw SM. Comparative chloroplast genomics reveals the evolution of Pinaceae genera and subfamilies. Genome Biol Evol. 2010;2:504–17.

224. Wu CS, Wang YN, Hsu CY, Lin CP, Chaw SM. Loss of different inverted repeat copies from the chloroplast genomes of Pinaceae and cupressophytes and influence of heterotachy on the evaluation of gymnosperm phylogeny. Genome Biol Evol. 2011;3:1284–95.

225. Male PJ, Bardon L, Besnard G, Coissac E, Delsuc F, Engel J, et al. Genome skimming by shotgun sequencing helps resolve the phylogeny of a pantropical tree family. Mol Ecol Resour. 2014;14:966–75.PubMed

226. Huang YY, Matzke AJ, Matzke M. Complete sequence and comparative analysis of the chloroplast genome of coconut palm (*Cocos nucifera*). PLoS One. 2013;8, e74736.

227. Bayly MJ, Rigault P, Spokevicius A, Ladiges PY, Ades PK, Anderson C, et al. Chloroplast genome analysis of Australian eucalypts--*Eucalyptus*, *Corymbia*, *Angophora*, *Allosyncarpia* and *Stockwellia* (Myrtaceae). Mol Phylogenet Evol. 2013;69:704–16.

228. Hirao T, Watanabe A, Kurita M, Kondo T, Takata K. Complete nucleotide sequence of the *Cryptomeria japonica* D. Don. chloroplast genome and comparative chloroplast genomics: diversified genomic structure of coniferous species. BMC Plant Biol. 2008;8:70.

229. Uthaipaisanwong P, Chanprasert J, Shearman JR, Sangsrakru D,

Yoocha T, Jomchai N, et al. Characterization of the chloroplast genome sequence of oil palm (*Elaeis guineensis* Jacq.). Gene. 2012;500:172–80.

230. Steane D, Jones R, Vaillancourt R. A set of chloroplast microsatellite primers for *Eucalyptus* (Myrtaceae). Mol Ecol Notes. 2005;5:538–41.

231. Tangphatsornruang S, Uthaipaisanwong P, Sangsrakru D, Chanprasert J, Yoocha T, Jomchai N, Tragoonrung S. Characterization of the complete chloroplast genome of *Hevea brasiliensis* reveals genome rearrangement, RNA editing sites and phylogenetic relationships. Gene. 2011;475:104–12.

232. Lee HL, Jansen RK, Chumley TW, Kim KJ. Gene relocations within chloroplast genomes of *Jasminum* and *Menodora*(Oleaceae) are due to multiple, overlapping inversions. Mol Biol Evol. 2007;24:1161–80.

233. Asif MH, Mantri SS, Sharma A, Srivastava A, Trivedi I, Gupta P, et al. Complete sequence and organisation of the *Jatropha curcas* (Euphorbiaceae) chloroplast genome. Tree Genet Genomes. 2010;6:941–52.

234. Guo W, Grewe F, Cobo-Clark A, Fan W, Duan Z, Adams RP, et al. Predominant and substoichiometric isomers of the plastid genome coexist within *Juniperus* plants and have shifted multiple times during cupressophyte evolution. Genome Biol Evol. 2014;6:580–90.

235. Cai Z, Penaflor C, Kuehl JV, Leebens-Mack J, Carlson JE, de Pamphilis CW, et al. Complete plastid genome sequences of *Drimys*, *Liriodendron*, and *Piper*: implications for the phylogenetic relationships of magnoliids. BMC Evol Biol. 2006;6:77.

236. Chen J, Hao Z, Xu H, Yang L, Liu G, Sheng Y, et al. The complete chloroplast genome sequence of the relict woody plant *Metasequoia glyptostroboides* Hu et Cheng. Front Plant Sci. 2015;6:447.

237. Ravi V, Khurana JP, Tyagi AK, Khurana P. The chloroplast genome of mulberry: complete nucleotide sequence, gene organization and comparative analysis. Tree Genet Genomes. 2006;3:49–59.

238. Straub SC, Moore MJ, Soltis PS, Soltis DE, Liston A, Livshultz T. Phylogenetic signal detection from an ancient rapid radiation: effects

of noise reduction, long-branch attraction, and model selection in crown clade Apocynaceae. Mol Phylogenet Evol. 2014;80:169–85.

239. Besnard G, Hernández P, Khadari B, Dorado G, Savolainen V. Genomic profiling of plastid DNA variation in the Mediterranean olive tree. BMC Plant Biol. 2011;11:80.

240. Yang M, Zhang X, Liu G, Yin Y, Chen K, Yun Q, et al. The complete chloroplast genome sequence of date palm (*Phoenix dactylifera* L.). PLoS One. 2010;5, e12762.

241. Fang MF, Wang YJ, Zu YM, Dong WL, Wang RN, Deng TT, Li ZH. The complete chloroplast genome of the Taiwan red pine *Pinus taiwanensis* (Pinaceae). Mitochondrial DNA. 2016;27:2732–3.

242. Vieira Ldo N, Faoro H, Rogalski M, Fraga HP, Cardoso RL, de Souza EM, et al. The complete chloroplast genome sequence of *Podocarpus lambertii*: genome structure, evolutionary aspects, gene content and SSR detection. PLoS One. 2014;9, e90618.

243. Okumura S, Sawada M, Park YW, Hayashi T, Shimamura M, Takase H, Tomizawa K. Transformation of poplar (*Populus alba*) plastids and expression of foreign proteins in tree chloroplasts. Transgenic Res. 2006;15:637–46.

244. Wang S, Shi C, Gao LZ. Plastid genome sequence of a wild woody oil species, *Prinsepia utilis*, provides insights into evolutionary and mutational patterns of Rosaceae chloroplast genomes. PLoS One. 2013;8, e73946.

245. Terakami S, Matsumura Y, Kurita K, Kanamori H, Katayose Y, Yamamoto T, Katayama H. Complete sequence of the chloroplast genome from pear (*Pyrus pyrifolia*): genome structure and comparative analysis. Tree Genet Genomes. 2012;8:841–54.

246. Alexander LW, Woeste KE. Pyrosequencing of the northern red oak (*Quercus rubra* L.) chloroplast genome reveals high quality polymorphisms for population management. Tree Genet Genomes. 2014;10:803–12.

247. Yang B, Li M, Ma J, Fu Z, Xu X, Chen Q, et al. The complete chloroplast genome sequence of *Sapindus mukorossi*. Mitochondrial DNA. 2016;27:1825–6.PubMed

248. Fajardo D, Senalik D, Ames M, Zhu H, Steffan SA, Harbut R, et al. Complete plastid genome sequence of *Vaccinium macrocarpon*:

structure, gene content, and rearrangements revealed by next generation sequencing. Tree Genet Genomes. 2013;9:489–98.

249. Yap JY, Rohner T, Greenfield A, Van Der Merwe M, McPherson H, Glenn W, et al. Complete chloroplast genome of the wollemi pine (*Wollemia nobilis*): structure and evolution. PLoS One. 2015;10, e0128126.

250. Shaw J, Lickey EB, Beck JT, Farmer SB, Liu W, Miller J, et al. The tortoise and the hare II: relative utility of 21 noncoding chloroplast DNA sequences for phylogenetic analysis. Am J Bot. 2005;92:142–66.

251. Ruhlman TA, Rajasekaran K, Cary JW. Expression of chloroperoxidase from *Pseudomonas pyrrocinia* in tobacco plastids for fungal resistance. Plant Sci. 2014;228:98–106.

252. Chakrabarti SK, Lutz KA, Lertwiriyawong B, Svab Z, Maliga P. Expression of the *cry9Aa2 B.t.* gene in tobacco chloroplasts confers resistance to potato tuber moth. Transgenic Res. 2006;15:481–8.

253. DeGray G, Rajasekaran K, Smith F, Sanford J, Daniell H. Expression of an antimicrobial peptide via the chloroplast genome to control phytopathogenic bacteria and fungi. Plant Physiol. 2001;127:852–62.

254. Chen PJ, Senthilkumar R, Jane WN, He Y, Tian Z, Yeh KW. Transplastomic *Nicotiana benthamiana* plants expressing multiple defence genes encoding protease inhibitors and chitinase display broad-spectrum resistance against insects, pathogens and abiotic stresses. Plant Biotechnol J. 2014;12:503–15.

255. Wang YP, Wei ZY, Zhang YY, Lin CJ, Zhong XF, Wang YL, et al. Chloroplast-expressed MSI-99 in tobacco improves disease resistance and displays inhibitory effect against rice blast fungus. Int J Mol Sci. 2015;16:4628–41.

256. McBride KE, Svab Z, Schaaf DJ, Hogan PS, Stalker DM, Maliga P. Amplification of a chimeric *Bacillus* gene in chloroplasts leads to an extraordinary level of an insecticidal protein in tobacco. Biotechnology (N Y). 1995;13:362–5.

257. Kota M, Daniell H, Varma S, Garczynski SF, Gould F, Moar WJ. Overexpression of the *Bacillus thuringiensis* (*Bt*) Cry2Aa2 protein in chloroplasts confers resistance to plants against susceptible and

Bt-resistant insects. Proc Natl Acad Sci U S A. 1999;96:1840–5.

258. Lee SB, Kwon HB, Kwon SJ, Park SC, Jeong MJ, Han SE, et al. Accumulation of trehalose within transgenic chloroplasts confers drought tolerance. Mol Breeding. 2003;11:1–13.
259. Ruiz ON, Hussein HS, Terry N, Daniell H. Phytoremediation of organomercurial compounds via chloroplast genetic engineering. Plant Physiol. 2003;132:1344–52.
260. Ruiz ON, Alvarez D, Torres C, Roman L, Daniell H. Metallothionein expression in chloroplasts enhances mercury accumulation and phytoremediation capability. Plant Biotechnol J. 2011;9:609–17.
261. Lutz KA, Knapp JE, Maliga P. Expression of bar in the plastid genome confers herbicide resistance. Plant Physiol. 2001;125:1585–90.
262. Ye GN, Hajdukiewicz PT, Broyles D, Rodriguez D, Xu CW, Nehra N, Staub JM. Plastid-expressed 5-enolpyruvylshikimate-3-phosphate synthase genes provide high level glyphosate tolerance in tobacco. Plant J. 2001;25:261–70.
263. Shimizu M, Goto M, Hanai M, Shimizu T, Izawa N, Kanamoto H, et al. Selectable tolerance to herbicides by mutated acetolactate synthase genes integrated into the chloroplast genome of tobacco. Plant Physiol. 2008;147:1976–83.
264. Iamtham S, Day A. Removal of antibiotic resistance genes from transgenic tobacco plastids. Nat Biotechnol. 2000;18:1172–6.
265. Dufourmantel N, Dubald M, Matringe M, Canard H, Garcon F, Job C, et al. Generation and characterization of soybean and marker-free tobacco plastid transformants over-expressing a bacterial 4-hydroxyphenylpyruvate dioxygenase which provides strong herbicide tolerance. Plant Biotechnol J. 2007;5:118–33.
266. Fouad WM, Altpeter F. Transplastomic expression of bacterial L-aspartate-alpha-decarboxylase enhances photosynthesis and biomass production in response to high temperature stress. Transgenic Res. 2009;18:707–18.
267. Lu Y, Rijzaani H, Karcher D, Ruf S, Bock R. Efficient metabolic pathway engineering in transgenic tobacco and tomato plastids with synthetic multigene operons. Proc Natl Acad Sci U S A. 2013;110:E623–32.

268. Ruiz ON, Daniell H. Engineering cytoplasmic male sterility via the chloroplast genome by expression of beta-ketothiolase. Plant Physiol. 2005;138:1232–46.

269. Yabuta Y, Tanaka H, Yoshimura S, Suzuki A, Tamoi M, Maruta T, Shigeoka S. Improvement of vitamin E quality and quantity in tobacco and lettuce by chloroplast genetic engineering. Transgenic Res. 2013;22:391–402.

270. Harada H, Maoka T, Osawa A, Hattan J, Kanamoto H, Shindo K, et al. Construction of transplastomic lettuce (*Lactuca sativa*) dominantly producing astaxanthin fatty acid esters and detailed chemical analysis of generated carotenoids. Transgenic Res. 2014;23:303–15.

271. Pengelly JJ, Forster B, von Caemmerer S, Badger MR, Price GD, Whitney SM. Transplastomic integration of a cyanobacterial bicarbonate transporter into tobacco chloroplasts. J Exp Bot. 2014;65:3071–80.

272. Sanz-Barrio R, Corral-Martinez P, Ancin M, Segui-Simarro JM, Farran I. Overexpression of plastidial thioredoxin f leads to enhanced starch accumulation in tobacco leaves. Plant Biotechnol J. 2013;11:618–27.

273. Elghabi Z, Karcher D, Zhou F, Ruf S, Bock R. Optimization of the expression of the HIV fusion inhibitor cyanovirin-N from the tobacco plastid genome. Plant Biotechnol J. 2011;9:599–608.

274. Agrawal P, Verma D, Daniell H. Expression of *Trichoderma reesei* beta-mannanase in tobacco chloroplasts and its utilization in lignocellulosic woody biomass hydrolysis. PLoS One. 2011;6, e29302.

275. Verma D, Jin S, Kanagaraj A, Singh ND, Daniel J, Kolattukudy PE, et al. Expression of fungal cutinase and swollenin in tobacco chloroplasts reveals novel enzyme functions and/or substrates. PLoS One. 2013;8, e57187.

276. Saxena B, Subramaniyan M, Malhotra K, Bhavesh NS, Potlakayala SD, Kumar S. Metabolic engineering of chloroplasts for artemisinic acid biosynthesis and impact on plant growth. J Biosci. 2014;39:33–41.

277. Yu LX, Gray BN, Rutzke CJ, Walker LP, Wilson DB, Hanson MR.

Expression of thermostable microbial cellulases in the chloroplasts of nicotine-free tobacco. J Biotechnol. 2007;131:362–9.

278. Petersen K, Bock R. High-level expression of a suite of thermostable cell wall-degrading enzymes from the chloroplast genome. Plant Mol Biol. 2011;76:311–21.

279. Arai Y, Shikanai T, Doi Y, Yoshida S, Yamaguchi I, Nakashita H. Production of polyhydroxybutyrate by polycistronic expression of bacterial genes in tobacco plastid. Plant Cell Physiol. 2004;45:1176–84.

280. Hasunuma T, Miyazawa SI, Yoshimura S, Shinzaki Y, Tomizawa KI, Shindo K, et al. Biosynthesis of astaxanthin in tobacco leaves by transplastomic engineering. Plant J. 2008;55:857–68.

281. Nakahira Y, Ishikawa K, Tanaka K, Tozawa Y, Shiina T. Overproduction of hyperthermostable beta-1,4-endoglucanase from the archaeon *Pyrococcus horikoshii* by tobacco chloroplast engineering. Biosci Biotechnol Biochem. 2013;77:2140–3.

282. Sherman A, Su J, Lin S, Wang X, Herzog RW, Daniell H. Suppression of inhibitor formation against FVIII in a murine model of hemophilia A by oral delivery of antigens bioencapsulated in plant cells. Blood. 2014;124:1659–68.

283. Sanz-Barrio R, Fernandez-San Millan A, Corral-Martinez P, Segui-Simarro JM, Farran I. Tobacco plastidial thioredoxins as modulators of recombinant protein production in transgenic chloroplasts. Plant Biotechnol J. 2011;9:639–50.

284. Farran I, McCarthy-Suarez I, Rio-Manterola F, Mansilla C, Lasarte JJ, Mingo-Castel AM. The vaccine adjuvant extra domain A from fibronectin retains its proinflammatory properties when expressed in tobacco chloroplasts. Planta. 2010;231:977–90.

285. Boyhan D, Daniell H. Low-cost production of proinsulin in tobacco and lettuce chloroplasts for injectable or oral delivery of functional insulin and C-peptide. Plant Biotechnol J. 2011;9:585–98.

286. Fernandez-San Millan A, Mingo-Castel A, Miller M, Daniell H. A chloroplast transgenic approach to hyper-express and purify human serum albumin, a protein highly susceptible to proteolytic degradation. Plant Biotechnol J. 2003;1:71–9.

287. Daniell H, Ruiz G, Denes B, Sandberg L, Langridge W. Optimization

of codon composition and regulatory elements for expression of human insulin like growth factor-1 in transgenic chloroplasts and evaluation of structural identity and function. BMC Biotechnol. 2009;9:33.

288. Wang X, Su J, Sherman A, Rogers GL, Liao G, Hoffman BE, et al. Plant-based oral tolerance to hemophilia therapy employs a complex immune regulatory response including LAP+CD4+ T cells. Blood. 2015;125:2418–27.

289. Su J, Sherman A, Doerfler PA, Byrne BJ, Herzog RW, Daniell H. Oral delivery of Acid Alpha Glucosidase epitopes expressed in plant chloroplasts suppresses antibody formation in treatment of Pompe mice. Plant Biotechnol J. 2015;13:1023–32.

290. Youm JW, Jeon JH, Kim H, Min SR, Kim MS, Joung H, et al. High-level expression of a human beta-site APP cleaving enzyme in transgenic tobacco chloroplasts and its immunogenicity in mice. Transgenic Res. 2010;19:1099–108.

291. Arlen PA, Falconer R, Cherukumilli S, Cole A, Cole AM, Oishi KK, Daniell H. Field production and functional evaluation of chloroplast-derived interferon-alpha2b. Plant Biotechnol J. 2007;5:511–25.

292. Ruhlman T, Ahangari R, Devine A, Samsam M, Daniell H. Expression of cholera toxin B-proinsulin fusion protein in lettuce and tobacco chloroplasts--oral administration protects against development of insulitis in non-obese diabetic mice. Plant Biotechnol J. 2007;5:495–510.

293. Leelavathi S, Reddy VS. Chloroplast expression of His-tagged GUS-fusions: a general strategy to overproduce and purify foreign proteins using transplastomic plants as bioreactors. Mol Breeding. 2003;11:49–58.

294. Lim S, Ashida H, Watanabe R, Inai K, Kim YS, Mukougawa K, et al. Production of biologically active human thioredoxin 1 protein in lettuce chloroplasts. Plant Mol Biol. 2011;76:335–44.

295. Nadai M, Bally J, Vitel M, Job C, Tissot G, Botterman J, Dubald M. High-level expression of active human alpha1-antitrypsin in transgenic tobacco chloroplasts. Transgenic Res. 2009;18:173–83.

296. Gisby MF, Mellors P, Madesis P, Ellin M, Laverty H, O›Kane S, et al. A synthetic gene increases TGF beta 3 accumulation by 75-fold

in tobacco chloroplasts enabling rapid purification and folding into a biologically active molecule. Plant Biotechnol J. 2011;9:618–28.

297. Farran I, Rio-Manterola F, Iniguez M, Garate S, Prieto J, Mingo-Castel AM. High-density seedling expression system for the production of bioactive human cardiotrophin-1, a potential therapeutic cytokine, in transgenic tobacco chloroplasts. Plant Biotechnol J. 2008;6:516–27.

298. Staub JM, Garcia B, Graves J, Hajdukiewicz PT, Hunter P, Nehra N, et al. High-yield production of a human therapeutic protein in tobacco chloroplasts. Nat Biotechnol. 2000;18:333–8.

299. Oey M, Lohse M, Scharff LB, Kreikemeyer B, Bock R. Plastid production of protein antibiotics against pneumonia via a new strategy for high-level expression of antimicrobial proteins. Proc Natl Acad Sci U S A. 2009;106:6579–84.

300. Koya V, Moayeri M, Leppla SH, Daniell H. Plant-based vaccine: mice immunized with chloroplast-derived anthrax protective antigen survive anthrax lethal toxin challenge. Infect Immun. 2005;73:8266–74.

301. Watson J, Koya V, Leppla SH, Daniell H. Expression of Bacillus anthracis protective antigen in transgenic chloroplasts of tobacco, a non-food/feed crop. Vaccine. 2004;22:4374–84.

302. Millan AFS, Ortigosa SM, Hervas-Stubbs S, Corral-Martinez P, Segui-Simarro JM, Gaetan J, et al. Human papillomavirus L1 protein expressed in tobacco chloroplasts self-assembles into virus-like particles that are highly immunogenic. Plant Biotechnol J. 2008;6:427–41.

303. Yacono MD, Farran I, Becher ML, Sander V, Sanchez VR, Martin V, et al. A chloroplast-derived *Toxoplasma gondii* GRA4 antigen used as an oral vaccine protects against toxoplasmosis in mice. Plant Biotechnol J. 2012;10:1136–44.

304. Kolotilin I, Kaldis A, Devriendt B, Joensuu J, Cox E, Menassa R. Production of a subunit vaccine candidate against porcine post-weaning diarrhea in high-biomass transplastomic tobacco. Plos One. 2012;7, e42405.

305. Arlen PA, Singleton M, Adamovicz JJ, Ding Y, Davoodi-Semiromi

A, Daniell H. Effective plague vaccination via oral delivery of plant cells expressing F1-V antigens in chloroplasts. Infect Immun. 2008;76:3640–50.

306. Molina A, Hervas-Stubbs S, Daniell H, Mingo-Castel AM, Veramendi J. High-yield expression of a viral peptide animal vaccine in transgenic tobacco chloroplasts. Plant Biotechnol J. 2004;2:141–53.

307. Lentz EM, Mozgovoj MV, Bellido D, Santos MJD, Wigdorovitz A, Bravo-Almonacid FF. VP8*antigen produced in tobacco transplastomic plants confers protection against bovine rotavirus infection in a suckling mouse model. J Biotechnol. 2011;156:100–7.

308. Daniell H, Lee SB, Panchal T, Wiebe PO. Expression of the native cholera toxin B subunit gene and assembly as functional oligomers in transgenic tobacco chloroplasts. J Mol Biol. 2001;311:1001–9.

309. Kang TJ, Loc NH, Jang MO, Jang YS, Kim YS, Seo JE, Yang MS. Expression of the B subunit of *E. coli* heat-labile enterotoxin in the chloroplasts of plants and its characterization. Transgenic Res. 2003;12:683–91.

310. Chebolu S, Daniell H. Stable expression of Gal/GalNAc lectin of *Entamoeba histolytica* in transgenic chloroplasts and immunogenicity in mice towards vaccine development for amoebiasis. Plant Biotechnol J. 2007;5:230–9.

311. Glenz K, Bouchon B, Stehle T, Wallich R, Simon MM, Warzecha H. Production of a recombinant bacterial lipoprotein in higher plant chloroplasts. Nat Biotechnol. 2006;24:76–7.

312. Rosales-Mendoza S, Alpuche-Solis AG, Soria-Guerra RE, Moreno-Fierros L, Martinez-Gonzalez L, Herrera-Diaz A, Korban SS. Expression of an *Escherichia coli* antigenic fusion protein comprising the heat labile toxin B subunit and the heat stable toxin, and its assembly as a functional oligomer in transplastomic tobacco plants. Plant J. 2009;57:45–54.

313. Soria-Guerra RE, Alpuche-Solis AG, Rosales-Mendoza S, Moreno-Fierros L, Bendik EM, Martinez-Gonzalez L, Korban SS. Expression of a multi-epitope DPT fusion protein in transplastomic tobacco plants retains both antigenicity and immunogenicity of all three components of the functional oligomer. Planta. 2009;229:1293–302.

314. Rubio-Infante N, Govea-Alonso DO, Alpuche-Solis AG, Garcia-Hernandez AL, Soria-Guerra RE, Paz-Maldonado LMT, et al. A chloroplast-derived C4V3 polypeptide from the human immunodeficiency virus (HIV) is orally immunogenic in mice. Plant Mol Biol. 2012;78:337–49.

315. Waheed MT, Thones N, Muller M, Hassan SW, Gottschamel J, Lossl E, et al. Plastid expression of a double-pentameric vaccine candidate containing human papillomavirus-16 L1 antigen fused with LTB as adjuvant: transplastomic plants show pleiotropic phenotypes. Plant Biotechnol J. 2011;9:651–60.

316. Gonzalez-Rabade N, McGowan EG, Zhou F, McCabe MS, Bock R, Dix PJ, et al. Immunogenicity of chloroplast-derived HIV-1 p24 and a p24-Nef fusion protein following subcutaneous and oral administration in mice. Plant Biotechnol J. 2011;9:629–38.

317. Zhou YX, Lee MYT, Ng JMH, Chye ML, Yip WK, Zee SY, Lam E. A truncated hepatitis E virus ORF2 protein expressed in tobacco plastids is immunogenic in mice. World J Gastroenterol. 2006;12:306–12.

318. Shao HB, He DM, Qian KX, Shen GF, Su ZL. The expression of classical swine fever virus structural protein E2 gene in tobacco chloroplasts for applying chloroplasts as bioreactors. C R Biol. 2008;331:179–84.

319. Tregoning JS, Nixon P, Kuroda H, Svab Z, Clare S, Bowe F, et al. Expression of tetanus toxin Fragment C in tobacco chloroplasts. Nucleic Acids Res. 2003;31:1174–9.

320. Morgenfeld M, Lentz E, Segretin ME, Alfano EF, Bravo-Almonacid F. Translational fusion and redirection to thylakoid lumen as strategies to enhance accumulation of human papillomavirus E7 antigen in tobacco chloroplasts. Mol Biotechnol. 2014;56:1021–31.

CHAPTER

6

CHLOROPLAST SIGNALING WITHIN, BETWEEN AND BEYOND CELLS

Krzysztof Bobik and Tessa M. Burch-Smith*

Department of Biochemistry, Cellular and Molecular Biology, University of Tennessee, Knoxville, TN, USA

The most conspicuous function of plastids is the oxygenic photosynthesis of chloroplasts, yet plastids are super-factories that produce a plethora of compounds that are indispensable for proper plant physiology and development. Given their origins as free-living prokaryotes, it is not surprising that plastids possess their own genomes whose expression is essential to plastid function. This semi-autonomous character of plastids requires the existence of sophisticated regulatory mechanisms that provide reliable communication between them and other cellular compartments. Such intracellular signaling is necessary for coordinating whole-cell responses to constantly varying environmental cues and cellular metabolic needs. This is achieved by plastids acting as receivers and transmitters of specific signals that coordinate expression of the nuclear and plastid genomes according to particular needs. In this review we will consider the so-called retrograde signaling occurring between plastids and nuclei, and between plastids and other organelles. Another important role of the plastid we will discuss is the involvement of plastid signaling in biotic and abiotic stress that, in addition to influencing retrograde signaling, has direct effects on several cellular compartments including the cell wall. We will also review recent evidence pointing to an intriguing function of chloroplasts in regulating intercellular symplasmic transport. Finally, we consider an intriguing yet less widely known aspect of plant biology, chloroplast signaling from the perspective of the entire plant. Thus, accumulating evidence highlights that chloroplasts, with their complex signaling pathways, provide a mechanism for exquisite regulation of plant development, metabolism and responses to the environment. As chloroplast processes are targeted for engineering for improved productivity the effect of such modifications on chloroplast signaling will have to be carefully considered in order to avoid unintended consequences on plant growth and development.

INTRODUCTION

According to the endosymbiotic theory plastids originated from free-living cyanobacteria that were engulfed by early eukaryotic cells. These cyanobacteria were retained by their hosts and have co-evolved with their host cells over 1.5 billion years to become an integral part of the modern plant cell (Yoon et al., 2004; Nakayama and Archibald, 2012). It is accepted that the successful stable integration occurred because of exceptional mutual benefits: the eukaryotic cell was able to establish an autotrophic lifestyle, while the engulfed cyanobacteria reached a pathogen-free asylum. Accumulating evidence suggests that this stable symbiosis between cyanobacteria and the eukaryotic cell was facilitated by infection of the latter with Chlamydiales pathogens (Ball et al., 2011, 2013).

The most conspicuous function of modern plastids is the sophisticated oxygenic photosynthesis performed by chloroplasts. However, plastids perform many other functions that are critical for proper plant development and physiology including the synthesis of amino acids, nucleotides and fatty acids, production of phytohormones, some vitamins and a multitude of secondary metabolites, as well nitrogen and sulfur assimilation. Many chloroplast secondary metabolites, besides being necessary for basic plant metabolic functions, are also important for interaction with the environment, as they function in plant defense against pathogen ingress and plant adaptation to stresses including heat, drought and high light. Thus, chloroplasts act as a hub in the cellular response to signals, generating a variety of signals that coordinate a fine-tuned and appropriate response to any given situation (Pfannschmidt and Yang, 2012).

The emerging view of the chloroplast is as a very dynamic signaling compartment. As a specific sensor of intra- and extracellular stimuli, chloroplasts constantly process and integrate a multitude of intracellular signals and pathways in order to sustain homeostasis at both the cellular and organismal levels (Figure 1A). An often over-looked aspect of cell biology is the physical interaction between organelles for coordination of signaling and metabolism (Figure 1B). This area is beginning to receive attention and we will examine these findings as they relate to chloroplasts and their roles in signaling. Given the dizzying array of signals that chloroplasts respond to and produce it is not surprising that there is considerable crosstalk between signaling pathways. This is particularly evident during responses to biotic and abiotic stress (Nakashima et al., 2014; Trotta et al., 2014; Zhou et al., 2015).

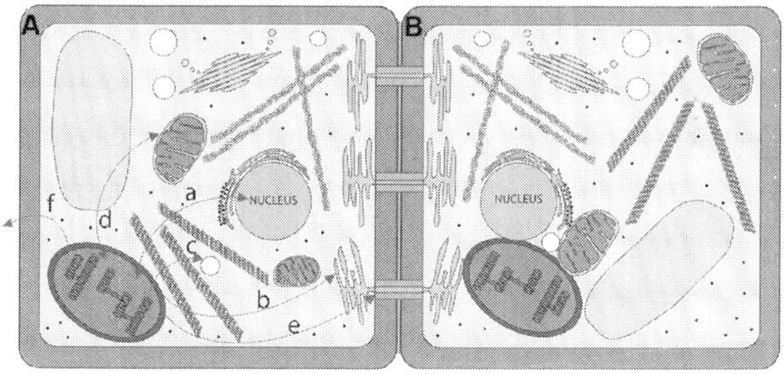

Figure 1. Routes for chloroplast signaling. (A) Chloroplasts generate signals that target multiple intercellular targets. (a) The majority of chloroplast proteins are encoded by the nucleus, and the import of those proteins into the chloroplast is anterograde signaling. In turn, several chloroplast products act as retrograde signals to regulate expression of nucleus-encoded genes. (b) Chloroplasts are metabolically coupled to the ER and it is likely that signals may move from the chloroplast to the ER. (c) Chloroplasts and peroxisomes are also closely associ-

ated, and numerous chloroplast products are substrates for peroxisomal pathways. (d) Mitochondria and chloroplasts are known to signal to each other. (e) Chloroplast signals regulate intercellular trafficking via plasmodesmata. It is not clear if this signaling is direct or involves retrograde signaling to the nucleus. (f) Chloroplasts produce volatile compounds that can signal to neighboring plants during pathogen attack. (B) The physical interaction between chloroplasts and various organelles may serve as a direct route for signaling.

Chloroplast signaling is not limited to exerting its effects within the cell. It is clear that chloroplast-derived signals can travel far beyond their site of production to induce changes in distal parts of the plant (e.g., Petrillo et al., 2014). Further, chloroplasts apparently regulate intercellular trafficking via the channels known as plasmodesmata. Through this action, chloroplasts could regulate almost all aspects of plant growth and development as it is becoming clear that not only do metabolites but also hormones, transcription factors and small RNA molecules use these channels for intercellular communication. Beyond whole-plant signaling, chloroplast signaling can impact entire ecosystem through production of volatile compounds.

While we focus here on the chloroplasts in mature leaves, it can be expected that other types of plastids, e.g., proplastids, etioplasts, or leucoplasts, participate in signaling networks in response to their unique developmental states and environmental conditions. Further, it is also likely that at any given time different subpopulations of chloroplasts within a cell are in various metabolic or physiologic states and are therefore likely to be involved in distinct signaling processes. Thus chloroplast signaling is complex, and dissecting crosstalk and feedback mechanisms remains a daunting task. With attempts to engineer chloroplasts for specialized or improved metabolic outputs, attention must be paid to how these adjustments may impact chloroplast behavior if unintended consequences are to be diminished. These consequences would not be limited to the chloroplasts, but could extend even to other plants cultivated in the vicinity of the engineered plants.

CHLOROPLASTS IN INTRACELLULAR SIGNALING

Chloroplast Signaling to the Nucleus

Over evolutionary time a significant number of the cyanobacterial genes were transferred to the host nucleus (Race et al., 1999; McFadden, 2001). These genes subsequently acquired sequences that function as transit peptides to enable import of their protein products back into plastids. The nucleus therefore exerts considerable control over chloroplast functions, and this nucleus-to-chloroplast signaling is termed anterograde signaling. Importantly, several genes encoding proteins that are integral components of photophosphorylation or photosynthesis were retained in the cyanobacterial genome (Allen, 1993; Race et al., 1999; Raven and Allen, 2003). Thus, in order to establish a stable eukaryotic plant cell it was necessary to synchronize the activities of both genomes. This has been achieved by creating a complex signaling system between the nucleus and plastids, able to transfer information and efficiently adjust gene expression in both compartments according to particular needs. Plastid derived-signals that regulate nuclear gene expression represent retrograde signaling. While anterograde signaling is well understood, it has been more challenging to unravel the molecular details of retrograde signaling.

Identification of molecules and signaling strategies underlying so-called retrograde signaling represents a long-standing quest in plant biology. Historically, Bradbeer et al. (1979) provided the first report describing the existence of communication between those organelles. They observed that the barley (*Hordeum vulgare*) *albostrians* chloroplast ribosome-deficient mutant had severely decreased chloroplast protein synthesis, in conjunction with depressed expression of nucleus-encoded chloroplast genes. This revolutionary discovery was in opposition to Ellis' "cytoplasmic control principle" that posited control of organellar protein synthesis by cytoplasmic components (Ellis, 1977). Bradbeer's discovery was soon confirmed by other researchers who treated young seedlings of mustard, *Arabidopsis*, pea or barley with lincomycin, chloramphenicol or streptomycin – inhibitors of plastid protein synthesis (Oelmuller et al., 1986; Susek et al., 1993; Yoshida et al., 1998; Sullivan and Gray,

1999). Besides the strategy of relying on systems with compromised chloroplast ribosome function, other approaches to perturbing distinct aspects of chloroplast function have also successfully interrogated chloroplast-to-nucleus signaling. The induction of carotenoid deficiency in genetic mutants or in plants treated with norflurazon (a herbicide that inhibits phytoene desaturase and blocks carotenoid synthesis), as well as in plants with compromised tetrapyrrole synthesis (resulting in accumulation of intermediates) caused suppressed expression of nucleus-encoded chloroplast genes (Johanningmeier and Howell, 1984; Mayfield and Taylor, 1984; Susek et al., 1993; Kropat et al., 1997, 2000; Strand et al., 2003; Zhang et al., 2011). These and other studies led to the realization that the plastid functional status can regulate the expression of photosynthesis associated nuclear genes (PhANGs; Surpin et al., 2002). Moreover, analysis of the *Arabidopsis* chloroplast ribosomal protein mutant *rps1* revealed that chloroplast translational capacity is a critical factor in developing heat tolerance. This is mediated by inducing expression of the heat stress transcription factor HsfA2, a key regulator of heat tolerance (Yu et al., 2012). Therefore, the functional status of chloroplasts also regulates nuclear genes involved in heat-tolerance. The involvement of plastid translation in retrograde signaling and plant development was recently discussed in detail (Tiller and Bock, 2014).

The intense search for factors mediating this chloroplast-to-nucleus communication has identified a set of plastid metabolic intermediates. Importantly, proteins with functions in both chloroplasts and nuclei have been recently identified and have been proposed to participate in retrograde signaling. These retrograde signals are expected to work by modifying the expression of nuclear genes in order to adapt plant development and physiology to constantly changing environmental conditions. Currently two major modes of retrograde signaling are distinguished and they are involved in so-called biogenic and operational type of control. Whereas the former includes signals responsible for chloroplast and photosynthesis biogenesis, the latter act in response to changing environmental cues in fully developed chloroplasts (Pogson et al., 2008). The best-studied target of retrograde signaling is represented by PhANGs, but plastids are also involved in tuning the expression of nuclear genes involved in response to a plethora of biotic and abiotic conditions. A recent meta-analysis of microarray studies of systems where high levels of retrograde signaling were induced has identified a core module of 39 nuclear genes that were subject to regulation in response to all signals examined (Glasser et al., 2014). The genes in this group, presumably representing the core retrograde-response module, are all known to be responsive to sugar, reactive oxygen species (ROS), abscisic acid (ABA) and/or auxin signaling pathways. Thus retrograde signaling may exploit a common component of these signaling pathways to mediate changes in gene expression.

THE GENOMES UNCOUPLED (GUN) MUTANTS – AIMING TO THE NUCLEUS WITH GUNS

Very helpful in deciphering the retrograde signaling phenomenon were mutants isolated from genetic screens. The *gun* (*genome uncoupled*) mutants escaped the pattern of suppressed PhANG expression despite defective chloroplast physiology or inhibited biogenesis. There are numerous excellent reviews of the role of *guns* in retrograde signaling (Woodson and Chory, 2008, 2012; Barajas-Lopez Jde et al., 2013; Chi et al., 2013). So far, six *gun* mutants have been identified and they can be classified according to pathways they belong to. Whereas the *gun1* mutant results from mutation in a gene encoding a chloroplast-localized pentatricopeptide repeat-containing protein (PPR), the *gun2-6* mutants are associated with tetrapyrroles synthesis (Susek et al., 1993; Mochizuki et al., 2001; Larkin et al., 2003; Strand et al., 2003; Koussevitzky et al., 2007). The exact role of GUN1 in PhANG regulation is far from understood, however, it is known to act upstream of ABSCISCIC ACID INSENSITIVE 4 (ABI4), an APETALA 2-type transcription factor that binds to the ACGT motif of light- and ABA-responsive elements (Koussevitzky et al., 2007). Interestingly, the expression of ABI4 was regulated by PTM, a chloroplast PHD-type transcription factor (Sun et al., 2011). The involvement of the key enzymes of the tetrapyrrole synthesis pathway in the *gun* phenotype led to detailed investigations of tetrapyrroles, especially Mg-protoIX, as putative retrograde signals. However no correlation between Mg-protoIX levels and retrograde signaling could be established (Matsui et al., 2008; Moulin et al., 2008). Interestingly, the *gun6* mutant identified heme as a strong candidate for mediating plastid-to-nucleus signaling (Woodson et al., 2011). Moreover, it was proposed that the impact of tetrapyrrole biosynthesis on nuclear gene expression is mediated by singlet oxygen ($^1O_2$)-induced signaling and feedback regulated 5-aminolevulinic acid (ALA) synthesis (Schlicke et al., 2014).

SAL1-PAP Chloroplast Retrograde Pathway

The detailed analysis of *sal1*, an *Arabidopsis* phosphonucleotidase mutant, has identified a known second messenger as acting in chloroplast-to-nucleus signaling. Estavillo et al. (2011) have demonstrated that the chloroplast and mitochondria-localized SAL1 phosphatase regulates the steady-state level of 3′-phosphoadenosine 5′-phosphate (PAP) by dephosphorylating it to an adenosine monophosphate (AMP). In the *sal1* mutant, or in response to drought stress or high light intensity, PAP levels increased, inducing expression of *ASCORBATE PEROXIDASE 2* and *EARLY LIGHT INDUCIBLE PROTEIN 2*, two nuclear genes whose expression is induced by high light stress (Harari-Steinberg et al., 2001; Caverzan et al., 2012). It has been proposed that PAP travels from chloroplasts to the nucleus where it regulates nuclear gene expression. Nucleus-localized exoribonucleases (XRNs) are likely targets of PAP, and by repressing their activity PAP may stimulate expression of high light and drought-responsive genes, leading to increased tolerance (Estavillo et al., 2011; Figure 2A).

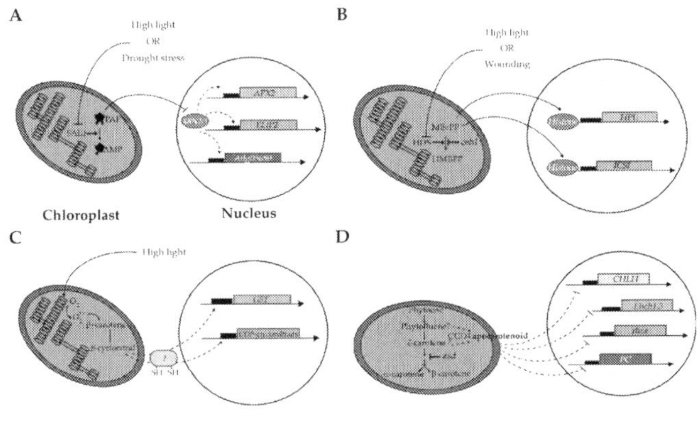

Figure2. Mechanisms of chloroplast-to-nucleus signaling. (A) Retrograde signaling by PAP. High light or drought stress inhibits SAL1 phosphatase and leads to the accumulation of PAP. PAP likely inhibits specific exoribonucleases

(XRNs) to modify nuclear genes expression. *APX2* and *ELIP2* stand for *ASCORBATE PEROXIDASE 2* and *EARLY LIGHT INDUCIBLE PROTEIN 2* genes, respectively. **(B)** Retrograde signaling by MEcPP. High light or wounding inhibits 1-hydroxy-2-methyl-2-(E)-butenyl-4-diphosphate synthase (HDS), leading to the subsequent accumulation of MEcPP. MEcPP affects nuclear gene expression via a mechanism proposed to involve chromatin remodeling by destabilizing DNA-histone interactions. *HPL* and *ICS1* stand for *HYDROPEROXIDE LYASE* and *ISOCHORISMATE SYNTHASE 1*genes, respectively. **(C)** Carotenoid-derivative β-cyclocitral mediates retrograde signaling. The ROS singlet oxygen induces formation of β-cyclocitral during high light treatment. β-cyclocitral's action on selected nuclear genes is proposed to involve proteins containing sulphydryl groups. The genes depicted are *GLUTATHIONE-S-TRANSPHERASE (GST)* and *UDP-glycosyltransferase.* **(D)** An unidentified apocarotenoid affects expression of nuclear genes. It is proposed that the putative signaling apocarotenoid accumulates in chloroplasts due to compromised ζ-carotene desaturase activity that results in accumulation of phytofluene and ζ-carotene, putative substrates for the carotenoid cleavage deoxygenase 4 (CCD4) enzyme that is prerequisite for the putative apocarotenoid synthesis. *CHLH, Lhcb1.3, rbcs* and *PC* stand for genes encoding the subunit H of the Mg-chelatase complex, light-harvesting complex *1.3* isoform, the Rubisco small subunit and plastocyanin, respectively.

Methylerythritol (MEcPP) Retrograde Pathway

Isoprenoid metabolism is a major biosynthetic pathway in plants (Cordoba et al., 2009). The Arabidopsis constitutively expressing HPL (ceh1) mutant displays enhanced expression of hydroperoxide lyase (HPL), a stress-inducible nuclear gene encoding a plastid-localized protein of the oxylipin pathway. The ceh1mutation disrupted a plastid-localized enzyme (HDS) that catalyzes conversion of methylerythritol cyclodiphosphate (MEcPP) to hydroxymethylbutenyl diphosphate (HMBPP; Xiao et al., 2012). The absence of CEH1 led to accumulation of MEcPP and induced the expression of a subset of stress-associated genes, including ISOCHORISMATE SYNTHASE 1 (a key plastidial enzyme in the salicylic acid (SA)-biosynthetic pathway) and HPL, but not ALLEN OXIDE SYNTHASE [AOS, encoding a plastid-localized protein of the jasmonic acid (JA)-biosynthetic pathway]. HDS-depleted plants with increased levels of MEcPP accumulated SA and displayed increased resistance to infection by biotrophic pathogens. Importantly, elevated MEcPP levels and increased expression of HPL are observed upon both wounding and high light treatment, demonstrating involvement of MEcPP in a retrograde pathway involved in abiotic stresses (operational control) distinct from the gun signaling pathway. Therefore MEcPP is a retrograde signal inducing targeted stress responses. The proposed mechanism of MEcPP action involves direct modification of chromatin remodeling by disruption of DNA-histone interactions (Xiao et al., 2012; Figure 2B).

Derivatives of Carotenoids as Signaling Molecules

Carotenoids are tetraterpenoid products of the isoprenoid biosynthetic pathway that also generates ABA and strigolactones (Ruyter-Spira et al., 2013; Giuliano, 2014). Carotenoids are constituents of the light harvesting complexes where they serve as accessory pigments to extend the absorption spectra of the chlorophylls, and they have critical protective roles as scavengers of singlet oxygen ($^1O_2$) generated by the PSII reaction center (Telfer, 2014). Carotenoid derivatives have recently been proposed to act as chloroplast-generated signaling molecules that link chloroplast activity and nuclear gene expression (Ramel et al., 2012; Avendano-Vazquez et al., 2014; Van Norman et al., 2014).

β-cyclocitral is a product of singlet oxygen-induced β-carotene oxidation. This volatile molecule contains an α,β-unsaturated carbonyl, designating it as a reactive electrophile species (RES). β-cyclocitral's accumulation in *Arabidopsis* leaves during high-light stress correlated with accumulation of singlet oxygen, supporting the notion of β-cyclocitral as an oxidation product of β-carotene (Ramel et al., 2012). Consistent with this, transcriptome analysis by DNA microarrays revealed that about 80% of β-cyclocitral-induced or repressed genes are also responsive to singlet oxygen overproduction in the *Arabidopsis fluorescent (flu)* mutant (op den Camp et al., 2003). Using qRT-PCR Ramel and coworkers demonstrated that all singlet oxygen marker genes tested are also induced by β-cyclocitral. Importantly, this effect seemed to be specific to β-cyclocitral as the overlap between gene expression changes induced by β-cyclocitral and other RES like methyl vinyl ketone or malondialdehyde (MDA) was smaller. Among the genes most strongly induced by β-cyclocitral were 10 glutathione-*S*-transferase (GST) genes and 12 UDP-glycosyltransferases (Figure 2C). Both groups of genes are involved in detoxification processes conferring tolerance to singlet oxygen in *Chlaydomonas reinhardtii* (Ledford et al., 2007). The protective effect of β-cyclocitral resulted in better PSII quantum efficiency and lower lipid peroxidation under high light stress (Ramel et al., 2012). Finally, the current model proposes β-cyclocitral as a stress molecule generated in chloroplasts under photooxidative stress that reprograms gene expression leading to stress acclimation. Thus, in addition to their roles in light harvesting and as antioxidants, carotenoids can also act as signaling molecules. The exact mechanism of β-cyclocitral action or its receptor in the nucleus are unknown, however, it is proposed that as an electrophile with a α,β-unsaturated carbonyl group it could react with electron donors such as proteins containing sulphydryl groups (Ramel et al., 2012, 2013). Very recently, elevated levels of β-cyclocitral were reported in *Arabidopsis* plastoglobule kinase mutants that have defective plastoglobule metabolism (Lundquist et al., 2013). The *abc1k1 abc1k3* double mutant shows rapid chlorosis under high light stress, confirming β-cyclocitral's, role in mediating stress responses.

In addition to β-cyclocitral, apocarotenoids are also potential chloroplast retrograde signaling components. Analysis of the *Arabidopsis ζ-carotene desaturase* mutant (*zds/clb5/spc1/pde181*) displaying arrested chloroplast biogenesis at a very early stage of development led to the conclusion that the accumulation of an uncharacterized apocarotenoid can act as a retrograde signal (Avendano-Vazquez et al., 2014). This apocarotenoid is likely generated by the activity of the carotenoid cleavage deoxygenase 4 (CCD4) enzyme on ζ-carotene (Figure 2D). Accumulation of this putative cleavage product was shown to modulate expression of many nuclear genes required for leaf development leading to a severe phenotype that included arrested chloroplast development and leaves with defective adaxial-abaxial patterning. Since neither ROS nor ABA nor strigolactone signaling pathways were responsible for the observed phenotypes, it was concluded that the putative phytofluene or ζ-carotene –derived apocaroteniod is part of a novel retrograde signaling pathway. Interestingly, the observed defects were restricted to primary leaves, underscoring the differences in developmental regulation between plastids in different organs (Avendano-Vazquez et al., 2014).

A third carotenoid derivative is implicated in plastid retrograde signaling, this time with respect to lateral root (LR) development (Van Norman et al., 2014). Reduced LR formation in *Arabidopsis* seedlings treated with norflurazon was observed and further investigation indicated that a β-carotene derivative is required for prebranch site formation (Van Norman et al., 2014). Extensive genetic analyses ruled out ABA and strigolactone as the carotene-derived signaling molecule involved in LR formation. Additionally, treatment of *Arabidopsis* seedlings with D15, a candidate inhibitor of carotenoid cleavage at the 9,10 position, resulted in a highly significant decrease in LR capacity, suggesting that this unknown apocarotenoid is likely cleaved at this position. Interestingly, even though reduced LR capacity and small albino shoots were observed in carotenoid biosynthesis mutants and in plants treated with norflurazon, plants treated with D15 had green shoots of comparable size to wild type. This demonstrates that the involvement of carotenoids in LR formation is separate from their photoprotective function (Van Norman et al., 2014). Intriguingly, carotenoid biosynthesis was found to occur in differentiated parts of the root at some distance from the oscillating zone where prebranch sites and eventually LRs formed. Thus the non-cell autonomous function of (apo) carotenoids seems to be required for development of LRs (Van Norman et al., 2014).

Chloroplast Proteins as Retrograde Signals

It is a broadly accepted paradigm that most nucleus-encoded chloroplast proteins reach this compartment due to the transit peptides located at their N-termini. Cleavage of the signal peptides after entry in to the chloroplast generates the functional chloroplast proteins. However, accumulating data show that some chloroplast proteins also act in the nucleus. While it is tempting to speculate that this is attributable to two-way protein movement between chloroplasts and cytoplasm, this has not been unambiguously demonstrated. However, a physiological function for this chloroplast-nucleus dual localization is apparent and it seems to be indispensable for proper plant response to pathogen attack and abiotic stress, and emphasizes the function of chloroplasts as signaling compartments.

One such protein, PTM (PHD type transcription factor with transmembrane domains) was shown to provide a physical link in signaling between chloroplasts and nucleus to regulate gene expression (Sun et al., 2011). This membrane-bound transcription factor (MTF) is localized to chloroplast outer envelope by four transmembrane domains at its C-terminus (Figure 3). The N-terminus of PTM contains a DNA-binding homeodomain box, a different transcription factors (DDT) domain and a plant homeodomain (PHD). Interestingly, a shorter variant of this protein, lacking the transmembrane domains, was detected in nuclear fractions. Notably, increased amounts of the shorter PTM variant were detected upon treatment with either norflurazon or lincomycin, and on exposure to high light. Through the application of protease inhibitors, it was demonstrated that the shorter form of PTM was the result of serine protease activity (Sun et al., 2011; Adam, 2015). According to the proposed model, chloroplast signals induce the intramembrane proteolytic cleavage of full length PTM, producing a soluble shorter variant (~58 kDa) that is released to the cytoplasm and finally transclocates to the nucleus where it binds, through its PHD domain, to the *ABI4*promoter to induce *ABI4* expression. ABI4, in turn binds to the *Lhcb* promoter, close to the CUF1 element and precludes binding of G-box-binding factors required for the expression of *Lhcb* and other PhANGs. This model explains the *gun* phenotype observed in *ptm* and *abi4* mutants. Moreover, the amount of processed PTM declined in the *gun1* mutant, suggesting a complex regulatory network.

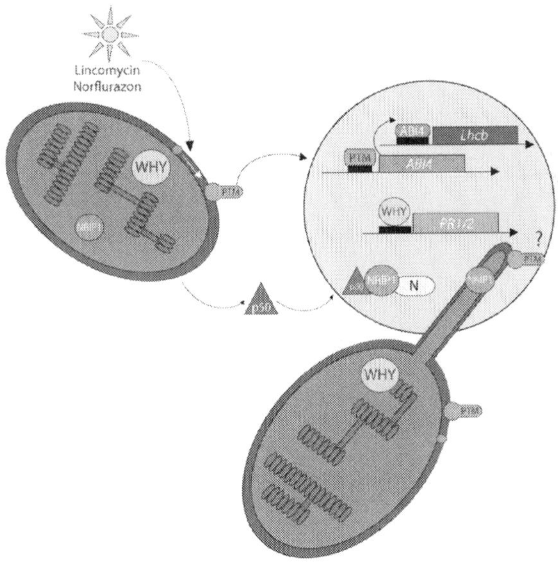

Figure 3. Chloroplast proteins as retrograde signals. A few chloroplast proteins have been implicated in directly modulating nuclear gene expression by their nuclear localization. These proteins may transit the cytoplasm by an un-

known mechanism. Alternatively, it has also been proposed that they may move through stromules to enter the nucleus. High light, lincomycin or norflurazon treatments induce a serine protease-dependent (blue dot) proteolytic-cleavage of the PTM, a chloroplast envelope-bound plant homeodomain (PHD) transcription factor. The cleavage product is found in the nucleus where it binds to promoter region of the ABI4 transcription factor. ABI4 in turn associates with the regulatory sequences of the *Lhcb* genes and prevents their transcription. The chloroplast protein Whirly1 also localizes to the nucleus and this is correlated with increased expression of *PATHOGENESIS RELATED GENE 1* and *2* (*PR1/2*). Upon TMV infection the chloroplast-localized NRIP1 is also detected in the nucleus where it interacts with the helicase domain of the TMV replicase (p50). Finally a trimeric complex of p50, NRIP1 and the N protein is localized to nucleus to provide resistance against the virus. It is suggested that NRIP1 may use stromules to translocate from the chloroplast to the nucleus.

WHIRLY1 provides another example of a chloroplast protein with a role in the nucleus. A WHIRLY1 fusion protein expressed in the plastid genome of tobacco, localized to both plastids and nuclei, and the two subpopulations were the same molecular size (Isemer et al., 2012). As a consequence WHIRLY1-regulated *PR* (*PATHOGENESIS RELATED*) genes were upregulated under normal growth conditions (Figure 3). WHIRLY1 is part of the transcriptionally active chromosome in plastids, and interestingly, another component of this complex, pTAC12 (HEMERA), also showed dual localization to chloroplasts and nuclei with unchanged molecular mass, excluding any proteolytic modification (Chen et al., 2010). However, unlike in the case of WHIRLY1, the dual localization of HEMERA in a single cell has not yet been demonstrated. Such a demonstration is important for ruling out the possibility that dual localization could be caused by fluctuating distribution resulting from specific cell types or developmental stages. It has been proposed that WHIRLY1 conveys information about the chloroplast redox state to the nucleus, and SA regulates this communication (Foyer et al., 2014). The proposed mechanism of WHIRLY1 action would be similar to that of NPR1 (See below). The mechanism and consequences of the proposed translocation of NRIP1 (another chloroplast-localized protein) to the nucleus is discussed later.

There are a few hypotheses that aim to explain the nuclear localization of chloroplast-targeted proteins and they invoke mechanisms enabling translocation of chloroplast-localized proteins to nuclei. One of them proposes permeabilization of the chloroplast outer membrane by an unknown mechanism. Another posits that stress-induced modification of chloroplasts results in the formation of stromules that contact nuclei could facilitate direct trafficking of chloroplast proteins (Caplan et al., 2008). In agreement with the former, a recent communication described GFP-fusion protein leakage from functional chloroplasts upon pathogen attack (Kwon et al., 2013). Interestingly, this occurrence was shown to be dependent on ROS. Evidence for the latter hypothesis of stromule involvement in chloroplast protein translocation is now being reported (see Stromules below). Regardless of the mechanism governing the distribution of chloroplast proteins to the cytoplasm and/or nucleus, this phenomenon may represent an important pathway for direct communication between chloroplasts and nuclei, and provides important insights in understanding the molecular basis of retrograde signaling.

CHLOROPLASTS AND INTER-ORGANELLAR SIGNALING

It has long been appreciated that there is metabolic crosstalk between organelles. Metabolic pathways often involve multiple organelles and metabolic intermediates may be transported across membranes by diffusion or by specialized transporters in an energy-dependent manner after traversing the cytosol. However, there is emerging evidence that direct physical contact between organelles may provide a major route for metabolic exchange (Figure 4). Such direct contact would also provide routes for inter-organelle signaling, although evidence for this is still limited. Thus, chloroplasts, besides producing signals that may travel long distances from plastids to their targets, may also communicate directly with other organelles through physical contacts. Here we consider evidence for chloroplast-organelle contacts and possible roles for these contacts in signaling.

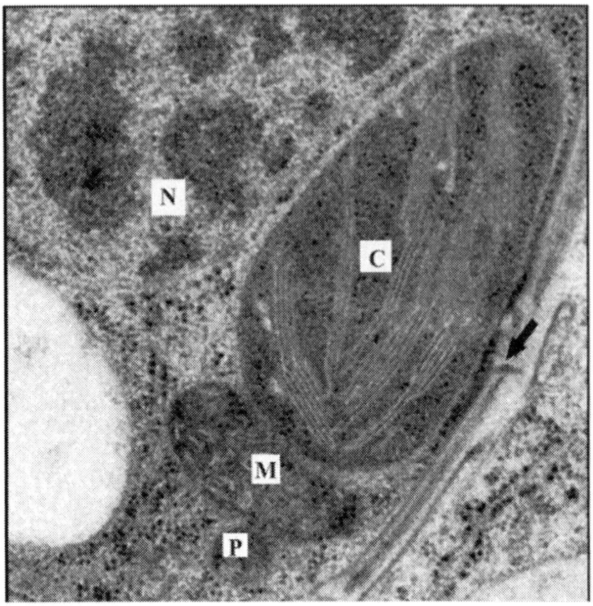

Figure 4. Arrangement of organelles in a leaf cell. Transmission electron microscopy images often reveal chloroplasts in close proximity with peroxisomes and mitochondria. Chloroplasts can also be observed near the nucleus and cell wall. Note the presence of a plasmodesma in the cell wall (arrow). Such arrangements of organelles would minimize distances that signals must traverse to arrive at their target. C, chloroplast; M, mitochondria; P, peroxisome; N, nucleus.

Stromules for Plastid-to-Plastid and Plastid-to-Nucleus Signaling

Rediscovered by Kohler et al. (1997), the observation of tubular protrusions from plastids changed our thinking about chloroplasts and possible pathways for signaling (Kohler et al., 1997). These so-called stromules are stroma-filled tubules enclosed by the inner and outer plastid envelope membranes, and are 0.4–0.8 μm in diameter and of variable length typically up to 65 μm (Gray et al., 2001). They are more abundant in non-green plastids than in chloroplasts. Stromules are distinct from the chloroplast protrusions (CPs) that form during stress (Holzinger et al., 2007a,b), and that are involved in the sequestration of Rubisco from the rest of the chloroplast body (Yamane et al., 2012).

Stromule formation is dependent on both intrinsic and extrinsic factors. The size of plastid, plastid identity, state of differentiation and density of plastids all determine stromule formation (Waters et al., 2004). The actin cytoskeleton has also been reported to be important for stromule formation and movement (Kwok and Hanson, 2003; Kwok and Hanson, 2004b; Gunning, 2005; Holzinger et al., 2007b), and the myosin XI motor is required for stromule formation in *Nicotiana benthamiana* chloroplasts (Natesan et al., 2009; Sattarzadeh et al., 2009). In contrast, it has recently been reported that isolated chloroplasts can form stromules (Brunkard et al., 2015b). Stromule formation is also temperature sensitive, and temperatures around 20°C appear optimal for stromule formation (Holzinger et al., 2007a), while lower temperatures inhibit their formation (Gray et al., 2012) and higher temperatures induce CP formation (Holzinger et al., 2007a). The light-dependence of stromule formation is somewhat controversial. There have been reports that light is not required for stromule formation (Kwok and Hanson, 2003; Gray et al., 2012), but recent findings demonstrate increased formation of stromules from *Arabidopsis* mesophyll chloroplasts during the day as compared to the night (Brunkard et al., 2015b). It is possible the differing results can be explained by differences in plant growth conditions and the types of plastids examined in each experiment. Besides light and temperature the hormone ABA, likely generated in response to environmental stresses, also induces stromule formation (Gray et al., 2012). Consistent with this, salt and osmotic stress also increased the fraction of stromule-bearing chloroplasts.

The most obvious consequences of stromules are significant enlargement of both plastid envelope surface and plastid volume. These modified plastid properties could affect the rate of plastid import and export, and also cause changes in plastid compartmentalization, respectively. Consistent with these presumed functions, it was shown that chlorophyll and thylakoid membranes are absent from stromules (Kohler et al., 1997; Holzinger et al., 2007a; Newell et al., 2012). The trafficking of plastid genomes or genetic material via stromules has also been ruled out (Newell et al., 2012). On the other hand proteins like GFP, aspartate aminotransferase and Rubisco complexes of a molecular weight around 550 kDa have been localized within stromules (Kwok and Hanson, 2004a). Moreover, interconnections between individual plastids via stromules have been reported, and photobleaching experiments and the use of photoconvertible proteins have demonstrated the transfer of proteins between them (Kohler et al., 2000; Hanson and Sattarzadeh, 2011, 2013, but see Schattat et al., 2012, 2015). Given that connections between plastids are rare, the biological significance of possible plastid-to-plastid trafficking remains unclear (Hanson and Sattarzadeh, 2013).

Stromules are involved in metabolic responses to stress, namely chloroplast autophagy in response to nutrient starvation (Ishida et al., 2008). Through observations of fluorescently tagged proteins and the *Arabidopsis atg5* autophagy mutant, it was demonstrated that plastid stromal proteins can be remobilized to the vacuole via the ATG-dependent authophagic pathway, without destroying the chloroplast. Similar observations have been reported from rice (Izumi et al., 2015). According to the proposed model stressed-induced autophagy sequesters stromules by forming an isolation membrane that eventually clips off a given stromule and its stromal contents, and then transports the cargo to the vacuole for degradation (Ishida et al., 2008). Interestingly, both stromules and protrusions were identified in potato tuber amyloplasts (Borucki et al., 2015). These protrusions, unlike stromules, are likely involved in starch accumulation in the parenchyma storage cells. The involvement of protrusions in accumulation of starch had been previously demonstrated (Langeveld et al., 2000). Thus stromules are involved in metabolism.

Plastids and stromules have been repeatedly observed in close proximity to other organelles including mitochondria, other plastids, ER, plasma membrane and nuclei. From the perspective of signaling, however, the most interesting seems to be the distinct distribution of plastids around nuclei, including concentration of stromules around and, most intriguingly, inside the nucleus (Collings et al., 2000; Kwok and Hanson, 2004c). Clusters of plastids with long stromules of 20–30 μm localized around the nucleus were observed in *N. tabacum* petioles of cotyledons (Kwok and Hanson, 2004c), as well as in the lower part of the hypocotyl where the plastids were preferentially arranged around the nucleus with long stromules of up to 100 μm extending to the cell periphery (Natesan et al., 2005). Such concentration of plastids around nucleus was also observed in petal cells and in shoot meristems (Kohler and Hanson, 2000; Kwok and Hanson, 2004c). Stromules have also been observed lying in grooves and invaginations of the nuclear membranes in tobacco epidermal cells (Kwok and Hanson, 2004c). Such direct connections were proposed to increase efficiency in plastid-nucleus communication. Stromules have also been implicated in chloroplast-to-ER signaling (Schattat et al., 2011a,b), and in intercellular signaling (Kwok and Hanson, 2004c) but this has not been explicitly tested or proven. These suggestions are based on interaction between stromules and ER or the plasma membrane, respectively, and further characterization of these interactions is warranted.

Stromules and stromule-nucleus contacts may have important roles in host–pathogen interactions. The induction of stromules and remobilization of chloroplasts to surround nuclei was observed in *N. benthamiana* leaves in response to infiltration with GV3101, common lab strain of *Agrobacterium tumefaciens* (Erickson et al., 2014). In addition, starch accumulated in GV3101-treated leaves and the levels of soluble sugars also increased. However, another lab strain LBA4404 did not produce these effects. The introduction of the *trans zeatin synthase* (*tsz*) gene from the GV3101 plasmid into LBA4404 led to the induction of stromules and other cellular changes typically observed on GV3101 infiltration. Indeed, direct application of cytokinin to leaves produced phenotypes similar to those obtained with GV3101. Thus, cytokinin may mediate the production of stromules and chloroplast movements during some plant–pathogen interactions.

The question of why stromule induction and chloroplast-nucleus associations occur during plant–pathogen interactions has recently been addressed. *Nicotiana* N RECEPTOR INTERACTING PROTEIN (NRIP)1 is a chloroplast rhodanese sulphurtransferase that is required for effector triggered immunity (ETI) against TMV mediated by the N innate immune receptor (Caplan et al., 2008). In the presence of the viral p50 helicase effector NRIP1 was recruited to the cytoplasm and nucleus where it formed a protein complex with the TIR domain of the N immune receptor (Figure 3). The interaction with N in the nucleus and cytoplasm is necessary for ETI. These results raised the intriguing questions of whether NRIP1was remobilized from the chloroplast and how this could be accomplished. Recent data suggest that stromules may provide a route for NRIP to traffic from the chloroplast to the nucleus. Co-expression of N and p50 induced formation of stroumles during

HR-PCD associated with this interaction (Caplan et al., 2015; Figure 5). Similarly, stromule induction was observed during ETI initiated in response to bacterial pathogens, and on treatment with the defense-related signaling molecules SA and the ROS H_2O_2. Through correlative EM-fluorescence microscopy, the authors provide convincing evidence that stromules contact the nucleus during N-mediated defense. In elegant experiments that fuse a nuclear exclusion sequence (NES) to NRIP, the authors provide quantitative evidence that NRIP1 is indeed trafficked from the chloroplast to the nucleus. H_2O_2 also traffics to the nucleus via stromules. In plants overexpressing chloroplast outer membrane protein CHLOROPLAST UNUSUAL POSITIONING 1 (CHUP1) stromule formation is abolished, suggesting that this membrane plays a role in stromule formation. This was confirmed in *chup* mutants and knockdown plants, where constitutive stromule formation was observed. The importance of stromules to HR-PCD and the defense response was underscored by the accelerated HR-PCD observed in those plants. These results are suggestive of a role for stromules in intracellular trafficking and possibly signaling during plant–pathogen interactions, and these possibilities warrant further examination. It will be exciting to test whether plastids and/or stromules are able to create hemifusion membranes with the nuclear membrane and whether other signaling molecules or metabolites use this route for chloroplast signaling.

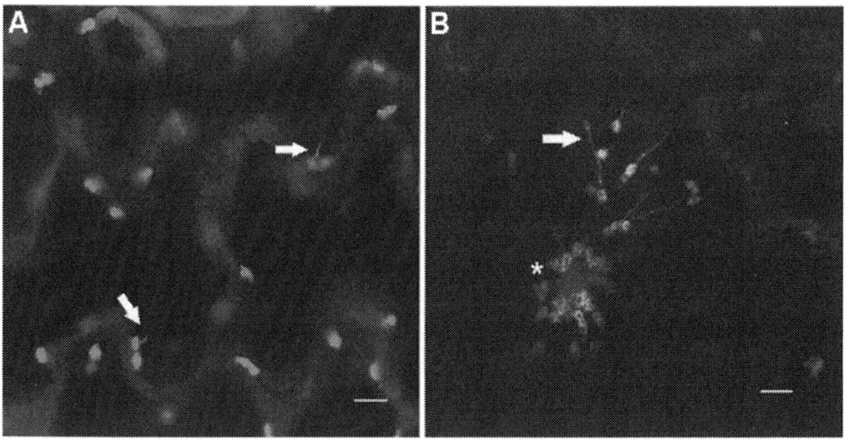

Figure 5. Chloroplast behavior during defense. **(A)** Stromules (arrows) are observed intermittently from chloroplasts in the epidermis of *Nicotiana benthamiana* leaves. **(B)** Upon infection with Tobacco mosaic virus, chloroplasts

cluster around the nucleus (asterisk), and stromule formation is induced (arrow). The chloroplasts shown are expressing NRIP1-CFP (Caplan et al., 2008, 2015). Note the NRIP1-CFP signal detected in the nucleus in TMV-infected chloroplasts, indicative of the translocation of NRIP1 from the chloroplasts to the nucleus. Images were collected on a Zeiss LSM 710 confocal laser scanning microscope and single focal plane images are shown. Scale bar is 10 μm.

A separate study raises the possibility that stromules may also function in intercellular and intracellular trafficking of pathogens. The chloroplast-localized chaperone heat shock cognate 70 kDa protein (cpHsc70-1) was identified as interacting with the AbMV movement protein (MP; Krenz et al., 2010). In uninfected tissues cpHsc70-1-YFP homooligomerization was demonstrated by bimolecular fluorescence complementation (BiFC). cpHsc70-1-YFP oligomers localized to chloroplasts and near the cell periphery. Intriguingly, cpHsc70-1-YFP localized to punctate structures in chloroplasts and filaments that stretched like a "string of pearls" between chloroplasts or to the cell periphery in AbMV-infected leaves. Reduced levels of the Hsc70 ortholog in *N. benthamiana* by virus induced gene silencing (VIGS) led to decreased intercellular trafficking of AbMV while viral replication was unaffected, suggesting that the chloroplast chaperone and stromules may have a role in viral trafficking that is independent of viral replication and accumulation (Krenz et al., 2010). Indeed, it has been proposed that the Hsc-70 chaperone in stromules may facilitate viral transport from chloroplasts to the cell periphery and plasmodesmata, from where intercellular spread could occur (Krenz et al., 2012). AbMV is a geminivirus that replicates in the plant nucleus, but AbMV is also found in plastids (Groning et al., 1987); therefore stromule-nucleus contacts could also provide a route for viral trafficking to nuclei in newly infected cells.

Another interesting question regarding the involvement of stromules in chloroplast signaling is when during evolution would such an innovation have arisen. The R gene *RESISTANCE TO POWDERY MILDEW (RPW)8.2* confers broad resistance to fungi that cause powdery mildew. RPW8.2 localizes to the extrahaustorial membrane (EHM) during fungal penetration of the host cell (Wang et al., 2009). Interestingly, RPW8.2 mutants have recently been reported to localize to an unknown membrane that appears to surround stromules (Wang et al., 2013a). This membrane has been termed the peristromule membrane (PSM) and it is proposed that they may share some as-yet-unknown characteristics with the EHM. Some RPW8.2 mutants localize to both the EHM and PSM, and while others localize to the nucleus. Notably, invasion by fungal haustoria induced the formation of stromules that purportedly connect plastids with haustoria. The authors speculate that stromules may represent an ancient interface for host–pathogen interactions. However, wild-type RPW8.2 was not observed labeling the PSM, and the authors speculate that this may be due to rapid cycling of RPW8.2 on and of the PSM (Wang et al., 2013a).

Together, these studies suggest that stromules could function in signaling in mediating responses to both biotic and abiotic stress, however, this remains to be demonstrated. Understanding the cellular functions of stromules will no doubt reveal intriguing aspects of plant cell biology and possibly identify new targets for engineering plants with modified responses to biotic and abiotic stresses.

Chloroplast-Endoplasmic Reticulum Contacts: Lipid Signaling

A conundrum in plant biology is that the enzymes for a given metabolic pathway are often found in different subcellular compartments. One striking example of this is the biogenesis of complex lipids in some plants including *Arabidopsis*. In these plants fatty acids synthesized *de novo* in the chloroplast by the prokaryotic pathway may be exported to the ER where they are assembled into lipids by the eukaryotic pathway before import back into chloroplasts (Benning, 2009). Despite concerted efforts to identify candidate transporters that would allow the substrates to translocate from the chloroplast to the ER, these transporters have not yet been discovered (Wang and Benning, 2012). This raises the intriguing possibility that metabolites may traffic directly from the chloroplast to the ER through membrane interactions. Moreover, direct physical contact sites between chloroplasts and ER, named plastid-associated membranes (PLAM), have long been reported by transmission electron microscopy in many plant and algal species (Cran and Dyer, 1973; Crotty and Ledbetter, 1973; Renaudin and Capdepon, 1977; McLean et al., 1988). More recently, confocal microscopy has also suggested chloroplast-ER membrane contact sites (MCS; Andersson et al., 2007; Tan et al., 2011). Indeed, these MCS are likely held together through strong protein–protein interactions since the ER remained associated with chloroplasts even after application of forces of 400 pN (Andersson et al., 2007). Once synthesized in the ER glycerolipids have to be imported into the chloroplasts. Mathematical modeling supports the import of diacylglycerol (DAG) from the ER to chloroplasts (Marechal and Bastien, 2014). Genetic analyses have identified the TGD complex as essential for import of 'eukaryotic' precursors into chloroplasts (Boudiere et al., 2014). TGD1-3 constitute a bacterial ABC transporter, while TGD4 apparently forms a β-barrel that localizes to both chloroplasts and ER (Wang and Benning, 2012). Thus transporters are indeed important mediators of lipid trafficking.

The DellaPena group has proposed an elegant hypothesis to explain chloroplast-ER membrane continuity (Mehrshahi et al., 2014). This group has developed a transorganellar assay to test whether non-polar metabolites, exemplified by tocopherols (vitamin E), located in the plastid envelope could directly access the lumen of the ER and modulate enzymes located there (Mehrshahi et al., 2013). To do this, tocopherol cyclase (TC) that is normally a chloroplast resident protein was retargeted to the ER in the background of the *vte1* mutant that does not make TC and tocopherols (Porfirova et al., 2002). Excitingly, this ER-localized TC complemented the *vte1* mutation and tocopherol levels were restored to almost wild-type levels in the rescued lines (Mehrshahi et al., 2013). This study revealed similar access to chloroplast-localized substrates for ER-localized γ-tocopherol methyltransferase (γTMT) and α-carotene ε-ring hydroxylase LUTEIN DEFICIENT1 (LUT1). Based on their successful transorganellar complementation assay, the authors propose that an exchange of non-polar metabolites between plastids and endoplasmic reticulum occurs, most likely due to direct contacts between those organelles at PLAM sites according to the membrane hemifusion model (Mehrshahi et al., 2014). The hemifusion model postulates that the fused membrane of the ER and chloroplast outer envelope would consist of the inner leaflet of the ER and the plastid outer envelope membranes, two envelopes with similar compositions of non-polar metabolites (Mehrshahi et al., 2014). A membrane of this nature would create an easy path for chloroplast-to-ER signaling by non-polar metabolites.

Chloroplast-Peroxisome Contacts

Peroxisomes are dynamic membrane-bound organelles of remarkable metabolic plasticity. They are found in all eukaryotic cells, and in plant cells are usually found in close association with mitochondria and chloroplasts (Figure 3). Peroxisomes are able to adjust their complement of enzymes in response to changes in environmental and developmental signals (reviewed in Goto-Yamada et al., 2015; Sandalio and Romero-Puertas, 2015). Peroxisomes were first recognized for their action in scavenging H_2O_2, however, it is clear that peroxisomes have much more extensive roles in cellular metabolism (Sandalio and Romero-Puertas, 2015). A subset of these reactions highlights the intimate metabolic coupling between chloroplasts and peroxisomes.

The photorespiratory pathway is the conversion of phosphoglycolate to CO_2 and 3PGA, via a complex series of reactions that takes place across three separate subcellular compartments: chloroplasts, peroxisomes, and mitochondria (Foyer et al., 2009). Severe photorespiratory conditions initiate ROS-dependent lipid peroxidation in the chloroplast that leads to activation of the lipoxygenase (LOX)-mediated reaction, which is one of the starting points in the oxylipin metabolic pathway. LOX also initiates the synthesis of many cell constituents and signaling molecules, including jasmonates via oxo-phytodienoic acid (OPDA; Weber et al., 1997; Montillet et al., 2004). OPDA is then transported into the peroxisomes where two rounds of beta-oxidation serve to modify the fatty-acid side-chain of the ring. It is believed that the transport of OPDA is carried out by both active and passive transport (Leon, 2013).

Recent work has revealed that in addition to close metabolic coupling, peroxisomes and chloroplasts physically interact with each other in a light-dependent manner. Using the newly developed femtosecond laser technology and confocal laser scanning microscopy, Oikawa and colleagues demonstrated that peroxisomes adopted an elliptical shape to increase their surface area and more tightly adhered to chloroplasts in light (Oikawa et al., 2015). A force of 61 fN nm^{-2} was needed to disrupt the chloroplast–peroxisome interaction in the light compared to 23 fN nm^{-2} required to so do in the dark. Interestingly, these changes in peroxisome shape and location depended on photosynthesis, but were independent of photorespiration or the activity of photoreceptors, and the actin cytoskeleton negatively regulated the interaction between the chloroplasts and peroxisomes (Oikawa et al., 2015). The chloroplast–peroxisome physical interaction is consistent with reports from other systems that suggest direct interaction between organelles is necessary for metabolite exchange (Binns et al., 2006; de Brito and Scorrano, 2008). Thus the trafficking of compounds like OPDA during JA synthesis could potentially contribute to signaling between plastids and peroxisomes.

The *Arabidopsis snowy cotyledon3 (sco3-1)* mutant also provides support for signaling between chloroplasts and peroxisomes. In seedlings, the *sco3-1* mutation interrupted chloroplast biogenesis, decreased chlorophyll accumulation, and disrupted thylakoids and, as a consequence, photosynthesis (Albrecht et al., 2010). This mutation also resulted in photoinhibition in mature leaves under high CO_2concentrations. It is quite interesting that the SCO3 protein is initially targeted to peroxisomes. Also interesting is that loss of SCO3 function led to cytoskeletal defects, specifically affecting microtubules. Thus, both the cytoskeleton and peroxisomes are necessary for normal chloroplast development. Further investigation of SCO3 function will illuminate the process of communication between chloroplasts and peroxisomes.

Chloroplast-Mitochondrion Contacts

Like chloroplasts, mitochondria are the end products of an endosymbiotic event, and they have also retained a portion of their ancestral genome (Woodson and Chory, 2008). Mitochondria-to-nucleus retrograde signaling is critical for coordinating expression of nuclear genes encoding mitochondrial proteins with expression of the mitochondrial genome (Woodson and Chory, 2008; Rhoads, 2011). Given their central roles in energy capture and utilization it is perhaps not surprising that chloroplasts and mitochondria exchange metabolites. Chloroplasts and mitochondria are also coupled by cellular redox status.

Mutant analyses have shed light on chloroplast-mitochondrion signaling. The expression of *alternative oxidase (AOX)*, a nucleus-encoded mitochondrial gene, seems to be regulated by chloroplasts as increased expression of *Arabidopsis* and soybean *AOX* has been observed upon high light treatment (Finnegan et al., 1997; Blanco et al., 2014). Intriguingly, the white leaves of the chloroplast ribosome-deficient barley mutant *albostrians* as well as photo-bleached leaves of wild type obtained upon treatment with norflurazon, displayed elevated levels of mitochondrial DNA and transcripts (Hedtke et al., 1999). Consistent with this observation, recent analyses of leaves with green/white variegation in 12 ornamental plants confirmed that chloroplast dysfunction leads to increased levels of mitochondrial DNA in white sectors (Toshoji et al., 2012). Communication between chloroplasts and mitochondria seems to be bidirectional as mutation of genes encoding mitochondrial proteins can have profound effects on chloroplasts (e.g., Xu et al., 2008; Burch-Smith et al., 2011).

Several possible routes for communication between chloroplasts and mitochondria are proposed. There is coordinated expression of nuclear genes encoding chloroplast and mitochondria proteins, and it is therefore likely that some of the chloroplast signaling to mitochondria is accomplished by modulating nuclear gene expression. Indeed, expression of the *MITOCHONDRIAL DYSFUNCTION STIMULON*(*MDS*) suite of mitochondria-associated genes was regulated in response to chloroplast perturbations that included increased ROS and NO production (Ng et al., 2014). These *MDS* genes carry a common regulatory motif in their promoters that mediates their induction in response to mitochondrial retrograde signals (De Clercq et al., 2013; Ng et al., 2013). Moreover, it was demonstrated that *Arabidopsis* ABI4 regulates both *Lhcb* and *AOX1A* genes, providing a molecular link for nucleus-coordinated chloroplast-mitochondria communication (Koussevitzky et al., 2007; Giraud et al., 2009). Analysis of the *Arabidopsis regulator of alternative oxidase 1* (*rao1)* mutant deficient in a nucleus-localized cyclin-dependent kinase E1 (CDKE1) that is a prerequisite for *AOX* induction, indicates it is another nucleus-localized sensor integrating mitochondrial and chloroplast retrograde signals (Blanco et al., 2014). Unlike wild type plants the *rao1(cdke1)* mutant was unable to induce *AOX1A* expression upon application of antimycin A or DCMU (that specifically targets chloroplasts). Further, the mutant displayed the *gun* phenotype upon induction of redox stress originating specifically from chloroplast photoelectron transport.

Another possible mechanism for chloroplast-to-mitochondrion signaling could involve dual targeting of proteins. The localization of proteins to multiple subcellular compartments is an ancient feature of land plants that can be observed in *Physcomitrella patens* (Xu et al., 2013b) and diatoms (Gile et al., 2015). In *Arabidopsis*, over 100 proteins are targeted to both chloroplasts and mitochondria (Carrie and Whelan, 2013). It is tempting to propose that, analogous to the situation where proteins have been shown to localize to both chloroplasts and nuclei, proteins may move from the chloroplast to the mitochondria. The regulated translocation of proteins from chloroplasts to mitochondria to modulate mitochondrial gene expression would mediate chloroplast signaling. Such translocation would be made much easier by direct contact between chloroplasts and mitochondria.

Third, there may be direct communication by physical interaction. In leaves, chloroplasts, mitochondria and peroxisomes have often been observed in close association, consistent with metabolic exchange among these organelles (Figure 4). It seems that the formation of this tri-organellar unit is regulated, with the chloroplast-peroxisome association being established first and then recruiting mitochondria (Oikawa et al., 2015). Application of the femtosecond laser pulses to chloroplast-mitochondria complexes in a variety of tissues should illuminate the biophysical characteristics of their association.

In all these instances of organelle interactions, it is apparent that more detailed analysis at the level of resolution provided by electron microscopy is needed. Such studies should incorporate state-of-the-art fixation techniques like tandem high-pressure freezing and freeze substitution in order to minimize artifact formation and to maintain intact the presumably delicate membrane extensions and contact sites. These techniques are becoming easier and less time-consuming (McDonald, 1999, 2014) and can easily be adopted for plant cell biology (Bobik et al., 2014). When coupled with fluorescence microscopy, this will be a powerful approach for ultrastructural interrogation, as exemplified by recent work from Caplan et al. (2015).

HORMONES AND REACTIVE MOLECULES AS CHLOROPLAST SIGNALS

As sessile organisms plants have evolved to cope with extreme environmental conditions and fluctuations. Some of the most common environmental challenges to plant survival include drought, flooding resulting in reduced oxygen availability, and temperature extremes. In addition, the photosynthetic machinery is sensitive to excessive light and such exposure results in oxidative stress at the cellular level. In addition to these abiotic stresses, pathogens pose a constant threat of disease. Plants have therefore evolved a complex suite of responses that are exquisitely fine-tuned to allow them to cope with these stresses. It is widely recognized that chloroplasts both sense and respond to environmental conditions. Indeed, the chloroplast-generated hormones SA, JA and ABA and other secondary messengers including ROS and reactive nitrogen species (RNS) as well as redox signals are critical components of the plant stress response. Therefore chloroplast signaling is indispensable for plant survival of abiotic and biotic stress. The roles of all these chloroplast-associated molecules in coping with biotic and abiotic stress have been extensively examined and several excellent reviews are available (Miller et al., 2008; Padmanabhan and Dinesh-Kumar, 2010; Kangasjarvi et al., 2012; Pfannschmidt and Yang, 2012; Trotta et al., 2014).

Salicylic Acid

Salicylic acid is best known for its role plant–pathogen interactions and particularly in plant defense. However, SA also has roles in plant developmental processes including germination, root and shoot growth and senescence, and also functions in abiotic stress responses (Rivas-San Vicente and Plasencia, 2011). A phenolic compound, SA is largely the product of the isochorismate pathway in the chloroplasts. Pathogen infection induces the production of SA by chloroplasts mainly through the action of the chloroplast-localized ISOCHORISMATE SYNTHASE (ICS)1 enzyme although there may be some contribution by ICS2 (Wildermuth et al., 2001; Garcion et al., 2008). Analysis of the *ics1 ics2* double mutant has revealed that there are other cellular sources of SA (Garcion et al., 2008), most likely via phenylpropanoid metabolism in the cytoplasm (Vlot et al., 2009). SA conjugates with glucose or a methyl side-group are the commonly active forms of SA. Indeed, methyl salicylate is a critical mediator of systemic acquired resistance (SAR; Park et al., 2007), and it may also function in ecological defense signaling when it becomes airborne (Shulaev et al., 1997).

Many SA-mediated defense responses rely on the action of the transcriptional activator NON-EXPRESSOR OF PR1 (NPR1; Fu and Dong, 2013). NPR1 interacts with TGA transcription factors, and it is believed to act with them as co-activators of defense gene expression

(Zhang et al., 1999; Zhou et al., 2000; Yan and Dong, 2014). The action of NPR1 is dependent on the cellular redox state. In the absence of SA, NPR1 oligomerizes in the cytoplasm, but on perception of SA, NPR1 is reduced and the oligomer disassembles into monomers that then relocate to the nucleus to modulate gene expression (Kinkema et al., 2000; Mou et al., 2003). The redox state of NPR1 is mediated by the glutathione and thioredoxin redox systems (Mhamdi et al., 2010; Han et al., 2013). In the nucleus, NPR3 and 4, two other SA-binding proteins (SABPs) that are closely related to NPR1, modulate NPR1's activity (Fu et al., 2012). It should be noted that there are also NPR1-independent pathways that mediate SA signaling (Rairdan et al., 2001; Shah, 2003; Uquillas et al., 2004; Blanco et al., 2005). Downstream of NPR1 the SA signaling pathway is well understood (Spoel et al., 2003; Fu and Dong, 2013).

Salicylic acid is important for basal defense as well as for ETI (Alazem and Lin, 2015), and application of SA or overexpression of its biosynthesis genes leads to increased pathogen resistance (Malamy et al., 1990; Ryals et al., 1996; Mur et al., 2008; Shah, 2009; Coll et al., 2011). Chloroplast Ca^{2+} signals are induced in a stress-specific manner, and this response is mediated by the calcium-sensing receptor (CAS; Nomura et al., 2012). CAS mediates the Ca^{2+} signals in ETI and in response to the presence of highly conserved pathogen associate molecular patterns (PAMPS). CAS, and thus CA^{2+}, regulates chloroplast SA biosynthesis and plants depleted of CAS failed to induce SA production in response to pathogen infection. In addition, expression of several nuclear defense-related genes was shown to be dependent on CAS, and the pattern of gene regulation was most similar to that observed in response to $^1O_2$.

Interestingly, exogenous application of SA induces closure of plasmodesmata (Wang et al., 2013b). This is mediated by PLASMODESMATA LOCALIZED PROTEIN (PDLP)5, and likely involves the action of a callose synthase (Lee et al., 2011; Wang et al., 2013b). Consistent with these findings, *Arabidopsis pdlp5*mutants have decreased resistance to bacterial pathogens. These results clearly demonstrate crosstalk between SA signaling and plasmodesmata, and illustrate how chloroplast signals can act to regulate intercellular trafficking via plasmodesmata (discussed below). It will be interesting to see if similar mechanisms are deployed for viral resistance since viruses use plasmodesmata for intercellular trafficking.

There has been some controversy over the SA-binding properties of NPR1. Binding assays using recombinant GST-NPR1 and tritiated-SA ([^{3}H]-SA) suggested that NPR1 did not bind SA (Fu et al., 2012). In contrast, equilibrium dialysis experiments revealed that NPR1 bound SA with a K_D similar to that observed for other receptor-ligand interactions for plant hormones (Wu et al., 2012). Recent results confirm that SA can indeed bind NPR1 (Manohar et al., 2014). However, it is clear that NPR1 is not the only protein that binds SA. The metabolic enzymes catalase (Chen et al., 1993), ascorbate peroxidase (Durner and Klessig, 1995) and methyl salicylate esterase (SABP2; Forouhar et al., 2005) have been shown to bind SA. Recent work reveals that SA potentially has numerous targets in the cell including, not unexpectedly, numerous chloroplast proteins.

Many of these SABPs have been identified by high-throughput approaches (Moreau et al., 2013; Manohar et al., 2014). By probing *Arabidopsis* protein microarrays with 4-azido SA (azSA), an SA analog, the Popescu lab identified numerous chloroplast-localized proteins with roles in photosynthesis and oxidative phosphorylation as proteins interacting with AzSA (Moreau et al., 2013). Two other interesting candidate SABPs were also identified: thimet metalloendopeptidase At5g65620 (TOP1) and its homolog encoded by *At5g10540* (TOP2), both of unknown cellular function. TOP1-GFP fusions localized to chloroplasts and TOP2 is likely cytosolic. Interestingly, *top1 top2* double mutants had compromised ETI and PCD to bacterial pathogens. Further, TOP1 and TOP2 dimerize in an SA- and redox-dependent manner (Westlake et al., 2015). However, TOP1 and TOP2 have distinct responses to the reductant DTT, suggesting they have different activities *in planta*. A role for TOP1 and TOP2 in the oxidative stress response was demonstrated by treating various *top* mutants with methyl viologen, a potent inducer of oxidative stress, but this function is likely restricted to early in plant development. Two additional high throughput screens have recently identified another 100 candidate SABPs (Manohar et al., 2014). Of these, nine were already known SABPs and the SA binding of nine of new candidate SABPs was verified. Notably, four of the new SABPs have roles in redox regulation, reiterating the interaction between these two pathways.

Jasmonates

Jasmonic acid is a lipid-derived hormone that is perhaps best known for its roles in insect herbivory and wounding, but also has roles in plant growth and development (Leon, 2013). The term jasmonates refers to a group of compounds that are derived from linoleic acid. JA is synthesized via the octadecanoid pathway and JA synthesis is initiated in the chloroplasts but is completed in the peroxisome. JA is then derivatized to yield a diverse array of metabolites that have different functions, ranging from storage to inactivation (reviewed in Leon, 2013; Wasternack and Hause, 2013). The most active of these compounds is a JA conjugate with isoleucine, *(+)-7-iso*-JA-Ile (Wasternack, 2014). JA biosynthesis and signaling pathways have been elucidated (Turner et al., 2002; Antico et al., 2012; Wasternack and Hause, 2013; Zhu, 2014). JA signaling in stress is closely liked with that of another hormone, ethylene (Kunkel and Brooks, 2002). Interestingly SA and JA/ET signaling are often antagonistic to each other (Robert-Seilaniantz et al., 2011). This highlights the crosstalk between chloroplast signals.

The final chloroplastic intermediate in JA-biosysnthesis is *cis*-(+)-12-oxophytodienoic acid (OPDA). OPDA is then translocated to the peroxisome, where its hydrocarbon chain is shortened by β-oxidation. It is not clear how OPDA is transported to the peroxisomes, but the process must be tightly regulated, given that OPDA itself is able to act as a signaling molecule. Indeed, OPDA has been implicated in tendril coiling (Blechert, 1999); in *Arabidopsis* seed germination (Dave et al., 2011); in tomato embryo development (Goetz et al., 2012), and fertility in *P. patens* (Stumpe et al., 2010). OPDA is also known to be an important signal for defense (Stintzi et al., 2001; Scalschi et al., 2015). Interestingly, *P. patens* does not make JA but instead uses OPDA for defense (Stumpe et al., 2010; Ponce De Leon et al., 2012). However, SA is used as a defense signal in this moss (Ponce De Leon et al., 2012). In the liverwort *Marchantia polymorpha*wounding induces OPDA production, and exogenous application limited *M. polymorpha* growth (Yamamoto et al., 2015). As in *P. patens*, JA was not detected in *M. polymorpha.* This finding was supported by OPDA regulating growth of *M. polymorpha* while JA could not. Thus, the production of lipid-derived signals by the chloroplast is an ancient feature of plants, and the production of JA may have developed more recently.

Crosstalk of RNA Silencing and Chloroplast Hormones

There is also crosstalk between the chloroplast-derived phytohormones and the RNAi machinery of plant cells. Several studies have reported increased expression of *RNA-DEPENDENT RNA POLYMERASE1 (RDR1)* on exogenous application of SA (Xie et al., 2001; Yu et al., 2003; Hunter et al., 2013), JA (Pandey and Baldwin, 2007) and ABA (Hunter et al., 2013). RDR1 is known to have a role in antiviral RNA silencing, in the production and amplification of virus-derived siRNAs (Donaire et al., 2008; Wang et al., 2010). JA interaction with SA is mostly antagonistic, and there are few genes whose expression is induced by both hormones (Pieterse et al., 2009). The finding RDR1 can be induced by multiple hormones suggests that there is crosstalk between hormones and the RNA silencing machinery to mediate stress responses. However, RDR1 does not seem to have a role in drought resistance (Hunter et al., 2013), although other components of the RNAi components have been demonstrated to function in stress tolerance (Earley et al., 2010; Li et al., 2012; Westwood et al., 2013). Together, these findings extend the role of chloroplast signaling in the stress response.

Abscisic Acid

Abscisic acid is one of the most important hormones mediating plant biotic and abiotic stress responses. ABA also has major roles in various plant physiological processes including stomatal movement and seed dormancy (Wensuo and Zhang, 2008; Rodriguez-Gacio Mdel et al., 2009; Kim et al., 2010). ABA is a sesquiterpenoid that is produced by the methylerythritol phosphate (MEP) pathway in plastids that produces carotenoids (Finkelstein, 2013). ABA levels in a cell are the result of both its synthesis and catabolism. Interestingly, several subcellular compartments are involved in ABA metabolism (Xu et al., 2013a). All the steps of the *de novo* ABA biosynthetic pathway occur in plastids except for the last two, which occur in the cytosol. The first committed step in ABA synthesis is the cleavage of the carotenoid xanthopyhll by 9-*cis*-epoxycarotenoid dioxygenase (NCED) to produce the C15 compound xanthoxin, that is transported from the plastids into the cytosol by an unknown mechanism, where it will be converted into ABA (Finkelstein, 2013). Once synthesized, ABA is transported from the sites of synthesis to sites action via the xylem and phloem and is thus made available to both roots and shoots. ABA catabolism occurs via one of two pathways: oxidation or conjugation to glucose to produce a glucosyl ester ABA-GE. ABA-GE is stored in the vacuole or ER until it mobilized under stress conditions. Because of its central roles in modulating responses to various stresses, ABA signaling is an attractive target for engineering plants with increased tolerance to those stresses.

The mechanisms governing perception of, and signaling by, ABA are being discovered (Yoshida et al., 2015). Members of the PYR/PYL/RCAR family of soluble proteins are ABA receptors (Ma et al., 2009; Park et al., 2009). This protein family has 14 members, almost all of which appear capable of forming an ABA-receptor complex that is able to activate the transcription of ABA-responsive genes (Kline et al., 2010). Downstream of the receptors, several kinases, including calcium dependent kinases (CDPKs), and phosphatases mediate ABA signaling, culminating in changes in nuclear gene expression (Finkelstein, 2013; Yoshida et al., 2015). The 26S proteasome is also important in mediating ABA signaling (Ludwikow, 2015).

Abscisic acid signaling is particularly important during drought, salinity and cold stress. During pathogen infection, ABA signaling is antagonistic to JA/Et signaling (Soosaar et al., 2005); and it can also antagonize SA signaling (Alazem and Lin, 2015). Indeed, crosstalk between SA, JA, and ABA signaling pathways during pathogen defense is well documented (Tuteja, 2007; Flors et al., 2008; Cao et al., 2011) and underscores the role of chloroplasts in integrating inputs for plant survival. Importantly, the H subunit of Mg chelatase was shown to be an ABA receptor, and moreover, the observations that ABA can repress *Lhcb* expression and the gun phenotype of the *abi4* mutant link this hormone to retrograde signaling (Shen et al., 2006; Koussevitzky et al., 2007).

Reactive Oxygen Species

Reactive oxygen species are formed by the reduction of molecular oxygen and the term ROS includes superoxide ($O_2^{\bullet -}$), hydroxyl, alkoxyl (·RO), and peroxyl radicals as well as non-radical molecules like hydrogen peroxide (H_2O_2) and singlet oxygen ($^1O_2$). It is well established that while large amounts of ROS are damaging, small amounts act as signaling molecules (Miller et al., 2008; Gill and Tuteja, 2010). In an effort to avoid the toxicity of ROS, plants have evolved multiple antioxidant systems. It is clear then that ROS signaling is complex and is often the outcome of the balance between production and scavenging.

In a typical plant cell ROS may be generated by a variety of subcellular compartments including chloroplasts, peroxisomes, mitochondria and the apoplast. In a photosynthesizing leaf most ROS is the product of the chloroplasts and peroxisomes with smaller contributions from mitochondria (Foyer and Noctor, 2005). During photosynthesis, in the reaction center of photosystem II (PSII), excited triplet center chlorophyll P680 interacts with oxygen to generate $^1O_2$. The acceptor side of PSI produces superoxide and hydrogen peroxide as electrons are transferred from reduced ferredoxin to molecular oxygen. Photorespiration in the peroxisomes to recycle Rubisco that has reacted with oxygen is a major source of H_2O_2 (Yoshida and Noguchi, 2011).

Plastid-generated ROS molecules have been shown to act as signals that modulate expression of nuclear genes. Different ROS molecules have been shown to induce expression of distinct suites of genes (Desikan et al., 2001; Vandenabeele et al., 2004; Vanderauwera et al., 2005; Laloi et al., 2007; Li et al., 2011; Balazadeh et al., 2012; Mor et al., 2014). This has made it possible to draw distinctions between the various ROS initiating responses to different stresses, (e.g., Gadjev et al., 2006). By virtue of their nature, ROS are known to interact with a variety of biological molecules. For example, as a strong electrophile, $^1O_2$can react spontaneously with many classes of biological molecules including proteins, lipids and nucleic acids. However, the question remains whether they act directly or indirectly through modification of other biomolecules.

Hydrogen peroxide is the least reactive ROS, and at high light intensities up to 5% of chloroplast-generated H_2O_2 has been detected outside the organelle (Mubarakshina et al., 2010). H_2O_2 may exit the chloroplasts via aquaporins (Bienert et al., 2007). Recently, H_2O_2 has also been shown to relocate from chloroplasts to the nucleus via stromules (Caplan et al., 2015). Indeed, chloroplast ROS production induces stromule formation (Brunkard et al., 2015b). These findings suggest that H_2O_2 may act directly on its targets to regulate their expression or behavior.

In contrast, $^1O_2$ is highly reactive, and, $^1O_2$ is involved in signaling pathways leading to cell death or to acclimation (Wagner et al., 2004;

Ledford et al., 2007). Yet, because of its very high reactivity and therefore short lifetime *in vivo*, $^1O_2$ is not considered to be a molecule directly involved in chloroplast-to-nucleus signaling. Studies in the *Arabidopsis fluorescent (flu)* mutant have shed light on $^1O_2$ signaling (Kim and Apel, 2013). The *flu* mutant accumulates protochlorophyllide and when plants are exposed to light they generate large amounts of $^1O_2$ (Meskauskiene et al., 2001). Increased $^1O_2$ production in *flu*chloroplasts was associated with induction of stress responses including dramatic changes in nuclear gene expression and enhanced accumulation of the stress hormones, SA, Et, and the oxylipins OPDA and JA (op den Camp et al., 2003). Interestingly, these $^1O_2$-induced changes were dependent on the chloroplastic EXECUTER1 (EX1) and EX2 proteins (Kim et al., 2012). To date, two direct protein targets of $^1O_2$ have been identified: one is β-carotene and the other is the D1 protein of PSII (Kim and Apel, 2013). Reaction of $^1O_2$ with β-carotene produces β-cyclocitral, shown to be involved in retrograde signaling (Ramel et al., 2012; see above). D1 is potent scavenger of $^1O_2$, and this activity leads to the destruction of D1 by protease activity. However, to date no biological activity for these D1 fragments in plants has been reported.

Plants possess several enzymes including catalase and ascorbate peroxidase that detoxify ROS (Miller et al., 2008). In addition plants have evolved antioxidant systems that allow them to not only cope with ROS production during stress but also transduce ROS signals. Such systems include tocopherols (vitamin E), ascorbate (vitamin C), and glutathione (a tripeptide thiol) and they all associate with chloroplasts. Glutathione is the most important antioxidant in plants (Zechmann, 2014) and it can act on ROS directly or through the ascorbate-glutathione system. Glutathione synthesis is initiated in chloroplasts (Wachter et al., 2005). The importance of glutathione highlights the importance of thiol-disulfide reactive proteins as antioxidants (Foyer and Noctor, 2003). Proteins including thioredxoxins, glutaredoxins, glutathione peroxidases are found in chloroplasts as well as other subcellular compartments, and these proteins convey the redox state of chloroplasts to the rest of the plant cell. The involvement of redox signaling in mediating plant stress responses are extensively documented (Foyer et al., 2009).

Nitric Oxide and Reactive Nitrogen Species

Nitric oxide (NO) is a small gaseous molecule whose role in signaling in plant and non-plant systems is well established. While NO is an important second messenger in plants, major aspects of NO synthesis and action remain undiscovered. NO has critical roles in normal plant physiology including seed dormancy, germination, the floral transition, and during stress responses (Wendehenne et al., 2004; Moreau et al., 2008; Sanz et al., 2015). In plants NO is synthesized in chloroplasts (Mandal et al., 2012; Tewari et al., 2013), and synthesis is closely linked to lipid metabolism, specifically that of oleic acid (Mandal et al., 2012).

In animals NO is synthesized by NITRIC OXIDE SYNTHASEs (NOSs) that oxidize L-arginine to NO (Stuehr, 2004). Curiously, NOS-like activity has long been described in plants but the identity of this enzyme remains elusive (Besson-Bard et al., 2008). In plants, nitrate reductase can reduce nitrate to NO (Desikan et al., 2002). Another enzyme implicated in NO synthesis is AtNOA1 (formerly AtNOS1). However, AtNOA1 is a GTPase that binds ribosomes, and it lacks NOS activity (Moreau et al., 2008). Nonetheless, *atnoa1* mutants displayed reduce levels of NO (Guo et al., 2003).

One of the primary mechanisms through which NO exerts its effects is by direct regulation of protein activity through nitrosylation of cysteine residues in a redox-dependent manner (Mengel et al., 2013). Nitration of tyrosine residues also occurs, and it is mediated by the peroxynitrite ($ONOO^-$) RNS formed on reaction of NO with O_2 (Corpas et al., 2013). Tyrosine nitration is viewed as a hallmark of plants under a variety of biotic and abiotic stresses (Corpas et al., 2013). While proteomic analyses have identified only relatively few nitrosylated proteins in *Arabidopsis* (Lindermayr et al., 2005; Romero-Puertas et al., 2008; Ortega-Galisteo et al., 2012; Camejo et al., 2013; Puyaubert et al., 2014), a recent computational approach identified more than 16, 000 potential protein target for nitrosylation (Chaki et al., 2014). Refined versions of this program should be very useful in streamlining investigation of NO function and effects in signaling.

There is substantial crosstalk between ROS and RNS signaling pathways (Lounifi et al., 2013). Indeed, even small amount of NO had profound effects on chloroplast PET and redox state (Vladkova et al., 2011; Misra et al., 2014). It is therefore not surprising that NO is important for plant defense (Zaninotto et al., 2006; Bellin et al., 2013). Further, NO signaling interacts with SA and JA signaling pathways (Wendehenne et al., 2004; Zhou et al., 2015). NO production has been proposed as a potential target for engineering plant fitness (Foresi et al., 2015). However, this approach should be carefully considered given the potentially deleterious effects of runaway NO production and the extensive crosstalk between NO and other stress signaling pathways.

Chloroplasts as Targets of Pathogen Effectors

The previous sections have highlighted the role of several chloroplast products in stress tolerance and plant defense. That chloroplasts have critical roles in pathogen defense is supported by the identification of numerous chloroplast proteins as direct targets of pathogen effectors (Table 1). Several of these targets are components of PET, suggesting that photosynthesis and chloroplast function are important for defense against pathogens. A series of experiments aimed at addressing the role of chloroplasts beyond phytohormone production in plant defense has been reported. VIGS of two genes encoding components of the PSII oxygen-evolving complex, *PsbQ* and *PsbO*, resulted in a three to four-fold increase in the number of Turnip mosaic virus (TuMV) infection foci (Manfre et al., 2011). A similar increase was observed when the chloroplast protease FtsH or Rubisco were depleted by VIGS (Manfre et al., 2011). These findings were further supported by the observation that treatment with lincomycin, and low-light conditions led to increased rates of TuMV infection. Importantly, it was demonstrated that these effects were independent of SA, as lincomycin-treated plants induced SA production to the same levels on TuMV infection as observed in non-treated controls. Given that TuMV replicates in association with the chloroplast outer envelope (Wei et al., 2010), it is perhaps not surprising that perturbing chloroplast function results in increased viral infection. However, chloroplast proteins are targets for other viruses, e.g., Tobacco mosaic virus (TMV), which are not known to replicate in association with chloroplast membranes. Interestingly, PsbO has been shown to interact the helicase domain of the TMV replicase, and VIGS of *PsbO* lead to a 10-fold increase in TMV infection in *N. benthamiana* (Abbink et al., 2002).

Table 1. Chloroplast proteins targeted by pathogen effectors.

Pathogen	Pathogen protein	Target	Reference
Viruses			
Alfalfa mosaic virus	Coat protein	PsbP	Balasubramaniam et al., 2014
Tobacco mosaic virus	Replicase (helicase)	NRIP1	Caplan et al., 2008
	Replicase	PsbO	Abbink et al., 2002
		AtpC	Bhat et al., 2013
		Rca	Bhat et al., 2013
Tomato mosaic virus	Movement protein	RbcS	Zhao et al., 2013
Plum pox virus	CI	PSI-K	Jimenez et al., 2006
Potato virus Y	HC-Pro	MinD	Jin et al., 2007
		XDS	Li et al., 2015
Soybean mosaic virus	P1	Rieske Fe/S	Shi et al., 2007
Sugarcane mosaic virus	HC-Pro	Ferredoxin	Cheng et al., 2008
Turnip mosaic virus	P3	Rubisco	Lin et al., 2011
Alternanthera mosaic virus	TGB3	PsbO	Jang et al., 2013
Bacteria			
Pseudomonas syringae pv Tomato DC3000	HopI1	Cytosolic Hsp70*	Jelenska et al., 2007; Jelenska et al., 2010
	HopN1	PsbO	Rodriguez-Herva et al., 2012
	HopU1	RNA binding proteins (CP-RBPs)	Fu et al., 2007
	HopK1	Unknown	Li et al., 2014
P. syringae pv pisi	AvrRps4	Unknown	Li et al., 2014
Fungi			
Melampsora larici-populina (rust fungus)	MLP10772 MLP124111	Unknown	Petre et al., 2015

CHLOROPLASTS AND INTERCELLULAR SIGNALING

The engulfment of cyanobacteria and subsequent evolution of the cellulosic cell wall surrounding the new plant cell likely created a physical barrier that hampered cell-to-cell communication and further development of multicellular, three-dimensional plants. It is important to highlight that modern chloroplasts are a sister group to the filament and heterocyst-forming cyanobacteria *Nostoc* and *Anabaena*(Falcon et al., 2010). Intercellular movement of molecules between the cells of *Anabaena cylindrica*filaments has been demonstrated (Mullineaux et al., 2008). Heterocysts are specialized cells that fix $N_2$and lack O_2-evolving PSII and typically they have thick cell walls with limited permeability to gases. Interestingly, heterocysts are connected to vegetative cells by microplasmodesmata (Falcon et al., 2010; Padmanabhan and Dinesh-Kumar, 2010). Thus, we propose that chloroplasts, as the successors of these ancient cyanobacteria, participate in regulation of the cell-to-cell connectivity ofmodern plant cells by regulating plasmodesmata biogenesis in contemporary plants. This hypothesis is supported by a number of reports showing that mutants with defective chloroplasts display aberrant trafficking and altered expression of cell wall-related genes (see below). The relationship between chloroplasts and plasmodesmata and thus cell wall becomes more palpable when one considers sugar transport from source to sink tissues through plasmodesmata.

Chloroplast Signaling and Symplasmic Transport

Communication between plant cells is enhanced by the presence of plasmodesmata, specialized channels traversing the cell wall between adjacent cells. Even though discovered more than 100 years ago, the structure and regulation of plasmodesmata are poorly understood. In contrast, recent advances in plant biology and genetics have helped reveal much about their function. Plasmodesmata are indispensable for proper plant development as they provide a route for exchange of metabolites including water, ions and photoassimilates, as well as information encoded by hormones, nucleic acids and proteins including transcription factors (reviewed in Gallagher et al., 2014; Brunkard et al., 2015a; Heinlein, 2015; Jackson, 2015). Plant viruses have evolved to take advantage of the intercellular connectivity afforded by plasmodesmata, and viruses use these channels for cell-to-cell spread during infections of their plant hosts. Recent studies on grafted plants resulted in a tantalizing hypothesis that plasmodesmata facilitate the movement of the entire chloroplast genomes between neighboring cells (Stegemann and Bock, 2009; Stegemann et al., 2012; Thyssen et al., 2012). However, the dimensions of plasmodesmata would constrain movement of intact organelles, and whether the translocated genomes use intact chloroplasts as vehicles or are transported through plasmodesmata as "naked" but specifically folded DNA remains to be elucidated.

Interestingly, genetic screens aimed at isolating genes regulating plasmodesmata have identified genes involved in the biogenesis and/or functioning of chloroplasts. The first such reported mutant was maize *sucrose export defective1 (sxd1;* Russin et al., 1996). The *sxd1* mutants failed to export photosynthate from sites of photosynthesis and have reduced intercellular transport due to accumulation of callose at plasmodesmata located at the bundle-sheath and vascular parenchyma interfaces (Botha et al., 2000). Callose is a polymer composed of β-1,3-linked glucose residues and callose accumulation at the necks of plasmodesmata is a major strategy for limiting plasmodesmal trafficking

(De Storme and Geelen, 2014). Notably, despite accumulating sugars and starch in source tissues, *sxd1* does not repress photosynthesis (Provencher et al., 2001). *SXD1* is the maize ortholog of *Arabidopsis VTE1*, encoding a chloroplast TC required for the production of the antioxidant vitamin E (Provencher et al., 2001; Porfirova et al., 2002). The *sxd1* mutant provided the first clue that chloroplast redox state may influence plasmodesmata. However, under optimal growth conditions *vte1* mutants were indistinguishable from wild-type plants, although they were more sensitive to photooxidative stress (Porfirova et al., 2002). *Arabidopsis* mutants with defects in tocopherol metabolism should be carefully examined for plasmodesmata-related changes in intercellular trafficking.

A genetic screen for *Arabidopsis thaliana* mutants with altered plasmodesmal function identified the *gfp arrested trafficking (gat)* mutants (Benitez-Alfonso et al., 2009). In these mutants GFP synthesized in the companion cells of the phloem failed to move via the plasmodesmata into the surrounding tissues as occurred in wild-type tissues; that is, *gat* mutants have decreased intercellular trafficking. The *gat1, 2, 4*, and *5* mutations are all seedling lethal, with development ceasing about 10 days after germination. *GAT1* encodes a plastid-localized thioredoxin-m3 (TRX-m3), and *gat1* roots accumulate higher levels of ROS than do wild-type roots. In addition, callose was found to accumulate at plasmodesmata in *gat1* roots. Similar observations have been made for the other *gat* mutants (Benitez-Alfonso and Jackson, 2009). Notably, approximately five percent of plasmodesmata in *gat1* seedling roots are occluded by electron-dense material (Benitez-Alfonso et al., 2009). Overexpression of *GAT1* results in the reciprocal phenotype of increased intercellular transport. Thus, like SXD1/VTE1, GAT1 likely functions in redox homeostasis involving the chloroplasts and perturbation of plastid redox state results in altered plasmodesmata. In addition, it has been proposed that altered metabolic flux in the *gat1* mutant may alter the redox state of TRX-m3 and ultimately influence plasmodesmata function (Benitez-Alfonso et al., 2009).

The Zambryski lab conducted a separate screen for *Arabidopsis* mutants with altered plasmodesmata-mediated intercellular trafficking (Kim et al., 2002; Burch-Smith and Zambryski, 2012). Several *increased size exclusion limit (ise)* mutants were identified by screening embryonically lethal mutants and monitoring their ability to traffic fluorescent dyes between cells. *ISE1* and *ISE2* have been mapped and cloned (Kobayashi et al., 2007; Stonebloom et al., 2009). In addition to increased plasmodesmal trafficking *ise1* and *ise2* embryos also contain increased numbers of plasmodesmata with multiple branches (Burch-Smith and Zambryski, 2010). Interestingly, nuclear *ISE2* encodes a chloroplast-localized DEVH-type RNA helicase, (Burch-Smith et al., 2011) while *ISE1* encodes a mitochondrial DEAD-box RNA helicase (Stonebloom et al., 2009). Further analyses in *Arabidopsis* and *N. benthamiana* revealed that loss of ISE1 and/or ISE2 leads to defective chloroplasts and ultimately leaves become yellow (Burch-Smith and Zambryski, 2010; Burch-Smith et al., 2011; Stonebloom et al., 2012). Gene expression analyses on *ise1* and *ise2* embryos by tiling microarrays revealed that the largest class of nuclear genes affected in both mutants, representing ~20% of the total, encode chloroplast-localized proteins (Burch-Smith et al., 2011). Importantly, genes encoding products for chlorophyll biosynthesis, photosynthetic light-harvesting reactions, and the Calvin cycle for carbon fixation were all affected in both mutants. Thus, the overlapping plasmodesmatal phenotypes of the *ise1* and *ise2* mutants are likely due to defective chloroplasts.

The changes in plasmodesmata and intercellular trafficking are likely mediated by altered redox signaling from chloroplasts and mitochondria. Using redox-sensitive fluorescent reporters the redox state of chloroplasts depleted of ISE1 or ISE2 was measured (Stonebloom et al., 2012). Significantly, chloroplasts in *ISE1*- or *ISE2*-silenced *N. benthamiana* leaves were more reduced than in non-silenced control leaves. These results are consistent those from the *gat1* mutant (Benitez-Alfonso et al., 2009). Thus, oxidized chloroplasts are associated with decreased intercellular trafficking while reduced chloroplasts increase plasmodesmata-mediated trafficking. Further experiments focused on identifying the redox pathways and signal transduction involved in regulating plasmodesmata will yield important insight into how the physiological state of the chloroplasts is communicated to the rest of the plant.

Redox Signals Travel between Cells

Recently the involvement of chloroplasts in remote control of nucleus-localized alternative splicing was reported. Specifically, the light-regulated redox state of plastoquinone is the source of a chloroplast-generated signal able to control alternative splicing (Petrillo et al., 2014). In the presence of a reduced or oxidized plastoquinone pool, the alternative splicing of selected nuclear genes is promoted or inhibited, respectively. Importantly, even though photosynthetic tissues generated it, the putative signal is able to travel to roots to control alternative splicing. Thus chloroplast redox signals can act non-cell-autonomously. The redox pathway leading to defective alternative splicing has not been identified; however, the redox-regulated protein kinases of chloroplasts are proposed to contribute to the signaling machinery through a process of phosphorylation of some other proteins (Petrillo et al., 2014). This result is consistent with the observation that an *Arabidopsis* plasma-membrane-localized thioredoxin, Trx h9 moved intercellularly (Meng et al., 2010), presumably via plasmodesmata. The potential for intercellular trafficking by redox signals greatly extends the reach of chloroplasts in signaling.

The question remains: how could chloroplasts regulate plasmodesmata and intercellular trafficking? It is possible that as-yet-unidentified chloroplasts signals could directly target plasmodesmata. Indeed, chloroplasts are often observed in the vicinity of plasmodesmata, and regardless of the reasons for this close proximity, this means signals would not have to travel across long distances to act (Figure 4). Studies in the algae *Chara australis* also suggest a relationship between chloroplast and cell wall structure (Foissner et al., 2015). It is also possible that chloroplast-to-nucelus retrograde signaling may be involved in regulating events at the cell wall. Indeed, disruption of chloroplast function often results in significant changes of expression of genes involved in cell wall synthesis and modification (Burch-Smith et al., 2011). Similarly, expression of genes with functions related to the cell wall was down regulated in plants accumulating β-cyclocitral upon high light treatment (Ramel et al., 2012). It is also perhaps significant that three of the 39 genes identified as the core response module of chloroplast retrograde signaling encode proteins involved in synthesizing or modifying cell walls (Glasser et al., 2014). Future experiments should determine the pathways used for chloroplast-to-plasmodesmata signaling.

CHLOROPLAST SIGNALING IN ECOSYSTEM FUNCTIONING – THE POWER OF SCENT

In addition to being the center for production of oxygen and sugars, chloroplasts are also the source of plethora of secondary metabolites. Chloroplast-produced biogenic volatile organic compounds (BVOCs) seem to represent a fascinating signaling system that enables a plant to communicate with not only itself (e.g., propagation of a systemic response), and with the entire ecosystem that includes plant–plant, plant–animal/insect interaction, but could potentially also impact atmospheric chemistry (Atkinson, 2000; Seybold, 2006; Heil and Silva Bueno, 2007; Heil, 2010).

Volatiles directly produced in chloroplasts include isoprene, monoterpenes, diterpenes, hemiterpenes and volatile carotenoid derivatives, all synthesized by the MEP pathway. Also generated by the chloroplasts are green leaf volatiles (GLVs), synthesized in the LOX-pathway that consists of C6 and C9 aldehydes, alcohols and their acetate esters, that are responsible for the 'mowed grass odor' (Dudareva et al., 2013). These volatiles have important physiological roles, for example, stabilization of thylakoid membranes, reduction of ROS, as well as an ecological function, emphasizing an often underestimated contribution of chloroplasts/plastids to fine regulation of the ecosystem (Velikova et al., 2012). It is estimated that each year about 500 Tg C of isoprene is emitted to the atmosphere. This significant amount has some consequences as the highly reactive isoprene molecule can oxidize OH groups to yield peroxyl radicals that in turn can convert NO into NO_2, producing O_3 (Thompson, 1992; Carter, 1994; Atkinson, 2000). Therefore isoprene as a reactive molecule plays a crucial role in establishing the content of atmospheric greenhouse gasses and pollutants (ozone, methane, secondary organic aerosols, etc.; Harrison et al., 2013).

Chloroplast-produced BVOCs have other ecological functions that are a consequence of their signaling propensity. One of these is the so-called priming effect that leads to enhanced and more effective defense response to pathogen or insect attack in plants pretreated with volatiles (Engelberth, 2004). Although they are released continuously, there is enhanced emission of volatiles upon biotic or abiotic stresses including drought, high light, parasite or herbivore activity (wounding), resulting in so-called HIPVs (herbivore induced plant volatiles) emissions that include monoterpenes and GLVs (Brilli et al., 2009; Bonaventure and Baldwin, 2010; Dicke and Baldwin, 2010; Niinemets, 2010, 2013; War et al., 2011; Heil, 2014). Leaves of attacked plants emit volatiles that can prime defense responses in intact leaves of the same plant as well as on neighboring plants. Moreover, as demonstrated in maize, the intact plants exposed to HIPVs (termed receiver plants) are able to store the information conveyed by volatiles and recall it upon herbivore attack to activate defense genes encoding proteinase inhibitors (Ali et al., 2013).

The molecular mechanism of volatile signaling is far from understood, however, an attractive scenario of volatile-driven epigenetic modification through DNA methylation has been demonstrated as a very reliable strategy for priming defense induction. Moreover, since the JA-signaling pathway is triggered upon herbivore attack (wounding), it is expected that induced JA formation as a downstream effect would be relevant to the HIPV-primed defense responses (Ali et al., 2013). However, some of the most important questions concern the perception and the very first steps in volatiles signaling in receiver plants. To this end it is proposed that volatiles, as hydrophobic compounds, could diffuse through the outer, lipophilic leaf layer (cuticle) or enter through stomata (Baldwin et al., 2006). Indeed, strong membrane depolarization as well as calcium influx were observed upon treatment of tomato plants with GLVs, two phenomena often associated with a signal transduction cascade in other systems (Zebelo et al., 2012). Rearrangements of cytoskeleton and expression of selected genes encoding, among others, protein kinases and transcription factors as well as changes in stomatal aperture were observed upon treatment with the monoterpenes camphor and menthol (Kriegs et al., 2010). However, the concentrations used in the assay are inconsistent with physiological levels, thus the observed effects are questionable. Elucidating the impact of BVOCs on gene expression and impacts on general cell biology are a likely future direction of this field.

Another intriguing aspect of chloroplast originating volatiles is that their release from attacked plants can serve as attractants of the offending herbivore's natural biological enemies, a phenomenon known as tritrophic chemical communication or tritrophic interaction (Heil, 2008; Mooney et al., 2012; Scala, 2013). Importantly, some volatiles also have allelopathic and insecticidal properties that could be deterrents for oviposition or can attract pollinators and/or seeds dispersers (Schulz et al., 2007; McCallum et al., 2011; Rodriguez et al., 2011; Reis et al., 2014). Together, these reports shed new light on chloroplasts as factories that generate volatile compounds that are important not only from the perspective of a single plant, but also the entire ecosystem. As new strategies for chloroplast biotechnology are deployed, it will be important to examine their effects on these volatiles so that susceptibility to herbivore predation and pathogen ingress are not inadvertently introduced through unintended modification of volatile production and perception.

CONCLUSION AND FUTURE DIRECTIONS

The era of transcriptomics initiated by the microarray platform and continued by Next-generation sequencing technology, together with metabolomics and proteomic approach provided by mass-spectrometry, established new possibilities and new directions in addressing questions in biology. Inevitably, due to these two strategies contemporary plant biology is progressing toward a more holistic view, which translates into studying not a single protein *per se*, but a single protein in the context of an entire cell, tissue, organ, individual plant and ultimately the entire ecosystem biology. Current progress in studying chloroplast biology shows their function goes far beyond photosynthesis and includes all aspects of plant biology, making these organelles an integral part and a full-fledged player in the cell. Understanding the signaling pathways chloroplasts and other plastids use to communicate with the rest of the plant cell under normal and stress conditions will provide new ideas for developing powerful strategies for engineering improved plants. Moreover, a thorough knowledge of chloroplast signaling will also allow the design of rational approaches that could minimize if not avoid unintended crosstalk and undesirable outcomes.

CONFLICT OF INTEREST STATEMENT

The authors declare that the research was conducted in the absence of any commercial or financial relationships that could be construed as a potential conflict of interest.

ACKNOWLEDGMENTS

We apologize to colleagues whose work we could not cite due to space limitations. We thank Dr. Elena Ganusova for helpful suggestions during manuscript preparation. This work was supported by University of Tennessee start-up funds and NSF award #1456761 to TB-S.

REFERENCES

1. Abbink, T. E., Peart, J. R., Mos, T. N., Baulcombe, D. C., Bol, J. F., and Linthorst, H. J. (2002). Silencing of a gene encoding a protein component of the oxygen-evolving complex of photosystem II enhances virus replication in plants. *Virology* 295, 307–319. doi: 10.1006/viro.2002.1332
2. Adam, Z. (2015). Plastid intramembrane proteolysis. *Biochim. Biophys. Acta* 1847, 910–914. doi: 10.1016/j.bbabio.2014.12.006
3. Alazem, M., and Lin, N. S. (2015). Roles of plant hormones in the regulation of host-virus interactions. *Mol. Plant Pathol.* 16, 529–540. doi: 10.1111/mpp.12204
4. Albrecht, V., Simkova, K., Carrie, C., Delannoy, E., Giraud, E., Whelan, J., et al. (2010). The cytoskeleton and the peroxisomal-targeted snowy cotyledon3 protein are required for chloroplast development in *Arabidopsis*. *Plant Cell* 22, 3423–3438. doi: 10.1105/tpc.110.074781
5. Ali, M., Sugimoto, K., Ramadan, A., and Arimura, G. (2013). Memory of plant communications for priming anti-herbivore responses. *Sci. Rep.* 3, 1872. doi: 10.1038/srep01872
6. Allen, J. F. (1993). Control of gene expression by redox potential and the requirement for chloroplast and mitochondrial genomes. *J. Theor. Biol.* 165, 609–631. doi: 10.1006/jtbi.1993.1210
7. Andersson, M. X., Goksor, M., and Sandelius, A. S. (2007). Membrane contact sites: physical attachment between chloroplasts and endoplasmic reticulum revealed by optical manipulation. *Plant Signal. Behav.* 2, 185–187. doi: 10.4161/psb.2.3.3973
8. Antico, C. J., Colon, C., Banks, T., and Ramonell, K. M. (2012). Insights into the role of jasmonic acid-mediated defenses against

necrotrophic and biotrophic fungal pathogens. *Front. Biol.* 7:48–56. doi: 10.1007/s11515-011-1171-1

9. Atkinson, R. (2000). Atmospheric chemistry of VOCs and NO. *Atmos. Environ.* 34, 2063–2101. doi: 10.1016/S1352-2310(99)00460-4
10. Avendano-Vazquez, A. O., Cordoba, E., Llamas, E., San Roman, C., Nisar, N., De La Torre, S., et al. (2014). An uncharacterized apocarotenoid-derived signal generated in zeta-carotene desaturase mutants regulates leaf development and the expression of chloroplast and nuclear genes in *Arabidopsis*. *Plant Cell* 26, 2524–2537. doi: 10.1105/tpc.114.123349
11. Balasubramaniam, M., Kim, B. S., Hutchens-Williams, H. M., and Loesch-Fries, L. S. (2014). The photosystem II oxygen-evolving complex protein PsbP interacts with the coat protein of *Alfalfa mosaic* virus and inhibits virus replication. *Mol. Plant Microbe Interact.* 27, 1107–1118. doi: 10.1094/MPMI-02-14-0035-R
12. Balazadeh, S., Jaspert, N., Arif, M., Mueller-Roeber, B., and Maurino, V. G. (2012). Expression of ROS-responsive genes and transcription factors after metabolic formation of H(2)O(2) in chloroplasts. *Front. Plant Sci.* 3:234. doi: 10.3389/fpls.2012.00234
13. Baldwin, I. T., Halitschke, R., Paschold, A., Von Dahl, C. C., and Preston, C. A. (2006). Volatile signaling in plant-plant interactions: "talking trees" in the genomics era. *Science* 311, 812–815. doi: 10.1126/science.1118446
14. Ball, S., Colleoni, C., Cenci, U., Raj, J. N., and Tirtiaux, C. (2011). The evolution of glycogen and starch metabolism in eukaryotes gives molecular clues to understand the establishment of plastid endosymbiosis. *J. Exp. Bot.* 62, 1775–1801. doi: 10.1093/jxb/erq411
15. Ball, S. G., Subtil, A., Bhattacharya, D., Moustafa, A., Weber, A. P., Gehre, L., et al. (2013). Metabolic effectors secreted by bacterial pathogens: essential facilitators of plastid endosymbiosis? *Plant Cell* 25, 7–21. doi: 10.1105/tpc.112.101329
16. Barajas-Lopez Jde, D., Blanco, N. E., and Strand, A. (2013). Plastid-to-nucleus communication, signals controlling the running of the plant cell. *Biochim. Biophys. Acta* 1833, 425–437. doi: 10.1016/j.bbamcr.2012.06.020
17. Bellin, D., Asai, S., Delledonne, M., and Yoshioka, H. (2013). Nitric

oxide as a mediator for defense responses. *Mol. Plant Microbe Interact.* 26, 271–277. doi: 10.1094/MPMI-09-12-0214-CR

18. Benitez-Alfonso, Y., Cilia, M., San Roman, A., Thomas, C., Maule, A., Hearn, S., et al. (2009). Control of *Arabidopsis* meristem development by thioredoxin-dependent regulation of intercellular transport. *Proc. Natl. Acad. Sci. U.S.A.* 106, 3615–3620. doi: 10.1073/pnas.0808717106
19. Benitez-Alfonso, Y., and Jackson, D. (2009). Redox homeostasis regulates plasmodesmal communication in *Arabidopsis* meristems. *Plant Signal. Behav.* 4, 655–659. doi: 10.1073/pnas.0808717106
20. Benning, C. (2009). Mechanisms of lipid transport involved in organelle biogenesis in plant cells. *Annu. Rev. Cell Dev. Biol.* 25, 71–91. doi: 10.1146/annurev.cellbio.042308.113414
21. Besson-Bard, A., Courtois, C., Gauthier, A., Dahan, J., Dobrowolska, G., Jeandroz, S., et al. (2008). Nitric oxide in plants: production and cross-talk with Ca2+ signaling. *Mol. Plant* 1, 218–228. doi: 10.1093/mp/ssm016
22. Bhat, S., Folimonova, S. Y., Cole, A. B., Ballard, K. D., Lei, Z., Watson, B. S., et al. (2013). Influence of host chloroplast proteins on *Tobacco mosaic* virus accumulation and intercellular movement. *Plant Physiol.* 161, 134–147. doi: 10.1104/pp.112.207860
23. Bienert, G. P., Moller, A. L., Kristiansen, K. A., Schulz, A., Moller, I. M., Schjoerring, J. K., et al. (2007). Specific aquaporins facilitate the diffusion of hydrogen peroxide across membranes. *J. Biol. Chem.* 282, 1183–1192. doi: 10.1074/jbc.M603761200
24. Binns, D., Januszewski, T., Chen, Y., Hill, J., Markin, V. S., Zhao, Y., et al. (2006). An intimate collaboration between peroxisomes and lipid bodies. *J. Cell Biol.* 173, 719–731. doi: 10.1083/jcb.200511125
25. Blanco, F., Garreton, V., Frey, N., Dominguez, C., Perez-Acle, T., Van Der Straeten, D., et al. (2005). Identification of NPR1-dependent and independent genes early induced by salicylic acid treatment in *Arabidopsis*. *Plant Mol. Biol.* 59, 927–944. doi: 10.1007/s11103-005-2227-x
26. Blanco, N. E., Guinea-Diaz, M., Whelan, J., and Strand, A. (2014). Interaction between plastid and mitochondrial retrograde signalling pathways during changes to plastid redox status. *Philos. Trans. R.*

Soc. Lond. B Biol. Sci. 369, 20130231. doi: 10.1098/rstb.2013.0231

27. Blechert, S. (1999). Structure-activity analyses reveal the existence of two separate groups of active octadecanoids in elicitation of the tendril-coiling response of *Bryonia dioica* Jacq. *Planta* 207, 470–479. doi: 10.1007/s004250050506
28. Bobik, K., Dunlap, J. R., and Burch-Smith, T. M. (2014). Tandem high-pressure freezing and quick freeze substitution of plant tissues for transmission electron microscopy. *J. Vis. Exp.* e51844. doi: 10.3791/51844
29. Bonaventure, G., and Baldwin, I. T. (2010). Transduction of wound and herbivory signals in plastids. *Commun. Integr. Biol.* 3, 313–317. doi: 10.4161/cib.3.4.11834
30. Borucki, W., Bederska, M., and Sujkowska-Rybkowska, M. (2015). Visualisation of plastid outgrowths in potato (*Solanum tuberosum* L.) tubers by carboxyfluorescein diacetate staining. *Plant Cell Rep.* 34, 853–860. doi: 10.1007/s00299-015-1748-2
31. Botha, C. E. J., Cross, R. H. M., Van Bel, A. J. E., and Peter, C. I. (2000). Phloem loading in the sucrose-export-defective (SXD-1) mutant maize is limited by callose deposition at plasmodesmata in bundle sheath-vascular parenchyma interface. *Protoplasma* 214, 65–72. doi: 10.1007/BF02524263
32. Boudiere, L., Michaud, M., Petroutsos, D., Rebeille, F., Falconet, D., Bastien, O., et al. (2014). Glycerolipids in photosynthesis: composition, synthesis and trafficking. *Biochim. Biophys. Acta* 1837, 470–480. doi: 10.1016/j.bbabio.2013.09.007
33. Bradbeer, J. W., Atkinson, Y. E., Borner, T., and Hagemann, R. (1979). Cytoplasmic synthesis of plastid polypeptides may be controlled by plastid-synthesized RNA. *Nature* 279, 816–817. doi: 10.1038/279816a0
34. Brilli, F., Ciccioli, P., Frattoni, M., Prestininzi, M., Spanedda, A. F., and Loreto, F. (2009). Constitutive and herbivore-induced monoterpenes emitted by *Populus x euroamericana* leaves are key volatiles that orient *Chrysomela populi* beetles. *Plant Cell Environ.*32, 542–552. doi: 10.1111/j.1365-3040.2009.01948.x
35. Brunkard, J. O., Runkel, A. M., and Zambryski, P. C. (2015a). The cytosol must flow: intercellular transport through plasmodesmata.

Curr. Opin. Cell Biol. 35, 13–20. doi: 10.1016/j.ceb.2015.03.003

36. Brunkard, J. O., Runkel, A. M., and Zambryski, P. C. (2015b). Chloroplasts extend stromules independently and in response to internal redox signals. *Proc. Natl. Acad. Sci. U.S.A.* 112, 10044–10049. doi: 10.1073/pnas.1511570112
37. Burch-Smith, T. M., Brunkard, J. O., Choi, Y. G., and Zambryski, P. C. (2011). Organelle-nucleus cross-talk regulates plant intercellular communication via plasmodesmata. *Proc. Natl. Acad. Sci. U.S.A.* 108, E1451–E1460. doi: 10.1073/pnas.1117226108
38. Burch-Smith, T. M., and Zambryski, P. C. (2010). Loss of INCREASED SIZE EXCLUSION LIMIT (ISE)1 or ISE2 increases the formation of secondary plasmodesmata. *Curr. Biol.* 20, 989–993. doi: 10.1016/j.cub.2010.03.064
39. Burch-Smith, T. M., and Zambryski, P. C. (2012). Plasmodesmata paradigm shift: regulation from without versus within. *Annu. Rev. Plant Biol.* 63, 239–260. doi: 10.1146/annurev-arplant-042811-105453
40. Camejo, D., Romero-Puertas Mdel, C., Rodriguez-Serrano, M., Sandalio, L. M., Lazaro, J. J., Jimenez, A., et al. (2013). Salinity-induced changes in S-nitrosylation of pea mitochondrial proteins. *J. Proteomics* 79, 87–99. doi: 10.1016/j.jprot.2012.12.003
41. Cao, F. Y., Yoshioka, K., and Desveaux, D. (2011). The roles of ABA in plant-pathogen interactions. *J. Plant Res.* 124, 489–499. doi: 10.1007/s10265-011-0409-y
42. Caplan, J. L., Kumar, A. S., Park, E., Padmanabhan, M. S., Hoban, K., Modla, S., et al. (2015). Chloroplast stromules function during innate immunity. *Dev. Cell* 34, 45–57. doi: 10.1016/j.devcel.2015.05.011
43. Caplan, J. L., Mamillapalli, P., Burch-Smith, T. M., Czymmek, K., and Dinesh-Kumar, S. P. (2008). Chloroplastic protein NRIP1 mediates innate immune receptor recognition of a viral effector. *Cell* 132, 449–462. doi: 10.1016/j.cell.2007.12.031
44. Carrie, C., and Whelan, J. (2013). Widespread dual targeting of proteins in land plants: when, where, how and why. *Plant Signal. Behav.* 8, e25034. doi: 10.4161/psb.25034
45. Carter, W. P. L. (1994). Development of ozone reactivity scales for

volatile organic compounds. *J. Air Waste Manag. Assoc.* 44, 881–899. doi: 10.1080/1073161X.1994.10467290

46. Caverzan, A., Passaia, G., Rosa, S. B., Ribeiro, C. W., Lazzarotto, F., and Margis-Pinheiro, M. (2012). Plant responses to stresses: role of ascorbate peroxidase in the antioxidant protection. *Genet. Mol. Biol.* 35, 1011–1019. doi: 10.1590/S1415-47572012000600016
47. Chaki, M., Kovacs, I., Spannagl, M., and Lindermayr, C. (2014). Computational prediction of candidate proteins for S-nitrosylation in *Arabidopsis thaliana*. *PLoS ONE* 9:e110232. doi: 10.1371/journal.pone.0110232
48. Chen, M., Galvao, R. M., Li, M., Burger, B., Bugea, J., Bolado, J., et al. (2010). *Arabidopsis* HEMERA/pTAC12 initiates photomorphogenesis by phytochromes. *Cell* 141, 1230–1240. doi: 10.1016/j.cell.2010.05.007
49. Chen, Z., Ricigliano, J. W., and Klessig, D. F. (1993). Purification and characterization of a soluble salicylic acid-binding protein from tobacco. *Proc. Natl. Acad. Sci. U.S.A.* 90, 9533–9537. doi: 10.1073/pnas.90.20.9533
50. Cheng, Y. Q., Liu, Z. M., Xu, J., Zhou, T., Wang, M., Chen, Y. T., et al. (2008). HC-Pro protein of sugar cane mosaic virus interacts specifically with maize ferredoxin-5 in vitro and in planta. *J. Gen. Virol.* 89, 2046–2054. doi: 10.1099/vir.0.2008/001271-0
51. Chi, W., Sun, X., and Zhang, L. (2013). Intracellular signaling from plastid to nucleus. *Annu. Rev. Plant Biol.* 64, 559–582. doi: 10.1146/annurev-arplant-050312-120147
52. Coll, N. S., Epple, P., and Dangl, J. L. (2011). Programmed cell death in the plant immune system. *Cell Death Differ.* 18, 1247–1256. doi: 10.1038/cdd.2011.37
53. Collings, D. A., Carter, C. N., Rink, J. C., Scott, A. C., Wyatt, S. E., and Allen, N. S. (2000). Plant nuclei can contain extensive grooves and invaginations. *Plant Cell* 12, 2425–2440. doi: 10.1105/tpc.12.12.2425
54. Cordoba, E., Salmi, M., and Leon, P. (2009). Unravelling the regulatory mechanisms that modulate the MEP pathway in higher plants. *J. Exp. Bot.* 60, 2933–2943. doi: 10.1093/jxb/erp190
55. Corpas, F. J., Palma, J. M., Del Rio, L. A., and Barroso, J. B.

(2013). Protein tyrosine nitration in higher plants grown under natural and stress conditions. *Front. Plant Sci.* 4:29. doi: 10.3389/fpls.2013.00029

56. Cran, D. G., and Dyer, A. F. (1973). Membrane continuity ans associations in the fern, *Dryopteris borreri*. *Protoplasma* 76, 103–108. doi: 10.1007/BF01279676
57. Crotty, W. J., and Ledbetter, M. C. (1973). Membrane continuities involving chloroplasts and other organelles in plant cells. *Science* 182, 839–841. doi: 10.1126/science.182.4114.839
58. Dave, A., Hernandez, M. L., He, Z., Andriotis, V. M., Vaistij, F. E., Larson, T. R., et al. (2011). 12-oxo-phytodienoic acid accumulation during seed development represses seed germination in *Arabidopsis*. *Plant Cell* 23, 583–599. doi: 10.1105/tpc.110.081489
59. de Brito, O. M., and Scorrano, L. (2008). Mitofusin 2 tethers endoplasmic reticulum to mitochondria. *Nature* 456, 605–610. doi: 10.1038/nature07534
60. De Clercq, I., Vermeirssen, V., Van Aken, O., Vandepoele, K., Murcha, M. W., Law, S. R., et al. (2013). The membrane-bound NAC transcription factor ANAC013 functions in mitochondrial retrograde regulation of the oxidative stress response in *Arabidopsis*. *Plant Cell*25, 3472–3490. doi: 10.1105/tpc.113.117168
61. Desikan, R., A-H-Mackerness, S., Hancock, J. T., and Neill, S. J. (2001). Regulation of the *Arabidopsis* transcriptome by oxidative stress. *Plant Physiol.* 127, 159–172. doi: 10.1104/pp.127.1.159
62. Desikan, R., Griffiths, R., Hancock, J., and Neill, S. (2002). A new role for an old enzyme: nitrate reductase-mediated nitric oxide generation is required for abscisic acid-induced stomatal closure in *Arabidopsis thaliana*. *Proc. Natl. Acad. Sci. U.S.A.* 99, 16314–16318. doi: 10.1073/pnas.252461999
63. De Storme, N., and Geelen, D. (2014). Callose homeostasis at plasmodesmata: molecular regulators and developmental relevance. *Front. Plant Sci.* 5:138. doi: 10.3389/fpls.2014.00138
64. Dicke, M., and Baldwin, I. T. (2010). The evolutionary context for herbivore-induced plant volatiles: beyond the 'cry for help'. *Trends Plant Sci.* 15, 167–175. doi: 10.1016/j.tplants.2009.12.002
65. Donaire, L., Barajas, D., Martinez-Garcia, B., Martinez-Priego, L.,

Pagan, I., and Llave, C. (2008). Structural and genetic requirements for the biogenesis of tobacco rattle virus-derived small interfering RNAs. *J. Virol.* 82, 5167–5177. doi: 10.1128/JVI.00272-08

66. Dudareva, N., Klempien, A., Muhlemann, J. K., and Kaplan, I. (2013). Biosynthesis, function and metabolic engineering of plant volatile organic compounds. *New Phytol.* 198, 16–32. doi: 10.1111/nph.12145
67. Durner, J., and Klessig, D. F. (1995). Inhibition of ascorbate peroxidase by salicylic acid and 2,6-dichloroisonicotinic acid, two inducers of plant defense responses. *Proc. Natl. Acad. Sci. U.S.A.* 92, 11312–11316. doi: 10.1073/pnas.92.24.11312
68. Earley, K., Smith, M., Weber, R., Gregory, B., and Poethig, R. (2010). An endogenous F-box protein regulates ARGONAUTE1 in *Arabidopsis thaliana*. *Silence* 1, 15. doi: 10.1186/1758-907X-1-15
69. Ellis, R. J. (1977). Protein synthesis by isolated chloroplasts. *Biochim. Biophys. Acta* 463, 185–215. doi: 10.1016/0304-4173(77)90008-8
70. Engelberth, J. (2004). Airborne signals prime plants against insect herbivore attack. *Proc. Natl. Acad. Sci. U.S.A.* 101, 1781–1785. doi: 10.1073/pnas.0308037100
71. Erickson, J. L., Ziegler, J., Guevara, D., Abel, S., Klosgen, R. B., Mathur, J., et al. (2014). Agrobacterium-derived cytokinin influences plastid morphology and starch accumulation in *Nicotiana benthamiana* during transient assays. *BMC Plant Biol.* 14:127. doi: 10.1186/1471-2229-14-127
72. Estavillo, G. M., Crisp, P. A., Pornsiriwong, W., Wirtz, M., Collinge, D., Carrie, C., et al. (2011). Evidence for a SAL1-PAP chloroplast retrograde pathway that functions in drought and high light signaling in *Arabidopsis*. *Plant Cell* 23, 3992–4012. doi: 10.1105/tpc.111.091033
73. Falcon, L. I., Magallon, S., and Castillo, A. (2010). Dating the cyanobacterial ancestor of the chloroplast. *ISME J.* 4, 777–783. doi: 10.1038/ismej.2010.2
74. Finkelstein, R. (2013). Abscisic acid synthesis and response. *Arabidopsis Book* 11, e0166. doi: 10.1199/tab.0166
75. Finnegan, P. M., Whelan, J., Millar, A. H., Zhang, Q., Smith, M. K., Wiskich, J. T., et al. (1997). Differential expression of the multigene

family encoding the soybean mitochondrial alternative oxidase. *Plant Physiol.* 114, 455–466. doi: 10.1104/pp.114.2.455

76. Flors, V., Ton, J., Van Doorn, R., Jakab, G., Garcia-Agustin, P., and Mauch-Mani, B. (2008). Interplay between JA, SA and ABA signalling during basal and induced resistance against *Pseudomonas syringae* and *Alternaria brassicicola*. *Plant J.* 54, 81–92. doi: 10.1111/j.1365-313X.2007.03397.x
77. Foissner, I., Sommer, A., and Hoeftberger, M. (2015). Photosynthesis-dependent formation of convoluted plasma membrane domains in *Chara* internodal cells is independent of chloroplast position. *Protoplasma* 252, 1085–1096. doi: 10.1007/s00709-014-0742-9
78. Foresi, N., Mayta, M. L., Lodeyro, A. F., Scuffi, D., Correa-Aragunde, N., Garcia-Mata, C., et al. (2015). Expression of the tetrahydrofolate-dependent nitric oxide synthase from the green alga *Ostreococcus tauri* increases tolerance to abiotic stresses and influences stomatal development in *Arabidopsis*. *Plant J.* 82, 806–821. doi: 10.1111/tpj.12852
79. Forouhar, F., Yang, Y., Kumar, D., Chen, Y., Fridman, E., Park, S. W., et al. (2005). Structural and biochemical studies identify tobacco SABP2 as a methyl salicylate esterase and implicate it in plant innate immunity. *Proc. Natl. Acad. Sci. U.S.A.* 102, 1773–1778. doi: 10.1073/pnas.0409227102
80. Foyer, C. H., Bloom, A. J., Queval, G., and Noctor, G. (2009). Photorespiratory metabolism: genes, mutants, energetics, and redox signaling. *Annu. Rev. Plant Biol.* 60, 455–484. doi: 10.1146/annurev.arplant.043008.091948
81. Foyer, C. H., Karpinska, B., and Krupinska, K. (2014). The functions of WHIRLY1 and REDOX-RESPONSIVE TRANSCRIPTION FACTOR 1 in cross tolerance responses in plants: a hypothesis. *Philos. Trans. R. Soc. Lond. B Biol. Sci.* 369, 20130226. doi: 10.1098/rstb.2013.0226
82. Foyer, C. H., and Noctor, G. (2003). Redox sensing snd signalling associated with reactive oxygen in chloroplasts, peroxisomes and mitochondria. *Physiol. Plant.* 119, 355–364. doi: 10.1034/j.1399-3054.2003.00223.x
83. Foyer, C. H., and Noctor, G. (2005). Redox homeostasis and

antioxidant signaling: a metabolic interface between stress perception and physiological responses. *Plant Cell* 17, 1866–1875. doi: 10.1105/tpc.105.033589

84. Fu, Z. Q., and Dong, X. (2013). Systemic acquired resistance: turning local infection into global defense. *Annu. Rev. Plant Biol.* 64, 839–863. doi: 10.1146/annurev-arplant-042811-105606
85. Fu, Z. Q., Guo, M., Jeong, B. R., Tian, F., Elthon, T. E., Cerny, R. L., et al. (2007). A type III effector ADP-ribosylates RNA-binding proteins and quells plant immunity. *Nature* 447, 284–288. doi: 10.1038/nature05737
86. Fu, Z. Q., Yan, S., Saleh, A., Wang, W., Ruble, J., Oka, N., et al. (2012). NPR3 and NPR4 are receptors for the immune signal salicylic acid in plants. *Nature* 486, 228–232. doi: 10.1038/nature11162
87. Gadjev, I., Vanderauwera, S., Gechev, T. S., Laloi, C., Minkov, I. N., Shulaev, V., et al. (2006). Transcriptomic footprints disclose specificity of reactive oxygen species signaling in *Arabidopsis*. *Plant Physiol.* 141, 436–445. doi: 10.1104/pp.106.078717
88. Gallagher, K. L., Sozzani, R., and Lee, C. M. (2014). Intercellular protein movement: deciphering the language of development. *Annu. Rev. Cell Dev. Biol.* 30, 207–233. doi: 10.1146/annurev-cellbio-100913-012915
89. Garcion, C., Lohmann, A., Lamodiere, E., Catinot, J., Buchala, A., Doermann, P., et al. (2008). Characterization and biological function of the ISOCHORISMATE SYNTHASE2 gene of *Arabidopsis*. *Plant Physiol.* 147, 1279–1287. doi: 10.1104/pp.108.119420
90. Gile, G. H., Moog, D., Slamovits, C. H., Maier, U. G., and Archibald, J. M. (2015). Dual organellar targeting of aminoacyl-tRNA synthetases in diatoms and cryptophytes. *Genome Biol. Evol.* 7, 1728–1742. doi: 10.1093/gbe/evv095
91. Gill, S. S., and Tuteja, N. (2010). Reactive oxygen species and antioxidant machinery in abiotic stress tolerance in crop plants. *Plant Physiol. Biochem.* 48, 909–930. doi: 10.1016/j.plaphy.2010.08.016
92. Giraud, E., Van Aken, O., Ho, L. H., and Whelan, J. (2009). The transcription factor ABI4 is a regulator of mitochondrial retrograde expression of ALTERNATIVE OXIDASE1a. *Plant Physiol.* 150, 1286–1296. doi: 10.1104/pp.109.139782

93. Giuliano, G. (2014). Plant carotenoids: genomics meets multi-gene engineering. *Curr. Opin. Plant Biol.* 19, 111–117. doi: 10.1016/j.pbi.2014.05.006
94. Glasser, C., Haberer, G., Finkemeier, I., Pfannschmidt, T., Kleine, T., Leister, D., et al. (2014). Meta-analysis of retrograde signaling in *Arabidopsis thaliana* reveals a core module of genes embedded in complex cellular signaling networks. *Mol. Plant* 7, 1167–1190. doi: 10.1093/mp/ssu042
95. Goetz, S., Hellwege, A., Stenzel, I., Kutter, C., Hauptmann, V., Forner, S., et al. (2012). Role of cis-12-oxo-phytodienoic acid in tomato embryo development. *Plant Physiol.* 158, 1715–1727. doi: 10.1104/pp.111.192658
96. Goto-Yamada, S., Mano, S., Yamada, K., Oikawa, K., Hosokawa, Y., Hara-Nishimura, I., et al. (2015). Dynamics of the light-dependent transition of plant peroxisomes. *Plant Cell Physiol.* 56, 1264–1271. doi: 10.1093/pcp/pcv081
97. Gray, J. C., Hansen, M. R., Shaw, D. J., Graham, K., Dale, R., Smallman, P., et al. (2012). Plastid stromules are induced by stress treatments acting through abscisic acid. *Plant J.* 69, 387–398. doi: 10.1111/j.1365-313X.2011.04800.x
98. Gray, J. C., Sullivan, J. A., Hibberd, J. M., and Hansen, M. R. (2001). Stromules: mobile protusions and interconnections between plastids. *Plant Biol. (Stuttg.)* 3, 223–233. doi: 10.1055/s-2001-15204
99. Groning, B. R., Abouzid, A., and Jeske, H. (1987). Single-stranded DNA from *Abutilon mosaic* virus is present in the plastids of infected *Abutilon sellovianum*. *Proc. Natl. Acad. Sci. U.S.A.* 84, 8996–9000. doi: 10.1073/pnas.84.24.8996
100. Gunning, B. E. (2005). Plastid stromules: video microscopy of their outgrowth, retraction, tensioning, anchoring, branching, bridging, and tip-shedding. *Protoplasma* 225, 33–42. doi: 10.1007/s00709-004-0073-3
101. Guo, F. Q., Okamoto, M., and Crawford, N. M. (2003). Identification of a plant nitric oxide synthase gene involved in hormonal signaling. *Science* 302, 100–103. doi: 10.1126/science.1086770
102. Han, Y., Mhamdi, A., Chaouch, S., and Noctor, G. (2013). Regulation of basal and oxidative stress-triggered jasmonic acid-related gene

expression by glutathione. *Plant Cell Environ.* 36, 1135–1146. doi: 10.1111/pce.12048

103. Hanson, M. R., and Sattarzadeh, A. (2011). Stromules: recent insights into a long neglected feature of plastid morphology and function. *Plant Physiol.* 155, 1486–1492. doi: 10.1104/pp.110.170852
104. Hanson, M. R., and Sattarzadeh, A. (2013). Trafficking of proteins through plastid stromules. *Plant Cell* 25, 2774–2782. doi: 10.1105/tpc.113.112870
105. Harari-Steinberg, O., Ohad, I., and Chamovitz, D. A. (2001). Dissection of the light signal transduction pathways regulating the two early light-induced protein genes in *Arabidopsis*. *Plant Physiol.* 127, 986–997. doi: 10.1104/pp.010270
106. Harrison, S. P., Morfopoulos, C., Dani, K. G., Prentice, I. C., Arneth, A., Atwell, B. J., et al. (2013). Volatile isoprenoid emissions from plastid to planet. *New Phytol.* 197, 49–57. doi: 10.1111/nph.12021
107. Hedtke, B., Wagner, I., Borner, T., and Hess, W. R. (1999). Inter-organellar crosstalk in higher plants: impaired chloroplast development affects mitochondrial gene and transcript levels. *Plant J.* 19, 635–643. doi: 10.1046/j.1365-313x.1999.00554.x
108. Heil, M. (2008). Indirect defence via tritrophic interactions. *New Phytol.* 178, 41–61. doi: 10.1111/j.1469-8137.2007.02330.x
109. Heil, M. (2010). Explaining evolution of plant communication by airborne signals. *Trends Ecol. Evol.* 25, 137–144. doi: 10.1016/j.tree.2009.09.010
110. Heil, M. (2014). Herbivore-induced plant volatiles: targets, perception and unanswered questions. *New Phytol.* 204, 297–306. doi: 10.1111/nph.12977
111. Heil, M., and Silva Bueno, J. C. (2007). Within-plant signaling by volatiles leads to induction and priming of an indirect plant defense in nature. *Proc. Natl. Acad. Sci. U.S.A.* 104, 5467–5472. doi: 10.1073/pnas.0610266104
112. Heinlein, M. (2015). Plasmodesmata: channels for viruses on the move. *Methods Mol. Biol.* 1217, 25–52. doi: 10.1007/978-1-4939-1523-1_2
113. Holzinger, A., Buchner, O., Lutz, C., and Hanson, M. R. (2007a).

Temperature-sensitive formation of chloroplast protrusions and stromules in mesophyll cells of *Arabidopsis thaliana. Protoplasma* 230, 23–30. doi: 10.1007/s00709-006-0222-y

114. Holzinger, A., Wasteneys, G. O., and Lutz, C. (2007b). Investigating cytoskeletal function in chloroplast protrusion formation in the arctic-alpine plant *Oxyria digyna. Plant Biol. (Stuttg.)* 9, 400–410. doi: 10.1055/s-2006-924727
115. Hunter, L. J., Westwood, J. H., Heath, G., Macaulay, K., Smith, A. G., Macfarlane, S. A., et al. (2013). Regulation of RNA-dependent RNA polymerase 1 and isochorismate synthase gene expression in *Arabidopsis. PLoS ONE* 8:e66530. doi: 10.1371/journal.pone.0066530
116. Isemer, R., Mulisch, M., Schafer, A., Kirchner, S., Koop, H. U., and Krupinska, K. (2012). Recombinant Whirly1 translocates from transplastomic chloroplasts to the nucleus. *FEBS Lett.* 586, 85–88. doi: 10.1016/j.febslet.2011.11.029
117. Ishida, H., Yoshimoto, K., Izumi, M., Reisen, D., Yano, Y., Makino, A., et al. (2008). Mobilization of rubisco and stroma-localized fluorescent proteins of chloroplasts to the vacuole by an ATG gene-dependent autophagic process. *Plant Physiol.* 148, 142–155. doi: 10.1104/pp.108.122770
118. Izumi, M., Hidema, J., Wada, S., Kondo, E., Kurusu, T., Kuchitsu, K., et al. (2015). Establishment of monitoring methods for autophagy in rice reveals autophagic recycling of chloroplasts and root plastids during energy limitation. *Plant Physiol.* 167, 1307–1320. doi: 10.1104/pp.114.254078
119. Jackson, D. (2015). Plasmodesmata spread their influence. *F1000Prime Rep.* 7, 25. doi: 10.12703/P7-25
120. Jang, C., Seo, E. Y., Nam, J., Bae, H., Gim, Y. G., Kim, H. G., et al. (2013). Insights into *Alternanthera mosaic* virus TGB3 functions: interactions with *Nicotiana benthamiana* PsbO correlate with chloroplast vesiculation and veinal necrosis caused by TGB3 over-expression. *Front. Plant Sci.* 4:5. doi: 10.3389/fpls.2013.00005
121. Jelenska, J., Van Hal, J. A., and Greenberg, J. T. (2010). *Pseudomonas syringae* hijacks plant stress chaperone machinery for virulence.

Proc. Natl. Acad. Sci. U.S.A. 107, 13177–13182. doi: 10.1073/pnas.0910943107

122. Jelenska, J., Yao, N., Vinatzer, B. A., Wright, C. M., Brodsky, J. L., and Greenberg, J. T. (2007). A J domain virulence effector of *Pseudomonas syringae* remodels host chloroplasts and suppresses defenses. *Curr. Biol.* 17, 499–508. doi: 10.1016/j.cub.2007.02.028
123. Jimenez, I., Lopez, L., Alamillo, J. M., Valli, A., and Garcia, J. A. (2006). Identification of a plum pox virus CI-interacting protein from chloroplast that has a negative effect in virus infection. *Mol. Plant Microbe Interact.* 19, 350–358. doi: 10.1094/MPMI-19-0350
124. Jin, Y., Ma, D., Dong, J., Li, D., Deng, C., Jin, J., et al. (2007). The HC-pro protein of *Potato virus* Y interacts with NtMinD of tobacco. *Mol. Plant Microbe Interact.* 20, 1505–1511. doi: 10.1094/MPMI-20-12-1505
125. Johanningmeier, U., and Howell, S. H. (1984). Regulation of light-harvesting chlorophyll-binding protein mRNA accumulation in *Chlamydomonas reinhardi*. Possible involvement of chlorophyll synthesis precursors. *J. Biol. Chem.* 259, 13541–13549.
126. Kangasjarvi, S., Neukermans, J., Li, S., Aro, E. M., and Noctor, G. (2012). Photosynthesis, photorespiration, and light signalling in defence responses. *J. Exp. Bot.* 63, 1619–1636. doi: 10.1093/jxb/err402
127. Kim, C., and Apel, K. (2013). Singlet oxygen-mediated signaling in plants: moving from flu to wild type reveals an increasing complexity. *Photosynth. Res.* 116, 455–464. doi: 10.1007/s11120-013-9876-4
128. Kim, C., Meskauskiene, R., Zhang, S., Lee, K. P., Lakshmanan Ashok, M., Blajecka, K., et al. (2012). Chloroplasts of *Arabidopsis* are the source and a primary target of a plant-specific programmed cell death signaling pathway. *Plant Cell* 24, 3026–3039. doi: 10.1105/tpc.112.100479
129. Kim, I., Hempel, F. D., Sha, K., Pfluger, J., and Zambryski, P. C. (2002). Identification of a developmental transition in plasmodesmatal function during embryogenesis in *Arabidopsis thaliana*. *Development* 129, 1261–1272.
130. Kim, T. H., Bohmer, M., Hu, H., Nishimura, N., and Schroeder,

J. I. (2010). Guard cell signal transduction network: advances in understanding abscisic acid, CO2, and Ca2+ signaling. *Annu. Rev. Plant Biol.* 61, 561–591. doi: 10.1146/annurev-arplant-042809-112226

131. Kinkema, M., Fan, W., and Dong, X. (2000). Nuclear localization of NPR1 is required for activation of PR gene expression. *Plant Cell* 12, 2339–2350. doi: 10.2307/3871233
132. Kline, K. G., Sussman, M. R., and Jones, A. M. (2010). Abscisic acid receptors. *Plant Physiol.* 154, 479–482. doi: 10.1104/pp.110.160846
133. Kobayashi, K., Otegui, M. S., Krishnakumar, S., Mindrinos, M., and Zambryski, P. (2007). INCREASED SIZE EXCLUSION LIMIT 2 encodes a putative DEVH box RNA helicase involved in plasmodesmata function during *Arabidopsis* embryogenesis. *Plant Cell* 19, 1885–1897. doi: 10.1105/tpc.106.045666
134. Kohler, R. H., Cao, J., Zipfel, W. R., Webb, W. W., and Hanson, M. R. (1997). Exchange of protein molecules through connections between higher plant plastids. *Science* 276, 2039–2042. doi: 10.1126/science.276.5321.2039
135. Kohler, R. H., and Hanson, M. R. (2000). Plastid tubules of higher plants are tissue-specific and developmentally regulated. *J. Cell Sci.*113(Pt 1), 81–89.
136. Kohler, R. H., Schwille, P., Webb, W. W., and Hanson, M. R. (2000). Active protein transport through plastid tubules: velocity quantified by fluorescence correlation spectroscopy. *J. Cell Sci.* 113(Pt 22), 3921–3930.
137. Koussevitzky, S., Nott, A., Mockler, T. C., Hong, F., Sachetto-Martins, G., Surpin, M., et al. (2007). Signals from chloroplasts converge to regulate nuclear gene expression. *Science* 316, 715–719. doi: 10.1126/science.%201140516
138. Krenz, B., Jeske, H., and Kleinow, T. (2012). The induction of stromule formation by a plant DNA-virus in epidermal leaf tissues suggests a novel intra– and intercellular macromolecular trafficking route. *Front. Plant Sci.* 3:291. doi: 10.3389/fpls.2012.00291
139. Krenz, B., Windeisen, V., Wege, C., Jeske, H., and Kleinow, T. (2010). A plastid-targeted heat shock cognate 70kDa protein

interacts with the *Abutilon mosaic* virus movement protein. *Virology* 401, 6–17. doi: 10.1016/j.virol.2010.02.011

140. Kriegs, B., Jansen, M., Hahn, K., Peisker, H., Samajova, O., Beck, M., et al. (2010). Cyclic monoterpene mediated modulations of *Arabidopsis thaliana* phenotype: effects on the cytoskeleton and on the expression of selected genes. *Plant Signal. Behav.* 5, 832–838. doi: 10.4161/psb.5.7.12032
141. Kropat, J., Oster, U., Rudiger, W., and Beck, C. F. (1997). Chlorophyll precursors are signals of chloroplast origin involved in light induction of nuclear heat-shock genes. *Proc. Natl. Acad. Sci. U.S.A.* 94, 14168–14172. doi: 10.1073/pnas.94.25.14168
142. Kropat, J., Oster, U., Rudiger, W., and Beck, C. F. (2000). Chloroplast signalling in the light induction of nuclear HSP70 genes requires the accumulation of chlorophyll precursors and their accessibility to cytoplasm/nucleus. *Plant J.* 24, 523–531. doi: 10.1046/j.1365-313x.2000.00898.x
143. Kunkel, B. N., and Brooks, D. M. (2002). Cross talk between signaling pathways in pathogen defense. *Curr. Opin. Plant Biol.* 5, 325–331. doi: 10.1016/S1369-5266(02)00275-3
144. Kwok, E. Y., and Hanson, M. R. (2003). Microfilaments and microtubules control the morphology and movement of non-green plastids and stromules in *Nicotiana tabacum*. *Plant J.* 35, 16–26. doi: 10.1046/j.1365-313X.2003.01777.x
145. Kwok, E. Y., and Hanson, M. R. (2004a). GFP-labelled Rubisco and aspartate aminotransferase are present in plastid stromules and traffic between plastids. *J. Exp. Bot.* 55, 595–604. doi: 10.1093/jxb/erh062
146. Kwok, E. Y., and Hanson, M. R. (2004b). In vivo analysis of interactions between GFP-labeled microfilaments and plastid stromules. *BMC Plant Biol.* 4:2. doi: 10.1186/1471-2229-4-2
147. Kwok, E. Y., and Hanson, M. R. (2004c). Plastids and stromules interact with the nucleus and cell membrane in vascular plants. *Plant Cell Rep.* 23, 188–195. doi: 10.1007/s00299-004-0824-9
148. Kwon, K. C., Verma, D., Jin, S., Singh, N. D., and Daniell, H. (2013). Release of proteins from intact chloroplasts induced by

reactive oxygen species during biotic and abiotic stress. *PLoS ONE* 8:e67106. doi: 10.1371/journal.pone.0067106

149. Laloi, C., Stachowiak, M., Pers-Kamczyc, E., Warzych, E., Murgia, I., and Apel, K. (2007). Cross-talk between singlet oxygen- and hydrogen peroxide-dependent signaling of stress responses in *Arabidopsis thaliana*. *Proc. Natl. Acad. Sci. U.S.A.* 104, 672–677. doi: 10.1073/pnas.0609063103
150. Langeveld, S. M., Van Wijk, R., Stuurman, N., Kijne, J. W., and De Pater, S. (2000). B-type granule containing protrusions and interconnections between amyloplasts in developing wheat endosperm revealed by transmission electron microscopy and GFP expression. *J. Exp. Bot.* 51, 1357–1361. doi: 10.1093/jexbot/51.349.1357
151. Larkin, R. M., Alonso, J. M., Ecker, J. R., and Chory, J. (2003). GUN4, a regulator of chlorophyll synthesis and intracellular signaling. *Science* 299, 902–906. doi: 10.1126/science.1079978
152. Ledford, H. K., Chin, B. L., and Niyogi, K. K. (2007). Acclimation to singlet oxygen stress in *Chlamydomonas reinhardtii*. *Eukaryot. Cell*6, 919–930. doi: 10.1128/EC.00207-06
153. Lee, J. Y., Wang, X., Cui, W., Sager, R., Modla, S., Czymmek, K., et al. (2011). A plasmodesmata-localized protein mediates crosstalk between cell-to-cell communication and innate immunity in *Arabidopsis*. *Plant Cell* 23, 3353–3373. doi: 10.1105/tpc.111.087742
154. Leon, J. (2013). Role of plant peroxisomes in the production of jasmonic acid-based signals. *Subcell. Biochem.* 69, 299–313. doi: 10.1007/978-94-007-6889-5_16
155. Li, G., Froehlich, J. E., Elowsky, C., Msanne, J., Ostosh, A. C., Zhang, C., et al. (2014). Distinct *Pseudomonas* type-III effectors use a cleavable transit peptide to target chloroplasts. *Plant J.* 77, 310–321. doi: 10.1111/tpj.12396
156. Li, H., Ma, D., Jin, Y., Tu, Y., Liu, L., Leng, C., et al. (2015). Helper component-proteinase enhances the activity of 1-deoxy–xylulose-5-phosphate synthase and promotes the biosynthesis of plastidic isoprenoids in *Potato virus* Y-infected tobacco. *Plant Cell Environ.* doi: 10.1111/pce.12526 [Epub ahead of print].
157. Li, T., Li, H., Zhang, Y. X., and Liu, J. Y. (2011). Identification

and analysis of seven H(2)O(2)-responsive miRNAs and 32 new miRNAs in the seedlings of rice (*Oryza sativa* L. ssp. indica). *Nucleic Acids Res.* 39, 2821–2833. doi: 10.1093/nar/gkq1047

158. Li, W., Cui, X., Meng, Z., Huang, X., Xie, Q., Wu, H., et al. (2012). Transcriptional regulation of *Arabidopsis* MIR168a and argonaute1 homeostasis in abscisic acid and abiotic stress responses. *Plant Physiol.* 158, 1279–1292. doi: 10.1104/pp.111.188789
159. Lin, L., Luo, Z., Yan, F., Lu, Y., Zheng, H., and Chen, J. (2011). Interaction between potyvirus P3 and ribulose-1,5-bisphosphate carboxylase/oxygenase (RubisCO) of host plants. *Virus Genes* 43, 90–92. doi: 10.1007/s11262-011-0596-6
160. Lindermayr, C., Saalbach, G., and Durner, J. (2005). Proteomic identification of S-nitrosylated proteins in *Arabidopsis*. *Plant Physiol.*137, 921–930. doi: 10.1104/pp.104.058719
161. Lounifi, I., Arc, E., Molassiotis, A., Job, D., Rajjou, L., and Tanou, G. (2013). Interplay between protein carbonylation and nitrosylation in plants. *Proteomics* 13, 568–578. doi: 10.1002/pmic.201200304
162. Ludwikow, A. (2015). Targeting proteins for proteasomal degradation-a new function of *Arabidopsis* ABI1 protein phosphatase 2C. *Front. Plant Sci.* 6:310. doi: 10.3389/fpls.2015.00310
163. Lundquist, P. K., Poliakov, A., Giacomelli, L., Friso, G., Appel, M., Mcquinn, R. P., et al. (2013). Loss of plastoglobule kinases ABC1K1 and ABC1K3 causes conditional degreening, modified prenyl-lipids, and recruitment of the jasmonic acid pathway. *Plant Cell* 25, 1818–1839. doi: 10.1105/tpc.113.111120
164. Ma, Y., Szostkiewicz, I., Korte, A., Moes, D., Yang, Y., Christmann, A., et al. (2009). Regulators of PP2C phosphatase activity function as abscisic acid sensors. *Science* 324, 1064–1068. doi: 10.1126/science.1172408
165. Malamy, J., Carr, J. P., Klessig, D. F., and Raskin, I. (1990). Salicylic Acid: a likely endogenous signal in the resistance response of tobacco to viral infection. *Science* 250, 1002–1004. doi: 10.1126/science.250.4983.1002
166. Mandal, M. K., Chandra-Shekara, A. C., Jeong, R. D., Yu, K., Zhu, S., Chanda, B., et al. (2012). Oleic acid-dependent modulation of NITRIC OXIDE ASSOCIATED1 protein levels regulates nitric

oxide-mediated defense signaling in *Arabidopsis*. *Plant Cell* 24, 1654–1674. doi: 10.1105/tpc.112.096768

167. Manfre, A., Glenn, M., Nunez, A., Moreau, R. A., and Dardick, C. (2011). Light quantity and photosystem function mediate host susceptibility to *Turnip mosaic* virus via a salicylic acid-independent mechanism. *Mol. Plant Microbe Interact.* 24, 315–327. doi: 10.1094/MPMI-08-10-0191
168. Manohar, M., Tian, M., Moreau, M., Park, S. W., Choi, H. W., Fei, Z., et al. (2014). Identification of multiple salicylic acid-binding proteins using two high throughput screens. *Front. Plant Sci.* 5:777. doi: 10.3389/fpls.2014.00777
169. Marechal, E., and Bastien, O. (2014). Modeling of regulatory loops controlling galactolipid biosynthesis in the inner envelope membrane of chloroplasts. *J. Theor. Biol.* 361, 1–13. doi: 10.1016/j.jtbi.2014.07.013
170. Matsui, A., Ishida, J., Morosawa, T., Mochizuki, Y., Kaminuma, E., Endo, T. A., et al. (2008). *Arabidopsis* transcriptome analysis under drought, cold, high-salinity and ABA treatment conditions using a tiling array. *Plant Cell Physiol.* 49, 1135–1149. doi: 10.1093/pcp/pcn101
171. Mayfield, S. P., and Taylor, W. C. (1984). Carotenoid-deficient maize seedlings fail to accumulate light-harvesting chlorophyll a/b binding protein (LHCP) mRNA. *Eur. J. Biochem.* 144, 79–84. doi: 10.1111/j.1432-1033.1984.tb08433.x
172. McCallum, E. J., Cunningham, J. P., Lucker, J., Zalucki, M. P., De Voss, J. J., and Botella, J. R. (2011). Increased plant volatile production affects oviposition, but not larval development, in the moth *Helicoverpa armigera*. *J. Exp. Biol.* 214, 3672–3677. doi: 10.1242/jeb.059923
173. McDonald, K. (1999). High-pressure freezing for preservation of high resolution fine structure and antigenicity for immunolabeling. *Methods Mol. Biol.* 117, 77–97. doi: 10.1385/1-59259-201-5:77
174. McDonald, K. L. (2014). Out with the old and in with the new: rapid specimen preparation procedures for electron microscopy of sectioned biological material. *Protoplasma* 251, 429–448. doi: 10.1007/s00709-013-0575-y

175. McFadden, G. I. (2001). Chloroplast origin and integration. *Plant Physiol.* 125, 50–53. doi: 10.1104/pp.125.1.50

176. McLean, B., Whatley, J. M., and Juniper, B. E. (1988). Continuity of chloroplast and endoplasmic reticulum membranes in *Chara* and *Equisetum*. *New Phytol.* 109, 59–65. doi: 10.1111/j.1469-8137.1988.tb00219.x

177. Mehrshahi, P., Johnny, C., and Dellapenna, D. (2014). Redefining the metabolic continuity of chloroplasts and ER. *Trends Plant Sci.* 19, 501–507. doi: 10.1016/j.tplants.2014.02.013

178. Mehrshahi, P., Stefano, G., Andaloro, J. M., Brandizzi, F., Froehlich, J. E., and Dellapenna, D. (2013). Transorganellar complementation redefines the biochemical continuity of endoplasmic reticulum and chloroplasts. *Proc. Natl. Acad. Sci. U.S.A.* 110, 12126–12131. doi: 10.1073/pnas.1306331110

179. Meng, L., Wong, J. H., Feldman, L. J., Lemaux, P. G., and Buchanan, B. B. (2010). A membrane-associated thioredoxin required for plant growth moves from cell to cell, suggestive of a role in intercellular communication. *Proc. Natl. Acad. Sci. U.S.A.* 107, 3900–3905. doi: 10.1073/pnas.0913759107

180. Mengel, A., Chaki, M., Shekariesfahlan, A., and Lindermayr, C. (2013). Effect of nitric oxide on gene transcription – S-nitrosylation of nuclear proteins. *Front. Plant Sci.* 4:293. doi: 10.3389/fpls.2013.00293

181. Meskauskiene, R., Nater, M., Goslings, D., Kessler, F., Op Den Camp, R., and Apel, K. (2001). FLU: a negative regulator of chlorophyll biosynthesis in *Arabidopsis thaliana*. *Proc. Natl. Acad. Sci. U.S.A.* 98, 12826–12831. doi: 10.1073/pnas.221252798

182. Mhamdi, A., Hager, J., Chaouch, S., Queval, G., Han, Y., Taconnat, L., et al. (2010). *Arabidopsis* GLUTATHIONE REDUCTASE1 plays a crucial role in leaf responses to intracellular hydrogen peroxide and in ensuring appropriate gene expression through both salicylic acid and jasmonic acid signaling pathways. *Plant Physiol.* 153, 1144–1160. doi: 10.1104/pp.110.153767

183. Miller, G., Shulaev, V., and Mittler, R. (2008). Reactive oxygen signaling and abiotic stress. *Physiol. Plant.* 133, 481–489. doi: 10.1111/j.1399-3054.2008.01090.x

184. Misra, A. N., Vladkova, R., Singh, R., Misra, M., Dobrikova, A. G., and Apostolova, E. L. (2014). Action and target sites of nitric oxide in chloroplasts. *Nitric Oxide* 39, 35–45. doi: 10.1016/j.niox.2014.04.003

185. Mochizuki, N., Brusslan, J. A., Larkin, R., Nagatani, A., and Chory, J. (2001). *Arabidopsis* genomes uncoupled 5 (GUN5) mutant reveals the involvement of Mg-chelatase H subunit in plastid-to-nucleus signal transduction. *Proc. Natl. Acad. Sci. U.S.A.* 98, 2053–2058. doi: 10.1073/pnas.98.4.2053

186. Montillet, J. L., Cacas, J. L., Garnier, L., Montane, M. H., Douki, T., Bessoule, J. J., et al. (2004). The upstream oxylipin profile of *Arabidopsis thaliana*: a tool to scan for oxidative stresses. *Plant J.* 40, 439–451. doi: 10.1111/j.1365-313X.2004.02223.x

187. Mooney, K. A., Pratt, R. T., and Singer, M. S. (2012). The tri-trophic interactions hypothesis: interactive effects of host plant quality, diet breadth and natural enemies on herbivores. *PLoS ONE* 7:e34403. doi: 10.1371/journal.pone.0034403

188. Mor, A., Koh, E., Weiner, L., Rosenwasser, S., Sibony-Benyamini, H., and Fluhr, R. (2014). Singlet oxygen signatures are detected independent of light or chloroplasts in response to multiple stresses. *Plant Physiol.* 165, 249–261. doi: 10.1104/pp.114.236380

189. Moreau, M., Lee, G. I., Wang, Y., Crane, B. R., and Klessig, D. F. (2008). AtNOS/AtNOA1 is a functional *Arabidopsis thaliana* cGTPase and not a nitric-oxide synthase. *J. Biol. Chem.* 283, 32957–32967. doi: 10.1074/jbc.M804838200

190. Moreau, M., Westlake, T., Zampogna, G., Popescu, G., Tian, M., Noutsos, C., et al. (2013). The *Arabidopsis* oligopeptidases TOP1 and TOP2 are salicylic acid targets that modulate SA-mediated signaling and the immune response. *Plant J.* 76, 603–614. doi: 10.1111/tpj.12320

191. Mou, Z., Fan, W., and Dong, X. (2003). Inducers of plant systemic acquired resistance regulate NPR1 function through redox changes. *Cell* 113, 935–944. doi: 10.1016/S0092-8674(03)00429-X

192. Moulin, M., Mccormac, A. C., Terry, M. J., and Smith, A. G. (2008). Tetrapyrrole profiling in *Arabidopsis* seedlings reveals that retrograde plastid nuclear signaling is not due to Mg-protoporphyrin

IX accumulation. *Proc. Natl. Acad. Sci. U.S.A.* 105, 15178–15183. doi: 10.1073/pnas.0803054105

193. Mubarakshina, M. M., Ivanov, B. N., Naydov, I. A., Hillier, W., Badger, M. R., and Krieger-Liszkay, A. (2010). Production and diffusion of chloroplastic H2O2 and its implication to signalling. *J. Exp. Bot.* 61, 3577–3587. doi: 10.1093/jxb/erq171
194. Mullineaux, C. W., Mariscal, V., Nenninger, A., Khanum, H., Herrero, A., Flores, E., et al. (2008). Mechanism of intercellular molecular exchange in heterocyst-forming cyanobacteria. *EMBO J.* 27, 1299–1308. doi: 10.1038/emboj.2008.66
195. Mur, L. A., Laarhoven, L. J., Harren, F. J., Hall, M. A., and Smith, A. R. (2008). Nitric oxide interacts with salicylate to regulate biphasic ethylene production during the hypersensitive response. *Plant Physiol.* 148, 1537–1546. doi: 10.1104/pp.108.124404
196. Nakashima, K., Yamaguchi-Shinozaki, K., and Shinozaki, K. (2014). The transcriptional regulatory network in the drought response and its crosstalk in abiotic stress responses including drought, cold, and heat. *Front. Plant Sci.* 5:170. doi: 10.3389/fpls.2014.00170
197. Nakayama, T., and Archibald, J. M. (2012). Evolving a photosynthetic organelle. *BMC Biol.* 10:35. doi: 10.1186/1741-7007-10-35
198. Natesan, S. K., Sullivan, J. A., and Gray, J. C. (2005). Stromules: a characteristic cell-specific feature of plastid morphology. *J. Exp. Bot.*56, 787–797. doi: 10.1093/jxb/eri088
199. Natesan, S. K., Sullivan, J. A., and Gray, J. C. (2009). Myosin XI is required for actin-associated movement of plastid stromules. *Mol. Plant* 2, 1262–1272. doi: 10.1093/mp/ssp078
200. Newell, C. A., Natesan, S. K., Sullivan, J. A., Jouhet, J., Kavanagh, T. A., and Gray, J. C. (2012). Exclusion of plastid nucleoids and ribosomes from stromules in tobacco and *Arabidopsis*. *Plant J.* 69, 399–410. doi: 10.1111/j.1365-313X.2011.04798.x
201. Ng, S., De Clercq, I., Van Aken, O., Law, S. R., Ivanova, A., Willems, P., et al. (2014). Anterograde and retrograde regulation of nuclear genes encoding mitochondrial proteins during growth, development, and stress. *Mol. Plant* 7, 1075–1093. doi: 10.1093/mp/ssu037
202. Ng, S., Ivanova, A., Duncan, O., Law, S. R., Van Aken, O., De

Clercq, I., et al. (2013). A membrane-bound NAC transcription factor, ANAC017, mediates mitochondrial retrograde signaling in *Arabidopsis*. *Plant Cell* 25, 3450–3471. doi: 10.1105/tpc.113.113985

203. Niinemets, U. (2010). The emission factor of volatile isoprenoids: stress, acclimation, and developmental responses. *Biogeosciences* 7, 2203–2223. doi: 10.5194/bg-7-2203-2010
204. Niinemets, U. (2013). Quantitative patterns between plant volatile emissions induced by biotic stresses and the degree of damage. *Front. Plant Sci.* 4:262. doi: 10.3389/fpls.2013.00262
205. Nomura, H., Komori, T., Uemura, S., Kanda, Y., Shimotani, K., Nakai, K., et al. (2012). Chloroplast-mediated activation of plant immune signalling in *Arabidopsis*. *Nat. Commun.* 3, 926. doi: 10.1038/ncomms1926
206. Oelmuller, R., Levitan, I., Bergfeld, R., Rajasekhar, V. K., and Mohr, H. (1986). Expression of nuclear genes as affected by treatments acting on the plastids. *Planta* 168, 482–492. doi: 10.1007/BF00392267
207. Oikawa, K., Matsunaga, S., Mano, S., Kondo, M., Yamada, K., Hayashi, M., et al. (2015). Physical interaction between peroxisomes and chloroplasts elucidated by in situ laser analysis. *Nat. Plants* 1, 15035. doi: 10.1038/NPLANTS.2015.35
208. op den Camp, R. G., Przybyla, D., Ochsenbein, C., Laloi, C., Kim, C., Danon, A., et al. (2003). Rapid induction of distinct stress responses after the release of singlet oxygen in *Arabidopsis*. *Plant Cell* 15, 2320–2332. doi: 10.1105/tpc.014662
209. Ortega-Galisteo, A. P., Rodriguez-Serrano, M., Pazmino, D. M., Gupta, D. K., Sandalio, L. M., and Romero-Puertas, M. C. (2012). S-Nitrosylated proteins in pea (*Pisum sativum* L.) leaf peroxisomes: changes under abiotic stress. *J. Exp. Bot.* 63, 2089–2103. doi: 10.1093/jxb/err414
210. Padmanabhan, M. S., and Dinesh-Kumar, S. P. (2010). All hands on deck-the role of chloroplasts, endoplasmic reticulum, and the nucleus in driving plant innate immunity. *Mol. Plant Microbe Interact.* 23, 1368–1380. doi: 10.1094/MPMI-05-10-0113
211. Pandey, S. P., and Baldwin, I. T. (2007). RNA-directed RNA polymerase 1 (RdR1) mediates the resistance of *Nicotiana attenuata*

to herbivore attack in nature. *Plant J.* 50, 40–53. doi: 10.1111/j.1365-313X.2007.03030.x

212. Park, S. W., Kaimoyo, E., Kumar, D., Mosher, S., and Klessig, D. F. (2007). Methyl salicylate is a critical mobile signal for plant systemic acquired resistance. *Science* 318, 113–116. doi: 10.1126/science.1147113
213. Park, S. Y., Fung, P., Nishimura, N., Jensen, D. R., Fujii, H., Zhao, Y., et al. (2009). Abscisic acid inhibits type 2C protein phosphatases via the PYR/PYL family of START proteins. *Science* 324, 1068–1071. doi: 10.1126/science.1173041
214. Petre, B., Saunders, D. G., Sklenar, J., Lorrain, C., Win, J., Duplessis, S., et al. (2015). Candidate effector proteins of the rust pathogen *Melampsora larici*-populina target diverse plant cell compartments. *Mol. Plant Microbe Interact.* 28, 689–700. doi: 10.1094/MPMI-01-15-0003-R
215. Petrillo, E., Godoy Herz, M. A., Fuchs, A., Reifer, D., Fuller, J., Yanovsky, M. J., et al. (2014). A chloroplast retrograde signal regulates nuclear alternative splicing. *Science* 344, 427–430. doi: 10.1126/science.1250322
216. Pfannschmidt, T., and Yang, C. (2012). The hidden function of photosynthesis: a sensing system for environmental conditions that regulates plant acclimation responses. *Protoplasma* 249(Suppl. 2), S125–S136. doi: 10.1007/s00709-012-0398-2
217. Pieterse, C. M., Leon-Reyes, A., Van Der Ent, S., and Van Wees, S. C. (2009). Networking by small-molecule hormones in plant immunity. *Nat. Chem. Biol.* 5, 308–316. doi: 10.1038/nchembio.164
218. Pogson, B. J., Woo, N. S., Forster, B., and Small, I. D. (2008). Plastid signalling to the nucleus and beyond. *Trends Plant Sci.* 13, 602–609. doi: 10.1016/j.tplants.2008.08.008
219. Ponce De Leon, I., Schmelz, E. A., Gaggero, C., Castro, A., Alvarez, A., and Montesano, M. (2012). *Physcomitrella patens* activates reinforcement of the cell wall, programmed cell death and accumulation of evolutionary conserved defence signals, such as salicylic acid and 12-oxo-phytodienoic acid, but not jasmonic acid, upon *Botrytis cinerea* infection. *Mol. Plant Pathol.* 13, 960–974. doi: 10.1111/j.1364-3703.2012.00806.x

220. Porfirova, S., Bergmuller, E., Tropf, S., Lemke, R., and Dormann, P. (2002). Isolation of an *Arabidopsis* mutant lacking vitamin E and identification of a cyclase essential for all tocopherol biosynthesis. *Proc. Natl. Acad. Sci. U.S.A.* 99, 12495–12500. doi: 10.1073/pnas.182330899

221. Provencher, L. M., Miao, L., Sinha, N., and Lucas, W. J. (2001). Sucrose export defective1 encodes a novel protein implicated in chloroplast-to-nucleus signaling. *Plant Cell* 13, 1127–1141. doi: 10.1105/tpc.13.5.1127

222. Puyaubert, J., Fares, A., Reze, N., Peltier, J. B., and Baudouin, E. (2014). Identification of endogenously S-nitrosylated proteins in *Arabidopsis* plantlets: effect of cold stress on cysteine nitrosylation level. *Plant Sci.* 21, 150–156. doi: 10.1016/j.plantsci.2013.10.014

223. Race, H. L., Herrmann, R. G., and Martin, W. (1999). Why have organelles retained genomes? *Trends Genet.* 15, 364–370. doi: 10.1016/S0168-9525(99)01766-7

224. Rairdan, G. J., Donofrio, N. M., and Delaney, T. P. (2001). Salicylic acid and NIM1/NPR1-independent gene induction by incompatible *Peronospora parasitica* in *Arabidopsis*. *Mol. Plant Microbe Interact.* 14, 1235–1246. doi: 10.1094/MPMI.2001.14.10.1235

225. Ramel, F., Birtic, S., Ginies, C., Soubigou-Taconnat, L., Triantaphylides, C., and Havaux, M. (2012). Carotenoid oxidation products are stress signals that mediate gene responses to singlet oxygen in plants. *Proc. Natl. Acad. Sci. U.S.A.* 109, 5535–5540. doi: 10.1073/pnas.1115982109

226. Ramel, F., Mialoundama, A. S., and Havaux, M. (2013). Nonenzymic carotenoid oxidation and photooxidative stress signalling in plants. *J. Exp. Bot.* 64, 799–805. doi: 10.1093/jxb/ers223

227. Raven, J. A., and Allen, J. F. (2003). Genomics and chloroplast evolution: what did cyanobacteria do for plants? *Genome Biol.* 4, 209. doi: 10.1186/gb-2003-4-3-209

228. Reis, S. L., Mantello, A. G., Rossete, E. A. G., Cardoso, A. M., and Beleboni, R. O. (2014). Insecticidal and repellent activity of typical monoterpenes from plant essential oils against *Callosobruchus maculatus*. *BMC Proc.* 8:115. doi: 10.1186/1753-6561-8-S4-P115

229. Renaudin, S., and Capdepon, M. (1977). Association of the

endoplasmic reticulum and the plastids in *Tozzia aplina* L. scale leaves. *J. Ultrastruct. Res.* 61, 303–308. doi: 10.1016/S0022-5320(77)80055-5

230. Rhoads, D. M. (2011). "Plant mitochondrial retrograde regulation," in *Plant Mitochondria*, ed. F. Kempken (New York, NY: Springer), 411–437. doi: 10.1007/978-0-387-89781-3_16
231. Rivas-San Vicente, M., and Plasencia, J. (2011). Salicylic acid beyond defence: its role in plant growth and development. *J. Exp. Bot.* 62, 3321–3338. doi: 10.1093/jxb/err031
232. Robert-Seilaniantz, A., Grant, M., and Jones, J. D. (2011). Hormone crosstalk in plant disease and defense: more than just jasmonate-salicylate antagonism. *Annu. Rev. Phytopathol.* 49, 317–343. doi: 10.1146/annurev-phyto-073009-114447
233. Rodriguez, A., San Andres, V., Cervera, M., Redondo, A., Alquezar, B., Shimada, T., et al. (2011). The monoterpene limonene in orange peels attracts pests and microorganisms. *Plant Signal. Behav.* 6, 1820–1823. doi: 10.4161/psb.6.11.16980
234. Rodriguez-Gacio Mdel, C., Matilla-Vazquez, M. A., and Matilla, A. J. (2009). Seed dormancy and ABA signaling: the breakthrough goes on. *Plant Signal. Behav.* 4, 1035–1049. doi: 10.4161/psb.4.11.9902
235. Rodriguez-Herva, J. J., Gonzalez-Melendi, P., Cuartas-Lanza, R., Antunez-Lamas, M., Rio-Alvarez, I., Li, Z., et al. (2012). A bacterial cysteine protease effector protein interferes with photosynthesis to suppress plant innate immune responses. *Cell. Microbiol.* 14, 669–681. doi: 10.1111/j.1462-5822.2012.01749.x
236. Romero-Puertas, M. C., Campostrini, N., Matte, A., Righetti, P. G., Perazzolli, M., Zolla, L., et al. (2008). Proteomic analysis of S-nitrosylated proteins in *Arabidopsis thaliana* undergoing hypersensitive response. *Proteomics* 8, 1459–1469. doi: 10.1002/pmic.200700536
237. Russin, W. A., Evert, R. F., Vanderveer, P. J., Sharkey, T. D., and Briggs, S. P. (1996). Modification of a specific class of plasmodesmata and loss of sucrose export ability in the sucrose export defective1 maize mutant. *Plant Cell* 8, 645–658. doi: 10.1105/tpc.8.4.645
238. Ruyter-Spira, C., Al-Babili, S., Van Der Krol, S., and Bouwmeester, H. (2013). The biology of strigolactones. *Trends Plant Sci.* 18, 72–

83. doi: 10.1016/j.tplants.2012.10.003

239. Ryals, J. A., Neuenschwander, U. H., Willits, M. G., Molina, A., Steiner, H. Y., and Hunt, M. D. (1996). Systemic acquired resistance. *Plant Cell* 8, 1809–1819. doi: 10.1105/tpc.8.10.1809

240. Sandalio, L. M., and Romero-Puertas, M. C. (2015). Peroxisomes sense and respond to environmental cues by regulating ROS and RNS signalling networks. *Ann. Bot.* doi: 10.1093/aob/mcv074 [Epub ahead of print].

241. Sanz, L., Albertos, P., Mateos, I., Sanchez-Vicente, I., Lechon, T., Fernandez-Marcos, M., et al. (2015). Nitric oxide (NO) and phytohormones crosstalk during early plant development. *J. Exp. Biol.* 66, 2857–2868. doi: 10.1093/jxb/erv213

242. Sattarzadeh, A., Krahmer, J., Germain, A. D., and Hanson, M. R. (2009). A myosin XI tail domain homologous to the yeast myosin vacuole-binding domain interacts with plastids and stromules in *Nicotiana benthamiana*. *Mol. Plant* 2, 1351–1358. doi: 10.1093/mp/ssp094

243. Scala, A. (2013). Green leaf volatiles: a plant's multifunctional weapon against herbivores and pathogens. *Int. J. Mol. Sci.* 14, 17781–17811. doi: 10.3390/ijms140917781

244. Scalschi, L., Sanmartin, M., Camanes, G., Troncho, P., Sanchez-Serrano, J. J., Garcia-Agustin, P., et al. (2015). Silencing of OPR3 in tomato reveals the role of OPDA in callose deposition during the activation of defense responses against *Botrytis cinerea*. *Plant J.* 81, 304–315. doi: 10.1111/tpj.12728

245. Schattat, M., Barton, K., Baudisch, B., Klosgen, R. B., and Mathur, J. (2011a). Plastid stromule branching coincides with contiguous endoplasmic reticulum dynamics. *Plant Physiol.* 155, 1667–1677. doi: 10.1104/pp.110.170480

246. Schattat, M., Barton, K., and Mathur, J. (2011b). Correlated behavior implicates stromules in increasing the interactive surface between plastids and ER tubules. *Plant Signal. Behav.* 6, 715–718. doi: 10.4161/psb.6.5.15085

247. Schattat, M. H., Barton, K. A., and Mathur, J. (2015). The myth of interconnected plastids and related phenomena. *Protoplasma* 252,

359–371. doi: 10.1007/s00709-014-0666-4

248. Schattat, M. H., Griffiths, S., Mathur, N., Barton, K., Wozny, M. R., Dunn, N., et al. (2012). Differential coloring reveals that plastids do not form networks for exchanging macromolecules. *Plant Cell* 24, 1465–1477. doi: 10.1105/tpc.111.095398
249. Schlicke, H., Hartwig, A. S., Firtzlaff, V., Richter, A. S., Glasser, C., Maier, K., et al. (2014). Induced deactivation of genes encoding chlorophyll biosynthesis enzymes disentangles tetrapyrrole-mediated retrograde signaling. *Mol. Plant* 7, 1211–1227. doi: 10.1093/mp/ssu034
250. Schulz, M., Kussmann, P., Knop, M., Kriegs, B., Gresens, F., Eichert, T., et al. (2007). Allelopathic monoterpenes interfere with *Arabidopsis thaliana* cuticular waxes and enhance transpiration. *Plant Signal. Behav.* 2, 231–239. doi: 10.4161/psb.2.4.4469
251. Seybold, S. J. (2006). Pine monoterpenes and pine bark beetles: a marriage of convenience for defense and chemical communications. *Phytochem. Rev.* 5, 143–178. doi: 10.1007/s11101-006-9002-8
252. Shah, J. (2003). The salicylic acid loop in plant defense. *Curr. Opin. Plant Biol.* 6, 365–371. doi: 10.1016/S1369-5266(03)00058-X
253. Shah, J. (2009). Plants under attack: systemic signals in defence. *Curr. Opin. Plant Biol.* 12, 459–464. doi: 10.1016/j.pbi.2009.05.011
254. Shen, Y. Y., Wang, X. F., Wu, F. Q., Du, S. Y., Cao, Z., Shang, Y., et al. (2006). The Mg-chelatase H subunit is an abscisic acid receptor. *Nature* 443, 823–826. doi: 10.1038/nature05176
255. Shi, Y., Chen, J., Hong, X., Chen, J., and Adams, M. J. (2007). A potyvirus P1 protein interacts with the rieske Fe/S protein of its host. *Mol. Plant Pathol.* 8, 785–790. doi: 10.1111/j.1364-3703.2007.00426.x
256. Shulaev, V., Silverman, P., and Raskin, I. (1997). Airborne signalling by methyl salicylate in plant pathogen resistance. *Nature* 385, 718–721. doi: 10.1038/385718a0
257. Soosaar, J. L., Burch-Smith, T. M., and Dinesh-Kumar, S. P. (2005). Mechanisms of plant resistance to viruses. *Nat. Rev. Microbiol.* 3, 789–798. doi: 10.1038/nrmicro1239
258. Spoel, S. H., Koornneef, A., Claessens, S. M., Korzelius, J. P., Van

Pelt, J. A., Mueller, M. J., et al. (2003). NPR1 modulates cross-talk between salicylate– and jasmonate-dependent defense pathways through a novel function in the cytosol. *Plant Cell* 15, 760–770. doi: 10.1105/tpc.009159

259. Stegemann, S., and Bock, R. (2009). Exchange of genetic material between cells in plant tissue grafts. *Science* 324, 649–651. doi: 10.1126/science.1170397

260. Stegemann, S., Keuthe, M., Greiner, S., and Bock, R. (2012). Horizontal transfer of chloroplast genomes between plant species. *Proc. Natl. Acad. Sci. U.S.A.* 109, 2434–2438. doi: 10.1073/pnas.1114076109

261. Stintzi, A., Weber, H., Reymond, P., Browse, J., and Farmer, E. E. (2001). Plant defense in the absence of jasmonic acid: the role of cyclopentenones. *Proc. Natl. Acad. Sci. U.S.A.* 98, 12837–12842. doi: 10.1073/pnas.211311098

262. Stonebloom, S., Brunkard, J. O., Cheung, A. C., Jiang, K., Feldman, L., and Zambryski, P. (2012). Redox states of plastids and mitochondria differentially regulate intercellular transport via plasmodesmata. *Plant Physiol.* 158, 190–199. doi: 10.1104/pp.111.186130

263. Stonebloom, S., Burch-Smith, T., Kim, I., Meinke, D., Mindrinos, M., and Zambryski, P. (2009). Loss of the plant DEAD-box protein ISE1 leads to defective mitochondria and increased cell-to-cell transport via plasmodesmata. *Proc. Natl. Acad. Sci. U.S.A.* 106, 17229–17234. doi: 10.1073/pnas.0909229106

264. Strand, A., Asami, T., Alonso, J., Ecker, J. R., and Chory, J. (2003). Chloroplast to nucleus communication triggered by accumulation of Mg-protoporphyrinIX. *Nature* 421, 79–83. doi: 10.1038/nature01204

265. Stuehr, D. J. (2004). Enzymes of the L-arginine to nitric oxide pathway. *J. Nutr.* 134, 2748S–2751S; discussion 2765S–2767S.

266. Stumpe, M., Gobel, C., Faltin, B., Beike, A. K., Hause, B., Himmelsbach, K., et al. (2010). The moss *Physcomitrella patens* contains cyclopentenones but no jasmonates: mutations in allene oxide cyclase lead to reduced fertility and altered sporophyte

morphology. *New Phytol.* 188, 740–749. doi: 10.1111/j.1469-8137.2010.03406.x

267. Sullivan, J. A., and Gray, J. C. (1999). Plastid translation is required for the expression of nuclear photosynthesis genes in the dark and in roots of the pea lip1 mutant. *Plant Cell* 11, 901–910. doi: 10.2307/3870823

268. Sun, X., Feng, P., Xu, X., Guo, H., Ma, J., Chi, W., et al. (2011). A chloroplast envelope-bound PHD transcription factor mediates chloroplast signals to the nucleus. *Nat. Commun.* 2, 477. doi: 10.1038/ncomms1486

269. Surpin, M., Larkin, R. M., and Chory, J. (2002). Signal transduction between the chloroplast and the nucleus. *Plant Cell* 14(Suppl.), S327–S338.

270. Susek, R. E., Ausubel, F. M., and Chory, J. (1993). Signal transduction mutants of *Arabidopsis* uncouple nuclear CAB and RBCS gene expression from chloroplast development. *Cell* 74, 787–799. doi: 10.1016/0092-8674(93)90459-4

271. Tan, X., Wang, Q., Tian, B., Zhang, H., Lu, D., and Zhou, J. (2011). A *Brassica napus* lipase locates at the membrane contact sites involved in chloroplast development. *PLoS ONE* 6:e26831. doi: 10.1371/journal.pone.0026831

272. Telfer, A. (2014). Singlet oxygen production by PSII under light stress: mechanism, detection and the protective role of beta-carotene. *Plant Cell Physiol.* 55, 1216–1223. doi: 10.1093/pcp/pcu040

273. Tewari, R. K., Prommer, J., and Watanabe, M. (2013). Endogenous nitric oxide generation in protoplast chloroplasts. *Plant Cell Rep.* 32, 31–44. doi: 10.1007/s00299-012-1338-5

274. Thompson, A. M. (1992). The oxidizing capacity of the earth's atmosphere: probable past and future changes. *Science* 256, 1157–1165. doi: 10.1126/science.256.5060.1157

275. Thyssen, G., Svab, Z., and Maliga, P. (2012). Cell-to-cell movement of plastids in plants. *Proc. Natl. Acad. Sci. U.S.A.* 109, 2439–2443. doi: 10.1073/pnas.1114297109

276. Tiller, N., and Bock, R. (2014). The translational apparatus of plastids and its role in plant development. *Mol. Plant* 7, 1105–1120. doi: 10.1093/mp/ssu022

277. Toshoji, H., Katsumata, T., Takusagawa, M., Yusa, Y., and Sakai, A. (2012). Effects of chloroplast dysfunction on mitochondria: white sectors in variegated leaves have higher mitochondrial DNA levels and lower dark respiration rates than green sectors. *Protoplasma* 249, 805–817. doi: 10.1007/s00709-011-0325-y

278. Trotta, A., Rahikainen, M., Konert, G., Finazzi, G., and Kangasjarvi, S. (2014). Signalling crosstalk in light stress and immune reactions in plants. *Philos. Trans. R. Soc. Lond. B Biol. Sci.* 369, 20130235. doi: 10.1098/rstb.2013.0235

279. Turner, J. G., Ellis, C., and Devoto, A. (2002). The jasmonate signal pathway. *Plant Cell* 14(Suppl.), S153–S164.

280. Tuteja, N. (2007). Abscisic Acid and abiotic stress signaling. *Plant Signal. Behav.* 2, 135–138. doi: 10.4161/psb.2.3.4156

281. Uquillas, C., Letelier, I., Blanco, F., Jordana, X., and Holuigue, L. (2004). NPR1-independent activation of immediate early salicylic acid-responsive genes in *Arabidopsis*. *Mol. Plant Microbe Interact.* 17, 34–42. doi: 10.1094/MPMI.2004.17.1.34

282. Vandenabeele, S., Vanderauwera, S., Vuylsteke, M., Rombauts, S., Langebartels, C., Seidlitz, H. K., et al. (2004). Catalase deficiency drastically affects gene expression induced by high light in *Arabidopsis thaliana*. *Plant J.* 39, 45–58. doi: 10.1111/j.1365-313X.2004.02105.x

283. Vanderauwera, S., Zimmermann, P., Rombauts, S., Vandenabeele, S., Langebartels, C., Gruissem, W., et al. (2005). Genome-wide analysis of hydrogen peroxide-regulated gene expression in *Arabidopsis* reveals a high light-induced transcriptional cluster involved in anthocyanin biosynthesis. *Plant Physiol.* 139, 806–821. doi: 10.1104/pp.105.065896

284. Van Norman, J. M., Zhang, J., Cazzonelli, C. I., Pogson, B. J., Harrison, P. J., Bugg, T. D., et al. (2014). Periodic root branching in *Arabidopsis* requires synthesis of an uncharacterized carotenoid derivative. *Proc. Natl. Acad. Sci. U.S.A.* 111, E1300–E1309. doi: 10.1073/pnas.1403016111

285. Velikova, V., Sharkey, T. D., and Loreto, F. (2012). Stabilization of thylakoid membranes in isoprene-emitting plants reduces formation of reactive oxygen species. *Plant Signal. Behav.* 7, 139–141. doi:

10.4161/psb.7.1.18521

286. Vladkova, R., Dobrikova, A. G., Singh, R., Misra, A. N., and Apostolova, E. (2011). Photoelectron transport ability of chloroplast thylakoid membranes treated with NO donor SNP: changes in flash oxygen evolution and chlorophyll fluorescence. *Nitric Oxide* 24, 84–90. doi: 10.1016/j.niox.2010.12.003
287. Vlot, A. C., Dempsey, D. A., and Klessig, D. F. (2009). Salicylic Acid, a multifaceted hormone to combat disease. *Annu. Rev. Phytopathol.* 47, 177–206. doi: 10.1146/annurev.phyto.050908.135202
288. Wachter, A., Wolf, S., Steininger, H., Bogs, J., and Rausch, T. (2005). Differential targeting of GSH1 and GSH2 is achieved by multiple transcription initiation: implications for the compartmentation of glutathione biosynthesis in the Brassicaceae. *Plant J.* 41, 15–30. doi: 10.1111/j.1365-313X.2004.02269.x
289. Wagner, D., Przybyla, D., Op Den Camp, R., Kim, C., Landgraf, F., Lee, K. P., et al. (2004). The genetic basis of singlet oxygen-induced stress responses of *Arabidopsis thaliana*. *Science* 306, 1183–1185. doi: 10.1126/science.1103178
290. Wang, W., Wen, Y., Berkey, R., and Xiao, S. (2009). Specific targeting of the *Arabidopsis* resistance protein RPW8.2 to the interfacial membrane encasing the fungal Haustorium renders broad-spectrum resistance to powdery mildew. *Plant Cell* 21, 2898–2913. doi: 10.1105/tpc.109.067587
291. Wang, W., Zhang, Y., Wen, Y., Berkey, R., Ma, X., Pan, Z., et al. (2013a). A comprehensive mutational analysis of the *Arabidopsis*resistance protein RPW8.2 reveals key amino acids for defense activation and protein targeting. *Plant Cell* 25, 4242–4261. doi: 10.1105/tpc.113.117226
292. Wang, X., Sager, R., Cui, W., Zhang, C., Lu, H., and Lee, J. Y. (2013b). Salicylic acid regulates Plasmodesmata closure during innate immune responses in *Arabidopsis*. *Plant Cell* 25, 2315–2329. doi: 10.1105/tpc.113.110676
293. Wang, X. B., Wu, Q., Ito, T., Cillo, F., Li, W. X., Chen, X., et al. (2010). RNAi-mediated viral immunity requires amplification of virus-derived siRNAs in *Arabidopsis thaliana*. *Proc. Natl. Acad. Sci. U.S.A.* 107, 484–489. doi: 10.1073/pnas.0904086107

294. Wang, Z., and Benning, C. (2012). Chloroplast lipid synthesis and lipid trafficking through ER-plastid membrane contact sites. *Biochem. Soc. Trans.* 40, 457–463. doi: 10.1042/BST20110752

295. War, A. R., Sharma, H. C., Paulraj, M. G., War, M. Y., and Ignacimuthu, S. (2011). Herbivore induced plant volatiles: their role in plant defense for pest management. *Plant Signal. Behav.* 6, 1973–1978. doi: 10.4161/psb.6.12.18053

296. Wasternack, C. (2014). Perception, signaling and cross-talk of jasmonates and the seminal contributions of the Daoxin Xie's lab and the Chuanyou Li's lab. *Plant Cell Rep.* 33, 707–718. doi: 10.1007/s00299-014-1608-5

297. Wasternack, C., and Hause, B. (2013). Jasmonates: biosynthesis, perception, signal transduction and action in plant stress response, growth and development. An update to the 2007 review in Annals of Botany. *Ann. Bot.* 111, 1021–1058. doi: 10.1093/aob/mct067

298. Waters, M. T., Fray, R. G., and Pyke, K. A. (2004). Stromule formation is dependent upon plastid size, plastid differentiation status and the density of plastids within the cell. *Plant J.* 39, 655–667. doi: 10.1111/j.1365-313X.2004.02164.x

299. Weber, H., Vick, B. A., and Farmer, E. E. (1997). Dinor-oxo-phytodienoic acid: a new hexadecanoid signal in the jasmonate family. *Proc. Natl. Acad. Sci. U.S.A.* 94, 10473–10478. doi: 10.1073/pnas.94.19.10473

300. Wei, T., Huang, T. S., Mcneil, J., Laliberte, J. F., Hong, J., Nelson, R. S., et al. (2010). Sequential recruitment of the endoplasmic reticulum and chloroplasts for plant potyvirus replication. *J. Virol.* 84, 799–809. doi: 10.1128/JVI.01824-09

301. Wendehenne, D., Durner, J., and Klessig, D. F. (2004). Nitric oxide: a new player in plant signalling and defence responses. *Curr. Opin. Plant Biol.* 7, 449–455. doi: 10.1016/j.pbi.2004.04.002

302. Wensuo, J., and Zhang, J. (2008). Stomatal movements and long-distabce signaling in plants. *Plant Signal. Behav.* 3, 772–777. doi: 10.4161/psb.3.10.6294

303. Westlake, T. J., Ricci, W. A., Popescu, G. V., and Popescu, S. C. (2015). Dimerization and thiol sensitivity of the salicylic acid binding thimet oligopeptidases TOP1 and TOP2 define their

functions in redox-sensitive cellular pathways. *Front. Plant Sci.* 6:327. doi: 10.3389/fpls.2015.00327

304. Westwood, J. H., Mccann, L., Naish, M., Dixon, H., Murphy, A. M., Stancombe, M. A., et al. (2013). A viral RNA silencing suppressor interferes with abscisic acid-mediated signalling and induces drought tolerance in *Arabidopsis thaliana. Mol. Plant Pathol.* 14, 158–170. doi: 10.1111/j.1364-3703.2012.00840.x
305. Wildermuth, M. C., Dewdney, J., Wu, G., and Ausubel, F. M. (2001). Isochorismate synthase is required to synthesize salicylic acid for plant defence. *Nature* 414, 562–565. doi: 10.1038/35107108
306. Woodson, J. D., and Chory, J. (2008). Coordination of gene expression between organellar and nuclear genomes. *Nat. Rev. Genet.* 9, 383–395. doi: 10.1038/nrg2348
307. Woodson, J. D., and Chory, J. (2012). Organelle signaling: how stressed chloroplasts communicate with the nucleus. *Curr. Biol.* 22, R690–R692. doi: 10.1016/j.cub.2012.07.028
308. Woodson, J. D., Perez-Ruiz, J. M., and Chory, J. (2011). Heme synthesis by plastid ferrochelatase I regulates nuclear gene expression in plants. *Curr. Biol.* 21, 897–903. doi: 10.1016/j.cub.2011.04.004
309. Wu, Y., Zhang, D., Chu, J. Y., Boyle, P., Wang, Y., Brindle, I. D., et al. (2012). The *Arabidopsis* NPR1 protein is a receptor for the plant defense hormone salicylic acid. *Cell Rep.* 1, 639–647. doi: 10.1016/j.celrep.2012.05.008
310. Xiao, Y., Savchenko, T., Baidoo, E. E., Chehab, W. E., Hayden, D. M., Tolstikov, V., et al. (2012). Retrograde signaling by the plastidial metabolite MEcPP regulates expression of nuclear stress-response genes. *Cell* 149, 1525–1535. doi: 10.1016/j.cell.2012.04.038
311. Xie, Z., Fan, B., Chen, C., and Chen, Z. (2001). An important role of an inducible RNA-dependent RNA polymerase in plant antiviral defense. *Proc. Natl. Acad. Sci. U.S.A.* 98, 6516–6521. doi: 10.1073/pnas.111440998
312. Xu, C., Moellering, E. R., Fan, J. B., and Benning, C. (2008). Mutation of a mitochondrial outer membrane protein affects chloroplast lipid biosynthesis. *Plant J.* 54, 163–175. doi: 10.1111/j.1365-313X.2008.03417.x
313. Xu, L., Carrie, C., Law, S. R., Murcha, M. W., and Whelan, J. (2013a).

Acquisition, conservation, and loss of dual-targeted proteins in land plants. *Plant Physiol.* 161, 644–662. doi: 10.1104/pp.112.210997

314. Xu, L., Law, S. R., Murcha, M. W., Whelan, J., and Carrie, C. (2013b). The dual targeting ability of type II NAD(P)H dehydrogenases arose early in land plant evolution. *BMC Plant Biol.* 13:100. doi: 10.1186/1471-2229-13-100

315. Yamamoto, Y., Ohshika, J., Takahashi, T., Ishizaki, K., Kohchi, T., Matusuura, H., et al. (2015). Functional analysis of allene oxide cyclase, MpAOC, in the liverwort *Marchantia polymorpha. Phytochemistry* 116, 48–56. doi: 10.1016/j.phytochem.2015.03.008

316. Yamane, K., Mitsuya, S., Taniguchi, M., and Miyake, H. (2012). Salt-induced chloroplast protrusion is the process of exclusion of ribulose-1,5-bisphosphate carboxylase/oxygenase from chloroplasts into cytoplasm in leaves of rice. *Plant Cell Environ.* 35, 1663–1671. doi: 10.1111/j.1365-3040.2012.02516.x

317. Yan, S., and Dong, X. (2014). Perception of the plant immune signal salicylic acid. *Curr. Opin. Plant Biol.* 20, 64–68. doi: 10.1016/j.pbi.2014.04.006

318. Yoon, H. S., Hackett, J. D., Ciniglia, C., Pinto, G., and Bhattacharya, D. (2004). A molecular timeline for the origin of photosynthetic eukaryotes. *Mol. Biol. Evol.* 21, 809–818. doi: 10.1093/molbev/msh075

319. Yoshida, K., and Noguchi, K. (2011). "Interaction between chloroplasts and mitochondria: activity, function and regulation of the mitochondrial respiratory system during photosynthesis," in *Plant Mitochondria*, ed. F. Kempken (New York, NY: Springer), 383–409. doi: 10.1007/978-0-387-89781-3_15

320. Yoshida, R., Sato, T., Kanno, A., and Kameya, T. (1998). Streptomycin mimics the cool temperature response in rice plants. *J. Exp. Bot.*49, 221–227. doi: 10.1093/jxb/49.319.221

321. Yoshida, T., Mogami, J., and Yamaguchi-Shinozaki, K. (2015). Omics approaches toward defining the comprehensive abscisic acid signaling network in plants. *Plant Cell Physiol.* 56, 1043–1052. doi: 10.1093/pcp/pcv060

322. Yu, D., Fan, B., Macfarlane, S. A., and Chen, Z. (2003). Analysis of the involvement of an inducible *Arabidopsis* RNA-dependent RNA

polymerase in antiviral defense. *Mol. Plant Microbe Interact.* 16, 206–216. doi: 10.1094/MPMI.2003.16.3.206

323. Yu, H. D., Yang, X. F., Chen, S. T., Wang, Y. T., Li, J. K., Shen, Q., et al. (2012). Downregulation of chloroplast RPS1 negatively modulates nuclear heat-responsive expression of HsfA2 and its target genes in *Arabidopsis*. *PLoS Genet* 8:e1002669. doi: 10.1371/journal.pgen.1002669
324. Zaninotto, F., La Camera, S., Polverari, A., and Delledonne, M. (2006). Cross talk between reactive nitrogen and oxygen species during the hypersensitive disease resistance response. *Plant Physiol.* 141, 379–383. doi: 10.1104/pp.106.078857
325. Zebelo, S. A., Matsui, K., Ozawa, R., and Maffei, M. E. (2012). Plasma membrane potential depolarization and cytosolic calcium flux are early events involved in tomato (*Solanum lycopersicon*) plant-to-plant communication. *Plant Sci.* 196, 93–100. doi: 10.1016/j.plantsci.2012.08.006
326. Zechmann, B. (2014). Compartment-specific importance of glutathione during abiotic and biotic stress. *Front. Plant Sci.* 5:566. doi: 10.3389/fpls.2014.00566
327. Zhang, Y., Fan, W., Kinkema, M., Li, X., and Dong, X. (1999). Interaction of NPR1 with basic leucine zipper protein transcription factors that bind sequences required for salicylic acid induction of the PR-1 gene. *Proc. Natl. Acad. Sci. U.S.A.* 96, 6523–6528. doi: 10.1073/pnas.96.11.6523
328. Zhang, Z. W., Yuan, S., Feng, H., Xu, F., Cheng, J., Shang, J., et al. (2011). Transient accumulation of Mg-protoporphyrin IX regulates expression of PhANGs – New evidence for the signaling role of tetrapyrroles in mature *Arabidopsis* plants. *J. Plant Physiol.* 168, 714–721. doi: 10.1016/j.jplph.2010.10.016
329. Zhao, J., Liu, Q., Zhang, H., Jia, Q., Hong, Y., and Liu, Y. (2013). The rubisco small subunit is involved in tobamovirus movement and Tm-2(2)-mediated extreme resistance. *Plant Physiol.* 161, 374–383. doi: 10.1104/pp.112.209213
330. Zhou, J., Jia, F., Shao, S., Zhang, H., Li, G., Xia, X., et al. (2015). Involvement of nitric oxide in the jasmonate-dependent basal

defense against root-knot nematode in tomato plants. *Front. Plant Sci.* 6:193. doi: 10.3389/fpls.2015.00193

331. Zhou, J. M., Trifa, Y., Silva, H., Pontier, D., Lam, E., Shah, J., et al. (2000). NPR1 differentially interacts with members of the TGA/OBF family of transcription factors that bind an element of the PR-1 gene required for induction by salicylic acid. *Mol. Plant Microbe Interact.* 13, 191–202. doi: 10.1094/MPMI.2000.13.2.191
332. Zhu, Z. (2014). Molecular basis for jasmonate and ethylene signal interactions in *Arabidopsis*. *J. Exp. Bot.* 65, 5743–5748. doi: 10.1093/jxb/eru349

CHAPTER

7

LIGHT-HARVESTING IN PHOTOSYSTEM I

Roberta Croce[1] and Herbert van Amerongen[2]

[1]Department of Physics and Astronomy, Faculty of Sciences, VU University Amsterdam, Amsterdam, The Netherlands

[2]Laboratory of Biophysics, Wageningen University, Wageningen, The Netherlands

ABSTRACT

This review focuses on the light-harvesting properties of photosystem I (PSI) and its LHCI outer antenna. LHCI consists of different chlorophyll *a*/*b* binding proteins called Lhca's, surrounding the core of PSI. In total, the PSI-LHCI complex of higher plants contains 173 chlorophyll molecules, most of which are there to harvest sunlight energy and to transfer the created excitation energy to the reaction center (RC) where it is used for charge separation. The efficiency of the complex is based on the capacity to deliver this energy to the RC as fast as possible, to minimize energy losses. The performance of PSI in this respect is remarkable: on average it takes around 50 ps for the excitation to reach the RC in plants, without being quenched in the meantime. This means that the internal quantum efficiency is close to 100 % which makes PSI the most efficient energy converter in nature. In this review, we describe the light-harvesting properties of the complex in relation to protein and pigment organization/composition, and we discuss the important parameters that assure its very high quantum efficiency. Excitation energy transfer and trapping in the core and/or Lhcas, as well as in the supercomplexes PSI-LHCI and PSI-LHCI-LHCII are described in detail with the aim of giving an overview of the functional behavior of these complexes.

Keywords: Excitation energy transfer Picosecond fluorescence Red forms Charge separation State transitions

INTRODUCTION

Photosystem I (PSI) is the multiprotein complex that reduces ferredoxin and oxidizes plastocyanin. It is composed of a core complex which contains around 100 chlorophylls *a* (Chls *a*) and all the cofactors of the electron transport chain and in most cases of an outer antenna system that increases the light-harvesting capacity. The core complex is conserved in all organisms performing oxygenic photosynthesis, while the outer antenna varies for different organisms. In plants, it is composed of Chl *a* and *b* binding proteins (Lhca's) belonging to the light-harvesting complex (Lhc) multigenic family and together they are called LHCI. In total, the PSI-LHCI complex of higher plants coordinates around 170 Chl molecules and 30 carotenoid molecules. In high-light conditions (2,000 $\mu E/m^2s$), this complex absorbs on average one photon per 600 μs. The structure of the PSI-LHCI complex of pea in which four Lhca's are associated with the core complex, is presented in Fig. 1. Structural details about the complexes can be found in Jordan et al. (2001) and Amunts et al. (2010), while the present review focuses on the light-harvesting process and the high energy conversion efficiency of this complex.

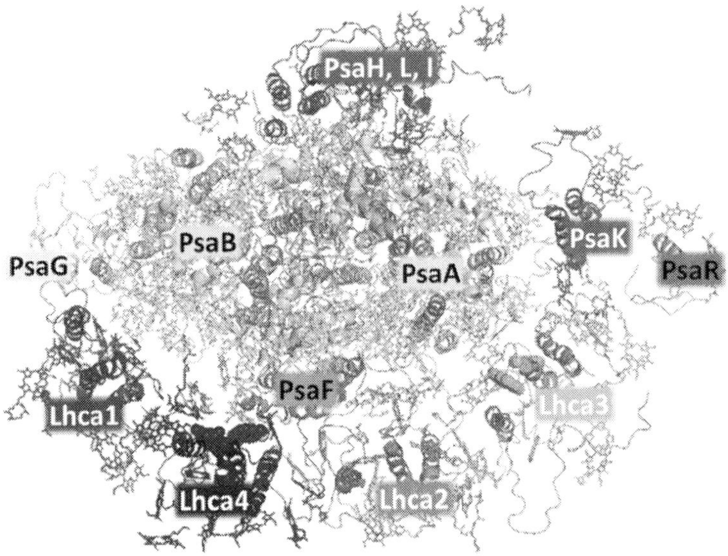

Figure 1: Structure of PSI-LHCI from pea (Amunts et al. 2010). *Top view* from the stromal side. The main subunits of core and antenna are indicated in figure. The Chls responsible for the red forms in Lhca4 and Lhca3 are presented in space-filled style

THE BASIS OF THE HIGH QUANTUM EFFICIENCY OF PSI

Photosystem I is known to be the most efficient light converter in nature (Nelson 2009), with a quantum efficiency (defined as the number of electrons produced per number of absorbed photons) that is close to 1. This fact is even more amazing, if we consider that PSI in plants contains around 200 pigments (Amunts et al. 2010). To achieve this high efficiency, it is necessary (1) that the energy is transferred very rapidly to the primary (electron) donor, (2) that the pigments in the complex are not being quenched, and (3) that the charge separation is to a large extent irreversible. In general, the published kinetic results on excitation trapping can be and have been modeled in different ways (see below), but all models have these three properties incorporated. In this review, we will mainly focus on excitation energy transfer (EET) and pay less attention to the charge-transfer processes. For the latter, we refer to an excellent review by Savikhin (2006).

The rate of energy transfer depends strongly on the distance between donor and acceptor of the excitation energy: the closer the pigments are the faster the transfer goes (although the mutual orientation also plays a role) and the higher the efficiency will be (Forster 1948). Down to a mutual center-to-center distance R between pigments of 1.5 nm, the transfer rate scales with R^{-6} according to the Förster equation whereas as shorter distances excitonic effects start to play a major role and excitations start to become more and more delocalized over the different pigments (see, e.g., van Amerongen et al. (2000)). However, if the pigments are getting too close, then an unwanted secondary effect called concentration quenching may occur, leading to a shortening of the excited-state lifetime, thereby decreasing the quantum efficiency (Beddard and Porter 1976). Very roughly, PSI of plants can be approximated by a cylinder of 12-nm diameter and 5-nm height, containing 170 Chls. This means that the pigment concentration in this system is 0.5 M. The excited-state lifetime of a diluted solution of Chls is around 6 ns, but it is below 100 ps at 0.5 M in lipid vesicles (Beddard et al. 1976). Apparently, PSI is able to avoid concentration quenching to keep the quantum efficiency close to 1.

What is the trick? It is the protein that keeps the pigments at the correct distance and geometry to facilitate fast energy transfer and to prevent excited-state quenching. In addition, the protein has a role in tuning the energy levels of the pigments (defining at which wavelength/color the maximum absorption occurs) whereas its vibrations (phonons) can couple to the electronic transitions of the pigments to broaden the absorption spectra and to allow energy transfer (both uphill and downhill) through the excited-state energy landscape (Van Amerongen et al. 2000).

But this is not yet all. When one reads about the energy transfer efficiency, it is nearly always written that EET should follow an energy gradient (from high-energy pigments to low-energy ones) to be efficient. Indeed, the picture used to exemplify photosynthetic energy transfer is commonly a deep funnel, where the energy is transferred between pigments of colors throughout the whole rainbow to end up on the primary donor which is the pigment with the lowest excited-state energy. This picture fits rather well with the antennae of cyanobacteria, the phycobilisomes, but it is clearly not a realistic representation of the situation in plants and green algae in which the most of the pigments are more or less isoenergetic. While it is correct for PSI that the primary electron donor (absorbing around 700 nm) is lower in energy than the bulk pigments (the maximum absorption of PSI is at 680 nm), it is also true that almost all PSI complexes contain Chls that absorb at energies below that of the primary donor, and they are responsible for the so-called red forms (Karapetyan 2006; Brecht et al. 2009). It was already shown in Croce et al. (1996) that at physiological temperatures a large part of the harvested energy reaches the red forms and thus needs to be transferred against the energy gradient in order to be used for charge separation. This is possible at the physiological temperatures at which these organisms live because thermal energy fills the energetic gap between donor and acceptor (Jennings et al. 2003). This means that the energy transfer pathways in PSI should be pictured more like a track for a roller coaster than like a descending road. Despite the presence of these pseudo traps, the system is extremely efficient. The role of these red forms in plants has not been completely elucidated yet, although it is clear that they extend the absorption capacity of the system to harvest solar energy in the near infrared, and thus provide an advantage in canopy or dense culture situations where the visible light is efficiently absorbed by the upper levels of the cells (Rivadossi et al. 2003). It has also been proposed that the red forms are important in photoprotection (Carbonera et al. 2005), and that they concentrate the excitation energy close to the reaction center (RC) (Trissl 1993). Although it should be mentioned that there are also red forms far away from the RC, and for example, the most red forms in plants are associated with LHCI (Croce et al. 1998). In the case of cyanobacteria, the red forms have a dual role which depends on the redox state of PSI: Karapetyan et al. (1999, 2006) and Schlodder et al. (2005) have shown with *Arthrospira platensis* that when the PSI RC is open, the energy absorbed by the red Chls migrates uphill to P700 at physiological temperatures thus increasing the absorption crosssection. If the PSI RC is closed, then the energy absorbed by the red Chls is dissipated, thus preventing PSI photodamage. The difference between plants and cyanobacteria is largely due to the location of the red forms: in higher plants, the red forms are mainly associated with the outer antenna (Croce et al.1998) and are distant from P700, while the red forms in the cyanobacterial core are supposed to be rather close to P700. This is supported by the observation that there is no energy transfer from LHCI to P700 in PSI of higher plants and algae at cryogenic temperatures, while energy migration from red Chls to P700 in PSI of cyanobacteria takes place even at cryogenic temperatures (Karapetyan 2006).

In the following, we will first describe the light-harvesting properties of the core and of the individual antenna complexes of higher plants before to move to the PSI-LHCI and PSI-LHCI-LHCII supercomplexes. A large part of the available data regarding the core complex has been obtained on cyanobacterial cores, and will only be briefly summarized here. Regarding LHCI and PSI-LHCI complexes, those of plants are clearly the best-studied ones, and the review will mainly focus on them. However, during recent years, information about PSI organization/composition/function of other eukaryotic organisms has emerged and the last paragraph focuses on the complex of Chlamydomonas reinhardtii, which is at present the best-studied alga system.

The core complex

The core complex of PSI (Fig. 2) is composed of 11–14 subunits depending on the organism, and it coordinates 96 Chls *a* and 22 β-carotene molecules in cyanobacteria (Fromme et al. 2001; Amunts et al. 2010). The main difference between PSI in cyanobacteria and higher plants is that the former occurs as a trimer, and the second one as a monomer. The pigments are mainly associated with the two largest subunits PsaA and PsaB, while the small subunits bind only a few Chls. For a detailed overview of the properties of the core subunits, the reader is referred to Jensen et al. (2007). The primary donor of PSI (P700) absorbs around 700 nm, below the energy of the bulk chlorophylls with average absorption around 680 nm. Nearly all PSI complexes also contain red forms (Karapetyan et al. 1999), but while in cyanobacteria the most red forms are associated with the core, in higher plants they are present in the outer antenna (Croce et al. 1998). The presence of red forms in the higher plant core is at present a point of discussion (Slavov et al. 2008). The absorption/emission of these forms varies for different organisms with emission maxima ranging from 720 to 760 nm (Gobets and van Grondelle 2001; Karapetyan 1998). Their number also varies and they are responsible for 3–10 % of the absorption in the region above 630 nm. Although it has been suggested that these forms originate from strongly interacting Chls (e.g., Gobets et al. 1994; Zazubovich et al. 2002), and several candidate pigments have been put forward (Zazubovich et al. 2002; Sener et al. 2002; Byrdin et al. 2002), it is still not exactly known which Chls are responsible for these forms. More in general, it should be noticed that all pigments in the core are very close together (see Fig. 2 bottom; average center-to-center distance between neighbors is around 10 Å), facilitating very efficient energy transfer. Indeed, many of the transfer steps between neighboring pigments were observed to take place with time constants between 100 and 200 fs (Du et al. 1993). The energy transfer to the red forms is slower and occurs in around 2–10 ps depending on the number of red forms in the different organisms (Savikhin et al. 2000; Hastings et al. 1995; Melkozernov et al. 2000a; Gobets and van Grondelle 2001; Gibasiewicz et al. 2001; Muller et al. 2003).

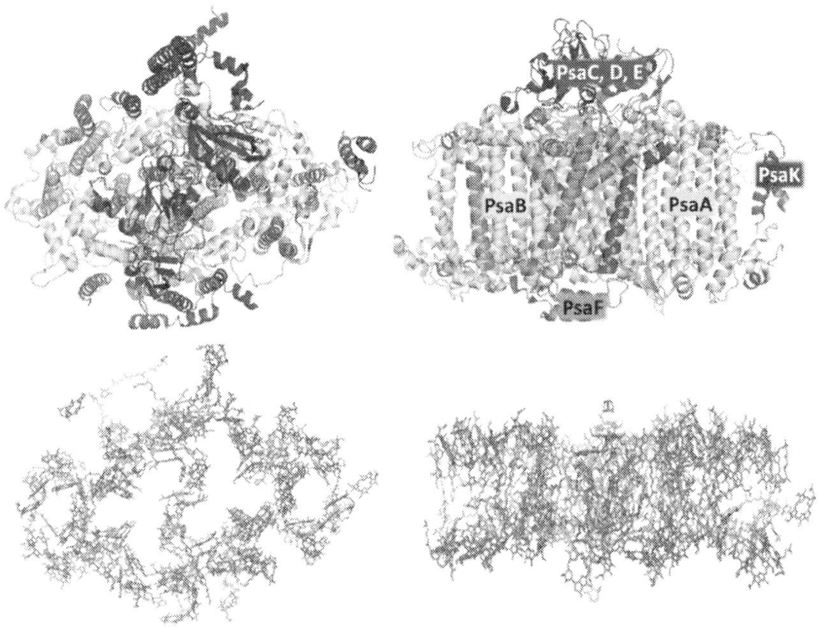

Figure 2: Structure of the cyanobacteria core (Jordan et al. 2001). *Top* protein organization. *Left*, *top view* from the stomal side. *Right*, *side view* the main proteins are indicated in figure, the color code for *left* and *right* is identical. *Bottom* pigment organization. Chlorophylls are in *green* with the exception of P700 which is in *red*. Carotenoids are in *yellow*. *Left*and *right* as in the *top panel*

This makes sense of course because there are only a few Chls responsible for this red-shifted absorption and many transfer steps are needed to reach them. It was shown that energy transfer and trapping in practically all PSI core complexes can be described with the same model which is composed of two parts: One part which represents the transfer from the bulk Chls to the primary donor and which is identical for all PSI species and other that depends on the different red-form contents and energy levels and thus is species-dependent. It was shown that the overall trapping time (the average time between the moment of absorption of a photon by one of the Chls in the core and the moment of charge separation) in the absence of red forms is around 18 ps (Gobets et al. 2001b). The presence of low-energy Chls slows down the trapping time; how much exactly depends on the number of red forms as mentioned above, but also on their excited-state energy levels: the more red forms there are and the lower their energy is, the longer it takes to transfer the excitations back from these Chls to pigments with higher energy, which is needed to reach the RC. For a comprehensive study in which different complexes were compared, we refer to Gobets et al. (2001b).

In summary, EET and trapping in the PSI core are very fast (20–40 ps), which means that the complex is very efficient in using sunlight despite the presence of chlorophylls that absorb at energies lower than the primary electron donor in the RC and partially slow down the EET. However, these red forms also broaden the absorption spectrum, apparently increasing the light-harvesting capacity.

Is charge separation in PS migration-limited or trap-limited?

There is a long-standing discussion whether the excitation energy trapping (i.e., the disappearance of an excitation due to charge separation) in the core of PSI is trap-limited, migration-limited (also called diffusion-limited) or something in between. If charge separation is migration-limited, then this means that the overall trapping time is dominated by the time it takes for an excitation to reach the primary donor P700 after which charge separation is so fast that the excitation cannot escape anymore into the antenna. On the other hand, when charge separation is trap-limited, EET is extremely fast, and an excitation might visit P700 many times before it gets trapped. However, experimentally it is very difficult to determine which model is the most appropriate for the core of PSI. Ultrafast fluorescence and transient absorption measurements have demonstrated that spectral equilibration occurs very rapidly, which at first sight may seem to argue against a migration-limited model. Savikhin et al. (2000) for instance observed spectral equilibration times of 0.53 and 2.3 ps, followed by charge separation from a spectrally equilibrated core with a time constant of 23.6 ps. However, it should be realized that spectral equilibration and spatial equilibration are not the same thing. It is very well possible that rapid spectral and spatial equilibration occur in the antenna surrounding the RC (see also below), but that subsequent EET to the primary donor is relatively slow, which corresponds to the so-called transfer-to-the-trap-limited model that was advocated in various studies (Gobets and van Grondelle 2001; Gobets et al. 2001b). However, as was pointed out by Savikhin (2006), it is often extremely difficult (if not impossible) to conclude from the experimental data alone, which model is correct. The main reason for adopting the transfer-to-the-trap-limited model is that the average distance between neighboring pigments in the surrounding antenna is much shorter than between the RC pigments and the antenna pigments. Although it is true that there are some "linker" pigments between RC and antenna, there are only two of them, one on each side of the RC, whereas most antenna pigments have several neighbors very close by.

In an illustrative modeling study by Gobets et al. (2003), the distances between pigments were explicitly taken into account. Use was made of the Förster equation for calculating interpigment EET to explain the overall trapping time of 18 ps in the absence of red forms. It was found that when an average hopping time of 150 fs (average lifetime of an excitation on a single pigment) was taken, right in the middle of the interval 100–200 fs mentioned before, a value of ~9 ps was found for the delivery time of an excitation to the primary donor. However, in that case a value of $n = 1.21$ is needed for the refractive index in the Förster equation to get a consistent description of the data, and this value seems rather low (Knox and van Amerongen 2002), although it has also been argued that for closely spaced pigments in PSI this may not be unrealistic (Damjanovic et al. 2002; Yang et al. 2003). Byrdin et al. (2002) used an approach where excitonic interactions were included to get a rather good description of the absorption, linear-dichroism, and circular-dichroism spectra of PSI from *Thermosynechococcus elongatus*. In order to get such a description, variations in the excited-state energy levels (site energies) of individual Chls were required and a certain assignment was chosen that led to the rather good simulated spectra. This assignment is certainly not unique, but the influence of variation of the site energies can be tested (see also below). For the energy-transfer calculations a hybrid approach was used, where transfer rates between pairs of pigments were calculated with the use of the Förster equation like Gobets and coworkers did but for each pair of pigments a weighted average was taken over the different exciton states in which the Chls were participating. The best description was obtained for an intrinsic charge-separation time of 0.9 ps^{-1}, and concomitantly, the charge-separation process was neither pure trap-limited nor transfer (-to-the-trap)-limited. More recently (Adolphs et al. 2010), the absorption, circular-dichroism, and linear-dichroism spectra were obtained with quantum-chemical/electrostatic calculations, i.e., with methods that had proven to lead to more or less correct site energies for simpler systems like the FMO complex and the calculated spectra showed rather good agreement with the experimental spectra. The results have not been used yet for describing the energy transfer properties, but it was concluded that the concentration of low-energy exciton states in the antenna is larger on one side of the RC, implying asymmetric delivery of excitation energy to the RC (Adolphs et al. 2010). The authors also proposed experiments to verify their calculations/predictions. Sener et al. (2002) also simulated EET transfer in PSI from *Thermosynechococcus elongatus* using a Förster-type approach and concluded that the overall transfer process does not depend very much on room-temperature fluctuations of the site energies, which were all chosen to fluctuate around a common average value.

Damjanovic et al. (2002) performed quantum-chemical calculations that showed substantial variations of the site energies of the Chls in PSI, leading to an overall absorption spectrum that was in reasonable agreement with the experimental one. However, these values did not lead to substantial changes in the overall diffusion time of excitations according to Sener et al. (2002). A very insightful modeling study is the one of Yang et al. (2003) in which excitonic interactions are not only used to calculate steady-state spectroscopic properties but are also included to model the excitation dynamics. The authors find that spectral and spatial equilibration outside the RC both occur within 5 ps, whereas the excitation transfer to the primary donor P700 is responsible for the largest contribution to the trapping time. Omitting the linker pigments in the simulations leads to somewhat slower transfer to the RC, but the overall trapping time is not changed substantially. Interestingly, the transfer from the antenna to P700 proceeds to a large extent via the other Chls in the RC and omitting those from the simulations slows down the transfer to P700 considerably. It is concluded that the combination of linker and RC pigments form a quasi-funnel structure that is highly optimized for efficient trapping. This trapping process is preceded by ultrafast "equilibration" in the antenna (within 5 ps), leading to a so-called transfer equilibrium state, and is followed by charge separation with a time constant between 0.9 and 1.7 ps. However, the actual value of the latter time constant does not influence the overall trapping time to a large extent, in contrast to the situation in trap-limited models. It should, however, be mentioned that not everyone agrees with these results; Muller et al. (2003) have for instance presented a transient absorption study in which it was concluded that charge separation in PSI with red forms is trap-limited. However, we are not aware of any theoretical studies so far that have been able to support this conclusion. On the other hand, it is also interesting to note that Holzwarth and coworkers have found indications that P700 might not be the primary electron donor (Holzwarth et al. 2006) and later work is in agreement with this proposal (Giera et al. 2010). Given the results of the calculations of Yang et al. this would imply that excitations reach the primary donor faster than was thought before. Finally, it is interesting to mention that recently ultrafast charge separation was observed with a time constant below 100 fs when photosystem I from *Synechocystis* was excited with spectrally broad 20 fs laser pulses centered at 720 nm. This is the fastest charge separation reported so far, and it does definitely not support a trap-limited scenario (Shelaev et al. 2010).

In conclusion, it seems most plausible that EET in the antenna system of the core occurs within a few ps (~5 ps) and is followed by far slower transfer to P700 (~20 ps) where charge separation occurs with an electron transfer time of ~1 ps. Although it seemed to be clear for a long time that P700 is the primary electron donor, this is not so certain anymore, meaning that transfer to the primary donor might be faster than was thought before.

The antenna complexes of PSI in higher plants

Biochemical and spectroscopic properties

A full characterization of the biochemical and spectroscopic properties of native Lhca complexes of *Arabidopsis thaliana*, which are present as functional dimers can be found in Wientjes and Croce (2011). The presence of an outer antenna system associated with PSI core in plants was first reported by Mullet et al. (1980). The first purification of LHCI complexes stems from 1983 by Haworth et al. (1983), who obtained an isolated fraction containing four polypeptides with molecular weights between 20 and 24 kDa. The four Lhca's correspond to the products of the Lhca1-4 genes. Two more Lhca genes were identified in the genome of *Arabidopsis thaliana*, Lhca5 and 6, but their expression level is always very low in all conditions tested (Ganeteg et al. 2004).

For a long time, it was believed that the LHCI antenna is composed of two complexes, called LHCI-730 and LHCI-680 based on their emission properties, with the former being enriched in Lhca1–Lhca4 and the latter in Lhca2 and Lhca3 (Lam et al. 1984; Bassi et al. 1985). However, while the properties of the Lhca1-4 heterodimer were studied on isolated and reconstituted complexes (Schmid et al. 1997; Knoetzel et al. 1992; Tjus et al. 1995; Croce et al. 2002), questions remained about the properties and the aggregation state of Lhca2 and Lhca3 due to the impossibility to purify them to homogeneity or even to reconstitute the dimer in vitro. Only recently all Lhcas were purified as two functional heterodimers, Lhca1/4 and Lhca2/3 (Wientjes and Croce 2011). They both emit in the red, with a maximum around 730 nm at low temperature. The absorption and emission spectra of the native dimers are reported in Fig. 3. This result is the final proof that the LHCI-680 complex that for a long time was believed to be part of the antenna of PSI is simply the result of a partial denaturation of the complexes with a loss of red forms: LHCI-680 complex does not exist in vivo.

The X-ray structure of the PSI-LHCI complex shows that each Lhca binds 13–14 Chls molecules (Ben-Shem et al. 2003), and the biochemical data indicate for both dimers a Chl *a*/*b* ratio of 3.7, meaning that they have lower affinity for Chl *b* than the complexes of PSII (LHCII has a Chl *a*/*b* ratio of 1.33). The dimers also bind five carotenoids each, mainly lutein and violaxanthin and substoichiometric amounts of β-carotene, while neoxanthin is not present at all, at variance with the antenna of PSII (Wientjes and Croce 2011).

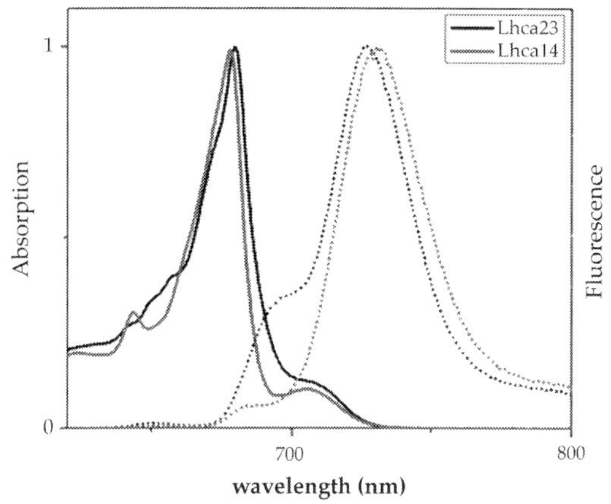

Figure 3: Absorption (*solid*) and fluorescence emission (*dot*) spectra of Lhca1/4 (*red*) and Lhca2/3 (*black*) native dimers at 77 K (Wientjes et al. 2011a)

The properties of the individual Lhca's have been studied by in vitro reconstitution of the complexes of tomato and *A. thaliana* (Schmid et al. 1997, 2002; Croce et al. 2002; Castelletti et al. 2003) because at present it is still not possible to obtain native preparations of pure Lhca monomers. The Lhca's seem to be stable in their dimeric form, while monomerization leads to the loss of some pigments. However, the properties of the reconstituted monomers were shown to be in agreement with the properties of the native dimers (Wientjes and Croce 2011). Although the properties of all individual monomers differ substantially from each other, it is interesting to notice that many spectral and biochemical properties of the dimer Lhca1+4 are very similar to those of Lhca2+3. For example, Chl *a*/*b* is 3.7 for both dimers whereas the Chl *a*/*b* ratios are 4.0 for Lhca1, 6.2 for Lhca3, 1.85 for Lhca2, and 2.3 for Lhca4 (Castelletti et al. 2003). Although the general structure and pigment coordination of Lhca complexes are very similar to those of the Lhcb antennae, which are mainly associated with PSII, Lhcas differ from Lhcbs because of the presence of low-energy absorption forms. The corresponding electronic transitions are responsible for fluorescence emission that is 50 nm red-shifted as compared to the emission of Lhcb complexes. Lam et al. (1984) observed for the first time emission of a purified fraction containing LHCI complexes that was peaking around 730 nm at 77 K, indicating that at least one of the complexes should contain red forms. The first candidate was Lhca4 (Bossmann et al. 1997; Zhang et al. 1997; Schmid et al. 1997) as suggested both by the analysis of plants lacking individual complexes and by in vitro reconstitution. Later it was shown that also Lhca3 emits above 725 nm and that Lhca1 and Lhca2 emit at 690 and 702 nm (Ganeteg et al. 2001; Croce et al. 2002; Schmid et al. 2002; Castelletti et al. 2003). This means that all Lhca's emit at energies below those of the antenna of PSII (680 nm). Lhca5 does not contain red forms and emits at 684 nm (Storf et al. 2005). Nothing is known about Lhca6, which has, however, a primary structure that is very similar to that of Lhca2 and probably has also similar spectroscopic properties.

The origin of the red forms in the Lhca complexes of higher plants was studied by mutation analysis and in vitro reconstitution (Morosinotto et al. 2002, 2005b; Croce et al. 2004; Mozzo et al. 2006). It was shown that the Chls that are responsible for the low-energy absorption in all Lhca's are Chls 603 and 609 (nomenclature from Liu et al. (2004), A5 and B5 according to Kuhlbrandt et al. (1994), these Chls are represented in space-fill style in Fig. 1), and that the difference in energy between the lowest energy state of the four complexes is due to variation in the interaction strength between these Chls. In Lhca3 and Lhca4 that harbor the most red forms, the ligand for Chl 603 is an asparagine, and it was shown that this residue is essential for stabilizing the most red form (Morosinotto et al. 2003). It was suggested that the presence of this asparagine maintains the correct geometry between the interacting Chls allowing for the formation of a charge-transfer (CT) state (Croce et al. 2007; Romero et al. 2009). More recently, a correlation between the presence of the asparagine as ligand for Chl 603 and the most red forms was also observed for the complexes of *Chlamydomonas reinhardtii* (Mozzo et al. 2010) and *Physcomitrella patens* (Alboresi et al. 2011), but it was suggested that this might not be the case in *Ostreococcus tauri* (Swingley et al. 2010). We would like to stress once more that the asparagine per se is not responsible for the red forms (and thus that the presence of an asparagine as ligand for a Chl is not a condition sufficient to induce red absorption), but Asn is necessary for maintaining the right geometry between the interacting Chls in the Lhca protein to allow for strong interaction, which is the reason for the red shift. Stark spectroscopy has shown that the red forms of Lhca4 originate from the mixing of the lowest excited state of a strongly coupled Chl dimer and a CT state (Romero et al. 2009), supporting earlier suggestions about the origin of these forms in Lhca complexes (Ihalainen et al. 2003) and in the core (Zazubovich et al. 2002; Vaitekonis et al. 2005).

In summary, four Lhca complexes (Lhca1–4), organized in two dimers (Lhca1–4, Lhca2–3), compose the outer antenna system of PSI in plants. The biochemical and spectroscopic properties of the dimers are very similar, and they both contain red forms (fluorescence maximum around 730 nm at 77 K) that originate from the mixing of the lowest excitonic state of a chlorophyll dimer (603(A5)/609(B5)) and a CT state.

Excitation Energy Transfer

Excitation energy transfer has been studied in reconstituted Lhca1 and Lhca4 and the native dimers of *Zea mays*, *A. thaliana*, and tomato (Melkozernov et al. 1998, 2000b, 2002; Gobets et al. 2001a; Gibasiewicz et al. 2005a; Wientjes et al. 2011a). It was shown that the equilibration in the Lhca4 monomeric complex occurs in <5 ps with EET from Chls *b* to Chls *a* occurring with time constants of 300 fs and 3 ps. All energy transfer processes, including carotenoid to Chl EET, seem to occur on a similar time scale as observed for the antenna complexes of PSII, as one might expect from the similar structure and pigment organization (Gibasiewicz et al. 2005a). The major difference is the fact that the lowest energy state is located on Chls 603/609 (instead of on Chls 610/611/612 (A1/B2/A2). The analysis of the Lhca1/4 and Lhca2/3 dimers shows that the transfer between monomers in the dimers is slower than the equilibration within a monomer, and it occurs in around 12 ps (Wientjes et al. 2011a). The average excited-state lifetime of the native complexes is 2.7 ns (Wientjes et al. 2011a), while shorter values were observed for the recombinant complexes (Melkozernov et al. 2000b; Ihalainen et al. 2005a; Passarini et al. 2010). The fluorescence decay is multiexponential for monomers and dimers, suggesting the presence of different conformations (Moya et al. 2001). The decay kinetics of Lhca complexes can be described with four components with lifetimes between 300 ps and 4 ns; the shortest component shows a spectrum with maximum at 690 nm and the longest one as a maximum at 720 nm (Wientjes et al. 2011a;

Passarini et al. 2010). These components correspond to different protein conformations as shown by single-molecule spectroscopy (Kruger et al. 2011). The equilibrium between the conformations can be changed by mutating particular residues in the proximity of the two interacting Chls responsible for the red forms: Small changes in the structure (e.g., the substitution of the asparagine by a glutamine) lead to a complete change in the equilibrium between the conformations (Wientjes et al. 2012). This means that they can easily switch from a "light-harvesting" state to a "quenched" state.

This property seems to be common to all members of the Lhc multigenic family (Moya et al. 2001; Kruger et al. 2011). Indeed, the light-harvesting complexes have been suggested to be involved in light-harvesting as well as in photoprotection (Ruban and Horton 1995). This means that they are able to optimize the absorption of photons and the transfer of excitation energy to the RC to maintain a very high quantum efficiency, but they are also able, when necessary, to quench their excited states and dissipate the excess energy as heat, thus preventing photodamage. When the Lhca's are connected to the core, they probably exist in their "light-harvesting" state to maximize the use of sunlight. One might speculate that the capacity of changing conformation becomes important when the antenna complexes are disconnected from the core and need to lower their excited-state population to minimize triplet formation which can lead to deleterious singlet oxygen formation, which can damage proteins, pigments, and lipids (Krieger-Liszkay et al. 2008). This idea seems to be supported by the fact that monomers (which have more structural degrees of freedom) are more often in the quenched conformation than dimers (Wientjes et al. 2011a; Passarini et al. 2010).

In conclusion, energy equilibration in monomeric Lhca complexes is very fast (5 ps) and occurs before equilibration between both monomers in a dimer. The complexes can exist in different conformations associated with different lifetimes and spectra.

PSI-LHCI supercomplex

Biochemical and structural characterization

In the PSI-LHCI supercomplex 4 Lhca's are associated with the core forming half a ring on the side of PsaF/J (Boekema et al. 2001; Ben-

Shem et al. 2003; Amunts et al. 2010). It is now generally accepted that one copy each of Lhca1-4 is present per supercomplex (Ballottari et al. 2004) and that each Lhca occupies a fixed position in the structure: The sequence going from the G pole (position of PsaG) of the core to that of K (position of PsaK) (Fig. 1), is Lhca1, Lhca4, Lhca2, and Lhca3 (Amunts et al. 2007; Wientjes et al. 2009). The composition of the outer antenna was found to be constant in all light conditions (Ballottari et al. 2007) and even in mutants lacking individual subunits, the place of the missing complex is not taken by any other Lhca (Klimmek et al. 2005; Morosinotto et al. 2005a; Wientjes et al. 2009), clearly indicating that the complexes are not interchangeable. The only exception is Lhca4 that in the Lhca4 KO mutant is partially substituted by Lhca5 (Wientjes et al. 2009) in agreement with the fact that in vitro Lhca5 is able to form a stable dimer with Lhca1 (Storf et al. 2005). This lowers the content of red forms in the complex as Lhca4 contains red forms, while Lhca5 does not, and may be of importance in specific light conditions. It has also been proposed that Lhca5 is interacting with Lhca2 and Lhca3 (Lucinski et al. 2006) and that Lhca5 and Lhca6 are necessary for the formation of the NADPH dehydrogenase-PSI supercomplex in *A. thaliana* (Peng et al. 2009). Although information about Lhca5 and Lhca6 is still lacking, their low expression levels in all tested conditions indicate that the basic PSI-LHCI unit in higher plants is only composed of the core complex and one copy each of Lhca1-4. The 3D structure has also shown that the PSI-LHCI supercomplex coordinates 173 Chl molecules in total. Around 100 of them are associated with the core as in cyanobacteria, 56 are associated with the Lhca complexes and the others are located in between the Lhca's and the core and are named "gap" pigments (Amunts et al. 2010). Interestingly, although the structure does not show tight protein–protein interactions between the subunits of the core and the outer antenna, their association appears to be very strong in plants at variance with the association of LHCII to the PSII core, which is rather weak (Wientjes et al. 2009).

In summary, the PSI-LHCI complex in plants is composed of the core plus 4 Lhca's. The number and organization of the Lhca's are identical in all growth conditions. The complex contains 173 Chls, most of which are associated with the core and the Lhcas, but a fraction is also located in between them.

Excitation energy transfer

A number of studies have investigated the light-harvesting process in the PSI-LHCI supercomplex of plants (Turconi et al. 1994; Croce et al. 2000; Ihalainen et al. 2002; Engelmann et al. 2006; Slavov et al. 2008; van Oort et al. 2008; Wientjes et al. 2011b). All measurements are characterized by the presence of two or three decay components. A fast component <10 ps represents excitation equilibration between the bulk pigments and the red most forms. A decay component in the range of 18–24 ps is normally considered to be associated with direct trapping in the core and a longer component of around 60–100 ps is thought to be due to trapping following excitation in the LHCI complexes. The average lifetime is similar to what was obtained by modeling (Sener et al. 2005). In order to extract details from the time-resolved measurements mainly two methods have been used. Target analysis, in which the complex is divided into several compartments, inside which the equilibration is considered to be very fast. The model fits the time-resolved data, while extracting the rate constants for energy transfer between the compartments. The spectra of the compartments are the second type of output from the fitting and should allow judging the quality of the fitting as they should match the steady-state emission spectra of the different PSI subcomplexes. This method has been used in Slavov et al. (2008). The other possibility is to analyze PSI complexes with different antenna size (Ihalainen et al. 2005b) and to excite at different wavelengths to vary the amount of excitation in the core and in the antenna. This method was used more recently (Wientjes et al. 2011b), measuring PSI-core, PSI-Lhca1/4, and PSI-Lhca1/4-Lhca2/3 upon excitation at 440 nm, which is more selective for the core and at 475 nm which excites preferentially the outer antenna complexes (because they contain Chl *b*, the Soret band of which is around 475 nm). In principle, both methods have their own pro's and contra's, but in the end they should lead to the same result. Unfortunately, the analysis of Slavov et al. was done before the Lhca2/3 dimer was fully characterized (Wientjes and Croce 2011), and thus the authors did not have the proper target spectra to validate their model. It would be very interesting to repeat the target analysis now that the spectra are available. In the following, we will summarize the results of Wientjes et al. (2011b), which represent the most recent PSI model, and put forward the points that still need clarification.

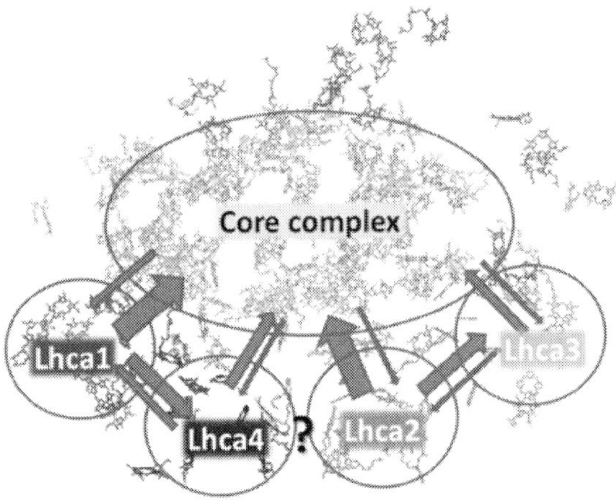

Figure 4: Schematic presentations of energy transfer and trapping in PSI-LHCI based on Wientjes et al. (2011b). Increasing thickness of the *arrows* indicates increasing rates. The transfer rate between Lhca2 and Lhca4 could not be estimated from the target analysis in that study, but based on structural data, it has been suggested to be similar to the intradimer transfer rates

Wientjes et al. observed that all Lhca's are transferring excitations directly to the core. The transfer from Lhca1 and Lhca2 (here named "blue" complexes) to the core is very fast and occurs in around 10 ps. These two complexes also transfer to the "red" Lhca's (Lhca3 and Lhca4) with a similar transfer rate. Lhca3 and Lhca4 transfer directly to the core but slower, in around 40 ps. This difference is due to the presence of the red most forms which are lower in energy than the accepting molecule in the core, thereby slowing down the transfer because of what we might call the "roller coaster" effect. It appears that in the end all Lhca's transfer a similar amount of excitations to the core (Wientjes et al. 2011b). To directly check the influence of the red forms on the trapping time, Wientjes et al. also measured a PSI-LHCI complex which is identical to that of the WT but in which Lhca4 had been substituted with Lhca5 that does not contain red forms. The fastest decay component becomes slower in the presence of Lhca5 (it goes from 20 to 26 ps), but the corresponding amplitude is strongly increased as compared to WT PSI (with Lhca4), whereas the amplitude of the slow component, which corresponds to a red spectrum, has concomitantly decreased. This clearly indicates that the transfer from the "blue" antenna Lhca5 to the core is extremely fast. This experiment also shows that the fast decay component commonly seen in the EET measurements of PSI, is not only due to the trapping from the core, but also from the "blue" antennae. The slow decay originates from Lhca4 and Lhca3. The data show that these red forms together slow down the transfer by a factor of two, in agreement with previous suggestions (Engelmann et al. 2006; Slavov et al. 2008). A scheme of the energy transfer in PSI-LHCI based on Wientjes et al. (2011b) is shown in Fig. 4.

In conclusion, PSI-LHCI in plants the trapping time is around 50 ps. The most red forms are associated with the outer antenna. All Lhca's transfer excitation energy to the core, the blue Lhca's (1 and 2) very rapidly and the red ones (Lhca3 and 4) somewhat slower.

PSI-LHCI-LHCII Supercomplex

In all conditions in which PSII is preferentially excited, part of the LHCII population moves to PSI to increase its antenna size, forming the PSI-LHCI-LHCII supercomplex (e.g., Lemeille and Rochaix 2010). This is considered to be a short-term acclimation mechanism that allows maintaining the excitation balance between the two photosystems upon rapid changes in light quality/quantity. However, it has recently been shown that the association of LHCII to PSI occurs also upon long-term acclimation, and it is in fact the most common state in *A. thaliana*(Wientjes et al. 2013). In normal light conditions (100 μmol/photons/m2) around 50 % of the PSI complexes is complemented by one LHCII trimer, while this value increases in low light and decreases in high light. This allows for the simultaneous regulation of the antenna size of PSI and PSII by just changing the expression level of the two main LHCII genes (Lhcb1 and Lhcb2) Ballottari et al. 2007) in order to maintain the excitation balance between the two photosystems (Wientjes et al. 2013).

The LHCII trimer is associated with the core on the opposite side of the Lhca's via the PsaH subunit (Lunde et al. 2000; Kouril et al. 2005). This complex is very sensitive to detergent, but it is stable in digitonian (Kouril et al. 2005; Pesaresi et al. 2009), and recently, it was purified to homogeneity (Galka et al. 2012). It was shown that the energy transfer from the LHCII trimer to the PSI core is extremely fast. Indeed, the presence of the trimer increases the antenna size of PSI by almost 25 %, while the increase in overall trapping time is only 6 ps (Wientjes et al. 2013), which indicates that there is a very good connection between the outer antenna and the core.

In summary, in most conditions, the PSI supercomplexes also bind one LHCII trimer in addition to the four Lhca's. EET from LHCII to PSI core is extremely fast, making LHCII a perfect light harvester for the system.

The PSI-LHCI complex of green algae

In recent years, the study of the PSI-LHCI supercomplex has been extended to organisms other than higher plants, revealing differences in the number and organization of the antenna complexes. An overview of the PSI antennae in the different organisms can be found in Busch and Hippler (2011). It seems that in mosses, green and red algae PSI-LHCI complexes with different antenna sizes are present. In the green alga *Chlamydomonas reinhardtii* there are nine Lhca genes (Elrad and Grossman 2004), and the largest purified supercomplex contains nine Lhca subunits per core (Drop et al. 2011) although smaller complexes have also been purified (Stauber et al. 2009). The additional (when compared to plants) 5 Lhca's form a second outer half ring around the core that is connected to the core via the 4 Lhca's forming the inner ring (Drop et al. 2011). The larger size of PSI of *C. reinhardtii* increases its light-absorption capacity but also slows down the excitation trapping. However, the fluorescence emission at low temperature peaks around 715 nm, which is 20 nm blue-shifted as compared to that of plant PSI (Bassi et al. 1992; Germano et al. 2002). Therefore, *C. reinhardtii* PSI contains red forms that on average are at higher energies than the ones in plants (Gibasiewicz et al. 2005b), and this speeds up the trapping process. In vitro reconstitution of the 9 Lhca's of *C. reinhardtii*has indicated that Lhca2, 4, and 9 are the antenna complexes that contain red pigments (Mozzo et al. 2010), but the exact number of red pigments in PSI of this alga is not known. Energy transfer and trapping in *C. reinhardtii* PSI-LHCI were investigated by two groups (Melkozernov et al. 2004; Ihalainen et al. 2005c). The results differ substantially, especially concerning the long decay component. This leads to different interpretations regarding the connection of the antenna with the core. However, considering the relative instability of the connection of part of the antenna to the supercomplex (Drop et al. 2011), it is possible that the sample properties were not the same in two studies.

In conclusion, PSI-LHCI is not only present in plants, but the antenna size and organization of the various complexes seem to vary for different organisms.

What next?

Many issues regarding energy transfer and trapping in PSI still need to be fully elucidated. This is mainly due to the high complexity of the system (the core alone contains around 100 Chls), which still represents a great challenge for modeling. In this respect an additional complication is represented by the red forms, which originate from excitonically coupled pigments but also have a strong charge-transfer character. Up to now the properties of these forms could not be reproduced in silico, thus limiting the possibility to study their properties and their effect on the kinetics via modeling.

Practically all studies addressing light-harvesting in PSI-LHCI have focused on the complex of higher plants with a few exceptions dealing with the complex from Chlamydomonas reinhardtii. However, the analysis of new organisms indicates that many different PSI-LHCI complexes exist in nature, varying in the number of antenna complexes and it their spectroscopic properties. This variability seems to be much more pronounced than in the case of PSII where LHCII trimers with properties similar to those of higher plants have been observed in many organisms, suggesting that the antenna complexes of PSI play a role in adaptation. This variability, on the other hand, provides the possibility to compare the functional behavior of PSI complexes which differ in antenna size and energy, in order to determine the robustness of the complex.

The comparison of all these complexes and of the environmental conditions in which these host organisms live would help in answering a long-standing question: what is the role of the red forms? Although we nowadays know a lot about their origin and their effect on the excitation trapping, we cannot answer this fundamental question yet. The possibility to produce plants or algae lacking red forms and to compare their growing capacity and their performance with those of the corresponding WT will form another strategy to unravel their physiological function. In principle, this is feasible because in vitro mutagenesis has clearly indicated which residues need to be changed to shift the red absorption of Lhca's to the blue.

Finally, in most organisms, the antenna of PSI is not only composed of Lhca, but also of LHCII. Although the PSI-LHCI-LHCII complex of higher plants has now been studied in some detail, very little information is available regarding this complex in other organisms. The case of *Chlamydomonas reinhardtii* is particularly interesting as it is generally believed that most of the LHCII moves to PSI in state 2. However, what does the PSI-LHCI-LHCII complex of *C. reinhardtii* look like and how is this large number of LHCII's associated with PSI? And finally, how efficient is the trapping in these large PSI-LHCI-LHCII supercomplexes?

ACKNOWLEDGMENTS

RC is supported by the ERC starting/consolidator grant number 281341 and by the Netherlands Organization for Scientific research (NWO) via a Vici grant.

REFERENCES

1. Adolphs J, Muh F, Madjet MA, Busch MS, Renger T (2010) Structure-based calculations of optical spectra of photosystem I suggest an asymmetric light-harvesting process. J Am Chem Soc 132(10):3331–3343. doi:10.1021/ja9072222
2. Alboresi A, Gerotto C, Cazzaniga S, Bassi R, Morosinotto T (2011) A red-shifted antenna protein associated with photosystem II in *Physcomitrella patens*. J Biol Chem 286(33):28978–28987. doi:10.1074/jbc.M111.226126
3. Amunts A, Drory O, Nelson N (2007) The structure of a plant photosystem I supercomplex at 3.4 angstrom resolution. Nature 447(7140):58–63
4. Amunts A, Toporik H, Borovikova A, Nelson N (2010) Structure determination and improved model of plant photosystem I. J Biol Chem 285(5):3478–3486
5. Ballottari M, Govoni C, Caffarri S, Morosinotto T (2004) Stoichiometry of LHCI antenna polypeptides and characterization of gap and linker pigments in higher plants photosystem I. Eur J Biochem 271(23–24):4659–4665
6. Ballottari M, Dall'Osto L, Morosinotto T, Bassi R (2007) Contrasting

behavior of higher plant photosystem I and II antenna systems during acclimation. J Biol Chem 282(12):8947–8958

7. Bassi R, Machold O, Simpson D (1985) Chlorophyll-proteins of two photosystem I preparations from maize. Carlsberg Res Commun 50:145–162
8. Bassi R, Soen SY, Frank G, Zuber H, Rochaix JD (1992) Characterization of chlorophyll a/b proteins of photosystem I from *Chlamydomonas reinhardtii*. J Biol Chem 267:25714–25721
9. Beddard GS, Porter G (1976) Concentration quenching in chlorophyll. Nature 260:366–367
10. Beddard GS, Carlin SE, Porter G (1976) Concentration quenching of chlorophyll fluorescence in bilayer lipid vesicles and liposomes. Chem Phys Lett 43:27–32
11. Ben-Shem A, Frolow F, Nelson N (2003) Crystal structure of plant photosystem I. Nature 426(6967):630–635
12. Boekema EJ, Jensen PE, Schlodder E, van Breemen JF, van Roon H, Scheller HV, Dekker JP (2001) Green plant photosystem I binds light-harvesting complex I on one side of the complex. Biochemistry 40(4):1029–1036
13. Bossmann B, Knoetzel J, Jansson S (1997) Screening of chlorina mutants of barley (*Hordeum vulgare* L.) with antibodies against light-harvesting proteins of PS I and PS II: absence of specific antenna proteins. Photosynth Res 52(2):127–136
14. Brecht M, Radics V, Nieder JB, Bittl R (2009) Protein dynamics-induced variation of excitation energy transfer pathways. Proc Natl Acad Sci USA 106(29):11857–11861. doi:10.1073/pnas.0903586106
15. Busch A, Hippler M (2011) The structure and function of eukaryotic photosystem I. Biochim Biophys Acta 1807(8):864–877. doi:10.1016/j.bbabio.2010.09.009
16. Byrdin M, Jordan P, Krauss N, Fromme P, Stehlik D, Schlodder E (2002) Light harvesting in photosystem I: modeling based on the 2.5-angstrom structure of photosystem I from *Synechococcus elongatus*. Biophys J 83(1):433–457
17. Carbonera D, Agostini G, Morosinotto T, Bassi R (2005) Quenching

of chlorophyll triplet states by carotenoids in reconstituted Lhca4 subunit of peripheral light-harvesting complex of photosystem I. Biochemistry 44(23):8337–8346

18. Castelletti S, Morosinotto T, Robert B, Caffarri S, Bassi R, Croce R (2003) Recombinant Lhca2 and Lhca3 subunits of the photosystem I antenna system. Biochemistry 42(14):4226–4234
19. Croce R, Zucchelli G, Garlaschi FM, Bassi R, Jennings RC (1996) Excited state equilibration in the photosystem I-light-harvesting I complex: p700 is almost isoenergetic with its antenna. Biochemistry 35:8572–8579
20. Croce R, Zucchelli G, Garlaschi FM, Jennings RC (1998) A thermal broadening study of the antenna chlorophylls in PSI-200, LHCI, and PSI core. Biochemistry 37:17355–17360
21. Croce R, Dorra D, Holzwarth AR, Jennings RC (2000) Fluorescence decay and spectral evolution in intact photosystem I of higher plants. Biochemistry 39:6341–6348
22. Croce R, Morosinotto T, Castelletti S, Breton J, Bassi R (2002) The Lhca antenna complexes of higher plants photosystem I. Biochim Biophys Acta Bioenerg 1556(1):29–40
23. Croce R, Morosinotto T, Ihalainen JA, Chojnicka A, Breton J, Dekker JP, van Grondelle R, Bassi R (2004) Origin of the 701-nm fluorescence emission of the Lhca2 subunit of higher plant photosystem I. J Biol Chem 279(47):48543–48549
24. Croce R, Chojnicka A, Morosinotto T, Ihalainen JA, van Mourik F, Dekker JP, Bassi R, van Grondelle R (2007) The low-energy forms of photosystem I light-harvesting complexes: spectroscopic properties and pigment–pigment interaction characteristics. Biophys J 93(7):2418–2428
25. Damjanovic A, Vaswani HM, Fromme P, Fleming GR (2002) Chlorophyll excitations in photosystem I of *Synechococcus elongatus*. J Phys Chem B 106(39):10251–10262
26. Drop B, Webber-Birungi M, Fusetti F, Kouril R, Redding KE, Boekema EJ, Croce R (2011) Photosystem I of *Chlamydomonas reinhardtii* contains nine light-harvesting complexes (Lhca) located on one side of the core. J Biol Chem 286(52):44878–44887. doi:10.1074/jbc.M111.301101

27. Du M, Xie XL, Jia YW, Mets L, Fleming GR (1993) Direct observation of ultrafast energy transfer in psi core antenna. Chem Phys Lett 201:535–542

28. Elrad D, Grossman AR (2004) A genome's-eye view of the light-harvesting polypeptides of *Chlamydomonas reinhardtii*. Curr Genet 45(2):61–75

29. Engelmann E, Zucchelli G, Casazza AP, Brogioli D, Garlaschi FM, Jennings RC (2006) Influence of the photosystem I-light harvesting complex I antenna domains on fluorescence decay. Biochemistry 45(22):6947–6955

30. Forster T (1948) Intermolecular energy transfer and fluorescence. Ann Phys Leipzig 2:55–75

31. Fromme P, Jordan P, Krauss N (2001) Structure of photosystem I. Biochim Biophys Acta 1507(1–3):5–31

32. Galka P, Santabarbara S, Khuong TT, Degand H, Morsomme P, Jennings RC, Boekema EJ, Caffarri S (2012) Functional analyses of the plant photosystem I-light-harvesting complex II supercomplex reveal that light-harvesting complex II loosely bound to photosystem II is a very efficient antenna for photosystem I in state II. Plant Cell 24(7):2963–2978. doi:10.1105/tpc.112.100339

33. Ganeteg U, Strand A, Gustafsson P, Jansson S (2001) The properties of the chlorophyll a/b-binding proteins Lhca2 and Lhca3 studied in vivo using antisense inhibition. Plant Physiol 127(1):150–158

34. Ganeteg U, Klimmek F, Jansson S (2004) Lhca5- an LHC-type protein associated with photosystem I. Plant Mol Biol 54(5):641–651

35. Germano M, Yakushevska AE, Keegstra W, van Gorkom HJ, Dekker JP, Boekema EJ (2002) Supramolecular organization of photosystem I and light-harvesting complex I in *Chlamydomonas reinhardtii*. FEBS Lett 525(1–3):121–125

36. Gibasiewicz K, Ramesh VM, Melkozernov AN, Lin S, Woodbury NW, Blankenship RE, Webber AN (2001) Excitation dynamics in the core antenna of PSI from *Chlamydomonas reinhardtii*CC 2696 at room temperature. J Phys Chem B 105(46):11498–11506

37. Gibasiewicz K, Croce R, Morosinotto T, Ihalainen JA, van Stokkum IHM, Dekker JP, Bassi R, van Grondelle R (2005a) Excitation

energy transfer pathways in Lhca4. Biophys J 88(3):1959–1969

38. Gibasiewicz K, Szrajner A, Ihalainen JA, Germano M, Dekker JP, van Grondelle R (2005b) Characterization of low-energy chlorophylls in the PSI-LHCI supercomplex from *Chlamydomonas reinhardtii*: a site-selective fluorescence study. J Phys Chem B 109(44):21180–21186
39. Giera W, Ramesh VM, Webber AN, van Stokkum I, van Grondelle R, Gibasiewicz K (2010) Effect of the P700 pre-oxidation and point mutations near A(0) on the reversibility of the primary charge separation in photosystem I from *Chlamydomonas reinhardtii*. Biochim Biophys Acta 1797(1):106–112. doi:10.1016/j.bbabio.2009.09.006
40. Gobets B, van Grondelle R (2001) Energy transfer and trapping in photosystem I. Biochim Biophys Acta 1057(1–3):80–99
41. Gobets B, Van Amerongen H, Monshouwer R, Kruip J, Rögner M, van Grondelle R, Dekker JP (1994) Polarized site-selected fluorescence spectroscopy of isolated photosystem I particles. Biochim Biophys Acta 1188:75–85
42. Gobets B, Kennis JTM, Ihalainen JA, Brazzoli M, Croce R, van Stokkum LHM, Bassi R, Dekker JP, Van Amerongen H, Fleming GR, van Grondelle R (2001a) Excitation energy transfer in dimeric light harvesting complex I: a combined streak-camera/fluorescence upconversion study. J Phys Chem B 105(41):10132–10139
43. Gobets B, van Stokkum IH, Rogner M, Kruip J, Schlodder E, Karapetyan NV, Dekker JP, van Grondelle R (2001b) Time-resolved fluorescence emission measurements of photosystem I particles of various cyanobacteria: a unified compartmental model. Biophys J 81(1):407–424
44. Gobets B, Valkunas L, van Grondelle R (2003) Bridging the gap between structural and lattice models: a parameterization of energy transfer and trapping in photosystem I. Biophys J 85(6):3872–3882
45. Hastings G, Reed LJ, Lin S, Blankenship RE (1995) Excited state dynamics in photosystem I: effects of detergent and excitation wavelength. Biophys J 69:2044–2055
46. Haworth P, Watson JL, Arntzen CJ (1983) The detection, isolation and characterization of a light-harvesting complex which is

specifically associated with photosystem I. Biochim Biophys Acta 724:151–158

47. Holzwarth AR, Muller MG, Niklas J, Lubitz W (2006) Ultrafast transient absorption studies on photosystem I reaction centers from *Chlamydomonas reinhardtii*. 2. Mutations near the P700 reaction center chlorophylls provide new insight into the nature of the primary electron donor. Biophys J 90(2):552–565
48. Ihalainen JA, Jensen PE, Haldrup A, van Stokkum IHM, van Grondelle R, Scheller HV, Dekker JP (2002) Pigment organization and energy transfer dynamics in isolated, photosystem I (PSI) complexes from Arabidopsis thaliana depleted of the PSI-G, PSI-K, PSI-L, or PSI-N subunit. Biophys J 83(4):2190–2201
49. Ihalainen JA, Ratsep M, Jensen PE, Scheller HV, Croce R, Bassi R, Korppi-Tommola JEI, Freiberg A (2003) Red spectral forms of chlorophylls in green plant PSI: a site-selective and high-pressure spectroscopy study. J Phys Chem B 107(34):9086–9093
50. Ihalainen JA, Croce R, Morosinotto T, van Stokkum IHM, Bassi R, Dekker JPX, van Grondelle R (2005a) Excitation decay pathways of Lhca proteins: a time-resolved fluorescence study. J Phys Chem B 109(44):21150–21158
51. Ihalainen JA, Klimmek F, Ganeteg U, van Stokkum IHM, van Grondelle R, Jansson S, Dekker JP (2005b) Excitation energy trapping in photosystem I complexes depleted in Lhca1 and Lhca4. FEBS Lett 579(21):4787–4791
52. Ihalainen JA, van Stokkum IHM, Gibasiewicz K, Germano M, van Grondelle R, Dekker JP (2005c) Kinetics of excitation trapping in intact photosystem I of *Chlamydomonas reinhardtii*and *Arabidopsis thaliana*. Biochim Biophys Acta Bioenerg 1706(3):267–275
53. Jennings RC, Zucchelli G, Croce R, Garlaschi FM (2003) The photochemical trapping rate from red spectral states in PSI-LHCI is determined by thermal activation of energy transfer to bulk chlorophylls. Biochim Biophys Acta Bioenerg 1557(1–3):91–98
54. Jensen PE, Bassi R, Boekema EJ, Dekker JP, Jansson S, Leister D, Robinson C, Scheller HV (2007) Structure, function and regulation of plant photosystem I. Biochim Biophys Acta Bioenerg 1767(5):335–352

55. Jordan P, Fromme P, Witt HT, Klukas O, Saenger W, Krauss N (2001) Three-dimensional structure of cyanobacterial photosystem I at 2.5 A resolution. Nature 411(6840):909–917

56. Karapetyan NV (1998) Organization and role of the long-wave chlorophylls in the photosystem I of the Cyanobacterium Spirulina. Membr Cell Biol 12:571–584

57. Karapetyan NV, Holzwarth AR, Rogner M (1999) The photosystem I trimer of cyanobacteria: molecular organisation, excitation dynamics and physiological significance. FEBS Lett 460:395–400

58. Karapetyan NV, Schlodder E, van Grondelle R, Dekker JP (2006) The long wavelength chlorophyll of photosystem I. In: Golbeck JH (ed) Photosystem I: the light-driven plastocyanin ferredoxin oxidoreductase, vol 24., Advances in photosynthesis and respirationSpringer, Dordrecht, pp 177–192

59. Klimmek F, Ganeteg U, Ihalainen JA, van Roon H, Jensen PE, Scheller HV, Dekker JP, Jansson S (2005) Structure of the higher plant light harvesting complex I: in vivo characterization and structural interdependence of the Lhca proteins. Biochemistry 44(8):3065–3073

60. Knoetzel J, Svendsen I, Simpson DJ (1992) Identification of the photosystem-I antenna polypeptides in barley: isolation of 3 pigment-binding antenna complexes. Eur J Biochem 206(1):209–215

61. Knox RS, van Amerongen H (2002) Refractive index dependence of the Forster resonance excitation transfer rate. J Phys Chem B 106(20):5289–5293. doi:10.1021/Jp013927

62. Kouril R, Zygadlo A, Arteni AA, de Wit CD, Dekker JP, Jensen PE, Scheller HV, Boekema EJ (2005) Structural characterization of a complex of photosystem I and light-harvesting complex II of Arabidopsis thaliana. Biochemistry 44(33):10935–10940

63. Krieger-Liszkay A, Fufezan C, Trebst A (2008) Singlet oxygen production in photosystem II and related protection mechanism. Photosynth Res 98(1–3):551–564

64. Kruger TP, Wientjes E, Croce R, van Grondelle R (2011) Conformational switching explains the intrinsic multifunctionality of plant light-harvesting complexes. Proc Natl Acad Sci USA

108(33):13516–13521. doi:10.1073/pnas.1105411108

65. Kuhlbrandt W, Wang DN, Fujiyoshi Y (1994) Atomic model of plant light-harvesting complex by electron crystallography. Nature 367:614–621
66. Lam E, Ortiz W, Malkin R (1984) Chlorophyll a/b proteins of photosystem I. FEBS Lett 168:10–14
67. Lemeille S, Rochaix JD (2010) State transitions at the crossroad of thylakoid signalling pathways. Photosynth Res 106(1–2):33–46. doi:10.1007/s11120-010-9538-8
68. Liu Z, Yan H, Wang K, Kuang T, Zhang J, Gui L, An X, Chang W (2004) Crystal structure of spinach major light-harvesting complex at 2.72 A resolution. Nature 428(6980):287–292
69. Lucinski R, Schmid VHR, Jansson S, Klimmek F (2006) Lhca5 interaction with plant photosystem I. FEBS Lett 580(27):6485–6488
70. Lunde C, Jensen PE, Haldrup A, Knoetzel J, Scheller HV (2000) The PSI-H subunit of photosystem I is essential for state transitions in plant photosynthesis. Nature 408(6812):613–615
71. Melkozernov AN, Schmid VHR, Schmidt GW, Blankenship RE (1998) Energy redistribution in heterodimeric light-harvesting complex LHCI-730 of photosystem I. J Phys Chem B 102:8183–8189
72. Melkozernov AN, Lin S, Blankenship RE (2000a) Excitation dynamics and heterogeneity of energy equilibration in the core antenna of photosystem I from the cyanobacterium *Synechocystis* sp. PCC 6803. Biochemistry 39:1489–1498
73. Melkozernov AN, Lin S, Schmid VHR, Paulsen H, Schmidt GW, Blankenship RE (2000b) Ultrafast excitation dynamics of low energy pigments in reconstituted peripheral light-harvesting complexes of photosystem I. FEBS Lett 471(1):89–92
74. Melkozernov AN, Schmid VHR, Lin S, Paulsen H, Blankenship RE (2002) Excitation energy transfer in the Lhca1 subunit of LHC I-730 peripheral antenna of photosystem I. J Phys Chem B 106(16):4313–4317
75. Melkozernov AN, Kargul J, Lin S, Barber J, Blankenship RE (2004) Energy coupling in the PSI-LHCI supercomplex from the green alga

Chlamydomonas reinhardtii. J Phys Chem B 108(29):10547–10555

76. Morosinotto T, Castelletti S, Breton J, Bassi R, Croce R (2002) Mutation analysis of Lhca1 antenna complex: low energy absorption forms originate from pigment–pigment interactions. J Biol Chem 277(39):36253–36261
77. Morosinotto T, Breton J, Bassi R, Croce R (2003) The nature of a chlorophyll ligand in Lhca proteins determines the far red fluorescence emission typical of photosystem I. J Biol Chem 278(49):49223–49229
78. Morosinotto T, Ballottari M, Klimmek F, Jansson S, Bassi R (2005a) The association of the antenna system to photosystem I in higher plants. J Biol Chem 280(35):31050–31058
79. Morosinotto T, Mozzo M, Bassi R, Croce R (2005b) Pigment–pigment interactions in Lhca4 antenna complex of higher plants photosystem I. J Biol Chem 280(21):20612–20619
80. Moya I, Silvestri M, Vallon O, Cinque G, Bassi R (2001) Time-resolved fluorescence analysis of the photosystem II antenna proteins in detergent micelles and liposomes. Biochemistry 40(42):12552–12561
81. Mozzo M, Morosinotto T, Bassi R, Croce R (2006) Probing the structure of Lhca3 by mutation analysis. Biochim Biophys Acta Bioenerg 1757(12):1607–1613
82. Mozzo M, Mantelli M, Passarini F, Caffarri S, Croce R, Bassi R (2010) Functional analysis of photosystem I light-harvesting complexes (Lhca) gene products of *Chlamydomonas reinhardtii*. Biochim Biophys Acta Bioenerg 1797(2):212–221
83. Muller MG, Niklas J, Lubitz W, Holzwarth AR (2003) Ultrafast transient absorption studies on photosystem I reaction centers from *Chlamydomonas reinhardtii*. 1. A new interpretation of the energy trapping and early electron transfer steps in photosystem I. Biophys J 85(6):3899–3922
84. Mullet JE, Burke JJ, Arntzen CJ (1980) A developmental study of photosystem I peripheral chlorophyll proteins. Plant Physiol 65:823–827
85. Nelson N (2009) Plant photosystem I: the most efficient nano-photochemical machine. J Nanosci Nanotechnol 9(3):1709–1713

86. Passarini F, Wientjes E, van Amerongen H, Croce R (2010) Photosystem I light-harvesting complex Lhca4 adopts multiple conformations: red forms and excited-state quenching are mutually exclusive. Biochim Biophys Acta Bioenerg 1797(4):501–508

87. Peng LW, Fukao Y, Fujiwara M, Takami T, Shikanai T (2009) Efficient operation of NAD(P)H dehydrogenase requires supercomplex formation with photosystem I via minor LHCI in arabidopsis. Plant Cell 21(11):3623–3640

88. Pesaresi P, Hertle A, Pribil M, Kleine T, Wagner R, Strissel H, Ihnatowicz A, Bonardi V, Scharfenberg M, Schneider A, Pfannschmidt T, Leister D (2009) Arabidopsis STN7 kinase provides a link between short- and long-term photosynthetic acclimation. Plant Cell 21(8):2402–2423

89. Rivadossi A, Zucchelli G, Garlaschi FM, Jennings RC (2003) The importance of PSI chlorophyll red forms in light-harvesting by leaves. Photosynth Res 60:209–215

90. Romero E, Mozzo M, van Stokkum IHM, Dekker JP, van Grondelle R, Croce R (2009) The origin of the low-energy form of photosystem I light-harvesting complex Lhca4: mixing of the lowest exciton with a charge-transfer state. Biophys J 96(5):L35–L37

91. Ruban AV, Horton P (1995) Regulation of non-photochemical quenching of chlorophyll fluorescence in plants. Aust J Plant Physiol 22:221–230

92. Savikhin S (2006) Ultrafast optical spectroscopy of photosystem I. In: Golbeck JH (ed) Photosystem I: the light-driven plastocyanin: ferredoxin oxidoreductase, vol 24., Advances in photosynthesis and respirationSpringer, Dordrecht, pp 155–175

93. Savikhin S, Xu W, Chitnis PR, Struve WS (2000) Ultrafast primary processes in PS I from *Synechocystis* sp. PCC 6803: roles of P700 and A(O). Biophys J 79:1573–1586

94. Schlodder E, Cetin M, Byrdin M, Terekhova IV, Karapetyan NV (2005) P700(+)- and (3)P700-induced quenching of the fluorescence at 760 nm in trimeric photosystem I complexes from the cyanobacterium *Arthrospira platensis*. Biochim Biophys Acta Bioenerg 1706(1–2):53–67

95. Schmid VHR, Cammarata KV, Bruns BU, Schmidt GW (1997) In

vitro reconstitution of the photosystem I light-harvesting complex LHCI-730: heterodimerization is required for antenna pigment organization. Proc Natl Acad Sci USA 94(14):7667–7672

96. Schmid VHR, Potthast S, Wiener M, Bergauer V, Paulsen H, Storf S (2002) Pigment binding of photosystem I light-harvesting proteins. J Biol Chem 277(40):37307–37314
97. Sener MK, Lu DY, Park SH, Schulten K, Fromme P (2002) Spectral disorder and excitation transfer dynamics in cyanobacterial photosystem I. Biophys J 82(1):292A
98. Sener MK, Jolley C, Ben-Shem A, Fromme P, Nelson N, Croce R, Schulten K (2005) Comparison of the light-harvesting networks of plant and cyanobacterial photosystem I. Biophys J 89(3):1630–1642
99. Shelaev IV, Gostev FE, Mamedov MD, Sarkisov OM, Nadtochenko VA, Shuvalov VA, Semenov AY (2010) Femtosecond primary charge separation in *Synechocystis* sp. PCC 6803 photosystem I. Biochim Biophys Acta 1797(8):1410–1420. doi:10.1016/j.bbabio.2010.02.026
100. Slavov C, Ballottari M, Morosinotto T, Bassi R, Holzwarth AR (2008) Trap-limited charge separation kinetics in higher plant photosystem I complexes. Biophys J 94(9):3601–3612
101. Stauber EJ, Busch A, Naumann B, Svatos A, Hippler M (2009) Proteotypic profiling of LHCI from *Chlamydomonas reinhardtii* provides new insights into structure and function of the complex. Proteomics 9(2):398–408
102. Storf S, Jansson S, Schmid VHR (2005) Pigment binding, fluorescence properties, and oligomerization behavior of Lhca5, a novel light-harvesting protein. J Biol Chem 280(7):5163–5168
103. Swingley WD, Iwai M, Chen Y, Ozawa S, Takizawa K, Takahashi Y, Minagawa J (2010) Characterization of photosystem I antenna proteins in the prasinophyte *Ostreococcus tauri*. Biochim Biophys Acta 1797(8):1458–1464. doi:10.1016/j.bbabio.2010.04.017
104. Tjus SE, Roobolboza M, Palsson LO, Andersson B (1995) Rapid isolation of photosystem-I chlorophyll-binding proteins by anion-exchange perfusion chromatography. Photosynth Res 45(1):41–49
105. Trissl HW (1993) Long-wavelength absorbing antenna pigments and heterogeneous absorption bands concentrate excitons and

increase absorption cross section. Photosynth Res 35:247–263

106. Turconi S, Weber N, Schweitzer G, Strotmann H, Holzwarth AR (1994) Energy transfer and charge separation kinetics in photosystem I. 2. Picosecond fluorescence study of various PSI particles and light-harvesting complex isolated from higher plants. Biochim Biophys Acta 1187:324–334
107. Vaitekonis S, Trinkunas G, Valkunas L (2005) Red chlorophylls in the exciton model of photosystem I. Photosynth Res 86(1–2):185–201
108. Van Amerongen H, Valkunas L, van Grondelle R (2000) Photosynthetic excitons. World Scientific Publishing, Singapore
109. van Oort B, Amunts A, Borst JW, van Hoek A, Nelson N, van Amerongen H, Croce R (2008) Picosecond fluorescence of intact and dissolved PSI-LHCI crystals. Biophys J 95(12):5851–5861
110. Wientjes E, Croce R (2011) The light-harvesting complexes of higher-plant photosystem I: lhca1/4 and Lhca2/3 form two red-emitting heterodimers. Biochem J 433(3):477–485. doi:10.1042/BJ20101538
111. Wientjes E, Roest G, Croce R (2012) From red to blue to far-red in Lhca4: how does the protein modulate the spectral properties of the pigments? Biochim Biophys Acta 1817(5):711–717. doi:10.1016/j.bbabio.2012.02.030
112. Wientjes E, van Amerongen H, Croce R (2013) LHCII is an antenna of both photosystems after long-term acclimation. Biochim Biophys Acta 1827(3):420–426. doi:10.1016/j.bbabio.2012.12.009
113. Wientjes E, Oostergetel GT, Jansson S, Boekema EJ, Croce R (2009) The role of Lhca complexes in the supramolecular organization of higher plant photosystem I. J Biol Chem 284(12):7803–7810
114. Wientjes E, van Stokkum IH, van Amerongen H, Croce R (2011a) Excitation-energy transfer dynamics of higher plant photosystem I light-harvesting complexes. Biophys J 100(5):1372–1380. doi:10.1016/j.bpj.2011.01.030
115. Wientjes E, van Stokkum IH, van Amerongen H, Croce R (2011b) The role of the individual Lhcas in photosystem I excitation energy trapping. Biophys J 101(3):745–754. doi:10.1016/j.bpj.2011.06.045

116. Yang M, Damjanovic A, Vaswani HM, Fleming GR (2003) Energy transfer in photosystem I of cyanobacteria *Synechococcus elongatus*: model study with structure-based semi-empirical Hamiltonian and experimental spectral density. Biophys J 85(1):140–158. doi:10.1016/S0006-3495(03)74461-0

117. Zazubovich V, Matsuzaki S, Johnson TW, Hayes JM, Chitnis PR, Small GJ (2002) Red antenna states of photosystem I from cyanobacterium Synechococcus elongatus: a spectral hole burning study. Chem Phys 275(1–3):47–59

118. Zhang H, Goodman HM, Jansson S (1997) Antisense inhibition of the photosystem I antenna protein Lhca4 in *Arabidopsis thaliana*. Plant Physiol 115(4):1525–1531.

CHAPTER

8

LIGHT HARVESTING IN PHOTOSYSTEM II

Herbert van Amerongen[1] and Roberta Croce[2]

[1]Laboratory of Biophysics, Wageningen University, Wageningen, The Netherlands

[2]Department of Physics and Astronomy, Faculty of Sciences, VU University Amsterdam, Amsterdam, The Netherlands

ABSTRACT

Water oxidation in photosynthesis takes place in photosystem II (PSII). This photosystem is built around a reaction center (RC) where sunlight-induced charge separation occurs. This RC consists of various polypeptides that bind only a few chromophores or pigments, next to several other cofactors. It can handle far more photons than the ones absorbed by its own pigments and therefore, additional excitations are provided by the surrounding light-harvesting complexes or antennae. The RC is located in the PSII core that also contains the inner light-harvesting complexes CP43 and CP47, harboring 13 and 16 chlorophyll pigments, respectively. The core is surrounded by outer light-harvesting complexes (Lhcs), together forming the so-called supercomplexes, at least in plants. These PSII supercomplexes are complemented by some "extra" Lhcs, but their exact location in the thylakoid membrane is unknown. The whole system consists of many subunits and appears to be modular, i.e., both its composition and organization depend on environmental conditions, especially on the quality and intensity of the light. In this review, we will provide a short overview of the relation between the structure and organization of pigment-protein complexes in PSII, ranging from individual complexes to entire membranes and experimental and theoretical results on excitation energy transfer and charge separation. It will become clear that time-resolved fluorescence data can provide invaluable information about the organization and functioning of thylakoid membranes. At the end, an overview will be given of unanswered questions that should be addressed in the near future.

KEYWORDS: Excitation energy transfer Picosecond fluorescence Thylakoid membrane Charge separation State transitions

INTRODUCTION

The photosynthetic light reactions of green plants, algae, and cyanobacteria take place in photosystems I and II (PSI and PSII). Light-induced charge separation in the reaction center (RC) of PSII leads to the oxidation of water, the reduction of plastoquinone and the formation of a proton gradient across the thylakoid membrane in which PSI and PSII are embedded, which is crucial for the production of ATP. PSII and PSI work in series and together they also drive $NADP^+$ to NADPH reduction with H_2O as electron donor (Nelson and Yocum 2006). Light-induced charge separation in the RC of PSII starts from the primary donor P680 and an electron proceeds via a pheophytin onto plastoquinone Q_A and subsequently to plastoquinone Q_B. The primary cation radical $P680^{+\cdot}$ has an E_m value of +1.25 V (Rappaport et al. 2009), far higher than the value of +0.80 for Chl in solution (Kobayashi et al. 2007) and this high value is ultimately responsible for the oxidation of water.

The RC of PSII itself only contains six chlorophylls a (Chls a) and two pheophytins but it is always present in the so-called core complex that also contains the pigment-proteins CP43 and CP47, providing additional 13 and 16 Chls a, respectively, together with several β-carotene molecules (see (Umena et al. 2011) for the most recent PSII core structure). Both antenna complexes feed excitation energy into the RC. These antenna Chls are on the one hand at a "safe" distance from the RC pigments, which are highly oxidizing after charge separation (see Fig. 1), preventing direct pigment oxidation in the antenna, and on the other hand close enough to perform efficient excitation energy transfer (EET).

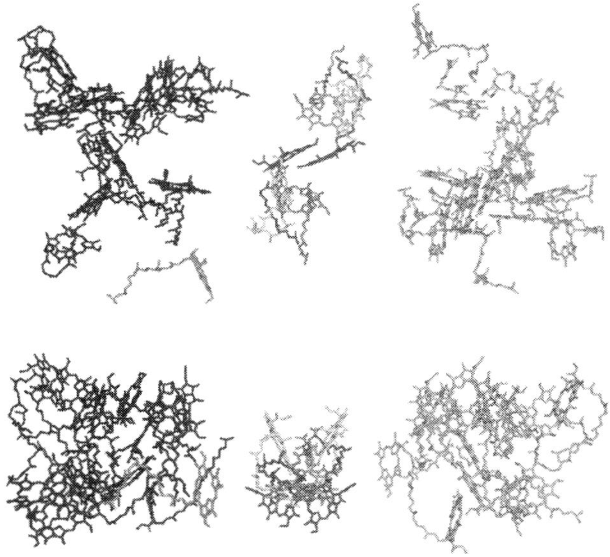

Figure 1: Chlorophyll organization in the core complex of PSII (Guskov et al. 2009). Chls P, red; Chls D1 and D2, orange; Chls z green; Pheos, yellow. The Chls of CP47 are in blue and those of CP43 in cyan. The phytol chains of the Chls are omitted for clarity. The upper figure shows a top view (from the stroma) and the lower figure provides a side view

The core consists of ~20 different subunits, and the pigment/protein ratio is low which makes it a rather expensive piece of machinery. To increase the absorption cross-section further in a cost-effective way, additional light-harvesting complexes have appeared during evolution. Plants and green algae possess membrane-embedded antennae with a high pigment/protein ratio (~1:2 regarding the mass), also called outer light-harvesting complexes (Lhcs) and to a large extent they are organized in larger supercomplexes, together with the PSII cores. Cyanobacteria, that appeared earlier in evolution contain membrane-associated phycobilisomes (see e.g., (Neilson and Durnford 2010)) with a pigment-to-protein ratio that is substantially lower (~1:5) although still higher than for the core complex. For recent studies of EET in/from phycobilisomes in vitro and in vivo the reader is referred to Tian et al. (Tian et al. 2011, 2012). The present review will focus on light harvesting in plants. The thylakoid membrane in plants is divided into grana, which are composed of stacks of membrane disks, and stroma lamellae, which connect the various grana in the choroplast (Mustardy and Garab 2003; Shimoni et al. 2005; Mustardy et al. 2008; Daum et al. 2010; Kouril et al. 2011). PSII is located in the grana (Andersson and Anderson 1980) whereas PSI is mainly present in the stroma lamellae (together with the ATP synthase). The thylakoid membrane is flexible and dynamic and able to respond to changes in environmental conditions by changing both composition and organization of the PSII supercomplexes (Anderson et al. 2008; Chuartzman et al. 2008; Goral et al. 2010). It has been shown that part of the grana membrane contains PSII arrays that consist of supercomplexes with different antenna sizes, but the abundance of the arrays seems to depend on the composition of PSII which for instance depends on the species analyzed and on the growth conditions (Boekema et al. 2000; Kouril et al.; Daum et al. 2010; Kirchhoff et al. 2007; Kouril et al. 2012; Kiss et al. 2008) (Kereiche et al. 2010; Kovacs et al. 2006; de Bianchi et al. 2008).

Only part of the PSII supercomplexes is embedded in these regular arrays, while another part is less organized. It is not exactly clear yet what the role of the arrays and the other parts is. But it is known that reorganizations in both arrays and other parts take place as a function of light quality and intensity (Wientjes et al. 2013; Kouril et al. 2012; Jahns and Holzwarth 2012; Betterle et al. 2009). In Fig. 2, a model of a plant PSII supercomplex is shown. It is composed of a PSII core together with the gene products of genes Lhcb1-6 in a well-defined arrangement. The largest supercomplexes contain a dimeric core, four LHCII (encoded by Lhcb1-3) trimers, two strongly bound (S) and two moderately strongly bound (M), and two monomeric copies each of CP29 (Lhcb4), CP26 (Lhcb5), and CP24 (Lhcb6). Supercomplexes of different sizes can be isolated (Caffarri et al. 2009), which is probably partly due to the solubilization process but it is also known that a sub-population of smaller supercomplexes is also observed in high light plants (see e.g., (Daum et al. 2010; Kouril et al. 2012)).

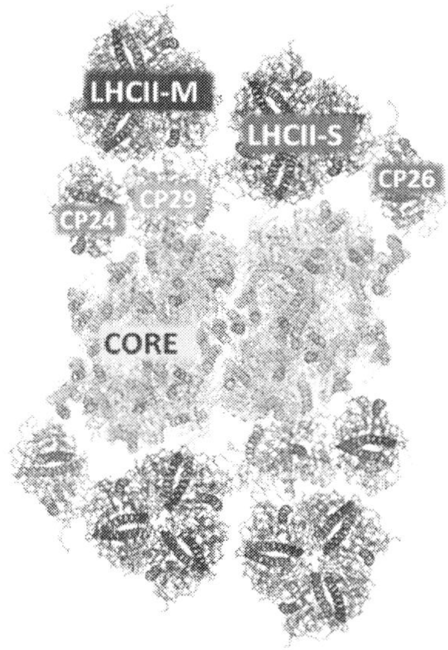

Figure 2: Model of the PSII supercomplex C2S2M2 from higher plants. Top-view for the stromal side on a C2S2M2 supercomplex from A. thaliana. Based on the projection map of C2S2M2 at 12 Å resolution, a model of the 3D structure of the supercomplex was reconstructed (Caffarri et al. 2009) using the crystal structures of PSII core (Guskov et al. 2009) and LHCII (Liu et al. 2004). For the minor antenna complexes, the structure of a monomer of LHCII was used while the pigment composition/occupancy was assigned based on the results of mutation analysis experiments on in vitro reconstituted complexes (Bassi et al. 1999; Remelli et al. 1999; Ballottari et al. 2009; Passarini et al. 2009)

The Lhc complexes are densely packed with Chl a and b pigments and the xanthophylls lutein (Lut), violaxanthin (Vx), and neoxanthin (Nx) (with the exception of CP24 that does not contain Nx) which are responsible for light absorption and EET. Xanthophyll excitations (xanthophylls are carotenoids which contain oxygen) are rapidly transferred, typically within one ps to the Chls that are in Van der Waals contact with these carotenoids. Chls b transfer excitations to Chls a, which have lower excited-state energy, and on average only a small fraction of the excitations (~5 %) is located on Chl b molecules, due to Boltzmann equilibration in the excited state. Via rapid EET between mainly Chls a the excitations end up in the RC (see (Croce and van Amerongen 2011) for a review). Some of the Chl a singlet excitations are transformed into

Chl a triplets, which can lead to the formation of destructive singlet oxygen molecules. Fortunately, most of these dangerous Chl triplets (>95 %) are scavenged by the carotenoids that are in Van der Waals contact with Chl a (Barzda et al. 1998; Lampoura et al. 2002; Mozzo et al. 2008a; Carbonera et al. 1992; van der Vos et al. 1991).

In this review, we will focus on the study of EET and CS in PSII, starting with the core, followed by outer antenna complexes and supercomplexes. A brief overview will then be given of results on thylakoid membranes, isolated from plants with varying antenna composition as a result of short- and long-term differences in light conditions. At the end, some unsolved problems will be presented together with suggestions for further research. We would also like to refer to other reviews from recent years for further information (Renger and Schlodder 2010; Vassiliev and Bruce 2008; Renger 2010; Van Amerongen et al. 2003; Minagawa and Takahashi 2004; Barber 2002; Muh et al. 2008; Renger and Renger 2008; Croce and van Amerongen 2011).

THE PSII CORE

In Fig. 3, the reconstructed picosecond fluorescence kinetics of the PSII core from Thermosynechococcus from two different studies are shown (Miloslavina et al. 2006; van der Weij-de Wit et al. 2011) and the results are nearly identical. Accurate data fitting requires five or more exponentials but two direct observations stand out. Charge separation occurs with an average time constant τ below 100 ps, leading to the relatively fast disappearance of the (fluorescence) signal. Since this time constant is much shorter than the normal excited-state lifetime of chlorophyll (typically several ns), this guarantees a high quantum efficiency of charge separation (φ_{CS}). The corresponding value is above 0.95, using the well-known relation $\varphi_{CS} = 1 - \tau/\tau_{Chl}$ (Croce and van Amerongen 2011), where τ_{Chl} is the average lifetime of the excited Chl in PSII in the absence of charge separation. The exact value for this parameter is unknown but a recent study led to a value of ~2 ns (Belgio et al. 2012). The kinetics also shows a small contribution of a long-lived component which is usually ascribed to the fact that charge separation is partly reversible. The amplitude and lifetime of this component depend on the competition between secondary charge separation in the RC (forward electron transfer from the primary electron acceptor) and back transfer of the electron from primary acceptor to primary donor.

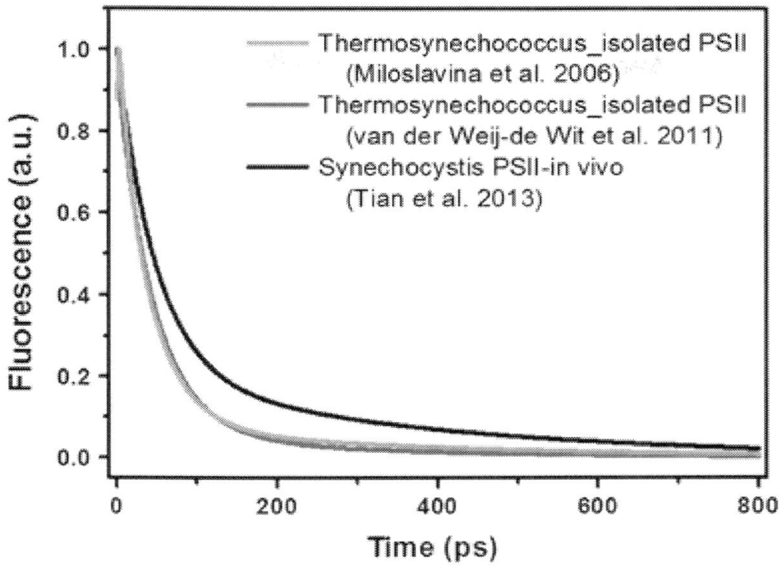

Figure 3: Picosecond kinetics of isolated PSII core complexes from Thermosynechococcus, reconstructed from (Miloslavina et al. 2006) (black solid) and (van der Weij-de Wit et al. 2011). The decay curve presented in (Miloslavina et al. 2006) was reconstructed based on the DAS shown in Fig. 7 of that work, and only $\tau 1$–$\tau 5$ are included in the calculation. The decay curve from (van der Weij-de Wit et al. 2011) was reconstructed based on the compartmental scheme shown in Fig. 6 in that article and the initial excitation fractions therein. Excitation wave lengths were 663 and 400 nm, respectively, but these differences are not expected to significantly influence the overall kinetics. The dotted line represents the fluorescence kinetics of PSII core in vivo for a Synechocystis mutant (excitation wavelength 400 nm) (Tian et al. 2013)

Although the kinetics in both studies is rather similar, the models that were used for the fitting differ considerably. It should be noted that the overall (average) trapping time τ of excitations can in good approximation be considered as the sum of two terms: $\tau = \tau_{mig} + \tau_{trap}$ (Van Amerongen et al. 2000; Broess et al. 2006). In a trap-limited model, the equilibration time (also called migration time τ_{mig}) of excitations over the photosystem is assumed to be much shorter than the overall trapping time, i.e., it can largely be neglected and thus $\tau = \tau_{trap}$. The best-known trap-limited model is the so-called exciton/radical pair equilibrium model (ERPE model) (van Grondelle 1985; Schatz et al. 1988, 1987), and it has widely been used to interpret all kinds of variations in fluorescence in photosynthesis. Besides primary charge separation, it also includes charge recombination and secondary charge separation (see above). In (Miloslavina et al. 2006), the data were fitted to a kind of trap-limited model and it was thus assumed that excitation equilibration in the core occurs on a time scale much faster than the overall trapping time. In good approximation τ_{trap} is equal to the intrinsic charge separation time of the primary donor τ_{iCS}, divided by the probability that the excitation is actually located on the primary donor. For N isoenergetic pigments, including the primary donor, $\tau_{trap} = N\tau_{iCS}$(when charge recombination is ignored). Taking for instance values of τ_{trap} = 60 ps and N = 35, one finds that τ_{iCS} = 1.7 ps.

However, the distances between the pigments in these complexes and the ones in the RC (Fig. 1) are so large that it was concluded in (van der Weij-de Wit et al. 2011) that the transfer time of excitations to the trap and therefore the contribution of τ_{mig} cannot be ignored. This means that the value of τ_{trap} should be smaller and concomitantly the same should be true for τ_{iCS}, which also comes out of the fitting (van der Weij-de Wit et al. 2011). Very recently, the picosecond fluorescence kinetics was obtained for the PSII core in vivo, by comparing the results of different mutants of Synechocystis PCC 6803 mutants (Tian et al. 2013). It turned out that the PSII core of this organism in vivo was somewhat slower than the one of Thermosynechococcus in vitro but again, the kinetics could be satisfactorily fitted with both a trap-limited and a migration-limited model. It is clear that comparing different fitting models cannot favor one trapping model above the other.

In a recent theoretical treatment Raszewski and Renger (Raszewski and Renger 2008) concluded that the trapping should be migration-limited: Transfer from CP43/CP47 occurs with time constants of 40–50 ps. The main reason for the slow transfer is the large distance between the pigments in the core antenna and those in the RC. As was mentioned above, this large distance is probably needed to avoid oxidation of the antenna pigments. The consequence of this slow EET is that the primary charge transfer time should be extremely fast, i.e., around 300 fs, accompanied by a very large initial drop in free energy to explain the overall time-resolved results. It should be noted that at least in isolated RC complexes such a fast charge separation time was not observed (Groot et al. 2005; Germano et al. 2004; van Mourik et al. 2004; Holzwarth et al. 2006; Prokhorenko and Holzwarth 2000; Andrizhiyevskaya et al. 2004; Wasielewski et al. 1990; Durrant et al. 1992; Pawlowicz et al. 2008) and one might wonder whether this is realistic. On the other hand, it is possible that isolated RC complexes are "slower" than the ones in vivo (see also below).

It is worthwhile to mention that the average lifetimes of core preparations from cyanobacteria are in general far shorter than for cores from plants (Raszewski and Renger 2008). Although this may be due to differences in the intrinsic properties of the cores, it is most likely related to problems associated with the isolation of core preparations from plants.

At the moment, there are still several unsolved issues with respect to PSII core kinetics. Both trap- and migration-limited models seem to have some intrinsic problem and maybe we should consider the possibility of coherent EET into the RC (Collini and Scholes 2009). Moreover, we do not know to which extent observed differences between in vivo and in vitro results are due to differences in biological species or to the fact that the thylakoid membrane might influence the primary EET and CT steps. Why do cores isolated from plants show slower kinetics than those of cyanobacteria? We have not mentioned any studies on the kinetics of isolated RCs but as was discussed in (Broess et al. 2006), they appear to be substantially slower than when embedded in larger systems. Therefore, it seems that the study of isolated complexes at the moment can only contribute to basic knowledge about charge separation mechanisms and pathways in PSII but they do not give realistic time constants.

Outer Antenna Complexes

The antenna complexes of PSII from higher plants are composed of members of the Lhc multigenic family. The structure of a monomeric subunit of trimeric LHCII (Liu et al. 2004; Standfuss et al. 2005) is given in Fig 4. Each monomer coordinates eight Chls a, six Chls b and four xanthophylls (one Nx, two Lut's and one Vx). The two Lut's are located at sites L1 and L2 in the center of the molecule while Nx and Vx are located at the periphery in sites N1 and V1, respectively (Croce et al. 1999; Caffarri et al. 2001; Ruban and Horton 1999). The average distance between the Chls is around 10 Å, which leads to excitonic interactions between the pigments, resulting in fast energy transfer within the complex.

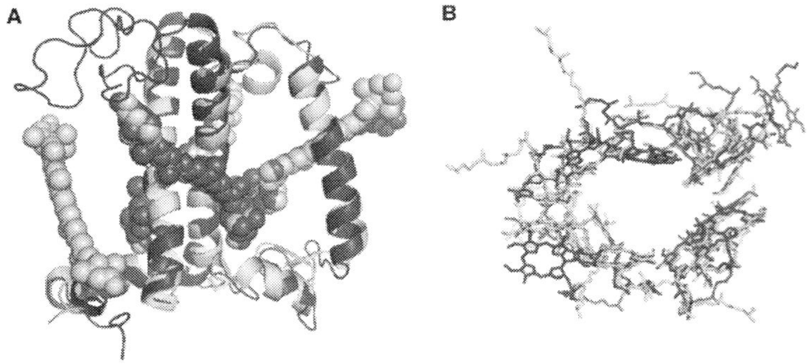

Figure 4: Overlap of the structural models of LHCII (Liu et al. 2004) and CP29 ((Pan et al. 2011)). **a**Side view (from within the membrane) on the protein backbone of LHCII (red) and CP29 (yellow) and the xanthophylls of LHCII (light blue) and CP29 (dark blue). Main differences are the lack of the N-terminal part of CP29 which apparently was cleaved off during crystallization and the lack of VX in CP29. For the rest, both proteins show very similar structures. **b**Top view showing that the Chl organization in LHCII (red) and CP29 (yellow) is rather similar although not identical

Based on sequence similarity, all members of the Lhc family are thought to have a similar structural arrangement (Green and Khlbrandt 1995) and most of the amino acids that bind Chl in LHCII are conserved in all family members (Bassi et al. 1997). This is now confirmed for CP29 and Lhca1-4 based on the comparison of the structures (Pan et al. 2011; Amunts et al. 2010). The structures of CP29 and monomeric LHCII are shown in Fig. 4. Nevertheless, individual complexes show different biochemical and spectroscopic properties (see e.g., (Pascal et al. 1999)), mainly due to the fact that the pigment composition is not identical (Sandona et al. 1998). Mutations of the putative Chl-binding residues followed by in vitro reconstitution (Plumley and Schmidt 1987) has allowed the characterization of the chromophores in most binding sites (Bassi et al. 1999; Remelli et al. 1999; Yang et al. 1999; Rogl and Kuhlbrandt 1999; Ballottari et al. 2009; Passarini et al. 2009). The four Chl-binding sites in the center (602, 603, 610 and 612) accommodate Chls a in all antenna complexes. Chls 602–603 absorb around 675 nm and Chls 610–612 absorb around 680 nm, representing the lowest energy state(s) of the system (Remelli et al. 1999; Rogl and Kuhlbrandt 1999).

The domain including helix C mainly coordinates Chls b (Remelli et al. 1999; Peterman et al. 1996). In all complexes, site L1 contains a Lut while L2 accommodates Lut in LHCII and CP26 but Vx in CP29 and CP24. Nx is present in the N1 site of all complexes apart from CP24 (Caffarri et al. 2007).

By combining the results of a large number of different studies on LHCII in the nineties (Visser et al. 1996; Savikhin et al. 1994a; Peterman et al. 1997; Croce et al. 2001; Connelly et al. 1997), it was concluded that (sub)picosecond EET leads to ps spectral equilibration and excitations become mainly localized on the peripheral Chl a pigments on the stromal part of the protein i.e., Chls 610–612 (Van Amerongen and van Grondelle 2001). From there, they can be transferred to neighboring complexes in the thylakoid membrane. Spatial equilibration within the trimers occurs on a slower time scale (tens of ps) as was concluded from several other studies (Savikhin et al. 1994b; Barzda et al. 2001; van Oort et al. 2007; Kwa et al. 1992; Novoderezhkin and van Grondelle 2010).

The results of the various time-resolved and steady-state spectroscopic studies were later modeled with the use of Redfield theory (Novoderezhkin et al. 2004, 2005) and led to a theoretical description of the data largely consistent with the crystal structure (Liu et al. 2004; Standfuss et al. 2005), demonstrating that within a few ps, the excitations are mainly localized on Chls 610–612.

More recent studies using 2-D electronic spectroscopy (Calhoun et al. 2009) are at least qualitatively in agreement with the modeling results of Novoderezhkin et al. (Novoderezhkin et al. 2005; Novoderezhkin and van Grondelle 2010) although it is not known whether the new models also lead to a correct description of for instance the linear-dichroism (LD) (Van Amerongen et al. 1994) and circular-dichroism (CD) spectra (Georgakopoulou et al. 2007). It is worth to point that in a very recent study, Müh and Renger were able to obtain rather satisfactory fits of all steady-state spectra of LHCII, demonstrating that not all site energies agree with those obtained before and that also the absolute LD spectra do not perfectly agree with the crystal structure (Muh and Renger 2012). Therefore, it seems that there is room for an additional round of improving both the structural model of LHCII and the understanding of its steady-state and time-resolved spectroscopic properties. At this point, it is also worth to mention that Zucchelli et al. (Zucchelli et al. 2012) recently calculated LHCII absorption spectra and obtained substantial variation for the monomeric subunits of three different trimers taken from the crystal structure (Liu et al. 2004; Standfuss et al. 2005), due to variations in the conformation of the Chl macrocycle and variations in the excitonic coupling strength between different Chls. Finally, it is worth mentioning that the (sub)ps transient absorption kinetics of the three gene products forming LHCII, Lhcb1, Lhcb2, and Lhcb3, are identical (Palacios et al. 2006).

EET in the minor antenna complexes (Cinque et al. 2000; Gradinaru et al. 1998, 2000; Salverda et al. 2003; Croce et al. 2003a, b; Marin et al. 2010, 2011) seems to occur along similar pathways as in LHCII. Also in these complexes equilibration occurs within a few ps, leading to excitation population mainly on Chls 610–612, the lowest energy pigments located on the stromal side at the periphery of the complex (Mozzo et al. 2008b).

PSII SUPER COMPLEXES

Obtaining homogeneous preparations of PSII supercomplexes is difficult because they disassemble quite easily (Wientjes et al. 2009; Caffarri et al. 2001). The largest supercomplex purified so far is C2S2M2 (Fig. 2) (Caffarri et al. 2009) and it is the most abundant complex in thylakoid membranes of Arabidopsis thaliana (Dekker and Boekema 2005; Kouril et al. 2012). The LHCII trimers differ somewhat in composition. The S trimer is composed of the products of the Lhcb1 and Lhcb2 genes and the M trimer in addition also contains the product of the Lhcb3 gene (Hankamer et al. 1997). Ordered arrays of C2S2, C2S2M, and C2S2M2 have been observed in membranes of different plants (Boekema et al. 2000; Daum et al. 2010; Yakushevska et al. 2001; Kouril et al. 2011). Smaller supercomplexes have also been purified but they are probably partly disassembled (Caffarri et al. 2009). Based on a projection map of the C2S2M2 supercomplex at 12 Å resolution (Caffarri et al. 2009) and the crystal structures of core and LHCII, a 3D supercomplex structure has been reconstructed (Fig. 2). Such a model can be used to visualize possible EET pathways (Croce and van Amerongen 2011).

Picosecond fluorescence measurements have been performed on four different PSII supercomplex preparations from A. thaliana (Caffarri et al. 2011). The smallest complex (C2S) contains a dimeric PSII core plus CP26, CP29 and one LHCII trimer. The largest complex (C2S2M2) corresponds to the structure in Fig. 2. The average fluorescence lifetime becomes longer upon increasing the antenna size from 109 ps for the dimeric core complex (~70 Chl amolecules) to 158 ps for C2S2M2 (~210 Chl a molecules), using a detergent concentration of 0.01 % α-DM. In 0.001 % α-DM the lifetimes decrease on average by around 20 ps. Plotting the average lifetimes versus the number of Chls a for the four supercomplex preparations and the core, shows that all values lie more or less on a straight line which evidently is not going through the origin as one might expect (Van Amerongen et al. 2003) indicating that the energy transfer and/or charge separation rates are reduced in complexes with decreased size, strongly suggesting that the antenna system is important for plant PSII integrity and functionality, which is also supported by biochemical results (Caffarri et al. 2011).

The kinetics were also simulated using coarse-grained modeling and the obtained parameters were used to illustrate various aspects of PSII functioning (Caffarri et al. 2011). It was for instance calculated that for the largest supercomplex the efficiency of charge separation is 89 %. In the presence of one open and one closed RC, the photochemical efficiency reduces to 78 %, which is much larger than the value of 45 % calculated when the cores are not connected into dimers. This demonstrates that a dimeric conformation increases the light-harvesting capacity by more than 70 % in the presence of one closed RC. This is an important property for PSII because of its slow turnover and it also suggests that the arrays of PSII that are observed in electron-microscopy measurements are advantageous when a substantial fraction of the RC's is closed.

In fact, the advantage of PSII units being connected to each other was already discussed many decades ago and it was experimentally determined that indeed many "photosynthetic units" (PSU's) are connected to each other (see e.g., (Clayton 1981)). Two popular models from those days were the puddle model, in which PSU's were not connected and the lake model, in which basically all PSU's were connected. Whereas for purple bacteria, the lake model is applicable, it was found that for plants, the situation was somewhere in between these extreme models (see e.g., also (Clayton 1981)), which is in agreement with the organization observed with electron-microscopy (see above).

Energy transfer and charge separation in PSII membranes

Grana membranes can be purified (the so-called BBY particles) that contain practically only PSII complexes (Berthold et al. 1981; Dunahay et al. 1984; Albertsson et al. 1981), although it is not completely understood how PSII is organized in these membranes. It had been suggested that C2S2 represents the supercomplex in high light, while C2S2M2 is the result of low-light growth (Daum et al. 2010). However, it was recently demonstrated that also in high light, C2S2M2 is still the main supercomplex in Arabidopsis (Kouril et al. 2012). In high light, the amount of LHCII trimers is lower than in low light, although in all cases the stoichiometry LHCII/core is higher than two (it is often between three and four) (Bailey et al. 2001; Anderson and Andersson 1988; Kouril et al. 2012), meaning that not all LHCII trimers are present in the supercomplexes but that there are also "extra" trimers. The location of these "extra" LHCII trimers, however, is still unknown and some of them might be located in the LHCII-only domains that were proposed by Boekema et al. (Boekema et al. 2000) although it should be emphasized that most of the "extra trimers" should be connected to PSII which is not necessarily the case for these LHCII-only domains. Part of these extra trimers attaches to PSI in the presence of light but detaches again in the dark and also during light-stress conditions (Wientjes et al. 2013). Recently, it was also found that in Arabidopsis plants, the amount of M trimers is decreasing when the grow-light intensity is increased from 100 to 800 μmol photons m^{-2} s^{-1}, whereas the amount of "extra" trimers remains the same. Decreasing on the other hand the intensity to 20 μmol photons m^{-2} s^{-1}, leads to an increase in the amount of "extra" trimers, whereas the amount of M trimers now remains unaltered (Kouril et al. 2012). For nearly all time-resolved studies in the literature, detailed information about the antenna composition is lacking.

In the past, various studies have been performed on BBY preparations (Berthold et al. 1981). The kinetics of these membranes were for instance described by a single lifetime of 210 ps (Schilstra et al. 1999) or with a major lifetime of 140 ps and a minor lifetime of 330 ps (Van Mieghem et al. 1992). More recently, two studies were done that showed average lifetimes in the order of 150–160 ps (Broess et al. 2006, 2008) and the results were interpreted with a coarse-grained model that uses the

C2S2M2 structure as a basis. Like in the ERPE model, it was assumed that primary charge separation (with rate k_{CS} or inverse rate/transfer time τ_{CS}) is reversible (first charge-separated state is ΔG lower in energy than the state in which the RC is excited in the Q_y state). Secondary charge separation (with rate k_{RP} or inverse rate/transfer time τ_{RP}) was supposed to be irreversible. EET was modeled by assuming hopping to occur between neighboring (monomeric) complexes with a rate called k_h (or inverse rate/hopping time τ_h) that was assumed to be the same for all hopping steps, whereas each rate was scaled with the number of pigments per complex. The basic difference with the earlier ERPE model is the fact that the supercomplex is used as a structural model to include EET steps and the fact that the hopping rate is not assumed to be infinitely fast. Using this model it was shown that different combinations of τ_{CS} and τ_H can describe the data nearly equally well (Broess et al. 2006), reminiscent of the data fitting results for core samples. Although it was not possible to extract more details about the charge transfer kinetics in the RC, it was possible to conclude that the BBY data could not be explained with published parameters for charge separation as obtained from time-resolved studies on cores by for instance Vasilliev et al. (Vassiliev et al. 2002) and Miloslavina et al. (Miloslavina et al. 2006) and other studies. Good resemblance could only be obtained when both the rate of charge separation and the drop in free energy upon charge separation were increased. It was also argued that previously published results on isolated PSII RC (Andrizhiyevskaya et al. 2004; Groot et al. 2005) were not in accordance with the BBY results. Again, this seems to indicate that the antenna system is important for plant PSII integrity and functionality.

In a follow-up study (Broess et al. 2008), time-resolved fluorescence measurements were performed on PSII membranes using two different excitation wavelengths, 420 and 483 nm. In this way, the relative number of excitations in core and outer antenna was varied, and the migration time from outer antenna to core was estimated to be 20–25 ps, much faster than one might expect based on earlier results on random aggregates of LHCII (Barzda et al. 2001). Therefore, it seems that the organization of the light-harvesting complexes in the supercomplexes/PSII membranes has been optimized in such a way that efficient EET takes place. However, at the moment detailed EET calculations are still lacking.

Energy transfer and charge separation in PSII in the thylakoid membrane

Isolated thylakoid membranes contain all complexes participating in the light reactions of photosynthesis but the large heterogeneity of the system and the presence of different complexes strongly complicate the interpretation of the time-resolved data. In general, the kinetics of thylakoids with open RCs are multi-exponential with lifetimes ranging from tens of picoseconds to values between 300 and 600 ps, and the average lifetime typically ranges from 300 to 400 ps (Engelmann et al. 2005; Leibl et al. 1989; Roelofs et al. 1992; Vasil'ev et al. 1998). However, interpretation of the individual lifetimes has remained ambiguous (for an overview see also (van Grondelle et al. 1994; Van Amerongen et al. 2003)). Recently, thylakoid membranes from A. thaliana with 4 LHCII trimers per RC were studied using various detection wavelengths to discriminate between PSI and PSII kinetics. Making use of two excitation wavelengths, it was possible to estimate the migration time from PSII outer antenna to core (van Oort et al. 2010). The fluorescence decay could be fitted very well with three lifetimes, in this particular case being 73, 251, and 531 ps (plus a very small contribution of a ns component) at all wavelengths with varying amplitudes. Shorter lifetimes mainly reflect spectral equilibration within individual complexes (see above) and are of less interest for the entire membrane. The three main lifetimes are sufficient to describe the data although they do not directly correspond to well-defined physical processes, and they are the result of different processes and heterogeneity in the membrane. Note that these lifetimes usually differ for different preparations, depending on for instance growth-light conditions and the state of the membrane (light- or dark-adapted, state 1 or state 2, in the presence or absence of nonphotochemical quenching (NPQ) and with open or closed RCs). The shortest of these three lifetimes (fitted with 73 ps in this case) is partly due to PSI whereas the other two are almost exclusively due to PSII as can be concluded from the shapes of the decay-associated spectra (van Oort et al. 2010). In this particular case, the thylakoids were dark-adapted and therefore in state 1 (no LHCII attached to PSI) with open RCs and in the absence of NPQ. It was estimated that $\tau_{trap} = 180$ ps and $\tau_{mig} = 150$ ps. This migration time is a factor of 4–5 longer than for the PSII membranes above, which contained 2.4–2.5 trimers per RC. Therefore, it is clear that the extra trimers are connected less well to the RCs. These results indicate that at the level of the thylakoid membrane trap-limited models are certainly not valid. At this point, it is also worth mentioning that different supercomplexes are functionally connected to each other and the domain size (how far does/can an excitation travel?) was estimated to be 12–24 LHCII trimers by Lambrev et al.(Lambrev et al. 2011).

In (Wientjes et al. 2013) it was studied for A. thaliana how the time-resolved fluorescence kinetics depends on the distribution of LHCII over PSI and PSII. In most light conditions some LHCII is attached to PSI (at most one LHCII trimer per PSI, on average around half a trimer). PSI and PSI-LHCII contribute only to the fastest (87 ps in this study) component to which also PSII contributes. Lifetimes of 0.26 and 0.54 ns are due to PSII and are very similar to the lifetimes reported above, namely 0.25, and 0.53 ns (van Oort et al. 2010) The longest lifetime is only observed in the presence of "extra" LHCII and is for instance not found for supercomplexes or PSII membranes with only 2.5 LHCII trimers per RC (see above). Upon relocation of LHCII from PSII to PSI the relative amplitude of the 87 ps component increases at the expense of the 0.26 and 0.54 ns components. This is explained by a decreased contribution of the "extra" LHCIIs to the "slow" PSII fluorescence decay, and an increased contribution to the ~87 ps component by PSI-LHCII, thereby shortening the average fluorescence lifetime of the thylakoids.

Where to go?

At the level of the individual pigment-protein complexes the functioning of the outer light-harvesting complexes of PSII seems to be relatively well understood ("Outer antenna complexes" section). When it comes to the PSII core, there is more uncertainty ("The PSII core" section,). Different labs are able to obtain very similar experimental results on the same samples but there is strong disagreement about the interpretation. Moreover, there seem to be differences between the "performance" of core complexes in vitro and in vivo and striking differences exist between core preparations from plants and cyanobacteria, although it is generally assumed that these cores are very similar. However, the cores in plants are surrounded by outer light-harvesting complexes, which is not the case in cyanobacteria. It is clear from the work on PSII supercomplexes that the intrinsic performance of the core of PSII is improving when the supercomplexes increase in size ("PSII supercomplexes" section). For PSII membranes, the performance seems to be even better and one might wonder to which extent the stacking of the membranes in the grana contributes to the overall performance. It is for instance still unknown how efficient EET between different membrane layers is: At the moment, the existing models mainly include EET within individual layers. It should, however, be noted that studies of Kirchhoff et al. (Kirchhoff et al. 2004) and Lambrev et al. (Lambrev et al. 2011) suggested that unstacking of the different membrane layers has no noticeable effect on excitation energy transfer, thereby implying that transfer between membrane layers is not very important. The modeling is not very sophisticated yet, which is partly due to the fact that also the structural models are not very accurate and good models should somehow also incorporate the structural variability of the membranes (in addition to heterogeneity): membranes are dynamic systems.

In thylakoid membranes where the average number of LHCII trimers can go up to four, depending on light conditions, the migration time is considerably slower, demonstrating that on the thylakoid level the charge separation process is definitely not trap-limited. It is still not known where the extra antenna complexes are located, but it is also not known to which extent they are disconnected and to which extent these complexes are quenched. There is a clear need for further studies on the grana organization and composition in different (light) conditions to enable more detailed modeling studies.

Finally, it will be very important to perform time-resolved studies in vivo, preferably at the single chloroplast level, using microscopic techniques. Only then will it be possible to see the "real" photosynthesis in action; after all, it is a very flexible and dynamic process and the chloroplast is continuously adapting to changing conditions.

ACKNOWLEDGMENTS

We thank Lijin Tian for providing Fig. 3. RC is supported by the ERC starting/consolidator Grant number 281341 and by the Netherland Organization for Scientific Research (NWO) via a Vici Grant.

REFERENCES

1. Albertsson PA, Andersson B, Larsson C, Akerlund HE (1981) Phase partition—a method for purification and analysis of cell organelles and membrane vesicles. Methods Biochem Anal 28:115–150
2. Amunts A, Toporik H, Borovikova A, Nelson N (2010) Structure determination and improved model of plant photosystem I. J Biol Chem 285(5):3478–3486
3. Anderson JM, Andersson B (1988) The dynamic photosynthetic membrane and regulation of solar-energy conversion. Trends Biochem Sci 13(9):351–355
4. Anderson JM, Chow WS, De Las Rivas J (2008) Dynamic flexibility in the structure and function of photosystem II in higher plant thylakoid membranes: the grana enigma. Photosynth Res 98(1–3):575–587
5. Andersson B, Anderson JM (1980) Lateral heterogeneity in the distribution of chlorophyll-protein complexes of the thylakoid membranes of spinach chloroplasts. Biochim Biophys Acta 593:427–440
6. Andrizhiyevskaya EG, Frolov D, van Grondelle R, Dekker JP (2004) On the role of the CP47 core antenna in the energy transfer and trapping dynamics of photosystem II. Phys Chem Chem Phys 6(20):4810–4819. doi:10.1039/b411977k
7. Bailey S, Walters RG, Jansson S, Horton P (2001) Acclimation of Arabidopsis thaliana to the light environment: the existence of separate low light and high light responses. Planta 213(5):794–801
8. Ballottari M, Mozzo M, Croce R, Morosinotto T, Bassi R (2009) Occupancy and functional architecture of the pigment binding sites of photosystem II antenna complex Lhcb5. J Biol Chem 284(12):8103–8113
9. Barber J (2002) Photosystem II: a multisubunit membrane protein that oxidises water. Curr Opin Struct Biol 12(4):523–530
10. Barzda V, Peterman EJG, van Grondelle R, Van Amerongen H (1998) The influence of aggregation on triplet formation in light-harvesting chlorophyll a/b pigment-protein complex II of green plants. Biochemistry 37:546–551

11. Barzda V, Gulbinas V, Kananavicius R, Cervinskas V, Van Amerongen H, van Grondelle R, Valkunas L (2001) Singlet-singlet annihilation kinetics in aggregates and trimers of LHCII. BiophysJ 80(5):2409–2421
12. Bassi R, Sandona D, Croce R (1997) Novel aspects of chlorophyll a/b-binding proteins. Physiol Plantarum 100:769–779
13. Bassi R, Croce R, Cugini D, Sandona D (1999) Mutational analysis of a higher plant antenna protein provides identification of chromophores bound into multiple sites. Proc Natl Acad Sci USA 96:10056–10061
14. Belgio E, Johnson MP, Juric S, Ruban AV (2012) Higher plant photosystem II light-harvesting antenna, not the reaction center, determines the excited-state lifetime-both the maximum and the nonphotochemically quenched. Biophys J 102(12):2761–2771. doi:10.1016/j.bpj.2012.05.004
15. Berthold DA, Babcock GT, Yocum CF (1981) A highly resolved, oxygen-evolving photosystem II preparation from spinach thylakoid membranes. EPR and electron-transport properties. FEBS Lett 134:231–234
16. Betterle N, Ballottari M, Zorzan S, de Bianchi S, Cazzaniga S, Dall'Osto L, Morosinotto T, Bassi R (2009) Light-induced dissociation of an antenna hetero-oligomer is needed for non-photochemical quenching induction. J Biol Chem 284(22):15255–15266
17. Boekema EJ, van Breemen JF, van Roon H, Dekker JP (2000) Arrangement of photosystem II supercomplexes in crystalline macrodomains within the thylakoid membrane of green plant chloroplasts. J Mol Biol 301(5):1123–1133
18. Broess K, Trinkunas G, van der Weij-de Wit CD, Dekker JP, van Hoek A, van Amerongen H (2006) Excitation energy transfer and charge separation in photosystem II membranes revisited. Biophys J 91(10):3776–3786
19. Broess K, Trinkunas G, van Hoek A, Croce R, van Amerongen H (2008) Determination of the excitation migration time in photosystem II—consequences for the membrane organization and charge separation parameters. Biochim Biophys Acta 1777(5):404–

409
20. Caffarri S, Croce R, Breton J, Bassi R (2001) The major antenna complex of photosystem II has a xanthophyll binding site not involved in light harvesting. J Biol Chem 276(38):35924–35933
21. Caffarri S, Passarini F, Bassi R, Croce R (2007) A specific binding site for neoxanthin in the monomeric antenna proteins CP26 and CP29 of Photosystem II. FEBS Lett 581(24):4704–4710
22. Caffarri S, Kouril R, Kereiche S, Boekema EJ, Croce R (2009) Functional architecture of higher plant photosystem II supercomplexes. Embo J 28:3052–3063
23. Caffarri S, Broess K, Croce R, van Amerongen H (2011) Excitation energy transfer and trapping in higher plant Photosystem II complexes with different antenna sizes. Biophys J 100(9):2094–2103. doi:10.1016/j.bpj.2011.03.049
24. Calhoun TR, Ginsberg NS, Schlau-Cohen GS, Cheng YC, Ballottari M, Bassi R, Fleming GR (2009) Quantum coherence enabled determination of the energy landscape in light-harvesting complex II. J Phys Chem B 113(51):16291–16295
25. Carbonera D, Giacometti G, Agostini G, Angerhofer A, Aust V (1992) ODMR of carotenoid and chlorophyll triplets in CP43 and CP47 complexes of spinach. Chem Phys Lett 194:275–281
26. Chuartzman SG, Nevo R, Shimoni E, Charuvi D, Kiss V, Ohad I, Brumfeld V, Reich Z (2008) Thylakoid membrane remodeling during state transitions in Arabidopsis. Plant Cell 20(4):1029–1039
27. Cinque G, Croce R, Holzwarth AR, Bassi R (2000) Energy transfer among CP29 chlorophylls: calculated Förster rates and experimental transient absorption at room temperature. BiophysJ 79:1706–1717
28. Clayton RK (1981) Photosynthesis: physical mechanism and chemical patterns. Cambridge University Press, Cambridge
29. Collini E, Scholes GD (2009) Coherent intrachain energy migration in a conjugated polymer at room temperature. Science 323(5912):369–373. doi:10.1126/science.1164016
30. Connelly JP, Muller MG, Hucke M, Gatzen G, Mullineaux CW, Ruban AV, Horton P, Holzwarth AR (1997) Ultrafast spectroscopy of trimeric light harvesting complex II from higher plants. J Phys

Chem B 101:1902–1909

31. Croce R, van Amerongen H (2011) Light-harvesting and structural organization of photosystem II: from individual complexes to thylakoid membrane. J Photochem Photobiol B 104(1–2):142–153. doi:10.1016/j.jphotobiol.2011.02.015
32. Croce R, Remelli R, Varotto C, Breton J, Bassi R (1999) The neoxanthin binding site of the major light harvesting complex (LHC II) from higher plants. FEBS Lett 456:1–6
33. Croce R, Muller MG, Bassi R, Holzwarth AR (2001) Carotenoid-to-chlorophyll energy transfer in recombinant major light- harvesting complex (LHCII) of higher plants I. Femtosecond transient absorption measurements. Biophys J 80(2):901–915
34. Croce R, Muller MG, Bassi R, Holzwarth AR (2003a) Chlorophyll b to chlorophyll a energy transfer kinetics in the CP29 antenna complex: a comparative femtosecond absorption study between native and reconstituted proteins. Biophys J 84(4):2508–2516
35. Croce R, Muller MG, Caffarri S, Bassi R, Holzwarth AR (2003b) Energy transfer pathways in the minor antenna complex CP29 of photosystem II: a femtosecond study of carotenoid to chlorophyll transfer on mutant and WT complexes. Biophys J 84(4):2517–2532
36. Daum B, Nicastro D, Austin J II, McIntosh JR, Kuhlbrandt W (2010) Arrangement of photosystem II and ATP synthase in chloroplast membranes of spinach and pea. Plant Cell 22(4):1299–1312
37. de Bianchi S, Dall'Osto L, Tognon G, Morosinotto T, Bassi R (2008) Minor antenna proteins CP24 and CP26 affect the interactions between photosystem II subunits and the electron transport rate in grana membranes of arabidopsis. Plant Cell 20(4):1012–1028
38. Dekker JP, Boekema EJ (2005) Supramolecular organization of thylakoid membrane proteins in green plants. Biochim Biophys Acta 1706:12–39
39. Dunahay TG, Staehelin LA, Seibert M, Ogilvie PD, Berg SP (1984) Structural, biochemical and biophysical characterization of four oxygen-evolving photosystem II preparations from spinach. Biochim Biophys Acta 764:179–193
40. Durrant JR, Hastings G, Joseph DM, Barber J, Porter G, Klug DR (1992) Subpicosecond equilibration of excitation energy in isolated

photosystem II reaction centers. Proc Natl Acad Sci USA 89:11632–11636

41. Engelmann ECM, Zucchelli G, Garlaschi FM, Casazza AP, Jennings RC (2005) The effect of outer antenna complexes on the photochemical trapping rate in barley thylakoid photosystem II. Biochim Biophys Acta 1706(3):276–286
42. Georgakopoulou S, van der Zwan G, Bassi R, van Grondelle R, van Amerongen H, Croce R (2007) Understanding the changes in the circular dichroism of light harvesting complex II upon varying its pigment composition and organization. Biochemistry 46(16):4745–4754
43. Germano M, Gradinaru CC, Shkuropatov AY, van Stokkum IH, Shuvalov VA, Dekker JP, van Grondelle R, van Gorkom HJ (2004) Energy and electron transfer in photosystem II reaction centers with modified pheophytin composition. Biophys J 86(3):1664–1672
44. Goral TK, Johnson MP, Brain APR, Kirchhoff H, Ruban AV, Mullineaux CW (2010) Visualizing the mobility and distribution of chlorophyll proteins in higher plant thylakoid membranes: effects of photoinhibition and protein phosphorylation. Plant J 62(6):948–959
45. Gradinaru CC, Pascal AA, van Mourik F, Robert B, Horton P, van Grondelle R, Van Amerongen H (1998) Ultrafast evolution of the excited states in the chlorophyll a/b complex CP29 from green plants studied by energy-selective pump- probe spectroscopy. Biochemistry 37:1143–1149
46. Gradinaru CC, van Stokkum IHM, Pascal AA, van Grondelle R, Van Amerongen H (2000) Identifying the pathways of energy transfer between carotenoids and chlorophylls in LHCII and CP29. A multicolor, femtosecond pump-probe study. J Phys Chem B 104:9330–9342
47. Green BR, Khlbrandt W (1995) Sequence conservation of light-harvesting and stress- response proteins in relation to the three-dimensional molecular structure of LHCII. Photosynth Res 44:139–148
48. Groot ML, Pawlowicz NP, van Wilderen LJGW, Breton J, van Stokkum IHM, van Grondelle R (2005) Initial electron donor and acceptor in isolated photosystem II reaction centers identified

with femtosecond mid-IR spectroscopy. Proc Natl Acad Sci USA 102(37):13087–13092

49. Guskov A, Kern J, Gabdulkhakov A, Broser M, Zouni A, Saenger W (2009) Cyanobacterial photosystem II at 2.9-angstrom resolution and the role of quinones, lipids, channels and chloride. Nat Struct Mol Biol 16(3):334–342
50. Hankamer B, Nield J, Zheleva D, Boekema E, Jansson S, Barber J (1997) Isolation and biochemical characterisation of monomeric and dimeric photosystem II complexes from spinach and their relevance to the organisation of photosystem II in vivo. Eur J Biochem 243:422–429
51. Holzwarth AR, Muller MG, Reus M, Nowaczyk M, Sander J, Rogner M (2006) Kinetics and mechanism of electron transfer in intact photosystem II and in the isolated reaction center: pheophytin is the primary electron acceptor. Proc Natl Acad Sci USA 103(18):6895–6900
52. Jahns P, Holzwarth AR (2012) The role of the xanthophyll cycle and of lutein in photoprotection of photosystem II. Biochim Biophys Acta 1817(1):182–193. doi:10.1016/j.bbabio.2011.04.012
53. Kereiche S, Kiss AZ, Kouril R, Boekema EJ, Horton P (2010) The PsbS protein controls the macro-organisation of photosystem II complexes in the grana membranes of higher plant chloroplasts. FEBS Lett 584(4):759–764
54. Kirchhoff H, Borinski M, Lenhert S, Chi LF, Buchel C (2004) Transversal and lateral exciton energy transfer in grana thylakoids of spinach. Biochemistry 43(45):14508–14516
55. Kirchhoff H, Haase W, Wegner S, Danielsson R, Ackermann R, Albertsson PA (2007) Low-light-induced formation of semicrystalline photosystem II arrays in higher plant chloroplasts. Biochemistry 46(39):11169–11176
56. Kiss AZ, Ruban AV, Horton P (2008) The PsbS protein controls the organization of the photosystem II antenna in higher plant thylakoid membranes. J Biol Chem 283(7):3972–3978
57. Kobayashi M, Ohashi S, Iwamoto K, Shiraiwa Y, Kato Y, Watanabe T (2007) Redox potential of chlorophyll d in vitro. Biochim Biophys Acta 1767(6):596–602

58. Kouril R, Oostergetel GT, Boekema EJ (2011) Fine structure of granal thylakoid membrane organization using cryo electron tomography. Biochim Biophys Acta 1807(3):368–374. doi:10.1016/j.bbabio.2010.11.007

59. Kouril R, Wientjes E, Bultema JB, Croce R, Boekema EJ (2012) High-light vs. low-light: effect of light acclimation on photosystem II composition and organization in Arabidopsis thaliana. Biochim Biophys Acta 1827(3):411–419. doi:10.1016/j.bbabio.2012.12.003

60. Kovacs L, Damkjaer J, Kereiche S, Ilioaia C, Ruban AV, Boekema EJ, Jansson S, Horton P (2006) Lack of the light-harvesting complex CP24 affects the structure and function of the grana membranes of higher plant chloroplasts. Plant Cell 18(11):3106–3120

61. Kwa SLS, Groeneveld FG, Dekker JP, van Grondelle R, Van Amerongen H, Lin S, Struve WS (1992) Steady-state and time-resolved polarized light spectroscopy of the green plant light-harvesting complex II. Biochim Biophys Acta 1101:143–146

62. Lambrev PH, Schmitt FJ, Kussin S, Schoengen M, Varkonyi Z, Eichler HJ, Garab G, Renger G (2011) Functional domain size in aggregates of light-harvesting complex II and thylakoid membranes. Biochim Biophys Acta 1807(9):1022–1031. doi:10.1016/j.bbabio.2011.05.003

63. Lampoura SS, Barzda V, Owen GM, Hoff AJ, Van Amerongen H (2002) Aggregation of LHCII leads to a redistribution of the triplets over the central xanthophylls in LHCII. Biochemistry 41(29):9139–9144

64. Leibl W, Breton J, Deprez J, Trissl HW (1989) Photoelectric study on the kinetics of trapping and charge stabilization in oriented PS II membranes. Photosynth Res 22:257–275

65. Liu Z, Yan H, Wang K, Kuang T, Zhang J, Gui L, An X, Chang W (2004) Crystal structure of spinach major light-harvesting complex at 2.72 A resolution. Nature 428(6980):287–292

66. Marin A, Passarini F, Croce R, van Grondelle R (2010) Energy transfer pathways in the CP24 and CP26 antenna complexes of higher plant photosystem II: a comparative study. Biophys J 99(12):4056–4065

67. Marin A, Passarini F, van Stokkum IH, van Grondelle R, Croce R

(2011) Minor complexes at work: light-harvesting by carotenoids in the photosystem II antenna complexes CP24 and CP26. Biophys J 100(11):2829–2838. doi:10.1016/j.bpj.2011.04.029

68. Miloslavina Y, Szczepaniak M, Muller MG, Sander J, Nowaczyk M, Rogner M, Holzwarth AR (2006) Charge separation kinetics in intact photosystem II core particles is trap-limited. A picosecond fluorescence study. Biochemistry 45(7):2436–2442
69. Minagawa J, Takahashi Y (2004) Structure, function and assembly of photosystem II and its light-harvesting proteins. Photosynth Res 82(3):241–263
70. Mozzo M, Dall'Osto L, Hienerwadel R, Bassi R, Croce R (2008a) Photoprotection in the antenna complexes of photosystem II—role of individual xanthophylls in chlorophyll triplet quenching. J Biol Chem 283(10):6184–6192
71. Mozzo M, Passarini F, Bassi R, van Amerongen H, Croce R (2008b) Photoprotection in higher plants: the putative quenching site is conserved in all outer light-harvesting complexes of photosystem II. Biochim Biophys Acta 1777(10):1263–1267
72. Muh F, Renger T (2012) Refined structure-based simulation of plant light-harvesting complex II: linear optical spectra of trimers and aggregates. Biochim Biophys Acta 1817(8):1446–1460. doi:10.1016/j.bbabio.2012.02.016
73. Muh F, Renger T, Zouni A (2008) Crystal structure of cyanobacterial photosystem II at 3.0 angstrom resolution: a closer look at the antenna system and the small membrane-intrinsic subunits. Plant Physiol Biochem 46(3):238–264
74. Mustardy L, Garab G (2003) Granum revisited. A three-dimensional model—where things fall into place. Trends Plant Sci 8(3):117–122
75. Mustardy L, Buttle K, Steinbach G, Garab G (2008) The three-dimensional network of the thylakoid membranes in plants: quasihelical model of the granum-stroma assembly. Plant Cell 20(10):2552–2557
76. Neilson JA, Durnford DG (2010) Evolutionary distribution of light-harvesting complex-like proteins in photosynthetic eukaryotes. Genome 53(1):68–78
77. Nelson N, Yocum CF (2006) Structure and function of photosystems

I and II. Annu Rev Plant Biol 57:521–565

78. Novoderezhkin VI, van Grondelle R (2010) Physical origins and models of energy transfer in photosynthetic light-harvesting. Phys Chem Chem Phys 12(27):7352–7365
79. Novoderezhkin V, Palacios MA, Van Amerongen H, van Grondelle R (2004) Energy-transfer dynamics in the LHCII complex of higher plants: modified redfield approach. J Phys Chem B 108(29):10363–10375
80. Novoderezhkin VI, Palacios MA, Van Amerongen H, van Grondelle R (2005) Excitation dynamics in the LHCII complex of higher plants: modeling based on the 2.72 angstrom crystal structure. J Phys Chem B 109(20):10493–10504
81. Palacios MA, Standfuss J, Vengris M, van Oort BF, van Stokkum IH, Kuhlbrandt W, van Amerongen H, van Grondelle R (2006) A comparison of the three isoforms of the light-harvesting complex II using transient absorption and time-resolved fluorescence measurements. Photosynth Res 88(3):269–285
82. Pan X, Li M, Wan T, Wang L, Jia C, Hou Z, Zhao X, Zhang J, Chang W (2011) Structural insights into energy regulation of light-harvesting complex CP29 from spinach. Nat Struct Mol Biol 18(3):309–315. doi:10.1038/nsmb.2008
83. Pascal A, Gradinaru C, Wacker U, Peterman E, Calkoen F, Irrgang KD, Horton P, Renger G, van Grondelle R, Robert B, Van Amerongen H (1999) Spectroscopic characterization of the spinach Lhcb4 protein (CP29), a minor light-harvesting complex of photosystem II. Eur J Biochem 262:817–823
84. Passarini F, Wientjes E, Hienerwadel R, Croce R (2009) Molecular basis of light harvesting and photoprotection in CP24 Unique Features of the most recent antenna complex. J Biol Chem 284(43):29536–29546
85. Pawlowicz NP, van Grondelle R, van Stokkum IH, Breton J, Jones MR, Groot ML (2008) Identification of the first steps in charge separation in bacterial photosynthetic reaction centers of Rhodobacter sphaeroides by ultrafast mid-infrared spectroscopy: electron transfer and protein dynamics. Biophys J 95(3):1268–1284
86. Peterman EJG, Hobe S, Calkoen F, van Grondelle R, Paulsen H, van

Amerongen H (1996) Low-temperature spectroscopy of monomeric and trimeric forms of reconstituted light-harvesting chlorophyll a/b complex. Biochim Biophys Acta 1273:171–174

87. Peterman EJG, Monshouwer R, van Stokkum IHM, van Grondelle R, Van Amerongen H (1997) Ultrafast singlet excitation transfer from carotenoids to chlorophylls via different pathways in light-harvesting complex II of higher plants. Chem Phys Lett 264:279–284
88. Plumley FG, Schmidt GW (1987) Reconstitution of chloroform a/b light-harvesting complexes: xanthophyll-dependent assembley and energy transfer. Proc Natl Acad Sci USA 84:146–150
89. Prokhorenko VI, Holzwarth AR (2000) Primary process and structure of the photosystem II reaction center: a photon echo study. J Phys Chem B 104:11563–11578
90. Rappaport F, Boussac A, Force DA, Peloquin J, Brynda M, Sugiura M, Un S, Britt RD, Diner BA (2009) Probing the coupling between proton and electron transfer in photosystem II core complexes containing a 3-fluorotyrosine. J Am Chem Soc 131(12):4425–4433
91. Raszewski G, Renger T (2008) Light harvesting in photosystem II core complexes is limited by the transfer to the trap: can the core complex turn into a photoprotective mode? J Am Chem Soc 130(13):4431–4446
92. Remelli R, Varotto C, Sandona D, Croce R, Bassi R (1999) Chlorophyll binding to monomeric light-harvesting complex. A mutation analysis of chromophore-binding residues. J Biol Chem 274(47):33510–33521
93. Renger G (2010) The light reactions of photosynthesis. Curr Sci 98(10):1305–1319
94. Renger G, Renger T (2008) Photosystem II: the machinery of photosynthetic water splitting. Photosynth Res 98(1–3):53–80
95. Renger T, Schlodder E (2010) Primary photophysical processes in photosystem II: bridging the gap between crystal structure and optical spectra. Chem Phys Chem 11(6):1141–1153
96. Roelofs TA, Lee CH, Holzwarth AR (1992) Global target analysis of picosecond chlorophyll fluorescence kinetics from pea chloroplasts. A new approach to the characterization of the primary processes in

photosystem II alfa- and beta-units. Biophys J 61:1147–1163

97. Rogl H, Kuhlbrandt W (1999) Mutant trimers of light-harvesting complex II exhibit altered pigment content and spectroscopic features. Biochemistry 38(49):16214–16222
98. Ruban AV, Horton P (1999) The xanthophyll cycle modulates the kinetics of nonphotochemical energy dissipation in isolated light-harvesting complexes, intact chloroplasts, and leaves of spinach. Plant Physiol 119:531–542
99. Salverda JM, Vengris M, Krueger BP, Scholes GD, Czarnoleski AR, Novoderezhkin V, Van Amerongen H, van Grondelle R (2003) Energy transfer in light-harvesting complexes LHCII and CP29 of spinach studied with three pulse echo peak shift and transient grating. BiophysJ 84(1):450–465
100. Sandona D, Croce R, Pagano A, Crimi M, Bassi R (1998) Higher plants light harvesting proteins. Structure and function as revealed by mutation analysis of either protein or chromophore moieties. Biochim Biophys Acta 1365:207–214
101. Savikhin S, Van Amerongen H, Kwa SLS, van Grondelle R, Struve WR (1994a) Low-temperature energy transfer in LHC-II trimers from the Chl a/b light-harvesting antenna of photosystem II. BiophysJ 66:1597–1603
102. Savikhin S, Zhu YW, Lin S, Blankenship RE, Struve WS (1994b) Femtosecond spectroscopy of chlorosome antennas from the green photosynthetic bacterium chloroflexus aurantiacus. J Phys Chem 98:10322–10334
103. Schatz GH, Brock H, Holzwarth AR (1987) Picosecond kinetics of fluorescence and absorbance changes in photosystem II particles excited at low photon density. Proc Natl Acad SciUSA 84:8414–8418
104. Schatz GH, Brock H, Holzwarth AR (1988) A kinetic and energetic model for the primary processes in photosystem II. Biophys J 54:397–405
105. Schilstra MJ, Nield J, Dorner W, Hankamer B, Carradus M, Barter LMC, Barber J, Klug DR (1999) Similarity between electron donor side reactions in the solubilized photosystem II-LHC II supercomplex and photosystem-II-containing membranes. Photosynth Res 60(2–

3):191–198

106. Shimoni E, Rav-Hon O, Ohad I, Brumfeld V, Reich Z (2005) Three-dimensional organization of higher-plant chloroplast thylakoid membranes revealed by electron tomography. Plant Cell 17(9):2580–2586
107. Standfuss R, van Scheltinga ACT, Lamborghini M, Kuhlbrandt W (2005) Mechanisms of photoprotection and nonphotochemical quenching in pea light-harvesting complex at 2.5A resolution. EMBO J 24(5):919–928
108. Tian L, van Stokkum IH, Koehorst RB, Jongerius A, Kirilovsky D, van Amerongen H (2011) Site, rate, and mechanism of photoprotective quenching in cyanobacteria. J Am Chem Soc 133(45):18304–18311. doi:10.1021/ja206414m
109. Tian L, van Stokkum IH, Koehorst RB, van Amerongen H (2012) Light harvesting and blue-green light induced non-photochemical quenching in two different C-phycocyanin mutants of synechocystis PCC 6803. J Phys Chem. doi:10.1021/jp309570u
110. Tian L, Farooq S, van Amerongen H (2013) Probing the picosecond kinetics of the photosystem II core complex in vivo. Phys Chem Chem Phys. doi:10.1039/c3cp43813a
111. Umena Y, Kawakami K, Shen JR, Kamiya N (2011) Crystal structure of oxygen-evolving photosystem II at a resolution of 1.9 A. Nature 473(7345):55–60. doi:10.1038/nature09913
112. Van Amerongen H, van Grondelle R (2001) Understanding the energy transfer function of LHCII, the major light-harvesting complex of green plants. J Phys Chem B 105(3):604–617
113. Van Amerongen H, Kwa SLS, van Bolhuis BM, van Grondelle R (1994) Polarized fluorescence and absorption of macroscopically aligned light harvesting complex II. Biophys J 67:837–847
114. Van Amerongen H, Valkunas L, van Grondelle R (2000) Photosynthenic excitons. World Scientific Publishing Co. Pte. Ltd, Singapore
115. Van Amerongen H, Dekker JP, Parson WW, Green BR (2003) Light-harvesting antennas in photosynthesis. Kluwer Academic, The Netherlands, pp 219–251

116. van der Vos R, Carbonera D, Hoff AJ (1991) Microwave and optical spectroscopy of carotenoid triplets in light-harvesting complex LHCII of spinach by absorbance-detected magnetic resonance. J Appl Magn Reson 2:179–202

117. van der Weij-de Wit CD, Dekker JP, van Grondelle R, van Stokkum IH (2011) Charge separation is virtually irreversible in photosystem II core complexes with oxidized primary quinone acceptor. J Phys Chem A 115(16):3947–3956. doi:10.1021/jp1083746

118. van Grondelle R (1985) Excitation energy transfer, trapping and annihilation in photosynthetic systems. Biochim Biophys Acta 811:147–195

119. van Grondelle R, Dekker JP, Gillbro T, Sundstr"m V (1994) Energy transfer and trapping in photosynthesis. Biochim Biophys Acta 1187:1–65

120. Van Mieghem FJE, Searle GFW, Rutherford AW, Schaafsma TJ (1992) The influence of the double reduction of Q(A) on the fluorescence decay kinetics of photosystem II. Biochim Biophys Acta 1100:198–206

121. van Mourik F, Groot ML, van Grondelle R, Dekker JP, van Stokkum IHM (2004) Global and target analysis of fluorescence measurements on photosystem 2 reaction centers upon red excitation. Phys Chem Chem Phys 6(20):4820–4824

122. van Oort B, van Hoek A, Ruban AV, van Amerongen H (2007) Equilibrium between quenched and nonquenched conformations of the major plant light-harvesting complex studied with high-pressure time-resolved fluorescence. J Phys Chem 111(26):7631–7637

123. van Oort B, Alberts M, de Bianchi S, Dall'Osto L, Bassi R, Trinkunas G, Croce R, van Amerongen H (2010) Effect of antenna-depletion in photosystern II on excitation energy transfer in Arabidopsis thaliana. Biophys J 98(5):922–931

124. Vasil'ev S, Wiebe S, Bruce D (1998) Non-photochemical quenching of chlorophyll fluorescence in photosynthesis. 5-Hydroxy-1,4-naphthoquinone in spinach thylakoids as a model for antenna based quenching mechanisms. Biochim Biophys Acta 1363:147–156

125. Vassiliev S, Bruce D (2008) Toward understanding molecular mechanisms of light harvesting and charge separation in photosystem

II. Photosynth Res 97(1):75–89

126. Vassiliev S, Lee CI, Brudvig GW, Bruce D (2002) Structure-based kinetic modeling of excited-state transfer and trapping in histidine-tagged photosystem II core complexes from synechocystis. Biochemistry 41(40):12236–12243
127. Visser HM, Kleima FJ, van Stokkum IHM, van Grondelle R, Van Amerongen H (1996) Probing the many energy-transfer processes in the photosynthetic light-harvesting complex II at 77 K using energy-selective sub- picosecond transient absorption spectroscopy. Chem Phys 210:297–312
128. Wasielewski MR, Johnson DG, Govindjee Preston C, Seibert M, Baltscheffsky M (1990) The primary charge-separation rate in isolated photosystem II reaction center complex. Current research in photosynthesis. Kluwer Academic Publishers, Dordrecht, pp 451–454
129. Wientjes E, Oostergetel GT, Jansson S, Boekema EJ, Croce R (2009) The role of Lhca complexes in the supramolecular organization of higher plant photosystem I. J Biol Chem 284(12):7803–7810
130. Wientjes E, van Amerongen H, Croce R (2013) LHCII is an antenna of both photosystems after long-term acclimation. Biochim Biophys Acta 1827(3):420–426. doi:10.1016/j.bbabio.2012.12.009
131. Yakushevska AE, Jensen PE, Keegstra W, van Roon H, Scheller HV, Boekema EJ, Dekker JP (2001) Supermolecular organization of photosystem II and its associated light-harvesting antenna in Arabidopsis thaliana. Eur J Biochem 268(23):6020–6028
132. Yang CH, Kosemund K, Cornet C, Paulsen H (1999) Exchange of pigment-binding amino acids in light-harvesting chlorophyll a/b protein. Biochemistry 38(49):16205–16213
133. Zucchelli G, Santabarbara S, Jennings RC (2012) The Q(y) absorption spectrum of the light-harvesting complex II as determined by structure-based analysis of chlorophyll macrocycle deformations. Biochemistry 51(13):2717–2736. doi:10.1021/bi201677q

CHAPTER

9

UNDERSTANDING THE ROLES OF THE THYLAKOID LUMEN IN PHOTOSYNTHESIS REGULATION

Sari Järvi, Peter J. Gollan and Eva-Mari Aro

Molecular Plant Biology, Department of Biochemistry, University of Turku, Turku, Finland

It has been known for a long time that the thylakoid lumen provides the environment for oxygen evolution, plastocyanin-mediated electron transfer, and photoprotection. More recently lumenal proteins have been revealed to play roles in numerous processes, most often linked with regulating thylakoid biogenesis and the activity and turnover of photosynthetic protein complexes, especially the photosystem II and NAD(P)H dehydrogenase-like complexes. Still, the functions of the majority of lumenal proteins in Arabidopsis thaliana are unknown. Interestingly, while the thylakoid lumen proteome of at least 80 proteins contains several large protein families, individual members of many protein families have highly divergent roles. This is indicative of evolutionary pressure leading to neofunctionalization of lumenal proteins, emphasizing the important role of the thylakoid lumen for photosynthetic electron transfer and ultimately for plant fitness. Furthermore, the involvement of anterograde and retrograde signaling networks that regulate the expression and activity of lumen proteins is increasingly pertinent. Recent studies have also highlighted the importance of thiol/disulfide modulation in controlling the functions of many lumenal proteins and photosynthetic regulation pathways.

INTRODUCTION

Photosystem (PS)I, PSII, and the light harvesting complexes (LHCI and LHCII), in concert with the cytochrome (cyt) b_6f, ATP synthase, and the NAD(P)H dehydrogenase-like (NDH) are responsible for light harvesting and transduction of solar energy into chemical energy via photosynthetic electron transport (PET). These multi-subunit pigment–protein complexes are embedded in the highly folded thylakoid membrane, which encloses a continuous internal compartment known as the thylakoid lumen. The linear electron transport (LET) chain represents the predominant pathway of PET. Three major thylakoid membrane protein complexes – PSII, cyt b_6f, and PSI – cooperate in LET in order to transport electrons from water molecules to oxidized nicotinamide adenine dinucleotide phosphate ($NADP^+$). Photosynthetic water-splitting occurs at the lumenal side of PSII at the oxygen-evolving complex (OEC). Hydrogen ions accumulating in the lumen as a result of water-splitting and cyt b_6f activity generate the proton motive force (*pmf*) that drives ATP synthesis. Lumenal proton concentration is also an important regulator of PET, triggering non-photochemical quenching (NPQ) of harvested energy and slowing down electron transfer in the cyt b_6f complex under acidic lumenal conditions. While LET generates both NADPH and ATP, cyclic electron transport (CET) around PSI produces *pmf* and thus ATP without reducing $NADP^+$(Heber and Walker, 1992). To that end, the main role of PSI CET is to balance the production of ATP and NADPH according to metabolic needs and to alleviate stromal over-reduction (Shikanai, 2007).

Although the photosynthetic apparatus and light-driven electron transport have been studied extensively, there remains a great deal to learn about factors that regulate PET according to the energy requirements of metabolic pathways and environmental cues. Recent characterizations of thylakoid lumen proteomes and analyses of the component proteins have revealed a range of novel proteins and protein families. Furthermore, the details of recent studies show that the lumen holds key factors for regulation and repair of the photosynthetic membrane, facilitating PET flexibility that is vital for efficient energy conversion. Here we review the current understanding of the functions of thylakoid lumen proteins in LET, CET, and PSII repair, and explore factors that regulate their expression, translocation, and activity (Figure 1; Table 1). Although uncharacterized lumen proteins have mainly been excluded from this review, their roles in PET regulation, retrograde signaling and/or acclimation are also likely to be vital for plant growth and development.

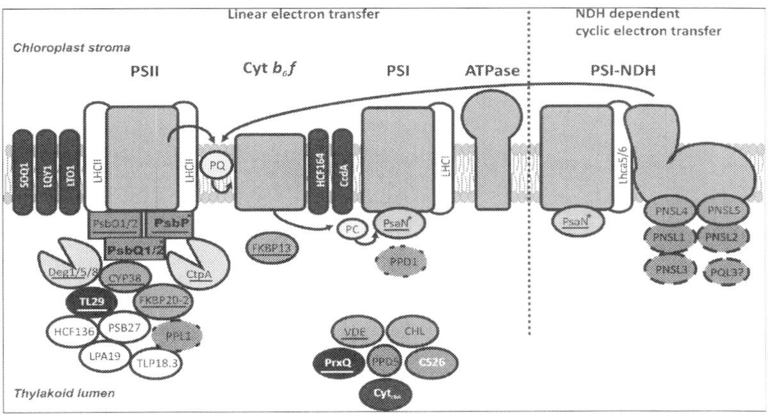

Figure 1. The majority of thylakoid lumen proteins with experimentally verified roles are involved in the function of either the PSII complex or the PSI–NDH supercomplex.The most abundant protein families in the thylakoid lumen are the OEC and OEC-like proteins (green), the immunophilins (blue), and proteases (yellow). In addition, the lumen proteome comprises peroxidases (black), photoprotective enzymes (purple), and several auxiliary proteins. The OEC proteins are proposed to function in water oxidation (square boxes), granal stacking (bolded), photosystem assembly (dotted outline), strigolactone biosynthesis (circle box), and NDH-dependent cyclic electron transfer (dashed outline). A high proportion of lumen proteins are thioredoxin targets (underlined). Regulation of thylakoid redox reactions involves membrane-embedded and soluble proteins (dark blue), and other lumen proteins are also implicated (white typeface). Lumen proteins with phosphorylation sites (asterisked) may be regulated by TLP18.3 phosphatase. Based on current knowledge, verified components of lumenal NDH subcomplex are not under post-translational regulation. No characterized lumenal proteins have so far been linked to the function of ATP synthase.

Table 1. Summary of characterized thylakoid lumen proteins in Arabidopsis.

Name	Gene code	Protein family/domain	Function/pathway	Signal peptide	Network	TRX target	Phosphorylation site	Reference
PSII								
PsbO1	At5g66570	OEC	Subunit of PSII OEC	Sec	C	Yes	–	Murakami et al. (2005)
PsbO2	At3g50820	OEC	Subunit of PSII OEC	Sec	C	Yes	–	Murakami et al. (2005)
PsbP1	At1g06680	OEC	Subunit of PSII OEC	Tat	C	Yes	Yes	Yi et al. (2007), Ifuku et al. (2008)
PsbQ1	At4g21280	OEC	Subunit of PSII OEC	Tat	C	–	–	Yi et al. (2006)
PsbQ2	At4g05180	OEC	Subunit of PSII OEC	Tat	C	–	Yes	Yi et al. (2006)
PSB27	At1g03600	PSB27	D1 processing	Tat	C	–	–	Chen et al. (2006)
LPA19	At1g05385	PSB27	D1 processing	Tat	–	–	–	Wei et al. (2010)
TLP18.3	At1g54780	Acid phosphatase	Degradation of D1, dimerization of PSII	Sec	C	–	–	Sirpio et al. (2007), Wu et al. (2011)
TL29/APX4	At4g09010	Peroxidase-like	Associated with PSII	Tat	C	Yes	–	Granlund et al. (2009b), Lundberg et al. (2011)
HCF136	At5g23120	–	Assembly of PSII	Tat	R	–	–	Meurer et al. (1998), Plucken et al. (2002)
CYP38	At3g01480	PPIase	Assembly of PSII	Sec	R	Yes	–	Fu et al. (2007), Sirpio et al. (2008), Vasudevan et al. (2012)
FKBP20-2	At3g60370	PPIase	Assembly of PSII	Tat	R	Yes	–	Lima et al. (2006)
PPL1	At3g55330	OEC, PsbP-like	Assembly of OEC	Tat	R	–	–	Ishihara et al. (2007)
CtpA	At5g46390	Peptidase	Processing of D1	Sec	R	Yes	–	Anbudurai et al. (1994)
CtpA	At4g17740	Peptidase	Processing of D1	Sec	R	Yes	–	Anbudurai et al. (1994)
CtpA	At3g57680	Peptidase	Processing of D1	No	–	–	–	Yin et al. (2008)
Deg1	At3g27925	Protease	Degradation of D1	Sec	R	Yes	–	Kapri-Pardes et al. (2007), Sun et al. (2007), Kley et al. (2011)
Deg5	At4g18370	Protease	Degradation of D1	Tat	–	Yes	–	Sun et al. (2007)
Deg8	At5g39830	Protease	Degradation of D1	Tat	R	–	–	Sun et al. (2007)
PSI								
PsaN	At5g64040	PSI	Subunit of PSI	Tat	C	Yes	Yes	Haldrup et al. (1999)

PPD1	At4g15510	OEC, PsbP-like	Assembly of PSI	Tat	R	–	–	Liu et al. (2012)
Cyt b6f								
FKBP13	At5g45680	PPIase	Assembly of cyt b6f	Tat	R	Yes	–	Gupta et al. (2002a), Gollan et al. (2011)
NDH								
PNSL1/PPL2	At2g39470	OEC, PsbP-like	Subunit of NDH	Tat	R	–	–	Ishihara et al. (2007)
PNSL2	At1g14150	OEC, PsbQ-like	Subunit of NDH	Tat	R	–	–	Suorsa et al. (2010), Yabuta et al. (2010)
PNSL3	At3g01440	OEC, PsbQ-like	Subunit of NDH	Tat	R	–	–	Suorsa et al. (2010), Yabuta et al. (2010)
PNSL4/ FKBP16-2	At4g39710	PPIase	Subunit of NDH	Tat	R	–	–	Peng et al. (2009)
PNSL5/ CYP20-2	At5g13120	PPIase	Subunit of NDH	Sec	–	–	–	Romano et al. (2004a), Shapiguzov et al. (2006), Peng et al. (2009),Sirpio et al. (2009)
PQL3	At2g01918	OEC, PsbQ-like	Subunit of NDH	Tat	–	–	–	Yabuta et al. (2010)
Photoprotection								
VDE	At1g08550	VDE, lipocalin	Xanthophyll cycle enzyme	Sec	–	Yes	–	Niyogi et al. (1998), Arnoux et al. (2009)
CHL	At3g47860	Lipocalin	Prevent lipid peroxidation	Sec	–	–	–	Levesque-Tremblay et al. (2009)
Others								
PETE2	At1g20340	Plastocyanin	Electron transport	Sec	C	–	–	Weigel et al. (2003), Pesaresi et al. (2009b)
PETE1	At1g76100	Plastocyanin	Electron transport	Sec	C	–	–	Weigel et al. (2003), Pesaresi et al. (2009b)
PRXQ	At3g26060	PRXQ-like	Antioxidant? Signaling?	No	–	Yes	–	Petersson et al. (2006)
PPD5	At5g11450	OEC, PsbP-like	Strigolactone biosynthesis	Tat	R	–	–	Roose et al. (2011)
Cyt c6A	At5g45040	Dithio-cytochrome	Oxidizing lumenal proteins	Sec	–	–	–	Gupta et al. (2002b), Weigel et al. (2003), Pesaresi et al. (2009b)
CS26	At3g03630	Cysteine synthase	S-sulfocysteine synthase	No	–	–	–	Bermudez et al. (2010), Bermudez et al. (2012)

DISTINCTIVE FEATURES OF THYLAKOID LUMEN

A decade ago, the thylakoid lumen was believed to be largely devoid of proteins, containing only the OEC proteins, the electron carrier plastocyanin (PC) and violaxanthin de-epoxidase (VDE). Proteomic and genomic studies have now revealed up to 80 proteins in *Arabidopsis thaliana* (*Arabidopsis*) to be localized in this compartment (Peltier et al., 2002; Schubert et al., 2002; Kieselbach and Schroder, 2003). All characterized lumenal proteins in *Arabidopsis* (Table 1) are nuclear-encoded and post-translationally transported into the chloroplast by the TOC/TIC (translocon at the outer/inner envelope of chloroplasts) system (Soll and Schleiff, 2004), while the secretory (Sec) and twin-arginine translocation (Tat) pathways import proteins into the lumen (discussed below; Albiniak et al., 2012). The thylakoid lumen is a constricted and crowded environment in which protein mobility is largely restricted; however, the dimensions of the thylakoid lumen are quite flexible (Mullineaux, 2008). Expansion of the lumenal space occurring in high light is linked to light-induced decrease in pH from around 7.0 in darkness to 5.8 and 6.5 in the light (Kramer et al., 1999; Cruz et al., 2001; Tikhonov, 2013) due the concomitant influx of anions upon acidification (Kirchhoff et al., 2011). The increase in lumenal space under light conditions is thought to allow protein diffusion that is important for PSII maintenance during photosynthetic activity (Kirchhoff et al., 2011). The lumenal pH controls the activities of many lumenal proteins, effectively functioning as a light-sensing on/off switch (discussed below).

PHOTOSYNTHETIC ELECTRON TRANSFER FROM A LUMENAL PERSPECTIVE

Water-Splitting

The OEC contains a Mn_4O_5Ca cluster that operates in water oxidation at the lumenal side of PSII. After storing four positive charges as a result of four successive electron transfer steps, the OEC oxidizes two water molecules and releases one oxygen molecule and four protons to the thylakoid lumen. Hence the OEC both liberates electrons for the electron transport chain and participates in acidification of the lumen. The OEC is supported by an extrinsic lumen protein complex, which reversibly associates with intrinsic PSII proteins. The OEC proteins are PsbO (also called OEC33), which is located proximal to the Mn_4O_5Ca cluster, PsbP (OEC23) and PsbQ (OEC17; Bricker et al., 2012). In *Arabidopsis* each OEC protein is encoded by two duplicate genes. The PsbO1 isoform exhibits higher oxygen-evolving activity than PsbO2 and accounts for around 90% of the total PsbO in WT plants (Murakami et al., 2005). PsbQ1 differs from PsbQ2 in a phosphorylatable serine residue that occurs in the latter (Reiland et al., 2009), while the PsbP1 is the major isoform of PsbP in the Columbia-0 ecotype since the *PSBP2* gene contains a frameshift that leads to truncated PsbP2 protein that is probably excluded from the thylakoids (Ifuku et al., 2008).

All three OEC proteins are required for maximal oxygen evolution, most likely because they sequester $Cl^-$and Ca^{2+} ions required for water-splitting (Miqyass et al., 2007; Popelkova and Yocum, 2007). Additionally, each of the OEC proteins appears to have a unique role in the integrity of PSII complexes. PsbQ is important for PSII stability, particularly under low light (Yi et al., 2006), while PsbP is required for assembly and/or stability of PSII and formation of PSII–LHCII supercomplexes (Yi et al., 2007; Ido et al., 2009). PsbQ stabilizes the interaction between PsbP and the membrane-bound PSII subunit PsbR (Suorsa et al., 2006; Allahverdiyeva et al., 2013). These results suggest that PsbP and PsbQ may coordinate the removal and/or reintegration of the Mn_4O_5Ca cluster with the disassembly and/or reassembly of PSII complexes during the PSII repair cycle (De Las Rivas et al., 2007). In addition, PsbP and PsbQ are linked to granal stacking (Dekker and Boekema, 2005), but evidence regarding the specific role of PsbQ in thylakoid architecture is contradictory (Yi et al., 2009). PsbO has been described as a GTPase that regulates PSII repair (Spetea et al., 2004; Lundin et al., 2007) and as a carbonic anhydrase (Lu et al., 2005), and has also demonstrated Ca^{2+} ion-binding activity (Heredia and De Las Rivas, 2003; Murray and Barber, 2006), although all of these features of PsbO are somewhat contentious and remain to be unequivocally demonstrated.

The Q Cycle and cyt b_6f

Part of electrons from PSII is shuttled to cyt b_6f via the so-called "Q cycle," which involves successive reduction and oxidation of the membrane-soluble electron- and proton-carrier plastoquinone (PQ). Each Q cycle pumps two protons from the stroma to the lumen, coupling the pH of the lumen to PET activity (Tikhonov, 2013). A subunit of the cyt b_6f complex known as the Rieske protein has a lumenal [2Fe-2S] cluster-binding domain that operates in electron transfer between cyt b_6 and cyt *f*. Rieske interacts with the lumenal immunophilin FKBP13 (Gupta et al., 2002a; Gollan et al., 2011), which was thought to regulate the assembly of the cyt b_6f complex from the stromal side (Gupta et al., 2002a), although recent results suggest that the FKBP13–Rieske interaction occurs in the thylakoid lumen (Gollan et al., 2011). The chaperone activity of FKBP13 is sensitive to redox regulation, as discussed below.

Plastocyanin and PSI

Electron transfer from cyt b_6f to PSI takes place in the lumen, and yet few lumen proteins appear to be directly involved. The major lumenal electron carrier is the copper-containing protein PC, comprising two isoforms in *Arabidopsis* that are encoded by the *PETE1* and *PETE2* genes, of which the latter is more highly expressed (Pesaresi et al., 2009b). A cyt *c* protein that operates as an alternative electron donor for PSI in cyanobacteria and green algae also occurs in the *Arabidopsis* thylakoid lumen (cyt c_{6A}), where it has been suggested to have a similar function (Gupta et al., 2002b); however, strong evidence suggests this is not the case in *Arabidopsis* (Weigel et al., 2003; Pesaresi et al., 2009b). PsaN is the only lumenal subunit of PSI (Kieselbach and Schroder, 2003). Mutants lacking PsaN are capable of assembling functional PSI complexes and growing photoautotrophically; however, restricted electron flow between PSII and PSI in mutant plants show that PsaN is necessary for efficient interaction between PSI and PC (Haldrup et al., 1999).

Cyclic Electron Transfer

In PSI CET, electrons are directed from ferredoxin (Fd) back into the Q cycle rather than to $NADP^+$. The commonly accepted role of CET is in adjusting stromal ATP:NADPH ratios in response to metabolic requirements; however, CET also operates to maintain the low lumenal pH required for NPQ and photosynthetic control of cyt b_6f to protect both

PSII and PSI, particularly under conditions where PSII is disengaged or inhibited (Kramer et al., 2004; Munekage et al., 2004; Joliot and Johnson, 2011). PSI CET proceeds by two partially redundant pathways; the major route is dependent on proton gradient regulation (PGR) proteins (Munekage et al., 2002; Okegawa et al., 2007) and the formation of a multi-protein CET supercomplex (DalCorso et al., 2008; Iwai et al., 2010), although some of the components remain to be discovered.

The second, minor route for PSI CET involves the membrane-intrinsic NADPH dehydrogenase-like (NDH) complex, which also forms a CET-specific supercomplex through association with PSI (Rumeau et al., 2007; Peng et al., 2009). Based on the structural similarities with mitochondrial complex I of the respiratory chain (Friedrich and Weiss, 1997), which oxidizes NADH and reduces ubiquinone in a process that is coupled to proton translocation across the mitochondrial inner membrane, the NDH-like complex is proposed to play a similar role in the thylakoid membrane. However, the physiological relevance, functional mechanism, and regulation of the chloroplast NDH-like complex have not been fully elucidated (Shikanai, 2007; Yamamoto et al., 2011), partly due to the fact that the abundance of NDH-like complexes in the thylakoid membrane is very low (Sazanov et al., 1998). Nevertheless, the complex is known to be stable only when associated with at least two copies of PSI, and the role of NDH in PSI CET and chlororespiration has been established (Peng and Shikanai, 2011).

The NDH-like complex is composed of at least 30 subunits and auxiliary proteins (Ifuku et al., 2011), and thus the PSI–NDH supercomplex is among the largest protein complexes in the thylakoid membrane. The subunits of NDH include both nuclear-encoded and plastid-encoded proteins, indicating strict control of expression, protein import and assembly processes. Based on characterization of *ndh* mutant *Arabidopsis* lines, the chloroplast NDH is postulated to comprise four subcomplexes, known as "A," "B," "membrane," and "lumen" subcomplexes (Ifuku et al., 2011), although detailed structural data for any of these is currently missing. The higher plant NDH is closely related to its cyanobacterial counterpart, with the major differences being the lumenal subcomplex and some auxiliary proteins that are characteristic to plant NDH (Peng et al., 2009; Battchikova et al., 2011). The lumenal subcomplex, which is vital for stability of subcomplex A, comprises a PsbP homolog (PPL2, also called PNSL1), a PsbQ homolog (PNSL2), and immunophilins FKBP16-2 (PNSL4) and CYP20-2 (PNSL5; Peng et al., 2009; Sirpio et al., 2009; Suorsa et al., 2010; Yabuta et al., 2010). Of these PNSL5/CYP20-2 is the sole contributor to cyclophilins (CYP)-mediated lumenal PPIase activity (Shapiguzov et al., 2006) and was initially found to co-migrate with LHCII (Romano et al., 2004a). Incorporation of subcomplex A into the thylakoid membrane is one of the final steps in formation of functional NDH, and it may be a reversible one that can disengage CET or accommodate NDH repair (Peng and Shikanai, 2011). It seems plausible that the lumenal subcomplex could regulate the assembly and/or (more likely) disassembly of NDH according to the conditions in the thylakoid lumen.

THE PSII ASSEMBLY AND REPAIR INVOLVE A LARGE ARRAY OF LUMENAL PROTEINS

Photosystem II biogenesis shares many components with the repair cycle occurring after photoinhibition of PSII. The D1 protein in the PSII reaction center is the major target of irreversible photodamage during photosynthesis under high light, leading to NPQ by photoinhibition-related quenching (qI); however, balanced damage and repair of PSII have been shown to occur at all light intensities (Tyystjärvi and Aro, 1996). Replacement of damaged D1 requires disassembly of PSII–LHCII complexes, PSII migration from crowded grana thylakoids to the stromal lamellae, D1 removal and replacement, reassembly and finally relocation of functional PSII (Baena-González and Aro, 2002). The lumenal components of PSII biogenesis/repair cycle are discussed below.

Degradation of the damaged D1 protein, carried out primarily by thylakoid-associated FtsH proteases, occurs in cooperation with the lumenal Deg1, Deg5, and Deg8 proteases (Kato et al., 2012). Deg1 is activated by homo-hexamerization in response to pH changes in the lumen (Kley et al., 2011), and interaction between Deg5 and Deg8 to form an active protease complex may also be pH-dependent (Sun et al., 2007). While activated, Deg proteases specifically degrade lumen-exposed loops of D1 (Kapri-Pardes et al., 2007; Sun et al., 2007). Deg1 has proteolytic activity against other lumenal proteins *in vitro*, including PC and PsbO, suggesting it may operate as a general protease in the thylakoid lumen (Chassin et al., 2002). In addition to proteolytic activity,

Deg1 assists PSII assembly through interaction with the reaction center protein D2 (Sun et al., 2010b). Interestingly, the thylakoid lumen acidic phosphatase TLP18.3 is also involved in the degradation of D1 protein, but also in dimerization of PSII (Sirpio et al., 2007; Wu et al., 2011). Interaction between TLP18.3 and Deg1 (Zienkiewicz et al., 2012) might regulate the protease through dephosphorylation (Spetea and Lundin, 2012). The D1 protein is the primary target of photodamage, but other PSII core proteins are also damaged and degraded, particularly in response to environmental stresses. Stromal Deg7 has been shown to be involved in the proteolysis of photodamaged D1, D2, CP47, and CP43 (Sun et al., 2010a), while stromal FtsH proteases and lumenal Deg1 mediate the degradation of LHCII proteins (Zienkiewicz et al., 2012; Luciński and Jackowski, 2013).

Newly synthesized D1 protein is co-translationally inserted into the thylakoid membrane and the core complex. The latter step is assisted by two lumenal proteins, immunophilin CYP38 and "high chlorophyll fluorescence 136" (HCF136), both of which are present already in the proteome of pre-chloroplastic etioplasts, presumably for prompt D1 assembly during thylakoid biogenesis (Meurer et al., 1998; Kanervo et al., 2008). The HCF136 protein is a prerequisite for the assembly of PSII reaction centers during complex biogenesis, while CYP38 assists the assembly of PSII core complexes during both biogenesis and repair (Meurer et al., 1998; Plucken et al., 2002; Fu et al., 2007; Sirpio et al., 2008). The C-terminal CYP-like domain of CYP38 interacts with the PSII apoprotein, CP47 (Vasudevan et al., 2012). Thus, CYP38 might assist the correct folding and integration of CP47 into the PSII core. An additional role for CYP38 may lie in the regulation of correct conformation of D1 and possibly also CP43 during PSII biogenesis and/or repair and as a negative regulator of the thylakoid protein phosphatase that dephosphorylates PSII core proteins (Fulgosi et al., 1998; Vener et al., 1999; Rokka et al., 2000; Fu et al., 2007; Sirpio et al., 2008). The lumenal immunophilin FKBP20-2 also has a role in PSII complex assembly by a yet unknown mechanism (Lima et al., 2006).

Processing of precursor D1 protein to the mature form by the C-terminal processing protease CtpA (Anbudurai et al., 1994; Yamamoto et al., 2001) is required for integration of the OEC complex to PSII (Roose and Pakrasi, 2004). The lumen proteome of *Arabidopsis* includes three CtpA homologs. Mutation in one of these genes (At3g57680) does not affect accumulation of the D1 precursor suggesting that there may be functional redundancy between the CtpA homologs (Yin et al., 2008). The lumenal homologs Psb27 and "low PSII accumulation 19" (LPA19) interact with the newly inserted D1 precursor and are involved in processing of nascent D1 during PSII biogenesis in *Arabidopsis* (Chen et al., 2006; Wei et al., 2010). The Psb27 homolog in cyanobacteria interacts with PSII to prevent premature assembly of the Mn_4O_5Ca cluster at the lumenal side of PSII (Roose and Pakrasi, 2008), suggesting that the timing of D1 maturation is important in the PSII assembly. The importance of a PsbP homolog PPL1 for the PSII repair cycle was shown by the slow recovery of PSII from photoinhibition in *ppl1* plants (Ishihara et al., 2007). Finally, the lumenal ascorbate peroxidase APX4/TL29 has been described as a lumen-located component and/or auxiliary protein of PSII (Granlund et al., 2009b), although according to its crystal structure its function is unlikely to involve peroxidase activity (Lundberg et al., 2011).

PSI ASSEMBLY IS DEPENDENT ON LUMENAL PSBP-LIKE PROTEIN PPD1

Compared to PSII, PSI is much more tolerant to, and/or very well protected from photoinhibition, as PSI photodamage exists *in vivo* only under specific conditions such as chilling temperature (Zhang and Scheller, 2004) or in the deficiency of PGR-dependent CET (Suorsa et al., 2012a, b). So far only one lumenal protein assisting PSI biogenesis, namely the PsbP-like protein PPD1, has been identified. PPD1 interacts directly with PSI reaction center proteins PsaA and PsaB and assists the folding and insertion of these two proteins into the thylakoid membrane (Liu et al., 2012). A lack of PPD1 leads to the loss of PSI and an inability to grow photoautotrophically (Liu et al., 2012).

PHOTOPROTECTION COMPONENTS IN THE THYLAKOID LUMEN

In naturally fluctuating light conditions, the energy harvested by LHCII can become unbalanced in relation to the capacity of stromal acceptors, thus saturating the electron transport chain and generating reactive oxygen species (ROS) that cause photodamage of membrane proteins (Nishiyama et al., 2006; Murata et al., 2007). In order to protect the photosynthetic machinery amidst natural light conditions, plants use energy dissipation mechanisms (NPQ) that are partially located in the thylakoid lumen (Niyogi et al., 1998).

Non-Photochemical Quenching

The major NPQ mechanism (qE) is rapid and reversible, involving dissipation of absorbed light energy as heat. This is predominantly achieved through production of the carotenoid zeaxanthin and reorganization of LHCII, both processes that are triggered by acidification of the thylakoid lumen. Upon protonation, lumenal VDE converts from a monomer to a dimer, opening access to the active site that facilitates the conversion of violaxanthin to zeaxanthin (Arnoux et al., 2009). Protonation of the PSII protein PsbS causes a structural rearrangement of PSII–LHCII supercomplexes (Li et al., 2000, 2004; Kereïche et al., 2010), although the exact role of PsbS in qE remains to be defined (Johnson and Ruban, 2011).

Lumen Response to Oxidative Stress

Cysteine synthesis 26 (CS26) is an *S*-sulfocysteine synthase and occurs in low abundance in the thylakoid lumen, but it has a vital role in detection of lumenal redox conditions, particularly in long photoperiods (Pesaresi et al., 2009a; Bermudez et al., 2010, 2012). A lack of CS26 led to strong photoinhibition and a systemic ROS response that was accompanied by reduced levels of OEC proteins and PSII assembly factors (Bermudez et al., 2012). CS26 was recently proposed as a ROS sensor through its sensitivity to thiosulfate accumulation in the lumen (Gotor and Romero, 2013). The "chloroplastic lipocalin" (CHL) is involved in photoprotection of thylakoid membrane lipids. CHL accumulates in the thylakoid lumen during environmental stress conditions such as drought and high light, as well as in paraquat and abscisic acid treatments, to protect the thylakoid membrane from peroxidation (Levesque-Tremblay et al., 2009).

LUMEN PROTEIN FAMILIES

Diverse Roles of the PsbP-Like and PsbQ-Like Proteins

The PsbP family has at least ten members in the *Arabidopsis* thylakoid lumen (Hall et al., 2010; Sato, 2010). Aside from the OEC protein PsbP, these are PPL1 and PPL2, involved in PSII repair and NDH stability, respectively (discussed above), and at least seven "PsbP domain" proteins (PPD1–7). An eighth (PPD8) is encoded, but has not been detected at the protein level. The role of PPD1 in PSI assembly has been discussed above, but the specific activities of other PPDs in the lumen remain a mystery in many respects. A homolog of PPD2 in the green alga *Chlamydomonas reinhardtii* is implicated in the generation of singlet oxygen signals (Brzezowski et al., 2012) and PPD5 knockout in *Arabidopsis* led to a reduction in NDH activity and is linked to production of the carotenoid-derived hormone strigolactone (Roose et al., 2011).

Similarly, multiple PsbQ-like proteins occur in the *Arabidopsis* lumen. PQL1 and PQL2 are lumenal subunits of NDH (see above), while a third (PQL3) is also required for NDH function, but has not been found in the proteome (Yabuta et al., 2010). The cyanobacterial ancestors of plant PsbP and PsbQ domains, called "cyanoP" and "cyanoQ," respectively, are involved in PSII oxygen evolution, but may have more of an auxiliary role in regulation of OEC structure and assembly. Notably, cyanoP is considerably more closely related, at least in sequence and structure, to PPL1 than to PsbP in plants (Sato, 2010; Jackson et al., 2012). Considering the few details about the PsbP- and PsbQ-like proteins known so far, it is tempting to speculate that expansion of these families in the lumen has provided opportunities for regulating the lumen-exposed parts of various photosynthetic complexes.

Lumenal Immunophilins Regulate the Assembly, Maintenance, and Turnover of Thylakoid Membrane Protein Complexes

The immunophilins include two unrelated protein families, the CYP and the FK506-binding proteins (FKBP), both of which are abundant in the thylakoid lumen proteome (He et al., 2004). Immunophilins are well known for their ability to rotate the peptide bond of a proline residue, known as PPIase activity, which has been linked to protein folding; however, a majority of the lumenal immunophilins does not show PPIase activity against synthetic peptides (Shapiguzov et al., 2006; Edvardsson et al., 2007). The best characterized of the lumen immunophilins is CYP38, which has an atypical CYP domain in the C-terminus and an N-terminal helical bundle, possibly for autoinhibition (Vasudevan et al., 2012). CYP38 does not show PPIase activity, but has a vital role in the assembly of PSII (Fu et al., 2007; Sirpio et al., 2008). Contrary to earlier observations (He et al., 2004; Romano et al., 2004b; Sirpio et al., 2008), CYP38 in *Arabidopsis* lacks a leucine zipper domain due to a frameshift in the coding sequence. The spinach ortholog of CYP38, called "thylakoid lumen PPIase of 40 kDa" (TLP40; 82% sequence identity to CYP38) is likely to possess a similar functional role to CYP38, but appears to behave differently to its *Arabidopsis*counterpart in that TLP40 has PPIase activity *in vitro* (Fulgosi et al., 1998; Vener et al., 1999). FKBP20-2 was also implicated in PSII assembly based on the observed increase of unassembled PSII monomers and dimers in the *fkbp20-2* knockout, suggesting a role in formation of PSII supercomplexes (Lima et al., 2006). As discussed earlier, FKBP16-2 and CYP20-2 take part in the lumenal NDH subcomplex (Peng et al., 2009), while another immunophilin, FKBP13, is linked to cyt b_6f regulation through interaction with Rieske (Gupta et al., 2002a; Gollan et al., 2011). In wheat, FKBP16-1 and FKBP16-3 may have a role in development of photosynthetic membranes through their interaction partners, the PsaL subunit of PSI and "thylakoid formation-1" (THF1, also called PSB29), respectively (Gollan et al., 2011).

The roles of most lumenal immunophilins remain unclear, although accumulating evidence indicates a primary role in the assembly and/or turnover of photosynthetic complexes. FKBP16-2, FKBP16-4, and CYP37 have been found both in the membrane-bound and lumen-soluble thylakoid proteomes (Peltier et al., 2002; Friso et al., 2004), suggesting that they may be involved in recruitment of lumen proteins to the membrane.

Pentapeptide Repeat Proteins in Thylakoid Lumen have Unknown Function

A lumenal pentapeptide repeat-containing (PPR) family has three members;

TL15, TL17, and TL20.3 (Schubert et al., 2002; Hall et al., 2010). The lumenal pentapeptide proteins TL15 and TL17 in *Arabidopsis*increase in abundance upon light adaptation (Granlund et al., 2009a) and are, together with TL20.3, putative targets of thioredoxin (TRX) reduction (Hall et al., 2010). In line with this, the crystal structure of TL15 has revealed an internal disulfide bridge (Ni et al., 2011). Cyanobacterial PPRs have diverse roles, two of which may be relevant in the thylakoid lumen; regulation of light-induced manganese ion import (Chandler et al., 2003) and galactolipid translocation (Black et al., 1995).

POST-TRANSLATIONAL MODIFICATIONS OF LUMEN PROTEINS

Regulation of Lumen Proteins by Reversible Phosphorylation

Phosphoproteomics studies have identified several phosphorylated proteins in the thylakoid lumen (Table 1), including the OEC proteins PsbP and PsbQ (Reiland et al., 2009) and lumen-exposed regions of the PSII subunits PsbR and CP47 (Reiland et al., 2009) and the PSI subunits PsaF (Sugiyama et al., 2008) and PsaN (Stael et al., 2012). Phosphorylation of photosynthetic proteins is thought to regulate assembly of the photosynthetic machinery in response to environmental conditions (Reiland et al., 2009). The recent discovery that PsaN phosphorylation is calcium-dependent may link PSI maintenance with dark-induced stromal Ca^{2+} flux (Stael et al., 2012). Despite these results, neither lumenal kinases, nor the physiological significance of phosphorylation events in the lumen have been found, while a single candidate for dephosphorylation activity is the membrane anchored TLP18.3 (Sirpio et al., 2007; Wu et al., 2011), although its substrates are unknown. The existence of any nucleotide-dependent processes in the lumen is contentious (Kieselbach and Schroder, 2003), although accumulating evidence suggests that ATP can be imported to the lumen by a membrane-embedded thylakoid ADP/ATP carrier (TAAC; Thuswaldner et al., 2007), where it is presumed to be available for protein phosphorylation. Recently TAAC was also described as a phosphosulfate channel in the plastid envelope (Gigolashvili et al., 2012). A nucleoside diphosphate kinase 3 (NDK3) found both in the thylakoid lumen and in mitochondria is capable of hydrolyzing ATP to generate GTP thought to be the substrate for GTPase activity of PsbO that is implicated in OEC dissociation for PSII repair cycle (Spetea and Lundin, 2012).

Redox Regulation through Disulfide Bridge Modulation

According to current knowledge, more than 40% of the lumen proteome may be regulated by redox reactions through modulation of disulfide bonds that control protein translocation and folding and/or enzyme activation (Hall et al., 2010). This observation places lumenal redox enzymes as powerful regulators of numerous processes. In comparison, less than 10% of stromal proteins are regulated by TRX, although at least 10 TRX isoforms exist in the stroma. Chloroplast redox enzymes have recently been thoroughly reviewed (Lindahl and Kieselbach, 2009; Hall et al., 2010), and will be discussed here only briefly.

The leading candidate for the source of disulfide reduction in the lumen is HCF164, an integral membrane enzyme with a lumenal TRX domain, thought to accept reducing equivalents from stromal TRX via the membrane-localized "cyt *c* defective A" (CcdA; Motohashi and Hisabori, 2006, 2010). HCF164 interacts with cyt *f* and the Rieske iron–sulfur protein and is required for assembly of the cyt b_6f complex (Lennartz et al., 2001), and is also capable of reducing PsaN (Motohashi and Hisabori, 2006). A similarly membrane-embedded TRX-like protein is the "suppressor of quenching 1" (SOQ1), thought to regulate NPQ through a previously uncharacterized pathway (Brooks et al., 2013). "Low quantum yield of photosystem II" (LQY1) is a thylakoid membrane-bound Zn finger protein with protein disulfide isomerase activity that interacts with PSII core complexes to modulate disulfide bond formation in PSII subunits during the PSII repair cycle (Lu et al., 2011). "Peroxiredoxin Q" (PRXQ) generally transfers reductants from TRX to hydrogen peroxide for detoxification; however, lumenal PRXQ does not appear to reduce hydrogen peroxide (Petersson et al., 2006).

Disulfide bond formation in the lumen requires an electron acceptor to oxidize thiol groups, although the mechanism for this is not clear. One prospect is lumenal oxygen that is released by water-splitting reactions (Buchanan and Luan, 2005). In an interesting development of this idea, CS26 was proposed to regulate thiol oxidation by production of *S*-sulfocysteine in the lumen (Bermudez et al., 2012). Another candidate thiol oxidase is the lumenal cyt c_{6A}, which is proposed to shuttle reducing equivalents between thiols and PC (Schlarb-Ridley et al., 2006). Recently the "lumen thiol oxidoreductase 1" (LTO1) protein was found to be a thylakoid membrane-localized enzyme with a lumenal TRX domain that was recently shown to catalyze disulfide bond formation in PsbO *in vitro* (Karamoko et al., 2011).

Although the mechanisms of thiol/disulfide modulation in the lumen remain unclear, important photosynthetic processes are redox-regulated. Disulfide bond formation is important for folding of PsbO1 and PsbO2, which are susceptible to proteolysis in their unfolded state (Hashimoto et al., 1997; Hall et al., 2010; Karamoko et al., 2011). VDE contains disulfides that are vital for its activity in NPQ (Hall et al., 2010). The substrate-binding/PPIase activity of FKBP13 is controlled by two disulfide bridges that can be reduced and oxidized *in vitro* by TRX (Gopalan et al., 2004, 2006) and LTO1 (Lu et al., 2013), respectively. This suggests that the interaction between FKBP13 and the Rieske iron–sulfur protein may be linked to redox state of the thylakoid (Gollan et al., 2011). Furthermore, homology between FKBP13 and FKBP16-2 infers similar redox sensitivity for the assembly of the lumenal NDH subcomplex (Gollan et al., 2012), although these possibilities have not been tested experimentally. The activity of lumen immunophilins FKBP20-2 and CYP38 may also be regulated by disulfide bond modulation (Lima et al., 2006; Fu et al., 2007; Sirpio et al., 2008). Identification of lumen TRX targets indicates that the PSII repair cycle and OEC assembly are under redox control (Hall et al., 2010). Finally, a lumen-exposed disulfide bridge is thought to regulate the activity of the membrane-bound LHCII kinase STN7 (Lemeille et al., 2009), although the redox factors responsible have not been found.

PROTEIN TRANSLOCATION INTO THYLAKOID LUMEN

Four separate methods of protein import into thylakoids are established; the "signal recognition particle-dependent" (SRP) method and the "spontaneous" method insert integral membrane proteins into the thylakoid membrane and are employed by many photosynthetic subunits (Michl et al., 1994; Kim et al., 1999). Lumen proteins are translocated from the chloroplast stroma by either the Sec pathway or the Tat pathway, depending on the signal peptide in the precursor of the passenger protein (Albiniak et al., 2012; Table 1).

The Sec system comprises three components; SecA binds the signal peptide in the passenger protein, hydrolyses ATP and threads the unfolded precursor through a fixed channel in the thylakoid membrane comprising SecE and SecY subunits (Yuan et al., 1994; Laidler et al., 1995; Schuenemann et al., 1999). Sec substrates include PsbO, PC, and VDE (Mori et al., 1999).

Unlike the Sec pathway, the Tat pathway operates independently of ATP hydrolysis, instead deriving energy from the *pmf* across the thylakoid membrane. The Tat system comprises three integral membrane subunits; "high chlorophyll fluorescence 106" (Hcf106) and cpTatC associate together to form a large, hetero-oligomeric complex in the thylakoid membrane, while Tha4 occurs in separate homo-oligomeric complexes. The signal peptides of Tat passengers conserve a central, basic "Arg-Arg" motif that is recognized by the Hcf106–cpTatC receptor complex which, in the presence of suitable *pmf*, then transiently associates with Tha4, which, according to the current model, forms the translocation pore to conduct the passenger protein through a membrane (Albiniak et al., 2012). According to their signal peptides, all PsbP and PsbQ proteins and their homologs in *Arabidopsis* are Tat substrates, as are all lumenal FKBPs (including FKBP16-2; Gollan et al., 2012). A compelling feature of the Tat pathway is its capacity to transport folded proteins and protein–cofactor complexes. In the homologous bacterial Tat system, this is a "quality control" mechanism that ensures proper protein folding and cofactor integration prior to protein export (Hynds et al., 1998; Berks et al., 2000). In plants the Tat pathway could similarly facilitate folding and assembly in the relatively stable environment of the chloroplast stroma to underwrite protein and cofactor integrity in the fluctuating conditions of the lumen (Muller and Klosgen, 2005). Furthermore, thylakoid import of folded proteins could abrogate the need for post-translational modifications such as phosphorylation in the lumen. It should be noted that important details of the Tat pathway in plants remain unclear, including (i) the physical mechanism of translocation, (ii) contributions of the *pmf* components, (iii) involvement of stromal chaperones, and (iv) the conformations, post-translational modifications and complex states of Tat passengers.

RESPONSE OF THE LUMEN PROTEOME TO ENVIRONMENTAL CUES

Transcription Regulation

The importance of retrograde signals emitted from the chloroplast, and from other sites in the plant cell, in regulating the nuclear expression of photosynthetic proteins is becoming clear (Foyer et al., 2012; Queval and Foyer, 2012). Similar signaling factors are likely to regulate expression of lumenal proteins, which are all encoded in the nucleus (Table 1), and yet elucidation of these signals has received little attention. A recent analysis of the expression profiles divided lumen proteins into two networks; a "constitutive factor" group that included predominantly PSI and PSII subunits and few PSII auxiliary proteins, and a "regulatory factor" group containing NDH subunits, as well as several proteins involved in PSII regulation (Ifuku et al., 2010).

Acclimations of the Lumen Proteome to Light and Temperature

Fifteen thylakoid lumen proteins displayed increased abundance in light-adapted *Arabidopsis* compared to dark-adapted plants indicating that their roles are related to photosynthetic activity (Granlund et al., 2009a). These include OEC proteins PsbP1 and PsbQ2, PSII auxiliary proteins HCF136 and PPL1 as well as major PC (PETE2). Additionally PPD5, two pentapeptide proteins and a group of other functionally uncharacterized thylakoid lumen proteins are up-regulated at the protein level in light compared to darkness (Granlund et al., 2009a). Notably, a majority of the proteins found in higher abundance in the light-adapted lumen are Tat substrates, suggesting that the Tat system may regulate the lumen proteome in response to prevailing light (and other stress) conditions according to the *pmf* that is generated.

Acclimation to low temperature affects the accumulation of eight thylakoid lumen proteins in *Arabidopsis*(Goulas et al., 2006). These include PsbO1/2, PsbP1/2 proteins, HCF136, NDH related immunophilin PNSL5/CYP20-2, and two FKBP proteins. The drastic increase in accumulation of PNSL5/CYP20-2, which occurs concomitantly with down-regulation of the Calvin–Benson cycle enzymes during cold acclimation, might be linked to the activation of NDH-dependent CET under such conditions. However, it should be noted that *Arabidopsis* is a cold-tolerant plant and a different response, e.g., in the accumulation of the NDH-like complex, could be present in rice or other cold-sensitive plant species.

The Importance of pH as a Regulator of Lumen Protein Activity

Light-induced protonation of the thylakoid lumen contributes the major portion of the *pmf* that drives ATP production; however, the acidic lumen is an important factor in many other processes, as reviewed above (Figure 2). Low pH is required to regulate electron transport, through qE activation and photosynthetic control of cyt b_6f (Bratt et al., 1995; Kramer et al., 2004; Li et al., 2004; Joliot and Johnson, 2011). pH-dependent oligomerization of Deg proteases connects thylakoid lumen pH to photoinhibition, recovery and the proteolytic breakdown of other lumenal proteins (Hall et al., 2010). Likewise, OEC is known to become inactivated by pH below 6.0 (Commet et al., 2012). Finally, the light- and dark-induced changes in thylakoid membrane architecture, and the internal dimensions of the thylakoids, are also linked to thylakoid lumen pH (Kirchhoff et al., 2011). The pH of the lumen is determined by the respective rates of electron transfer and ATP synthase activity, and regulation of these processes is used to maintain stromal homeostasis (Kramer et al., 2004; Joliot and Johnson, 2011). It stands to reason that other lumenal activities may also be regulated according to metabolic requirements through controlled changes in thylakoid lumen pH.

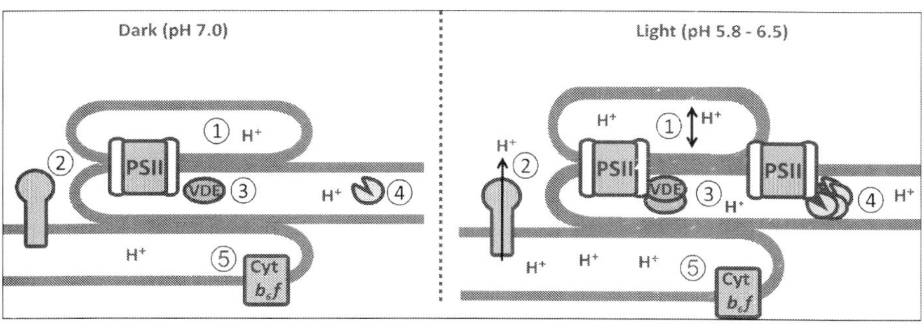

FIGURE 2. Lumenal processes under pH regulation. (1) Light-induced expansion of the lumen volume facilitating plastocyanin migration

and (2) enhancement of ATP synthase activity; (3) Light-induced activation of VDE and PsbS by protonation for photoprotection; (4) Deg oligomerization for PSII repair; and (5) induction of photosynthetic control via cyt b_6f.

CONCLUDING REMARKS

The thylakoid lumen not only provides the environment for oxygen evolution, PC-mediated electron transfer and zeaxanthin formation, but also houses factors that are important for the biogenesis, maintenance and turnover of photosynthetic protein complexes, activity of the NDH-like complex and, based on recent findings, even various signaling cascades. Indeed, most characterized lumenal proteins are linked to the PSII and NDH-like complexes, while only few are associated with PSI or cyt b_6f complexes and none have functions related to ATP synthase (Figure 1). A striking feature of the thylakoid lumen proteome is the presence of large protein families such as the OEC-like proteins and immunophilins, suggesting that neofunctionalization of lumenal protein homologs in regulation of photosynthetic complexes has driven the evolution of the lumen proteome. It is evident that lumenal proteins are imported, regulated and degraded directly by changes in the lumenal conditions that reflect the metabolic requirements of the plant. Several novel retrograde and anterograde signaling networks regulating expression and activity of lumen proteins according to environmental cues are likely to be revealed during forthcoming years. To that end, the multitude of photosynthetic regulatory proteins located in the thylakoid lumen should be carefully considered when identifying targets for improving photosynthetic reactions through genetic modifications and/or selection.

CONFLICT OF INTEREST STATEMENT

The authors declare that the research was conducted in the absence of any commercial or financial relationships that could be construed as a potential conflict of interest.

ACKNOWLEDGMENT

This research was supported by the Academy of Finland project 118637.

REFERENCES

1. Albiniak, A. M., Baglieri, J., and Robinson, C. (2012). Targeting of lumenal proteins across the thylakoid membrane. J. Exp. Bot. 63, 1689–1698. doi: 10.1093/jxb/err444
2. Allahverdiyeva, Y., Suorsa, M., Rossi, F., Pavesi, A., Kater, M. M., Antonacci, A., et al. (2013). Arabidopsis plants lacking PsbQ and PsbR subunits of the oxygen-evolving complex show altered PSII super-complex organization and short-term adaptive mechanisms. Plant J. 75, 671–684. doi: 10.1111/tpj.12230
3. Anbudurai, P. R., Mor, T. S., Ohad, I., Shestakov, S. V., and Pakrasi, H. B. (1994). The ctpA gene encodes the C-terminal processing protease for the D1 protein of the photosystem II reaction center complex. Proc. Natl. Acad. Sci. U.S.A. 91, 8082–8086. doi: 10.1073/pnas.91.17.8082
4. Arnoux, P., Morosinotto, T., Saga, G., Bassi, R., and Pignol, D. (2009). A structural basis for the pH-dependent xanthophyll cycle in Arabidopsis thaliana. Plant Cell 21, 2036–2044. doi: 10.1105/tpc.109.068007
5. Baena-González, E., and Aro, E. M. (2002). Biogenesis, assembly and turnover of photosystem II units. Philos. Trans. R. Soc. Lond. B Biol. Sci. 357, 1451–1460. doi: 10.1098/rstb.2002.1141
6. Battchikova, N., Eisenhut, M., and Aro, E. M. (2011). Cyanobacterial NDH-1 complexes: novel insights and remaining puzzles. Biochim. Biophys. Acta 1807, 935–944. doi: 10.1016/j.bbabio.2010.10.017
7. Berks, B. C., Sargent, F., and Palmer, T. (2000). The Tat protein export pathway. Mol. Microbiol. 35, 260–274. doi: 10.1046/j.1365-2958.2000.01719.x
8. Bermudez, M. A., Galmes, J., Moreno, I., Mullineaux, P. M., Gotor, C., and Romero, L. C. (2012). Photosynthetic adaptation to length of day is dependent on S-sulfocysteine synthase activity in the thylakoid lumen. Plant Physiol. 160, 274–288. doi: 10.1104/pp.112.201491
9. Bermudez, M. A., Paez-Ochoa, M. A., Gotor, C., and Romero, L. C. (2010). Arabidopsis S-sulfocysteine synthase activity is essential for chloroplast function and long-day light-dependent redox control. Plant Cell 22, 403–416. doi: 10.1105/tpc.109.071985

10. Black, K., Buikema, W. J., and Haselkorn, R. (1995). The hglK gene is required for localization of heterocyst-specific glycolipids in the cyanobacterium Anabaena sp. strain PCC 7120. J. Bacteriol. 177, 6440–6448.

11. Bratt, C. E., Arvidsson, P., Carlsson, M., and Åkerlund, H. (1995). Regulation of violaxanthin de-epoxidase activity by pH and ascorbate concentration. Photosynth. Res. 45, 169–175. doi: 10.1007/BF00032588

12. Bricker, T. M., Roose, J. L., Fagerlund, R. D., Frankel, L. K., and Eaton-Rye, J. J. (2012). The extrinsic proteins of Photosystem II. Biochim. Biophys. Acta 1817, 121–142. doi: 10.1016/j.bbabio.2011.07.006

13. Brooks, M. D., Sylak-Glassman, E. J., Fleming, G. R., and Niyogi, K. K. (2013). A thioredoxin-like/beta-propeller protein maintains the efficiency of light harvesting in Arabidopsis. Proc. Natl. Acad. Sci. U.S.A. 110, E2733–E2740. doi: 10.1073/pnas.1305443110

14. Brzezowski, P., Wilson, K. E., and Gray, G. R. (2012). The PSBP2 protein of Chlamydomonas reinhardtii is required for singlet oxygen-dependent signaling. Planta 236, 1289–1303. doi: 10.1007/s00425-012-1683-1

15. Buchanan, B. B., and Luan, S. (2005). Redox regulation in the chloroplast thylakoid lumen: a new frontier in photosynthesis research. J. Exp. Bot. 56, 1439–1447. doi: 10.1093/jxb/eri158

16. Chandler, L. E., Bartsevich, V. V., and Pakrasi, H. B. (2003). Regulation of manganese uptake in Synechocystis 6803 by RfrA, a member of a novel family of proteins containing a repeated five-residues domain. Biochemistry 42, 5508–5514. doi: 10.1021/bi027113a

17. Chassin, Y., Kapri-Pardes, E., Sinvany. G., Arad, T., and Adam, Z. (2002). Expression and characterization of the thylakoid lumen protease DegP1 from Arabidopsis. Plant Physiol. 130, 857–864. doi: 10.1104/pp.007922

18. Chen, H., Zhang, D., Guo, J., Wu, H., Jin, M., Lu, Q., et al. (2006). A Psb27 homologue in Arabidopsis thaliana is required for efficient repair of photodamaged photosystem II. Plant Mol. Biol. 61, 567–575. doi: 10.1007/s11103-006-0031-x

19. Commet, A., Boswell, N., Yocum, C. F., and Popelka, H. (2012). pH optimum of the photosystem II H_2O oxidation reaction: effects of PsbO, the manganese-stabilizing protein, Cl^- retention, and deprotonation of a component required for O_2 evolution activity. Biochemistry 51, 3808–3818. doi: 10.1021/bi201678m

20. Cruz, J. A., Sacksteder, C. A., Kanazawa, A., and Kramer, D. M. (2001). Contribution of electric field ($\Delta\psi$) to steady-state transthylakoid proton motive force (pmf) in vitro and in vivo. Control of pmf parsing into $\Delta\psi$ and ΔpH by ionic strength. *Biochemistry* 40, 1226–1237. doi: 10.1021/bi0018741

21. DalCorso, G., Pesaresi, P., Masiero, S., Aseeva, E., Schunemann, D., Finazzi, G., et al. (2008). A complex containing PGRL1 and PGR5 is involved in the switch between linear and cyclic electron flow in Arabidopsis. Cell 132, 273–285. doi: 10.1016/j.cell.2007.12.028

22. Dekker, J. P., and Boekema, E. J. (2005). Supramolecular organization of thylakoid membrane proteins in green plants. Biochim. Biophys. Acta 1706, 12–39. doi: 10.1016/j.bbabio.2004.09.009

23. De Las Rivas, J., Heredia, P., and Roman, A. (2007). Oxygen-evolving extrinsic proteins (PsbO, P, Q, R): bioinformatic and functional analysis. Biochim. Biophys. Acta 1767, 575–582. doi: 10.1016/j.bbabio.2007.01.018

24. Edvardsson, A., Shapiguzov, A., Petersson, U. A., Schroder, W. P., and Vener, A. V. (2007). Immunophilin AtFKBP13 sustains all peptidyl-prolyl isomerase activity in the thylakoid lumen from Arabidopsis thaliana deficient in AtCYP20-2. Biochemistry 46, 9432–9442. doi: 10.1021/bi700426q

25. Foyer, C. H., Neukermans, J., Queval, G., Noctor, G., and Harbinson, J. (2012). Photosynthetic control of electron transport and the regulation of gene expression. J. Exp. Bot. 63, 1637–1661. doi: 10.1093/jxb/ers013

26. Friedrich, T., and Weiss, H. (1997). Modular evolution of the respiratory NADH:ubiquinone oxidoreductase and the origin of its modules. J. Theor. Biol. 187, 529–540. doi: 10.1006/jtbi.1996.0387

27. Friso, G., Giacomelli, L., Ytterberg, A. J., Peltier, J. B., Rudella, A. Sun, Q., et al. (2004). In-depth analysis of the thylakoid membrane proteome of Arabidopsis thaliana chloroplasts: new proteins, new

functions, and a plastid proteome database. Plant Cell 16, 478–499. doi: 10.1105/tpc.017814

28. Fu, A., He, Z., Cho, H. S., Lima, A., Buchanan, B. B., and Luan, S. (2007). A chloroplast cyclophilin functions in the assembly and maintenance of photosystem II in Arabidopsis thaliana. Proc. Natl. Acad. Sci. U.S.A. 104, 15947–15952. doi: 10.1073/pnas.0707851104
29. Fulgosi, H., Vener, A. V., Altschmied, L., Herrmann, R. G., and Andersson, B. (1998). A novel multi-functional chloroplast protein: identification of a 40 kDa immunophilin-like protein located in the thylakoid lumen. EMBO J. 17, 1577–1587. doi: 10.1093/emboj/17.6.1577
30. Gigolashvili, T., Geier, M., Ashykhmina, N., Frerigmann, H., Wulfert, S., Krueger, S., et al. (2012). The Arabidopsis thylakoid ADP/ATP carrier TAAC has an additional role in supplying plastidic phosphoadenosine 5′-phosphosulfate to the cytosol. Plant Cell 24, 4187–4204. doi: 10.1105/tpc.112.101964
31. Gollan, P. J., Bhave, M., and Aro, E. M. (2012). The FKBP families of higher plants: exploring the structures and functions of protein interaction specialists. FEBS Lett. 586, 3539–3547. doi: 10.1016/j.febslet.2012.09.002
32. Gollan, P. J., Ziemann, M., and Bhave, M. (2011). PPIase activities and interaction partners of FK506-binding proteins in the wheat thylakoid. Physiol. Plant. 143, 385–395. doi: 10.1111/j.1399-3054.2011.01503.x
33. Gopalan, G., He, Z., Balmer, Y., Romano, P., Gupta, R., Heroux, A., et al. (2004). Structural analysis uncovers a role for redox in regulating FKBP13, an immunophilin of the chloroplast thylakoid lumen. Proc. Natl. Acad. Sci. U.S.A. 101, 13945–13950. doi: 10.1073/pnas.0405240101
34. Gopalan, G., He, Z., Battaile, K. P., Luan, S., and Swaminathan, K. (2006). Structural comparison of oxidized and reduced FKBP13 from Arabidopsis thaliana. Proteins 65, 789–795. doi: 10.1002/prot.21108
35. Gotor, C., and Romero, L. C. (2013). S-sulfocysteine synthase function in sensing chloroplast redox status. Plant Signal. Behav. 8, e23313. doi: 10.4161/psb.23313

36. Goulas, E., Schubert, M., Kieselbach, T., Kleczkowski, L. A., Gardestrom, P., Schroder, W., et al. (2006). The chloroplast lumen and stromal proteomes of Arabidopsis thaliana show differential sensitivity to short- and long-term exposure to low temperature. Plant J. 47, 720–734. doi: 10.1111/j.1365-313X.2006.02821.x

37. Granlund, I., Hall, M., Kieselbach, T., and Schroder, W. P. (2009a). Light induced changes in protein expression and uniform regulation of transcription in the thylakoid lumen of Arabidopsis thaliana. PLoS ONE 4:e5649. doi: 10.1371/journal.pone.0005649.

38. Granlund, I., Storm, P., Schubert, M., Garcia-Cerdan, J. G., Funk, C., and Schroder, W. P. (2009b). The TL29 protein is lumen located, associated with PSII and not an ascorbate peroxidase. Plant Cell Physiol. 50, 1898–1910. doi: 10.1093/pcp/pcp134

39. Gupta, R., Mould, R. M., He, Z., and Luan, S. (2002a). A chloroplast FKBP interacts with and affects the accumulation of Rieske subunit of cytochrome bf complex. Proc. Natl. Acad. Sci. U.S.A. 99, 15806–15811. doi: 10.1073/pnas.222550399

40. Gupta, R., He, Z., and Luan, S. (2002b). Functional relationship of cytochrome c6 and plastocyanin in Arabidopsis. Nature 417, 567–571. doi: 10.1038/417567a

41. Haldrup, A., Naver, H., and Scheller, H. V. (1999). The interaction between plastocyanin and photosystem I is inefficient in transgenic Arabidopsis plants lacking the PSI-N subunit of photosystem I. Plant J. 17, 689–698. doi: 10.1046/j.1365-313X.1999.00419.x

42. Hall, M., Mata-Cabana, A., Akerlund, H. E., Florencio, F. J., Schroder, W. P., Lindahl, M., et al. (2010). Thioredoxin targets of the plant chloroplast lumen and their implications for plastid function. Proteomics 10, 987–1001.

43. Hashimoto, A., Ettinger, W. F., Yamamoto, Y., and Theg, S. M. (1997). Assembly of newly imported oxygen-evolving complex subunits in isolated chloroplasts: sites of assembly and mechanism of binding. Plant Cell 9, 441–452.

44. He, Z., Li, L., and Luan, S. (2004). Immunophilins and parvulins. Superfamily of peptidyl prolyl isomerases in Arabidopsis. *Plant Physiol.* 134, 1248–1267. doi: 10.1104/pp.103.031005

45. Heber, U., and Walker, D. (1992). Concerning a dual function of coupled cyclic electron transport in leaves. Plant Physiol. 100, 1621–1626. doi: 10.1104/pp.100.4.1621

46. Heredia, P., and De Las Rivas, J. (2003). Calcium-dependent conformational change and thermal stability of the isolated PsbO protein detected by FTIR spectroscopy. Biochemistry 42, 11831–11838. doi: 10.1021/bi034582j

47. Hynds, P. J., Robinson, D., and Robinson, C. (1998). The sec-independent twin-arginine translocation system can transport both tightly folded and malfolded proteins across the thylakoid membrane. J. Biol. Chem. 273, 34868–34874. doi: 10.1074/jbc.273.52.34868

48. Ido, K., Ifuku, K., Yamamoto, Y., Ishihara, S., Murakami, A., Takabe, K., et al. (2009). Knockdown of the PsbP protein does not prevent assembly of the dimeric PSII core complex but impairs accumulation of photosystem II supercomplexes in tobacco. Biochim. Biophys. Acta 1787, 873–881. doi: 10.1016/j.bbabio.2009.03.004

49. Ifuku, K., Endo, T., Shikanai, T., and Aro, E. M. (2011). Structure of the chloroplast NADH dehydrogenase-like complex: nomenclature for nuclear-encoded subunits. Plant Cell Physiol. 52, 1560–1568. doi: 10.1093/pcp/pBR098

50. Ifuku, K., Ishihara, S., and Sato, F. (2010). Molecular functions of oxygen-evolving complex family proteins in photosynthetic electron flow. J. Integr. Plant. Biol. 52, 723–734. doi: 10.1111/j.1744-7909.2010.00976.x

51. Ifuku, K., Ishihara, S., Shimamoto, R., Ido, K., and Sato, F. (2008). Structure, function, and evolution of the PsbP protein family in higher plants. Photosynth. Res. 98, 427–437. doi: 10.1007/s11120-008-9359-1

52. Ishihara, S., Takabayashi, A., Ido, K., Endo, T., Ifuku, K., and Sato, F. (2007). Distinct functions for the two PsbP-like proteins PPL1 and PPL2 in the chloroplast thylakoid lumen of Arabidopsis. Plant Physiol. 145, 668–679. doi: 10.1104/pp.107.105866

53. Iwai, M., Takizawa, K., Tokutsu, R., Okamuro, A., Takahashi, Y., and Minagawa, J. (2010). Isolation of the elusive supercomplex that drives cyclic electron flow in photosynthesis. Nature 464, 1210–

1213. doi: 10.1038/nature08885

54. Jackson, S. A., Hinds, M. G., and Eaton-Rye, J. J. (2012). Solution structure of CyanoP from Synechocystis sp. PCC 6803: new insights on the structural basis for functional specialization amongst PsbP family proteins. Biochim. Biophys. Acta 1817, 1331–1338. doi: 10.1016/j.bbabio.2012.02.032
55. Johnson, M. P., and Ruban, A. V. (2011). Restoration of rapidly reversible photoprotective energy dissipation in the absence of PsbS protein by enhanced ΔpH. J. Biol. Chem. 286, 19973–19981. doi: 10.1074/jbc.M111.237255
56. Joliot, P., and Johnson, G. N. (2011). Regulation of cyclic and linear electron flow in higher plants. Proc. Natl. Acad. Sci. U.S.A. 108, 13317–13322. doi: 10.1073/pnas.1110189108
57. Kanervo, E., Singh, M., Suorsa, M., Paakkarinen, V., Aro, E., Battchikova, N., et al. (2008). Expression of protein complexes and individual proteins upon transition of etioplasts to chloroplasts in pea (Pisum sativum). Plant Cell Physiol. 49, 396–410. doi: 10.1093/pcp/pcn016
58. Kapri-Pardes, E., Naveh, L., and Adam, Z. (2007). The thylakoid lumen protease Deg1 is involved in the repair of photosystem II from photoinhibition in Arabidopsis. Plant Cell 19, 1039–1047. doi: 10.1105/tpc.106.046573
59. Karamoko, M., Cline, S., Redding, K., Ruiz, N., and Hamel, P. P. (2011). Lumen thiol oxidoreductase1, a disulfide bond-forming catalyst, is required for the assembly of photosystem II in Arabidopsis. Plant Cell 23, 4462–4475. doi: 10.1105/tpc.111.089680
60. Kato, Y., Sun, X., Zhang, L., and Sakamoto, W. (2012). Cooperative D1 degradation in the photosystem II repair mediated by chloroplastic proteases in Arabidopsis. Plant Physiol. 159, 1428–1439. doi: 10.1104/pp.112.199042
61. Kereïche, S., Kiss, A. Z., Kouøil, R., Boekema, E. J., and Horton, P. (2010). The PsbS protein controls the macro-organisation of photosystem II complexes in the grana membranes of higher plant chloroplasts. FEBS Lett. 584, 759–764. doi: 10.1016/j.febslet.2009.12.031

62. Kieselbach, T., and Schroder, W. P. (2003). The proteome of the chloroplast lumen of higher plants. Photosynth. Res. 78, 249–264. doi: 10.1023/B:PRES.0000006913.86689.f1

63. Kim, S. J., Jansson, S., Hoffman, N. E., Robinson, C., and Mant, A. (1999). Distinct "assisted" and "spontaneous" mechanisms for the insertion of polytopic chlorophyll-binding proteins into the thylakoid membrane. J. Biol. Chem. 274, 4715–4721. doi: 10.1074/jbc.274.8.4715

64. Kirchhoff, H., Hall, C., Wood, M., Herbstova, M., Tsabari, O., Nevo, R., et al. (2011). Dynamic control of protein diffusion within the granal thylakoid lumen. Proc. Natl. Acad. Sci. U.S.A. 108, 20248–20253. doi: 10.1073/pnas.1104141109

65. Kley, J., Schmidt, B., Boyanov, B., Stolt-Bergner, P. C., Kirk, R., Ehrmann, M., et al. (2011). Structural adaptation of the plant protease Deg1 to repair photosystem II during light exposure. Nat. Struct. Mol. Biol. 18, 728–731. doi: 10.1038/nsmb.2055

66. Kramer, D., Sacksteder, C., and Cruz, J. (1999). How acidic is the lumen. Photosynth. Res. 60, 151–163. doi: 10.1023/A:1006212014787

67. Kramer, D. M., Avenson, T. J., and Edwards, G. E. (2004). Dynamic flexibility in the light reactions of photosynthesis governed by both electron and proton transfer reactions. Trends Plant Sci. 9, 349–357. doi: 10.1016/j.tplants.2004.05.001

68. Laidler, V., Chaddock, A. M., Knott, T. G., Walker, D., and Robinson, C. (1995). A SecY homolog in Arabidopsis thaliana. Sequence of a full-length cDNA clone and import of the precursor protein into chloroplasts. J. Biol. Chem. 270, 17664–17667.

69. Lemeille, S., Willig, A., Depege-Fargeix, N., Delessert, C., Bassi, R., and Rochaix, J. D. (2009). Analysis of the chloroplast protein kinase Stt7 during state transitions. PLoS Biol. 7:e1000045. doi: 10.1371/journal.pbio.1000045.

70. Lennartz, K., Plucken, H., Seidler, A., Westhoff, P., Bechtold, N., and Meierhoff, K. (2001). HCF164 encodes a thioredoxin-like protein involved in the biogenesis of the cytochrome b(6)f complex in Arabidopsis. Plant Cell 13, 2539–2551.

71. Levesque-Tremblay, G., Havaux, M., and Ouellet, F. (2009). The chloroplastic lipocalin AtCHL prevents lipid peroxidation and

protects Arabidopsis against oxidative stress. Plant J. 60, 691–702. doi: 10.1111/j.1365-313X.2009.03991.x

72. Li, X.-P., Björkman, O., Shih, C., Grossman, A. R., Rosenquist, M., Jansson, S., et al. (2000). A pigment-binding protein essential for regulation of photosynthetic light harvesting. Nature 403, 391–395. doi: 10.1038/35000131

73. Li, X. P., Gilmore, A. M., Caffarri, S., Bassi, R., Golan, T., Kramer, D., et al. (2004). Regulation of photosynthetic light harvesting involves intrathylakoid lumen pH sensing by the PsbS protein. J. Biol. Chem. 279, 22866–22874. doi: 10.1074/jbc.M402461200

74. Lima, A., Lima, S., Wong, J. H., Phillips, R. S., Buchanan, B. B., and Luan, S. (2006). A redox-active FKBP-type immunophilin functions in accumulation of the photosystem II supercomplex in Arabidopsis thaliana. Proc. Natl. Acad. Sci. U.S.A. 103, 12631–12636. doi: 10.1073/pnas.0605452103

75. Lindahl, M., and Kieselbach, T. (2009). Disulphide proteomes and interactions with thioredoxin on the track towards understanding redox regulation in chloroplasts and cyanobacteria. J. Proteomics 72, 416–438. doi: 10.1016/j.jprot.2009.01.003

76. Liu, J., Yang, H., Lu, Q., Wen, X., Chen, F., Peng, L., et al. (2012). PsbP-domain protein1, a nuclear-encoded thylakoid lumenal protein, is essential for photosystem I assembly in Arabidopsis. Plant Cell 24, 4992–5006. doi: 10.1105/tpc.112.106542

77. Lu, Y., Hall, D. A., and Last, R. L. (2011). A small zinc finger thylakoid protein plays a role in maintenance of photosystem II in Arabidopsis thaliana. Plant Cell 23, 1861–1875. doi: 10.1105/tpc.111.085456

78. Lu, Y., Wang, H. R., Li, H., Cui, H. R., Feng, Y. G., and Wang, X. Y. (2013). A chloroplast membrane protein LTO1/AtVKOR involving in redox regulation and ROS homeostasis. Plant Cell Rep. 32, 1427–1440. doi: 10.1007/s00299-013-1455-9

79. Lu, Y. K., Theg, S. M., and Stemler, A. J. (2005). Carbonic anhydrase activity of the photosystem II OEC33 protein from pea. Plant Cell Physiol. 46, 1944–1953. doi: 10.1093/pcp/pci209

80. Luciński, R., and Jackowski, G. (2013). AtFtsH heterocomplex-mediated degradation of apoproteins of the major light harvesting

complex of photosystem II (LHCII) in response to stresses. J. Plant Physiol. 170, 1082–1089. doi: 10.1016/j.jplph.2013.03.008

81. Lundberg, E., Storm, P., Schroder, W. P., and Funk, C. (2011). Crystal structure of the TL29 protein from Arabidopsis thaliana: an APX homolog without peroxidase activity. J. Struct. Biol. 176, 24–31. doi: 10.1016/j.jsb.2011.07.004
82. Lundin, B., Hansson, M., Schoefs, B., Vener, A. V., and Spetea, C. (2007). The Arabidopsis PsbO2 protein regulates dephosphorylation and turnover of the photosystem II reaction centre D1 protein. Plant J. 49, 528–539. doi: 10.1111/j.1365-313X.2006.02976.x
83. Meurer, J., Plucken, H., Kowallik, K. V., and Westhoff, P. (1998). A nuclear-encoded protein of prokaryotic origin is essential for the stability of photosystem II in Arabidopsis thaliana. EMBO J. 17, 5286–5297. doi: 10.1093/emboj/17.18.5286
84. Michl, D., Robinson, C., Shackleton, J. B., Herrmann, R. G., and Klosgen, R. B. (1994). Targeting of proteins to the thylakoids by bipartite presequences: CFoII is imported by a novel, third pathway. EMBO J. 13, 1310–1317.
85. Miqyass, M., van Gorkom, H. J., and Yocum, C. F. (2007). The PSII calcium site revisited. Photosynth Res. 92, 275–287. doi: 10.1007/s11120-006-9124-2
86. Mori, H., Summer, E. J., Ma, X., and Cline, K. (1999). Component specificity for the thylakoidal Sec and delta pH-dependent protein transport pathways. J. Cell Biol. 146, 45–56.
87. Motohashi, K., and Hisabori, T. (2006). HCF164 receives reducing equivalents from stromal thioredoxin across the thylakoid membrane and mediates reduction of target proteins in the thylakoid lumen. J. Biol. Chem. 281, 35039–35047. doi: 10.1074/jbc.M605938200
88. Motohashi, K., and Hisabori, T. (2010). CcdA is a thylakoid membrane protein required for the transfer of reducing equivalents from stroma to thylakoid lumen in the higher plant chloroplast. Antioxid. Redox Signal. 13, 1169–1176. doi: 10.1089/ars.2010.3138
89. Muller, M., and Klosgen, R. B. (2005). The Tat pathway in bacteria and chloroplasts (review). Mol. Membr. Biol. 22, 113–121. doi: 10.1080/09687860500041809
90. Mullineaux, C. W. (2008). Factors controlling the mobility of

photosynthetic proteins. Photochem. Photobiol. 84, 1310–1316. doi: 10.1111/j.1751-1097.2008.00420.x

91. Munekage, Y., Hashimoto, M., Miyake, C., Tomizawa, K., Endo, T., Tasaka, M., et al. (2004). Cyclic electron flow around photosystem I is essential for photosynthesis. Nature 429, 579–582. doi: 10.1038/nature02598

92. Munekage, Y., Hojo, M., Meurer, J., Endo, T., Tasaka, M., and Shikanai, T. (2002). PGR5 is involved in cyclic electron flow around photosystem I and is essential for photoprotection in Arabidopsis. Cell 110, 361–371. doi: 10.1016/S0092-8674(02)00867-X

93. Murakami, R., Ifuku, K., Takabayashi, A., Shikanai, T., Endo, T., and Sato, F. (2005). Functional dissection of two Arabidopsis PsbO proteins: PsbO1 and PsbO2. FEBS J. 272, 2165–2175. doi: 10.1111/j.1742-4658.2005.04636.x

94. Murata, N., Takahashi, S., Nishiyama, Y., and Allakhverdiev, S. I. (2007). Photoinhibition of photosystem II under environmental stress. Biochim. Biophys. Acta 1767, 414–421. doi: 10.1016/j.bbabio.2006.11.019

95. Murray, J. W., and Barber, J. (2006). Identification of a calcium-binding site in the PsbO protein of photosystem II. Biochemistry 45, 4128–4130. doi: 10.1021/bi052503t

96. Ni, S., McGookey, M. E., Tinch, S. L., Jones, A. N., Jayaraman, S., Tong, L., et al. (2011). The 1.7 A resolution structure of At2g44920, a pentapeptide-repeat protein in the thylakoid lumen of Arabidopsis thaliana. Acta Crystallogr. Sect. F Struct. Biol. Cryst. Commun. 67, 1480–1484. doi: 10.1107/S1744309111037432

97. Nishiyama, Y., Allakhverdiev, S. I., and Murata, N. (2006). A new paradigm for the action of reactive oxygen species in the photoinhibition of photosystem II. Biochim. Biophys. Acta 1757, 742–749. doi: 10.1016/j.bbabio.2006.05.013

98. Niyogi, K. K., Grossman, A. R., and Bjorkman, O. (1998). Arabidopsis mutants define a central role for the xanthophyll cycle in the regulation of photosynthetic energy conversion. Plant Cell 10, 1121–1134.

99. Okegawa, Y., Long, T. A., Iwano, M., Takayama, S., Kobayashi, Y.,

Covert, S. F., et al. (2007). A balanced PGR5 level is required for chloroplast development and optimum operation of cyclic electron transport around photosystem I. Plant Cell Physiol. 48, 1462–1471. doi: 10.1093/pcp/pcm116

100. Peltier, J. B., Emanuelsson, O., Kalume, D. E., Ytterberg, J., Friso, G., Rudella, A., et al. (2002). Central functions of the lumenal and peripheral thylakoid proteome of Arabidopsis determined by experimentation and genome-wide prediction. Plant Cell 14, 211–236. doi: 10.1105/tpc.010304

101. Peng, L., Fukao, Y., Fujiwara, M., Takami, T., and Shikanai, T. (2009). Efficient operation of NAD(P)H dehydrogenase requires supercomplex formation with photosystem I via minor LHCI in Arabidopsis. Plant Cell 21, 3623–3640. doi: 10.1105/tpc.109.068791

102. Peng, L., and Shikanai, T. (2011). Supercomplex formation with photosystem I is required for the stabilization of the chloroplast NADH dehydrogenase-like complex in Arabidopsis. Plant Physiol. 155, 1629–1639. doi: 10.1104/pp.110.171264

103. Pesaresi, P., Hertle, A., Pribil, M., Kleine, T., Wagner, R., Strissel, H., et al. (2009a). Arabidopsis STN7 kinase provides a link between short- and long-term photosynthetic acclimation. Plant Cell 21, 2402–2423. doi: 10.1105/tpc.108.064964

104. Pesaresi, P., Scharfenberg, M., Weigel, M., Granlund, I., Schroder, W. P., Finazzi, G., et al. (2009b). Mutants, overexpressors, and interactors of Arabidopsis plastocyanin isoforms: revised roles of plastocyanin in photosynthetic electron flow and thylakoid redox state. Mol. Plant. 2, 236–248. doi: 10.1093/mp/ssn041

105. Petersson, U. A., Kieselbach, T., Garcia-Cerdan, J. G., and Schroder, W. P. (2006). The Prx Q protein of Arabidopsis thaliana is a member of the luminal chloroplast proteome. FEBS Lett. 580, 6055–6061. doi: 10.1016/j.febslet.2006.10.001

106. Plucken, H., Muller, B., Grohmann, D., Westhoff, P., and Eichacker, L. A. (2002). The HCF136 protein is essential for assembly of the photosystem II reaction center in Arabidopsis thaliana. FEBS Lett. 532, 85–90. doi: 10.1016/S0014-5793(02)03634-7

107. Popelkova, H., and Yocum, C. F. (2007). Current status of the role

of Cl(-) ion in the oxygen-evolving complex. Photosynth. Res. 93, 111–121. doi: 10.1007/s11120-006-9121-5

108. Queval, G., and Foyer, C. H. (2012). Redox regulation of photosynthetic gene expression. Philos. Trans. R. Soc. Lond. B Biol. Sci. 367, 3475–3485. doi: 10.1098/rstb.2012.0068

109. Reiland, S., Messerli, G., Baerenfaller, K., Gerrits, B., Endler, A., Grossmann, J., et al. (2009). Large-scale Arabidopsis phosphoproteome profiling reveals novel chloroplast kinase substrates and phosphorylation networks. Plant Physiol. 150, 889–903. doi: 10.1104/pp.109.138677

110. Rokka, A., Aro, E. M., Herrmann, R. G., Andersson, B., and Vener, A. V. (2000). Dephosphorylation of photosystem II reaction center proteins in plant photosynthetic membranes as an immediate response to abrupt elevation of temperature. Plant Physiol. 123, 1525–1536. doi: 10.1104/pp.123.4.1525

111. Romano, P. G., Edvardsson, A., Ruban, A. V., Andersson, B., Vener, A. V., Gray, J. E., et al. (2004a). Arabidopsis AtCYP20-2 is a light-regulated cyclophilin-type peptidyl-prolyl cis–trans isomerase associated with the photosynthetic membranes. Plant Physiol. 134, 1244–1247. doi: 10.1104/pp.104.041186

112. Romano, P. G., Horton, P., and Gray, J. E. (2004b). The Arabidopsis cyclophilin gene family. Plant Physiol. 134, 1268–1282. doi: 10.1104/pp.103.022160

113. Roose, J. L., Frankel, L. K., and Bricker, T. M. (2011). Developmental defects in mutants of the PsbP domain protein 5 in Arabidopsis thaliana. PLoS ONE 6:e28624. doi: 10.1371/journal.pone.0028624

114. Roose, J. L., and Pakrasi, H. B. (2004). Evidence that D1 processing is required for manganese binding and extrinsic protein assembly into photosystem II. J. Biol. Chem. 279, 45417–45422. doi: 10.1074/jbc.M408458200

115. Roose, J. L., and Pakrasi, H. B. (2008). The Psb27 protein facilitates manganese cluster assembly in photosystem II. J. Biol. Chem. 283, 4044–4050. doi: 10.1074/jbc.M708960200

116. Rumeau, D., Peltier, G., and Cournac, L. (2007). Chlororespiration and cyclic electron flow around PSI during photosynthesis and plant stress response. Plant Cell Environ. 30, 1041–1051. doi:

10.1111/j.1365-3040.2007.01675.x

117. Sato, N. (2010). Phylogenomic and structural modeling analyses of the PsbP superfamily reveal multiple small segment additions in the evolution of photosystem II-associated PsbP protein in green plants. Mol. Phylogenet. Evol. 56, 176–186. doi: 10.1016/j.ympev.2009.11.021
118. Sazanov, L. A., Burrows, P. A., and Nixon, P. J. (1998). The plastid ndh genes code for an NADH-specific dehydrogenase: isolation of a complex I analogue from pea thylakoid membranes. Proc. Natl. Acad. Sci. U.S.A. 95, 1319–1324. doi: 10.1073/pnas.95.3.1319
119. Schlarb-Ridley, B. G., Nimmo, R. H., Purton, S., Howe, C. J., and Bendall, D. S. (2006). Cytochrome c(6A) is a funnel for thiol oxidation in the thylakoid lumen. FEBS Lett. 580, 2166–2169. doi: 10.1016/j.febslet.2006.03.052
120. Schubert, M., Petersson, U. A., Haas, B. J., Funk, C., Schroder, W. P., and Kieselbach, T. (2002). Proteome map of the chloroplast lumen of Arabidopsis thaliana. J. Biol. Chem. 277, 8354–8365. doi: 10.1074/jbc.M108575200
121. Schuenemann, D., Amin, P., Hartmann, E., and Hoffman, N. E. (1999). Chloroplast SecY is complexed to SecE and involved in the translocation of the 33-kDa but not the 23-kDa subunit of the oxygen-evolving complex. J. Biol. Chem. 274, 12177–12182. doi: 10.1074/jbc.274.17.12177
122. Shapiguzov, A., Edvardsson, A., and Vener, A. V. (2006). Profound redox sensitivity of peptidyl-prolyl isomerase activity in Arabidopsisthylakoid lumen. FEBS Lett. 580, 3671–3676. doi: 10.1016/j.febslet.2006.05.054
123. Shikanai, T. (2007). Cyclic electron transport around photosystem I: genetic approaches. Annu. Rev. Plant. Biol. 58, 199–217. doi: 10.1146/annurev.arplant.58.091406.110525
124. Sirpio, S., Allahverdiyeva, Y., Suorsa, M., Paakkarinen, V., Vainonen, J., Battchikova, N., et al. (2007). TLP18.3, a novel thylakoid lumen protein regulating photosystem II repair cycle. *Biochem. J.* 406, 415–425. doi: 10.1042/BJ20070460
125. Sirpio, S., Holmstrom, M., Battchikova, N., and Aro, E. M. (2009). AtCYP20-2 is an auxiliary protein of the chloroplast NAD(P)

H dehydrogenase complex. FEBS Lett. 583, 2355–2358. doi: 10.1016/j.febslet.2009.06.031

126. Sirpio, S., Khrouchtchova, A., Allahverdiyeva, Y., Hansson, M., Fristedt, R., Vener, A. V., et al. (2008). AtCYP38 ensures early biogenesis, correct assembly and sustenance of photosystem II. Plant J. 55, 639–651. doi: 10.1111/j.1365-313X.2008.03532.x

127. Soll, J., and Schleiff, E. (2004). Protein import into chloroplasts. Nat. Rev. 5, 198–208. doi: 10.1038/nrm1333

128. Spetea, C., Hundal, T., Lundin, B., Heddad, M., Adamska, I., and Andersson, B. (2004). Multiple evidence for nucleotide metabolism in the chloroplast thylakoid lumen. Proc. Natl. Acad. Sci. U.S.A. 101, 1409–1414. doi: 10.1073/pnas.0308164100

129. Spetea, C., and Lundin, B. (2012). Evidence for nucleotide-dependent processes in the thylakoid lumen of plant chloroplasts – an update. FEBS Lett. 586, 2946–2954. doi: 10.1016/j.febslet.2012.07.005

130. Stael, S., Rocha, A. G., Wimberger, T., Anrather, D., Vothknecht, U. C., and Teige, M. (2012). Cross-talk between calcium signalling and protein phosphorylation at the thylakoid. J. Exp. Bot. 63, 1725–1733. doi: 10.1093/jxb/err403

131. Sugiyama, N., Nakagami, H., Mochida, K., Daudi, A., Tomita, M., Shirasu, K., et al. (2008). Large-scale phosphorylation mapping reveals the extent of tyrosine phosphorylation in Arabidopsis. Mol. Syst. Biol. 4, 193. doi: 10.1038/msb.2008.32

132. Sun, X., Fu, T., Chen, N., Guo, J., Ma, J., Zou, M., et al. (2010a). The stromal chloroplast Deg7 protease participates in the repair of photosystem II after photoinhibition in Arabidopsis. Plant Physiol. 152, 1263–1273. doi: 10.1104/pp.109.150722

133. Sun, X., Ouyang, M., Guo, J., Ma, J., Lu, C., Adam, Z., et al. (2010b). The thylakoid protease Deg1 is involved in photosystem-II assembly in Arabidopsis thaliana. Plant J. 62, 240–249. doi: 10.1111/j.1365-313X.2010.04140.x

134. Sun, X., Peng, L., Guo, J., Chi, W., Ma, J., Lu, C., et al. (2007). Formation of DEG5 and DEG8 complexes and their involvement in the degradation of photodamaged photosystem II reaction center D1 protein in Arabidopsis. Plant Cell 19, 1347–1361. doi: 10.1105/

tpc.106.049510

135. Suorsa, M., Grieco, M., Jarvi, S., Gollan, P. J., Kangasjarvi, S., Tikkanen, M., et al. (2012a). PGR5 ensures photosynthetic control to safeguard photosystem I under fluctuating light conditions. Plant Signal. Behav. 8, e22741. doi: 10.4161/psb.22741

136. Suorsa, M., Jarvi, S., Grieco, M., Nurmi, M., Pietrzykowska, M., Rantala, M., et al. (2012b). PROTON GRADIENT REGULATION5 is essential for proper acclimation of Arabidopsis photosystem I to naturally and artificially fluctuating light conditions. Plant Cell 24, 2934–2948. doi: 10.1105/tpc.112.097162

137. Suorsa, M., Sirpio, S., Allahverdiyeva, Y., Paakkarinen, V., Mamedov, F., Styring, S., et al. (2006). PsbR, a missing link in the assembly of the oxygen-evolving complex of plant photosystem II. J. Biol. Chem. 281, 145–150. doi: 10.1074/jbc.M510600200

138. Suorsa, M., Sirpio, S., Paakkarinen, V., Kumari, N., Holmstrom, M., and Aro, E. M. (2010). Two proteins homologous to PsbQ are novel subunits of the chloroplast NAD(P)H dehydrogenase. Plant Cell Physiol. 51, 877–883. doi: 10.1093/pcp/pcq070

139. Thuswaldner, S., Lagerstedt, J. O., Rojas-Stutz, M., Bouhidel, K., Der, C., Leborgne-Castel, N., et al. (2007). Identification, expression, and functional analyses of a thylakoid ATP/ADP carrier from Arabidopsis. J. Biol. Chem. 282, 8848–8859. doi: 10.1074/jbc.M609130200

140. Tikhonov, A. N. (2013). pH-dependent regulation of electron transport and ATP synthesis in chloroplasts. Photosynth. Res. doi: 10.1007/s11120-013-9845-y

141. Tyystjärvi, E., and Aro, E. M. (1996). The rate constant of photoinhibition, measured in lincomycin-treated leaves, is directly proportional to light intensity. Proc. Natl. Acad. Sci. U.S.A. 93, 2213–2218. doi: 10.1073/pnas.93.5.2213

142. Vasudevan, D., Fu, A., Luan, S., and Swaminathan, K. (2012). Crystal structure of Arabidopsis cyclophilin38 reveals a previously uncharacterized immunophilin fold and a possible autoinhibitory mechanism. Plant Cell 24, 2666–2674. doi: 10.1105/tpc.111.093781

143. Vener, A. V., Rokka, A., Fulgosi, H., Andersson, B., and Herrmann, R. G. (1999). A cyclophilin-regulated PP2A-like protein phosphatase

in thylakoid membranes of plant chloroplasts. Biochemistry 38, 14955–14965. doi: 10.1021/bi990971v

144. Wei, L., Guo, J., Ouyang, M., Sun, X., Ma, J., Chi, W., et al. (2010). LPA19, a Psb27 homolog in Arabidopsis thaliana, facilitates D1 protein precursor processing during PSII biogenesis. J. Biol. Chem. 285, 21391–21398. doi: 10.1074/jbc.M110.105064

145. Weigel, M., Varotto, C., Pesaresi, P., Finazzi, G., Rappaport, F., Salamini, F., et al. (2003). Plastocyanin is indispensable for photosynthetic electron flow in Arabidopsis thaliana. J. Biol. Chem. 278, 31286–31289. doi: 10.1074/jbc.M302876200

146. Wu, H. Y., Liu, M. S., Lin, T. P., and Cheng, Y. S. (2011). Structural and functional assays of AtTLP18.3 identify its novel acid phosphatase activity in thylakoid lumen. Plant Physiol. 157, 1015–1025. doi: 10.1104/pp.111.184739

147. Yabuta, S., Ifuku, K., Takabayashi, A., Ishihara, S., Ido, K., Ishikawa, N., et al. (2010). Three PsbQ-like proteins are required for the function of the chloroplast NAD(P)H dehydrogenase complex in Arabidopsis. Plant Cell Physiol. 51, 866–876. doi: 10.1093/pcp/pcq060

148. Yamamoto, H., Peng, L., Fukao, Y., and Shikanai, T. (2011). An Src homology 3 domain-like fold protein forms a ferredoxin binding site for the chloroplast NADH dehydrogenase-like complex in Arabidopsis. Plant Cell 23, 1480–1493. doi: 10.1105/tpc.110.080291

149. Yamamoto, Y., Inagaki, N., and Satoh, K. (2001). Overexpression and characterization of carboxyl-terminal processing protease for precursor D1 protein: regulation of enzyme-substrate interaction by molecular environments. J. Biol. Chem. 276, 7518–7525. doi: 10.1074/jbc.M008877200

150. Yi, X., Hargett, S. R., Frankel, L. K., and Bricker, T. M. (2006). The PsbQ protein is required in Arabidopsis for photosystem II assembly/stability and photoautotrophy under low light conditions. J. Biol. Chem. 281, 26260–26267. doi: 10.1074/jbc.M603582200

151. Yi, X., Hargett, S. R., Frankel, L. K., and Bricker, T. M. (2009). The PsbP protein, but not the PsbQ protein, is required for normal thylakoid architecture in Arabidopsis thaliana. FEBS Lett. 583,

2142–2147. doi: 10.1016/j.febslet.2009.05.048

152. Yi, X., Hargett, S. R., Liu, H., Frankel, L. K., and Bricker, T. M. (2007). The PsbP protein is required for photosystem II complex assembly/stability and photoautotrophy in Arabidopsis thaliana. J. Biol. Chem. 282, 24833–24841. doi: 10.1074/jbc.M705011200

153. Yin, S., Sun, X., and Zhang, L. (2008). An Arabidopsis ctpA homologue is involved in the repair of photosystem II under high light. Chin. Sci. Bull. 53, 1021–1026. doi: 10.1007/s11434-008-0153-4

154. Yuan, J., Henry, R., McCaffery, M., and Cline, K. (1994). SecA homolog in protein transport within chloroplasts: evidence for endosymbiont-derived sorting. Science 266, 796–798. doi: 10.1126/science.7973633

155. Zhang, S., and Scheller, H. V. (2004). Photoinhibition of photosystem I at chilling temperature and subsequent recovery in Arabidopsis thaliana. Plant Cell Physiol. 45, 1595–1602. doi: 10.1093/pcp/pch180

156. Zienkiewicz, M., Ferenc, A., Wasilewska, W., and Romanowska, E. (2012). High light stimulates Deg1-dependent cleavage of the minor LHCII antenna proteins CP26 and CP29 and the PsbS protein in Arabidopsis *thaliana*. *Planta* 235, 279–288. doi: 10.1007/s00425-011-1505-x

CHAPTER

10

PLASTOQUINONE AND UBIQUINONE IN PLANTS: BIOSYNTHESIS, PHYSIOLOGICAL FUNCTION AND METABOLIC ENGINEERING

Miaomiao Liu and Shanfa Lu

Institute of Medicinal Plant Development, Chinese Academy of Medical Sciences and Peking Union Medical College, Beijing, China

Plastoquinone (PQ) and ubiquinone (UQ) are two important prenylquinones, functioning as electron transporters in the electron transport chain of oxygenic photosynthesis and the aerobic respiratory chain, respectively, and play indispensable roles in plant growth and development through participating in the biosynthesis and metabolism of important chemical compounds, acting as antioxidants, being involved in plant response to stress, and regulating gene expression and cell signal transduction. UQ, particularly UQ_{10}, has also been widely used in people's life. It is effective in treating cardiovascular diseases, chronic gingivitis and periodontitis, and shows favorable impact on cancer treatment and human reproductive health. PQ and UQ are made up of an active benzoquinone ring attached to a polyisoprenoid side chain.

Biosynthesis of PQ and UQ is very complicated with more than thirty five enzymes involved. Their synthetic pathways can be generally divided into two stages. The first stage leads to the biosynthesis of precursors of benzene quinone ring and prenyl side chain. The benzene quinone ring for UQ is synthesized from tyrosine or phenylalanine, whereas the ring for PQ is derived from tyrosine. The prenyl side chains of PQ and UQ are derived from glyceraldehyde 3-phosphate and pyruvate through the 2-*C*-methyl-D-erythritol 4-phosphate pathway and/or acetyl-CoA and acetoacetyl-CoA through the mevalonate pathway. The second stage includes the condensation of ring and side chain and subsequent modification. Homogentisate solanesyltransferase, 4-hydroxybenzoate polyprenyl diphosphate transferase and a series of benzene quinone ring modification enzymes are involved in this stage. PQ exists in plants, while UQ widely presents in plants, animals and microbes. Many enzymes and their encoding genes involved in PQ and UQ biosynthesis have been intensively studied recently. Metabolic engineering of UQ_{10} in plants, such as rice and tobacco, has also been tested. In this review, we summarize and discuss recent research progresses in the biosynthetic pathways of PQ and UQ and enzymes and their encoding genes involved in side chain elongation and in the second stage of PQ and UQ biosynthesis. Physiological functions of PQ and UQ played in plants as well as the practical application and metabolic engineering of PQ and UQ are also included.

INTRODUCTION

Plastoquinone (PQ) and ubiquinone (UQ) are two important prenylquinones functioning as electron transporters in plants. They are involved in photophosphorylation and oxidative phosphorylation located in chloroplast thylakoids and mitochondrial inner membrane, respectively (Swiezewska, 2004). PQ and UQ are both made up of an active benzoquinone ring attached to a polyisoprenoid side chain. The length of polyisoprenoid side chain determines the type of PQ and UQ. Difference between PQ and UQ in chemical structure mainly lies in diverse substituent groups of benzoquinone ring (Figure 1). Plant PQ and UQ usually include nine or ten units of isoprenoid side chain. For instance, the main PQ and UQ in *Arabidopsis thaliana* have nine such units, known as PQ_9 and UQ_9, respectively. PQ and UQ are localized in different organelles of plant cells. PQ is located on the thylakoids of chloroplasts, while UQ is located on the inner membrane of mitochondria. The locations of PQ and UQ are consistent with their roles in photophosphorylation and oxidative phosphorylation. Lifespan of PQ and UQ is very short in cells. The half-time of PQ and UQ in spinach cells is about 15 and 30 h, respectively (Wanke et al., 2000). Therefore, to maintain the concentration stable and dynamic balance for normal plant photosynthesis and respiration, PQ and UQ need to be continuously synthesized in cells. In addition to the significance of PQ and UQ in plants, UQ_{10} has also been used in people's life. Significant progress has been made recently in PQ and UQ biosynthetic pathways and genes associated with PQ and UQ production. The biosynthesis and functions of UQ in *Escherichia coli, Saccharomyces cerevisiae* and animals (Clarke, 2000; Meganathan, 2001; Tran and Clarke, 2007; Bentinger et al., 2010; Aussel et al., 2014), UQ_{10} production in plants (Parmar et al., 2015) and metabolic engineering of UQ_{10} in microbes (de Dieu Ndikubwimana and Lee, 2014) have been reviewed. Here we mainly summarize and discuss recent advances in biosynthetic pathways, key enzymes and their encoding genes, physiological functions, and metabolic engineering of PQ and UQ in plants.

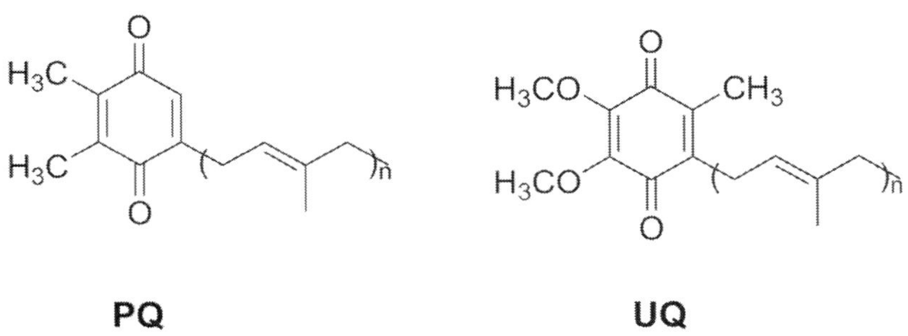

Figure 1. The chemical structures of PQ and UQ. PQ, plastoquinone; UQ, ubiquinone.

BIOSYNTHETIC PATHWAYS OF PQ AND UQ

The biosynthetic pathways of PQ and UQ can be generally divided into two stages (Figures 2 and 3). The first stage leads to the biosynthesis of precursors of benzene quinone ring and prenyl side chain. The second stage includes the condensation of ring and side chain and subsequent modification. The benzene quinone ring precursor for UQ is 4-hydroxybenzoic acid (4HB) synthesized from tyrosine or phenylalanine under the catalysis of various enzymes known as phenylalanine ammonia-lyase (PAL), cinnamic acid 4-hydroxylase (C4H), 4-coumarate CoA ligase (4CL), and other unknown enzymes. The benzene quinone ring precursor for PQ is homogentisic acid (HGA). It is synthesized from tyrosine under the catalysis of tyrosine aminotransferase (TAT) and 4-hydroxyphenylpyruvate reductase (HPPR). The prenyl side chains of PQ and UQ are derived from glyceraldehyde 3-phosphate (G3P) and pyruvate through the 2-*C*-methyl-D-erythritol 4-phosphate (MEP) pathway and/or acetyl-CoA and acetoacetyl-CoA through the mevalonate (MVA) pathway. The universal isoprene precursor isopentenyl diphosphate (IPP, C_5) and its isomer dimethylallyl diphosphate (DMAPP) synthesized through the MEP and MVA pathways are converted into intermediate diphosphate precursors, including geranyl diphosphate (GPP, C_{10}), farnesyl diphosphate (FPP, C_{15}), and geranylgeranyl diphosphate (GGPP, C_{20}). Enzymes catalyzing this conversion are a group of polyprenyl diphosphate synthases (PPSs), including geranyl diphosphate synthase (GPPS), farnesyl diphosphate synthase (FPPS), and geranylgeranyl diphosphate synthase (GGPPS) (Ma et al., 2012; Zhang and Lu, 2016). Both PQ and UQ share the prenyl side chain precursors. Side chain may be elongated to solanesyl diphosphate (SPP, C_{45}) and decaprenyl diphospate (DPS, C_{50}) under the catalysis of solanesyl diphosphate synthase (SPS) and decaprenyl diphosphate synthase (DPS), respectively. In the second stage, SPP is attached to HGA by homogentisate solanesyltransferase (HST) to produce intermediate 2-dimethyl-plastoquinone, which is then catalyzed by a methytransferase to form the end-product PQ_9 in plants. The condensation of PHB and the corresponding prenyl side chain is catalyzed by PHB polyprenyltransferase (PPT). It leads to the formation of 3-polyprenyl-4-hydroxybenzoate. After three methylations, three hydroxylations and one decarboxylation, ubiquinonol-n and ubiquinone-n are formed eventually. Enzymes catalyzing these modification steps are currently not well understood.

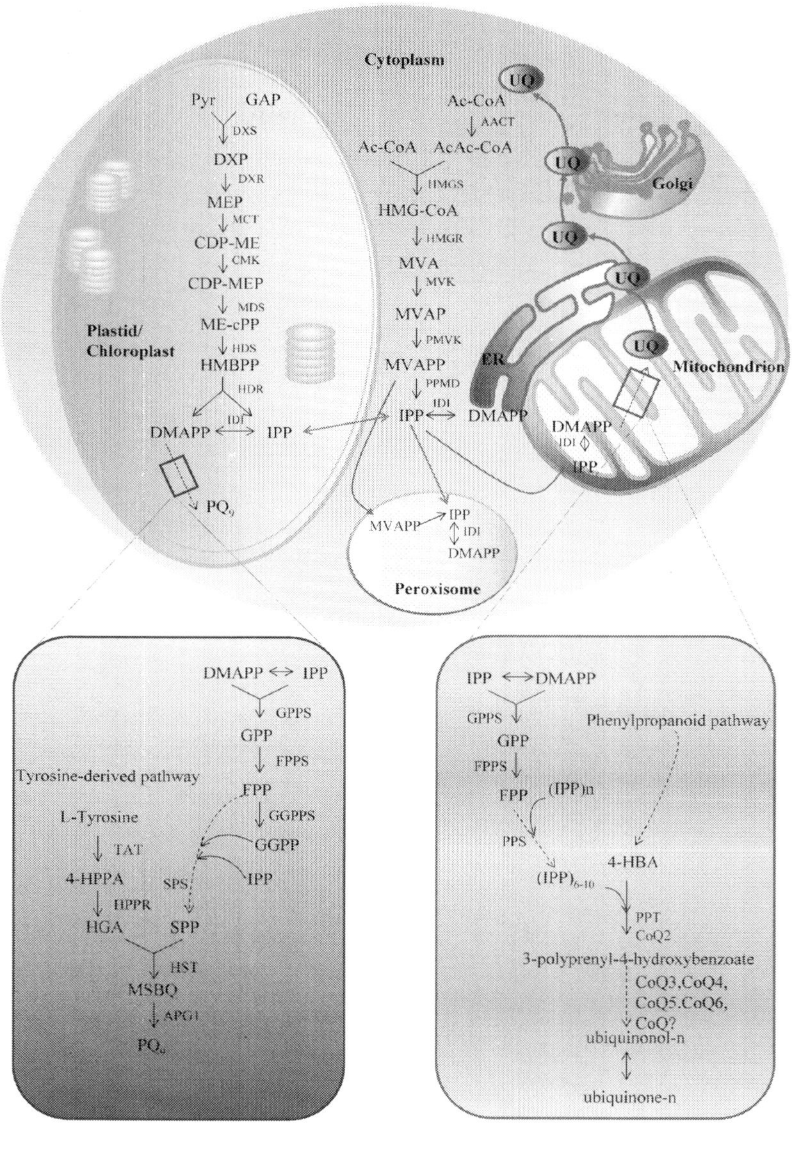

Figure 2. The biosynthetic pathways of PQ_9 **and UQ.** 4-HBA, 4-hydroxybenzoate acid; 4-HPPA, 4-hydroxyphenylpyruvic acid; AACT, acetoacetyl-CoA thiolase; AcAc-CoA, acetyloacetyl-coenzyme A; Ac-CoA, acetyl-coenzyme A; APG1, MSBQ/MPBQ methyltransferase; CDP-ME, 4-diphosphocytidyl-2C-methyl-D-erythritol; CDP-MEP, 4-diphosphocytidyl-2C-methyl-D-erythritol 2-phosphate; CMK,

4-diphosphocytidyl-2-*C*-methyl-D-erythritol kinase; Coq?, unknown enzymes; Coq3 and Coq5, *O*-methyltransferase and *C*-methyltransferase; Coq6, monooxygenase; DMAPP, dimethylallyl pyrophosphate; DMAPP, dimethylallyl pyrophosphate; DPS, decaprenyl diphosphate synthese; DXP, 1-deoxy-D-xylulose-5 phosphate; DXR, 1-deoxy-D-xylulose 5-phosphate reductoisomerase; DXS, 1-deoxy-D-xylulose-5-phosphate; FPP, farnesyl diphosphate; FPPS, farnesyl diphosphate synthase; GAP, D-glyceraldehyde 3-phosphate; GGPP, geranylgeranyl diphosphate; GGPPS, geranylgeranyl diphosphate synthase; GPP, geranyl diphosphate; GPPS, geranyl diphosphate synthase; HDR, 1-hydroxy-2-methyl-2-(E)-butenyl 4-diphosphate reductase; HDS, 1-hydroxy-2-methyl-2-(E)-butenyl 4-diphosphate synthase; HGA, homogentisate acid; HMBPP, 1-hydroxy-2-methyl-2-(E)-butenyl 4-diphosphate; HMG-CoA, 3-hydroxy-3-methylglutaryl-coenzyme A; HMGR, 3-hydroxy-3-methylglutaryl-CoA reductase; HMGR, hydroxymethylglutaryl-CoA reductase; HMGS, 3-hydroxy-3-methylglutaryl-CoA synthase; HMGS, hydroxymethylglutaryl-CoA synthetase; HPPR, 4-hydroxyphenylpyruvate reductase; HST, homogentisate solanesyltransferase; IDI, isopentenyl-diphosphate *delta*-isomerase; IPP, isopentenyl diphosphate isomerase; IPP, isopentenyl diphosphate; MCT, 2C-methyl-D-erythritol 4-phosphate cytidyl transferase; MDS, 2C-methyl-D-erythritol 2,4-cyclodiphosphate synthase; ME-cPP, 2C-methyl-D-erythritol 2,4-cyclodiphosphate; MEP, 2C-methyl-D-erythritol-4-phosphate; MSBQ, methyl-solanesylbenzoquinone; MVA, mevalonic acid; MVAP, mevalonate-5-phosphate; MVAPP, mevalonate-5-diphosphate; MVK, mevalonate kinase; PMVK, phosphomevalonate kinase; PPMD, diphosphomevalonate decarboxylase; PPS, diphosphate synthese; PPT/Coq2, 4-hydroxybenzoate polyprenyl transferase; Pyr, pyruvate; SPP, solanesyl diphosphate; SPS, solanesyl diphosphate synthase; TAT, tyrosine aminotransferase.

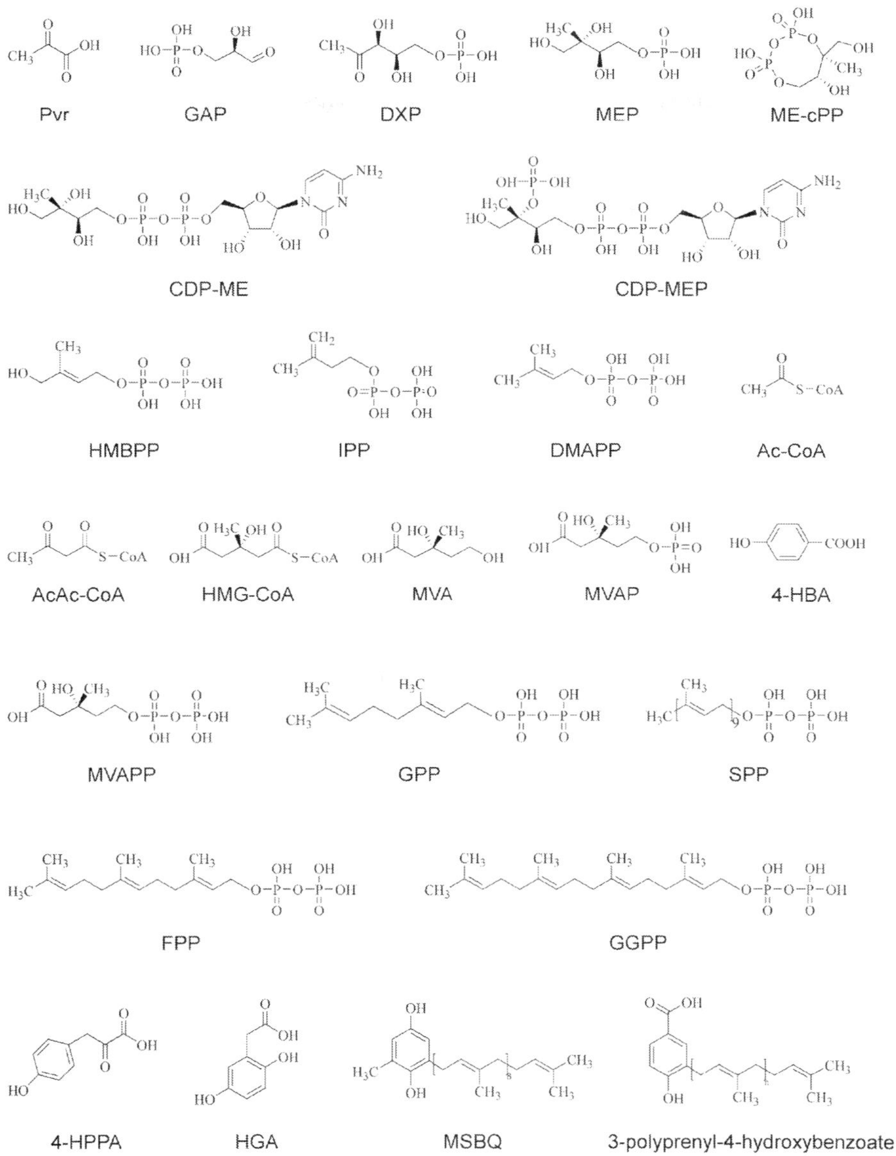

FIGURE 3. The chemical structures of various compounds in PQ and UQ biosynthetic pathways. 4-HBA, 4-hydroxybenzoate acid; 4-HPPA, 4-hydroxyphenylpyruvic acid; AcAc-CoA, acetyloacetyl-coenzyme A; Ac-CoA, acetyl-coenzyme A; CDP-ME, 4-diphosphocytidyl-2C-methyl-D-erythritol; CDP-MEP, 4-diphosphocytidyl-2C-methyl-D-erythritol 2-phosphate; DMAPP, dimethylallyl pyrophosphate; DMAPP,

dimethylallyl pyrophosphate; DXP, 1-deoxy-D-xylulose-5 phosphate; DXS, 1-deoxy-D-xylulose-5-phosphate; FPP, farnesyl diphosphate; GAP, D-glyceraldehyde 3-phosphate; GGPP, geranylgeranyl diphosphate; GPP, geranyl diphosphate; HGA, homogentisate acid; HMBPP, 1-hydroxy-2-methyl-2-(E)-butenyl 4-diphosphate; HMG-CoA, 3-hydroxy-3-methylglutaryl-coenzyme A; IPP, isopentenyl diphosphate; ME-cPP, 2C-methyl-D-erythritol 2,4-cyclodiphosphate; MEP, 2C-methyl-D-erythritol-4-phosphate; MSBQ, methyl-solanesyl-benzoquinone; MVA, mevalonic acid; MVAP, mevalonate-5-phosphate; MVAPP, mevalonate-5-diphosphate; Pyr, pyruvate; SPP, solanesyl diphosphate.

ENZYMES AND THEIR ENCODING GENES INVOLVED IN PQ AND UQ BIOSYNTHESIS

Recently, genes involved in the MEP and MVA pathways have been intensively studied. Chemical compounds synthesized through these pathways are precursors for many terpenoids and their derivatives. They are not specific to PQ and UQ production. Here, we are not going to summarize the advances in these genes. Instead, genes involved in side chain elongation and in the second stage of PQ and UQ biosynthesis are reviewed in detail (Table 1).

Table 1. Genes involved in side chain elongation and in the second stage of PQ and UQ biosynthesis in photosynthetic organisms.

Gene	Enzyme	Organism	Function	Reference
AtSPS1, AtSPS2	Solanesyl diphosphate synthase	Arabidopsis thaliana	PQ9	Hirooka et al., 2003, 2005; Jun et al., 2004; Block et al., 2013; Ksas et al., 2015
AtSPS3	Solanesyl diphosphate synthase	Arabidopsis thaliana	UQ9	Hsieh et al., 2011; Ducluzeau et al., 2012; Block et al., 2013
AtPPT1/ HRL1	4-hydroxybenzoate polyprenyltransferase	Arabidopsis thaliana	UQ9	Devadas et al., 2002; Okada et al., 2004; Dutta et al., 2015
AtCoQ3	O-methyltransferase	Arabidopsis thaliana	UQ9	Avelange-Macherel and Joyard, 1998
AtCoQ5	C-methyltransferase	Arabidopsis thaliana	UQ9	Hayashi et al., 2014

AtCOQ6	Monooxygenase	Arabidopsis thaliana	UQ	Hayashi et al., 2014
AtHST	Homogentisate solanesyltransferase	Arabidopsis thaliana	PQ9	Sadre et al., 2006; Tian et al., 2007; Chao et al., 2014
APG1/ VTE3	MSBQ/MPBQ methyltransferase	Arabidopsis thaliana	PQ9 and tocopherol	Cheng et al., 2003; Motohashi et al., 2003; Naqvi et al., 2011
BoCOQ5	C-methyltransferase	Brassica oleracea	UQ9	Zhou et al., 2009
BrCoQ3	O-methyltransferase	Brassica rapa	UQ	Wang et al., 2011
CrHST	Homogentisate solanesyltransferase	Chlamy-domonas reinhardtii	PQ9 and tocochr-omanol	Venkatesh et al., 2006
CrAPG1	MSBQ/MPBQ methyltransferase	Chlamy-domonas reinhardtii	PQ and tocopherol	Cheng et al., 2003
EsCoQ3	O-methyltransferase	Eutrema salsugineum	UQ	Yang et al., 2013
HbSDS	Solanesyl diphosphate synthase	Hevea brasiliensis	PQ9	Phatthiya et al., 2007
LsAPG1	MSBQ/MPBQ methyltransferase	Lactuca sativa	tocopherol	Tang et al., 2016
OsSPS1	Solanesyl diphosphate synthase	Oryza sativa	UQ9	Ohara et al., 2010; Block et al., 2013
OsSPS2, OsSPS3	Solanesyl diphosphate synthase	Oryza sativa	PQ9	Ohara et al., 2010; Block et al., 2013
OsPPT1a	4-hydroxybenzoate Polyprenyltransferase	Oryza sativa	UQ9	Ohara et al., 2006
SISPS	Solanesyl diphosphate synthase	Tomato	PQ9	Jones et al., 2013
SIDPS	Decaprenyl diphosphate synthase	Tomato	UQ10	Jones et al., 2013
VTE3	MPBQ methyltransferase	Peanut	methyl-transferase	Liu et al., 2013
APG1	MSBQ/MPBQ methyltransferase	Synechocystis PCC6803	PQ and tocopherol	Shintani et al., 2002
Slr0926	4-hydroxybenzote solanesyltransferase	Synechocystis PCC6803	PQ9	Sadre et al., 2012

Polyprenyl Diphosphate Synthase (PPS)

Polyprenyl diphosphate synthases, also known as prenyltransferases

(PTs) and isoprenyl diphosphate synthases (IDSs), are a group of enzymes widely distributed in organisms. They catalyze the formation of polyprenyl diphosphates with various chain lengths through consecutive condensation of IPP and are key enzymes involved in the biosynthesis of isoprenoid compounds, including monoterpenes, diterpenes, triterpenoids, carotenoids, natural rubber and many derivatives, such as PQ and UQ, vatamin E, prenylflavonoids, and shikonnin (Kellogg and Poulter, 1997).

Polyprenyl diphosphate synthases possess seven common conserved domains, I–VII, of which domain II is characterized with the first aspartate-rich motif (FARM), $DDX_{2-4}D$, while domain VI is characterized with the second aspartate-rich motif (SARM), DDXXD (Wang and Ohnuma, 1999; Phatthiya et al., 2007). These aspartate-rich motifs are involved in substrate binding and catalysis via chelating Mg^{2+}, a cofactor required for enzyme activity (Wang and Ohnuma, 1999). Based on the chain length of final products, PPSs can be divided into three subfamilies: short- (C_{15}–C_{25}), medium- (C_{30}–C_{35}) and long-chain (C_{40}–C_{50}) PTs (Hemmi et al., 2002). The most common PQ and UQ in plants have a C_{45} or C_{50} prenyl side chain moiety. For instance, *Oryza sativa* PQ_9 and UQ_9 contain a C_{45} prenyl side chain (Ohara et al., 2010), while cauliflower and pea UQ_{10} has a C_{50} prenyl side chain (Mattila and Kumpulainen, 2001). Thus, long chain PPSs, such as SPS catalyzing the formation of solanesyl diphosphate (SPP, C_{45}) and decaprenyl diphosphate synthase (DPS) involved in decaprenyl diphosphate (DPP, C_{50}) production, are significant to PQ and UQ biosynthesis in plants.

Database mining of fully sequenced genomes showed that *Chlamydomnas reinhardtii, Cyanidioschyzon merolae, Cucumis sativus, Vitis vinifera* and *Hordeum vulgare* contained two *SPSs, Physcomitrella patens, Arabidopsis, Glycine max, Oryza sativa* and *Zea mays* had three, and *Brachypodium distachyon*contained four (Block et al., 2013). Among the three *Arabidopsis SPS* genes, *AtSPS1* (*At1g78510*) and *AtSPS2* (*At1g17050*) are highly expressed in leaves with the level of *AtSPS1* transcripts higher than *AtSPS2*(Hirooka et al., 2003, 2005), whereas *AtSPS3* (*At2g34630*) is ubiquitously expressed with peaks in seeds and shoot apical meristems (Ducluzeau et al., 2012). Although AtSPS1 had been shown previously to be localized in the ER (Hirooka et al., 2003, 2005; Jun et al., 2004), recent analysis have clearly demonstrated that both AtSPS1 and AtSPS2 is targeted exclusively to plastids and contribute to the biosynthesis of PQ_9 (Block et al., 2013). Overexpression of *AtSPS1* resulted in the accumulation of PQ_9 and its derivative plastochromanol-8 (PC_8) (Ksas et al., 2015). Moreover, the enzymatic activity of AtSPS1 and AtSPS2 is stimulated by a member of the lipid-associated fibrillin protein family, fibrillin 5 (FNB5-B), which physically binds to the hydrophobic solanesyl moiety and helps to release the moiety from the enzymes in *Arabidopsis* cells (Kim et al., 2015). AtSPS3 had been shown previously to be targeted to plastids and to contribute to gibberellin biosynthesis (Bouvier et al., 2000). However, recent results suggest that it is actually dual targeted to mitochondria and plastids and appears very likely bifunctional (Ducluzeau et al., 2012). AtSPS3 is able to complement a yeast *coq1* knockout lacking mitochondrial hexaprenyl diphosphate synthase. Silence of *AtSPS3* using RNAi technology led to 75–80% reduction of the UQ pool size. *AtSPS3* overexpression resulted in a 40% increase in UQ content. No significant alternation of PQ levels was observed in *AtSPS3* silenced or overexpressing lines. Therefore, AtSPS3 seems to be the main contributor to SPS activity required for UQ_9 biosynthesis in *Arabidopsis* (Hsieh et al., 2011; Ducluzeau et al., 2012).

Similarly, three *SPS* genes exist in *O. sativa* (Ohara et al., 2010; Block et al., 2013). *OsSPS1* is highly expressed in roots, whereas *OsSPS2* is highly expressed in both leaves and roots. TargetP prediction and transient expression of GFP fusion protein showed the localization of OsSPS1 in mitochondria and OsSPS2 in plastids. Recombinant proteins of both OsSPS1 and OsSPS2 catalyzed the formation of solanesyl diphosphates. The enzyme activity of OsSPS1 was stronger than OsSPS2. OsSPS1 complemented the yeast *coq1* disruptant and produced UQ_9 in yeast cells, whereas OsSPS2 weakly complemented the growth defect of the *coq1* mutant (Ohara et al., 2010). The results suggest that OsSPS1 and OsSPS2 are involved in the supply of solanesyl diphosphate for UQ_9 production in mitochondria and PQ_9 biosynthesis in chloroplasts, respectively. *OsSPS3* is less studied compared with *OsSPS1* and *OsSPS2*. Since it shows high homology with OsSPS2 (Block et al., 2013), OsSPS3 appears to be also involved in PQ_9 formation.

Analysis of the fully sequenced tomato genomes showed that *Solanum lycopersicum* contained two long-chain *PPSs* genes (Block et al.,

2013). Jones et al. (2013) cloned and termed them *SlSPS* and *SlDPS*, respectively. *SlSPS* is targeted to plastids, whereas the fluorescence signal of SlDPS-GFP may resemble the mitochondrial localization reported for rice OsSPS1 (Ohara et al., 2010; Jones et al., 2013). In *E. coli*, SlSPS and SlDPS extend the side chain of endogenous UQ to nine and ten isoprene units, respectively (Jones et al., 2013). Overexpression of *SlSPS* elevated the content of PQ in immature tobacco leaves. Silence of *SlSPS* resulted in photobleached phenotype and accumulated phytoene. *SlSPS* and *SlDPS* could not complement silencing of each other. SlDPS can use GPP, FPP or GGPP in SPP and DPP biosynthesis. Silence of SlDPS did not affect leaf appearance, but impacted on primary metabolism (Jones et al., 2013). The roles of *SlSPS* and *SlDPS* in PQ and UQ biosynthesis need to be further investigated.

Although long chain PPSs play significant roles in PQ and UQ biosynthesis, the corresponding genes have only been identified from various plants, such as *Arabidopsis* (Hirooka et al., 2003, 2005; Jun et al., 2004; Ducluzeau et al., 2012; Block et al., 2013), rice (Ohara et al., 2010), tomato (Jones et al., 2013), and *Hevea brasiliensis* (Phatthiya et al., 2007). Greater efforts are required for molecular cloning and functional characterization of the genes in other plant species and utilization of plant source long chain *PPS*s in PQ and UQ production through biotechnology.

Homogentisate Solanesyltransferase (HST)

HSTs catalyze the condensation of HGA and SPP to form 2-demethylplastoquinol-9 leading to PQ_9biosynthesis (Sadre et al., 2010). Although HSTs are members of the homogentisate PT family, they differ from other homogentisate PT members, such as the homogentisate phytyltransferases (HPTs) isolated from *Synechocystis* (slr1736) and *Arabidopsis* (VTE2), in enzyme properties (Collakova and DellaPenna, 2001; Savidge et al., 2002). HPTs showed the highest enzyme activities with phytyl diphosphate, whereas HSTs had the highest activities with solanesyl diphosphate and were hardly active with phytyl diphosphate (Sadre et al., 2006).

Compared with other genes involved in the upstream of PQ biosynthesis, HSTs are less studied, although they play an important role in the biosynthesis of PQ_9, GAs and carotenoids, of which carotenoids are precursors of ABA and strigolactones. Genes encoding HSTs have only reported in *C. reinhardtii* and *Arabidopsis* (Sadre et al., 2006; Venkatesh et al., 2006; Tian et al., 2007; Sadre et al., 2010; Chao et al., 2014). The deduced *C. reinhardtii* CrHST and *Arabidopsis AtHST* proteins contain a chloroplast targeting sequence (Sadre et al., 2006; Tian et al., 2007). Confocal images of the AtHST-GFP fusion protein showed that AtHST protein was located in the chloroplast, most likely on the envelope membrane (Tian et al., 2007). Overexpression of *AtHST* caused modest elevation of PQ_9 levels (Sadre et al., 2006). The *pds2*mutant with an in-frame 6 bp deletion in *AtHST* showed an albino phenotype and the mutation can be functionally complemented by constitutive expression of *AtHST* (Norris et al., 1995; Tian et al., 2007). Similarly, T-DNA insertion mutant of *HST* gene displayed albino, dwarf and early flowering phenotypes with chloroplast development arrested, chlorophyll (Chl) absent and stomata closure defected (Chao et al., 2014). The GA and ABA levels were very low in the mutant. Exogenous GA could partially rescue the dwarf phenotype and the root development defects. Exogenous ABA could rescue the stomata closure defects (Chao et al., 2014). The expression levels of many genes involved in flowering time regulation and PQ, Chl, GA, ABA and carotenoid biosynthesis were changed in the mutant, suggesting the key roles of *AtHST*played in chloroplast development and plant hormone biosynthesis (Chao et al., 2014). *HST* genes in other plant species remain to be isolated and characterized.

4-Hydroxybenzoate Polyprenyl Diphosphate Transferase (PPT)

PPT is a rate-limiting enzyme in UQ biosynthesis. It catalyzes the condensation of benzoquinone ring and polyisoprenoid side chain to form the intermediate, 3-polyprenyl-4-hydroxybenzoate. All of the reported eukaryotic PPTs contain a mitochondrial targeting signal at N-terminal sequence. Most PPTs have a broad specificity for prenyl diphosphates, accepting substrates of different chain lengths, but show a high specificity for 4HB (Forsgren et al., 2004; Okada et al., 2004; Ohara et al., 2006). *PPT* genes have been identified from various organisms, such as *E. coli* (Siebert et al., 1992), *S. cerevisiae* (Ashby et al., 1992), *Schizosaccharomyces pombe* (Uchida et al., 2000), human (Forsgren et al., 2004), and two plant species, including *Arabidopsis* (Okada et al., 2004) and rice (Ohara et al., 2006). PPTs are members of the polyprenyl diphosphate transferase family. Although proteins in this family have sequence homology and relatively close phylogenetic relationship, only a small subset catalyzing the condensation of 4HB and prenyl chain are involved in UQ biosynthesis (Ohara et al., 2006). Other members may be involved in other metabolic pathways. For instance, *Cyanobacteria* hydroxybenzoate solanesyltransferase, Slr0926, which exhibits highly specific for 4-hybroxybenzoate and a broad specificity with regard to the prenyl donor substrate, including SPP and a number of shorter-chain prenyl diphosphates, is actually involved in PQ_9 biosynthesis (Sadre et al., 2012). *Lithospermum erythrorhizon*4HB geranyltransferases (*LePGT1* and *LePGT2*) with a strict specificity for GPP as a prenyl substrate are involved in shikonin biosynthesis and not relevant to UQ formation (Heide and Berger, 1989; Yazaki et al., 2002; Ohara et al., 2006, 2009).

Arabidopsis AtPPT1 is the first *PPT* gene identified in plants (Okada et al., 2004). It is predominantly expressed in the flower cluster. The deduced AtPPT1 protein is localized in mitochondria and can complement the yeast *coq2* disruptant. AtPPT1 has broad substrate specificity in terms of the prenyl donor. The T-DNA insertion mutant of *AtPPT1* shows arrest of embryo development at an early stage of zygotic embryogenesis (Okada et al., 2004). The other mutant, known as *hypersensitive response-like lesions 1* (*hrl1*), was identified in an ethyl methanesulfonate (EMS)-mutagenesis screen (Devadas et al., 2002). The *hrl1* mutant spontaneously develops HR-like lesions and shows enhanced resistance against bacterial pathogens (Devadas et al., 2002). Positional cloning and subsequent DNA sequencing showed that the mutant had a single base change in an exon of *AtPPT1*. The mutation results in a leucine to phenylalanine substitution at position 228. Leucine 228 is not a part of the active site of the enzyme but is conserved across PPT sequences from various organisms (Dutta et al., 2015). Overexpression of *HRL1* in *A. thaliana* leads to elevated UQ and decreased ubiquinol levels (Dutta et al., 2015).

The *O. sativa* genome contains three *PPT* genes, including *OsPPT1a, OsPPT1b* and *OsPPT1c*. However, only OsPPT1a was found to be expressed (Ohara et al., 2006). The deduced OsPPT1a proteins contain a putative mitochondrial sorting signal at the N-terminus. Consistently, GFP-PPT fusion proteins are mainly localized in mitochondria (Ohara et al., 2006). Same as AtPPT1 and other PPT proteins, OsPPT1a can complement the yeast *coq2* mutant, accepts prenyl diphosphates of various chain lengths as prenyl donors, and shows strict substrate specificity for the aromatic substrate 4HB as a prenyl acceptor (Ohara et al., 2006).

The Modification Enzyme of PQ Benzene Quinone Ring

2-demethylplastoquinol formed through the condensation of HGA and SPP by HSTs is a key intermediate in PQ synthesis. This intermediate can be converted to the final product PQ under the catalyzing of a modification enzyme, termed albino or pale green 1 (APG1) or methy-phytyl-benzoquinone (MPBQ)/methyl-solanesyl-benzoquinone (MSBQ) methyltransferase (Shintani et al., 2002; Motohashi et al., 2003; Cheng et al., 2003).

Arabidopsis APG1 was identified through characterization of the Ds-tagged albino or pale green mutant 1 (*apg1*) (Motohashi et al., 2003). This mutant lacks PQ, cannot survive beyond the seedling stage when germinated on soil, and contains decreased numbers of lamellae with reduced levels of chlorophyll (Motohashi et al., 2003). The insertion of Ds transposon disrupts a gene encoding a 37 kDa polypeptide precursor of the chloroplast inner envelope membrane. The 37 kDa protein had partial sequence similarity to the *S*-adenosylmethionine-dependent methyltransferase (Motohashi et al., 2003). Because of the lack of PQ in apg1 mutant and the putative methyltrasferase activity of the 37 kDa protein, APG1 appears to be involved in the methylation step of PQ biosynthesis (Motohashi et al., 2003). Almost at the same time, using a combined genomic, genetic and biochemical approach, Cheng et al. (2003) isolated and characterized the *Arabidopsis VTE3* (vitamin E defective) locus. This locus is actually identical to APG1. *In vitro* enzyme assays showed that *VTE3* was the plant functional equivalent of MPBQ/MSBQ methyltransferase from *Synechocystis* sp PCC6803 (Shintani et al., 2002).

In cyanobacterium, various genes involved in the PQ biosynthetic pathway and the vitamin E biosynthetic pathway are highly conserved. The biosynthesis of PQ and vitamin E share common precursor, HGA, which is prenylated with different substrates in the biosynthesis of different products. HGA is prenylated with phytyl diphosphate (PDP) or GGPP in tocochromanol formation, whereas it is prenylated with

SPP in PQ_9 biosynthesis. Thus, the MPBQ/MSBQ methyltransferase identified from *Synechocystis* sp PCC6803 has a dual function in the final methylation step of PQ and vitamin E biosynthesis (Shintani et al., 2002). Most genes involved in PQ and vitamin E synthesis are homologous, but MPBQ/MSBQ methyltransferase from *Synechocystis* sp PCC6803 and VTE3 from *Arabidopsis* are highly divergent in primary sequence (Cheng et al., 2003). The orthologs of *Synechocystis* MPBQ/MSBQ methyltransferase exist in *C. reinhardtii* and *Thalassiosira pseudonana* but absent from vascular and non-vascular plants. VTE3 orthologs exist in *C. reinhardtii* and vascular and non-vascular plants but absent from cyanobacteria (Cheng et al., 2003). It suggests that VTE3 is evolutionarily originated from archea rather than cyanobacteria. Interestingly, two mutants of *VTE3, vte3-1* and *vte3-2*, which show partial or total disruption of MPBQ/MSBQ methyltransferase activity, respectively, have different effects on PQ and vitamin E biosynthesis. *vte3-1* mainly impairs the methylation of tocopherol substrates. It has little effect on the methylation of MSBQ to PQ. On the contrary, *vte3-2* completely disrupts MPBQ/MSBQ methyltransferase (Shintani et al., 2002). The underlying mechanism remains to be elucidated.

Although genes encoding MPBQ/MSBQ methyltransferase are largely unknown in plant species other than *Arabidopsis*, engineering of *VTE3* has been tested (Van Eenennaam et al., 2003; Naqvi et al., 2011). Seed-specific expression of *Arabidopsis VTE3* in transgenic soybean reduced seed delta-tocopherol from 20 to 2%. Coexpression of *Arabidopsis VTE3* and *VTE4* (gamma-tocopherol methyltransferase gene) resulted in a greater than eightfold increase of alpha-tocopherol and an up to fivefold increase in vitamin E activity in transgenic soybean seeds (Van Eenennaam et al., 2003). Simultaneous expression of *Arabidopsis* -hydroxyphenylpyruvate dioxygenase and *VTE3* in transgenic corn kernels triples the tocopherol content (Naqvi et al., 2011). It is currently unknown whether it is possible to increase the content of PQ in plants through *VTE3* overexpression.

Modification Enzymes of UQ Benzene Quinone Ring

Compared with PQ benzene quinine ring modification, the process of UQ aromatic ring modification is complex. It includes three methylations (two *O*-methylations and one *C*-methylation), three hydroxylations, and one decarboxylation. UQ biosyntheses in prokaryotes and eukaryotes are similar. The difference lies in the reaction order of modifications. In eukaryotes, proposed modifications start with hydroxylation, followed by *O*-methylation, decarboxylation, two additional hydroxylations, *C*-methylation and one additional *O*-methylation (Tran and Clarke, 2007), whereas the reaction order of modifications in prokaryotes is decarboxylation, three hydroxylations, *O*-methylation, *C*-methylation and then an additional *O*-methylation (Meganathan, 2001).

Hydroxylation

The modification of UQ aromatic ring requires a total of three hydroxylations. The first hydroxylation of 3-polyprenyl-4-hydroxybenzoate in eukaryotes occurs before decarboxylation (Goewert et al., 1977). Two additional hydroxylations occur after decarboxylation in eukaryotes. Enzymes involved in hydroxylation are cytochrome P450 monooxygenases, known as COQ6 and COQ7 in yeast (Olson and Rudney, 1983). Genes encoding these enzymes have been identified from various organisms, such as *S. cerevisiae*(Marbois and Clarke, 1996; Kawamukai, 2000; Ozeir et al., 2011), rat (Jonassen et al., 1996), *Caenorhabditis elegans* (Ewbank et al., 1997), human (Vajo et al., 1999; Lu T.T. et al., 2013), and *E. coli*(Hajj Chehade et al., 2013). Homologs of the enzymes involved in UQ aromatic ring hydroxylations have also been found in various plants, such as *Arabidopsis* (Lange and Ghassemian, 2003) and alga (Blanc et al., 2010, 2012). *Arabidopsis* contains a COQ6 homolog, but lacks COQ7 (Hayashi et al., 2014). Although AtCOQ6 cannot complement the yeast *coq6* disruptant, addition of a mitochondrial targeting signal to AtCOQ6 will enable the production of UQ_{10} (Hayashi et al., 2014). Except AtCOQ6, other plant COQ6 and COQ7 homologs were identified based on genome sequence search and computational annotation, their roles in UQ biosynthesis need to be further confirmed using experimental approaches.

Methylation

The three methylations of UQ aromatic ring include two *O*-methylations and one *C*-methylation. The two *O*-methylation steps are catalyzed by COQ3, whereas the *C*-methylation step is catalyzed by COQ5 in yeast (Clarke et al., 1991; Barkovich et al., 1997; Poon et al., 1999; Hsu et al., 2000; Baba et al., 2004). COQ3 homologs have been identified in various organisms, such as *E. coli* (Hsu et al., 1996), yeast (Clarke et al., 1991; Poon et al., 1999; Hsu et al., 2000), rat (Marbois et al., 1994a,b) and human (Jonassen and Clarke, 2000). The gene encoding plant COQ3 was first identified in *Arabidopsis* (Avelange-Macherel and Joyard, 1998). *AtCOQ3* gene product is localized within mitochondrial membranes and can complement *Saccharomyces cerevisiae* or *Schizosaccharomyces pombe coq3* disruptant (Avelange-Macherel and Joyard, 1998; Hayashi et al., 2014). With the decoding of plant genomes, COQ3 has also been found in other plant species, such as *Brassica rapa* (Wang et al., 2011), *Eutrema salsugineum* (Yang et al., 2013). Similar to COQ3, COQ5 is associated with other COQ proteins on the inner mitochondrial membrane at the matrix side (Baba et al., 2004; Nguyen et al., 2014). *COQ5* genes have been identified in yeast (Barkovich et al., 1997; Baba et al., 2004), human (Chen et al., 2013; Nguyen et al., 2014), broccoli (Zhou et al., 2009) and *Arabidopsis* (Hayashi et al., 2014). *S. cerevisiae* COQ5 displays a typical class I *S*-adenosyl methionine-dependent methyltransferase crystal structure (Dai et al., 2014). The broccoli BoCOQ5 can complement a yeast *coq5* mutant and increases cellular UQ levels in bacteria (Zhou et al., 2009). Similarly, AtCOQ5 can complement the *Schizosaccharomyces pombe coq5* disruptant (Hayashi et al., 2014).

Decarboxylation

The proposed modifications of UQ aromatic ring include a decarboxylation step. It occurs before the three hydroxylation steps in prokaryotes, whereas in eukaryotes, decarboxylation occurs after a hydroxylation step and an *O*-methylation step. It has been shown that *ubiX* and *ubiD* genes are involved in the decarboxylation of UQ aromatic ring in bacteria (Meganathan, 2001; Gulmezian et al., 2007; Aussel et al., 2014). *ubiX* encodes a flavin mononucleotide (FMN)-binding protein with no decarboxylase activity detected *in vitro* (Gulmezian et al., 2007; Aussel et al., 2014), and UbiX proteins are metal-independent and require dimethylallyl-monophosphate as substrate (White et al., 2015). During the biosynthesis of UQ, UbiX acts as a flavin PT, producing a flavin-derived cofactor required for the decarboxylase activity of UbiD (Payne et al., 2015; White et al., 2015). *pad1* and *fdc1* are two fungal genes related to bacterial *ubiX*and *ubiD* (Mukai et al., 2010; Lin et al., 2015; Payne et al., 2015). Although Pad1 and Fdc1 proteins are homologous with UbiX and UbiD, respectively, and UbiX can activate Fdc1, they are not essential for UQ synthesis in yeast (Mukai et al., 2010; Lin et al., 2015). Thus, the decarboxylation step is largely unknown in eukaryotes. Feeding the isolated mitochondria from potato tubers with ^{14}C-labeled IPP, 4HB and *S*-adenosylmethionine shows the accumulation of methoxy-4-hydroxy-5-decaprenylbenzoate. It indicates the occurrence of decarboxylation after the first hydroxylation and subsequent *O*-methylation in plants; however, the immediate methoxy-4-hydroxy-5-decaprenylbenzoate cannot be detected in potato tubers (Lütke-Brinkhaus et al., 1984; Lütke-Brinkhaus and Kleinig, 1987), and no plant genes and enzymes involved in the decarboxylation of UQ aromatic ring have been identified.

In addition to the enzymes mentioned above, various other enzymes, such as yeast COQ4, COQ8, and COQ9, are potentially involved in UQ benzene quinone ring modification, although their actual functions are currently unknown (Tran and Clarke, 2007). Moreover, it seems that modification enzymes involved in modification of UQ benzene quinone ring form a multi-subunit complex. The interaction among subunits guarantees the normal function of enzymes.

PHYSIOLOGICAL FUNCTIONS OF PQ AND UQ IN PLANTS

Both PQ and UQ are functionally important electron transporters in plants. PQ is involved in the electron transport chain of oxygenic photosynthesis, whereas UQ works exclusively as an electron carrier in the aerobic respiratory chain (Cramer et al., 2011; de Dieu Ndikubwimana and Lee, 2014; Parmar et al., 2015). In addition to their basal functions in photophosphorylation and oxidative phosphorylation, PQ and UQ also play indispensable roles in plant growth and development through participating in the biosynthesis or metabolism of various important chemical compounds, acting as antioxidants, being involved in plant response to stress, and regulating gene expression and cell signal transduction.

Involved in Biosynthesis or Catabolism of Chemical Compounds

It has been shown that PQ and UQ are involved in the biosynthesis or metabolism of various important chemical compounds in plants. For instance, PQ participates in the biosynthesis of carotenoids (Norris et al., 1995), abscisic acid (ABA) (Rock and Zeevaart, 1991) and gibberellin (GA) (Nievelstein et al., 1995), whereas UQ is involved in branch-chain amino acid metabolism (Ishizaki et al., 2006; Araújo et al., 2010).

Carotenoids are C_{40} tetraterpenoids functioned as accessory light-harvesting pigments in photosynthetic tissues. In non-photosynthetic tissues, such as fruits and flowers, high levels of carotenoids often bring intense orange, yellow and red colors (Pfander, 1992). During carotenoid biosynthesis, the phytoene desaturation reaction is a rate-limiting step. A certain quinone/hydroquinone balance is necessary for optimal phytoene desaturation (Mayer et al., 1990). In an anaerobic environment, the oxidized quinones rather than reduced quinones are involved in the desaturation of phytoene (Mayer et al., 1990). *pds1* and *pds2* are two *Arabidopsis* mutants showing albino phenotype (Norris et al., 1995). The mutations affect phytoene desaturation and cause accumulation of phytoene, but they are not occurred in the phytoene desaturation enzyme. Analysis of *pds1* and *pds2* shows that *pds1* is 4-hydroxypheylpyruvate dioxygenase deficient (Norris et al., 1998), whereas *pds2* is deficient in HST, a critical enzyme involved in PQ biosynthesis (Tian et al., 2007). Both of the mutations lead to plastoquionone/tocopherol absence from different aspects in *Arabidopsis*, providing conclusive evidence that PQ is an essential component in phytoene desaturation (Nievelstein et al., 1995; Norris et al., 1995).

Since the plant hormone ABA is synthesized by oxidative cleavage of epoxy-carotenoids (Rock and Zeevaart, 1991), it is reasonable that PQ is also important for ABA biosynthesis. In a T-DNA insertion mutant of *HST* gene (*pds2-1*), not only PQ but also carotenoids, ABA and GA_3 levels are dramatically reduced (Chao et al., 2014). PQ works as the co-factor of phytoene desaturase and ζ-carotene desaturase and is the immediate electron acceptor in carotenoid and ABA biosynthesis (Chao et al., 2014). Disruption of *HST* gene results in PQ content decrease, which subsequently affects carotenoid and ABA biosynthesis. On the other hand, the biosynthesis of PQ, carotenoid, ABA and GA shares the common precursor, GGPP (Ma et al., 2012; Chao et al., 2014; Du et al., 2015; Zhang and Lu, 2016). In the *pds2-1* mutant, expression of GA biosynthesis genes, such as *GA1, GA2*, and *GA3*, is significantly down-regulated (Chao et al., 2014). Consistently, in the *Arabidopsis* phytoene desaturase gene mutant (*pds3*), gibberellin biosynthesis is impaired (Qin et al., 2007). It indicates that PQ may affect the biosynthesis of other chemical compounds through negative feedback regulation or other indirect mechanisms.

The relationship between UQ and amino acids lies in two aspects: (1) The precursors of UQ biosynthesis are derived from amino acids, including phenylalanine and tyrosine; and (2) UQ is involved in catabolism of some branched-chain amino acids in mitochondria. It is well known that mitochondrion is an integration point of cellular metabolism and signaling. Amino acids are not only metabolized in peroxisomes but also broken down in mitochondria, which provide carbon skeletons for biosynthesis of many important compounds, such as vitamins, amino acids, and lipids (Sweetlove et al., 2007). In *Arabidopsis*, leucine is catabolized to form isovaleryl-CoA in mitochondrial matrix, and then the intermediate of the leucine catabolic pathway, isovaleryl-CoA, is dehydrogenated to 3-methylcrotonyl-CoA by isovaleryl-CoA dehydrogenase (IVD). This process occurs on the matrix face of the inner mitochondrial membrane and an electron is transferred through the electron-transfer flavoprotein/electron-transfer falvoprotein:ubiquinone oxidoreductase (ETF/ETFQO) system, first to flavoprotein and then to flavoprotein:ubiquione oxidoreductase (Ishizaki et al., 2006; Araújo et al., 2010). UQ is the final acceptor of electrons in the decomposition of leucine. Similarly, electrons produced during the catabolism of lysine can also be channeled to the mitochondrial electron transport chain (Araújo et al., 2010).

Act as Antioxidants and Involved in Plant Response to Stress

Plastoquinone and ubiquinone can scavenge free radicals to prevent lipid peroxidation, protein oxidation and DNA damage in plant response to biotic and abiotic stresses. They exert antioxidant activity in the reduced forms, plastoquinol and uibiquinol, located in chloroplast thylakoid membrane and mitochondrial membrane, respectively. Analysis of the antioxidant effect of reduced PQ in isolated spinach thylakoid membranes showed that the reduced PQ acted as a scavenger of toxic oxygen species generated in the thylakoid membranes under strong illumination stress (Hundal et al., 1995). Reduced PQ inhibits lipid peroxidation and pigment bleaching, whereas oxidized PQ plays an opposite role (Hundal et al., 1995). In PQ-depleted spinach PSII membranes, exogenously added plastoquinol serves as an efficient scavenger of singlet oxygen (Yadav et al., 2010). Similarly, in *C. reinhardtii* cells, the level of reduced PQ markedly increased under high-light stress. When pyrazolate, an inhibitor of PQ and tocopherol biosynthesis, was added, the content of reduced PQ quickly decreases (Kruk et al., 2005). Further analyzing the turnover of plastoquinol showed that, due to scavenging of singlet oxygen, the reduced PQ underwent high turnover rate under high-light conditions (Kruk and Trebst, 2008). Moreover, the redox state of PQ pool was found to be an upstream master switch associated with programmed cell death in *Arabidopsis* leaves in response to excess excitation energy and may be play a central role in the light acclimation of diatoms (Mühlenbock et al., 2008; Lepetit et al., 2013). PSII photoinhibition occurred as a consequence of more reduced PQ pool (Darwish et al., 2015).

In addition to abiotic stress, PQ is also involved in plant response to biotic stress. When *Solanum nigrum*was treated with the pathogen *Phytophthora infestans*-derived elicitor, reactive oxygen species (ROS) production, lipid peroxidation and lipoxygenase activity were elevated (Maciejewska et al., 2002). These events were accompanied by a significant increase in PQ level. The increase of PQ level was more significant in plants growing in darkness than under continuous light. It suggests that PQ may be involved in maintaining a tightly controlled balance between the accumulation of ROS and antioxidant activity (Maciejewska et al., 2002). *Mesembryanthemum crystalinum* performs C_3 and CAM carbon metabolism. Analysis of *M. crystalinum* plants infected with pathogen *Botrytis cinere* showed that the redox state of PQ pool modifies plant response to biotic stress and hypersensitive-like response is accelerated when PQ pool is in the reduced state (Nosek et al., 2015).

Ubiquinone is an obligatory element of mitochondrial functions in both animals and plants. The antioxidant activity of UQ has been extensively characterized in animals (Littarru and Tiano, 2007, 2010). It prevents DNA damage and cell membrane lipid peroxidation through the elimination of ROS. Same as PQ, UQ also has two forms: the reduced type (ubiquinol) and the oxidized one (ubiquinone), of which ubiquinol is the form exerting antioxidant activity. Overexpression of yeast *coq2* (*p*-hydroxybenzoate poliprenyltransferase) in tobacco resulted in the increase of UQ in transgenic lines (Ohara et al., 2004). Transgenics with the higher UQ level showed the greater tolerance to oxidative stresses caused by methyl viologen or high salinity (Ohara et al., 2004). Analysis of the suspension-cultured *Chorispora bungeana*cells showed that the redox transition of UQ played key roles in adaptation of cellular regulations under chilling stress (Chang et al., 2006). In addition, the redox state of UQ determines the levels of ROS and plays a key regulatory role in *Arabidopsis* basal resistance against bacterial pathogens and in response to high oxidative stress environments (Dutta et al., 2015).

Actually, the PQ and UQ pools play a dual role: (1) reducing O_2 to superoxide by semiquinone; and (2) reducing superoxide to hydrogen peroxide by hydroquinone. In plant cells, the predominant ROS involved in plant defense includes superoxide and hydrogen peroxide, which are distributed in different pools. Moreover, ROS are generated in two ways. One is elicited by external stresses, such as environmental stresses and biotic stresses. The other way is produced through metabolic processes in the cells, such as the electron transport chains in mitochondria and chloroplasts (Mubarakshina and Ivanov, 2010; Tripathy and Oelmuller, 2012). An unknown interaction may be existed between different pools in modulation of ROS generation and plant response to stress.

Regulate Cell Signal Transduction and Gene Expression

Plastoquinone and ubiquinone may regulate cell signaling and gene expression indirectly through the generation of hydrogen peroxide, an important signaling molecule in plant resistance and cell metabolism. They can also directly regulate the expression of genes involved in cell metabolism. For instance, UQ_{10}influences the expression of hundreds of human genes involved in different cellular pathways. Among them, seventeen are functionally connected by signaling pathways of G-protein coupled receptor, JAK/STAT, integrin and β-arrestin (Schmelzer et al., 2007). These UQ_{10}-inducible genes possess a common promoter framework with binding domains of transcription factor families EVII, HOXF, HOXC and CLOX (Schmelzer et al., 2007). Moreover, UQ_{10}–mediated gene-gene network are involved in inflammation, cell differentiation, and peroxisome proliferator-activated receptor signaling (Schmelzer et al., 2011).

In plants, Chla/b-binding protein complex II (LHC II) and NADPH dehydrogenase complex are two important protein complexes in photosynthesis. The cytochrome b_6f deficient mutant of *lemna perpusilla*maintains a low level of the light-harvesting chl a/b-binding protein complex II (LHC II) at low-light intensities (Yang et al., 2001). Inhibiting the reduction of PQ pool increases the level of LHC II in the mutant at both low- and high-light intensities, whereas the level of LHC II is increased in wild-type plants only under high-light conditions (Yang et al., 2001). It suggests that the redox state of PQ is an important signal-transducing component in plant photoacclimation process (Yang et al., 2001). Analysis of gene expression using DNA microarray technology showed that 663 genes were differentially expressed in *A. thaliana* under low, medium, high and excessive irradiances, of which 50 genes were reverted by 3-(3,4-dichlorophenyl)-1,1-dimethylurea (DCMU), an inhibitor of the photosynthetic electron transport chain (Adamiec et al., 2008). It indicates that the expression of the 50 genes is regulated by the redox state of PQ pool. Hierarchical clustering and promoter motif analysis showed that the promoter regions of PQ-regulated genes contain conserved *cis*-acting elements involved in signal transduction from the redox state of the PQ pool (Adamiec et al., 2008). Active NADPH dehydrogenase complex is necessary for cyclic electron transport in photosystem I (PSI) and respiration. In glucose-treated cyanobacteria *Synechocystis*sp. strain PCC 6803 cells, NADPH dehydrogenase complex activity was inhibited and the cyclic PSI rate was decreased. In contrast, when the cells were treated with DCMU, the activity of NADPH dehydrogenase was significantly stimulated (Ma et al., 2008). Glucose treatment causes partial reduction of the PQ pool, whereas DCMU results in overoxidation. Differential responses of enzyme activity and cyclic PSI rate to glucose and DCMU treatments indicate that the redox state of PQ pool controls the NADPH dehydrogenase complex activity and further influences on cyclic PSI (Ma et al., 2008). In addition to PQ, UQ is also involved in signal transduction in plants. Comparative analysis of hypersensitive tobacco *Nicotiana tabacum L.* variety *Samsun NN* treated with UQ_{10} and TMV and those treated with TMV only showed that UQ_{10} and TMV-treated tobbaco had less number of lesions and TMV and greater change of plant hormone levels, including the decrease of ABA and increase of IAA level (Rozhnova and Gerashchenkov, 2006). It indicates that UQ_{10} has a protective antiviral effect through controlling plant hormonal status (Rozhnova and Gerashchenkov, 2008). On the other hand, it has been reported that the ROS level generated by UQ redox state is a threshold for successful basal resistance response in plants (Dutta et al., 2015). In plant defenses, ROS acts as signaling molecules directly or mediates the generation of phytoalexins or serves as a source for activation of further defenses indirectly (Kovtun et al., 2000; Thoma et al., 2003; Mur et al., 2008). Plants employ both pathogen-associated molecular pattern (PAMP)-triggered immunity (PTI) and effector-triggered immunity (ETI) in basal and R gene-mediated defense response. ROS is induced rapidly and transiently and then mediates signaling during PTI and ETI (Frederickson Matika and Loake, 2013). The induction of ROS is considered as a defining hallmark of identification and subsequent defense activation against pathogens (Torres et al., 2006; Dutta et al., 2015). Moreover, UQ is involved in the mitochondrial glycerol-3-phosphate shuttle for redox homeostasis in plants (Shen et al., 2007) and serves as mitochondrial permeability transition pore in cell metabolism (Amirsadeghi et al., 2007; Reape et al., 2007).

UTILIZATION AND METABOLIC ENGINEERING OF PQ AND UQ

Utilization of PQ and UQ

Plastoquinone is specific to plants. It has not been directly utilized for human. However, various synthesized PQ derivatives, such as SkQ1 (plastoquinonyl-decyl-triphenylphosphonium), SkQR1 (the rhodamine-containing analog of SkQ1) and SkQ3 (methylplastoquinonyl-decyl-triphenylphosphonium), were reported to show antioxidant and protonophore activity (Skulachev et al., 2011). They are able to penetrate cell membranes and potentially used in anti-aging treatment (Anisimov et al., 2008, 2011; Obukhova et al., 2009). SkQ1 is currently under clinical trials for glaucoma treatment in Russia (Iomdina et al., 2015). The phase 2 clinical trial indicates that SkQ1 is safe and efficacious in treating dry eye signs and symptoms (Petrov et al., 2016). In plants, SkQ1 and SkQ3 can retard the senescence of *Arabidopsis*rosette leaves and their death, increase the vegetative period, and improve crop structure of wheat (Dzyubinskaya et al., 2013). In addition, SkQ1 effectively suppresses the development of p50-induced PCD in tobacco plants through inhibiting ROS production (Solovieva et al., 2013; Samuilov and Kiselevsky, 2015). The role of SkQ1 and SkQ3 played in cells is mainly based on its antioxidant activity. Compared with PQ, the practical application of UQ, particularly UQ_{10}, has attracted more attention. $UQ_{10}$is effective in treating cardiovascular diseases, particularly in preventing and treating hypertension, hyperlipidemia, coronary artery disease and heart failure (Tran et al., 2001; Moludi et al., 2015). In the past decades, many studies have reported the remarkable clinical benefits of UQ_{10} and illuminated its antioxidant activity as the basis of pathology (Lee et al., 2012). Moreover, UQ_{10} controls energy metabolism and regulates cell death via redox signaling, indicating its potential in cancer treatment (Chai et al., 2010). Moderate UQ_{10} levels have favorable impact on breast cancer (Folkers et al., 1993; Lockwood et al., 1994; Premkumar et al., 2008). In addition, UQ_{10} can enhance viral immunity and affect the development of AIDS (Folkers et al., 1988, 1991). UQ_{10} is also closely related to human reproductive health. Moderate UQ_{10} supplement can effectively reduce the risk of spontaneous abortion and develop pre-eclampsia in pregnant women (Palan et al., 2004; Teran et al., 2009). Exogenous administration of UQ_{10} can increase sperm cell motility and the mean pregnancy rate. Its positive role in the treatment of male infertility also relies on the antioxidant properties and bioenergetics (Balercia et al., 2004; Mancini et al., 2005; Balercia et al., 2009; Mancini and Balercia, 2011; Safarinejad, 2012). As a good immunomdulator, UQ_{10} has been used to treat chronic gingivitis and periodontitis (Chatterjee et al., 2012; Hans et al., 2012). A series of toothpaste containing UQ_{10} are developed and sold in the market. Other UQ_{10}-containing chemical products include cleanser, cosmetic products, and healthy foods. Similar to PQ derivatives, the role of UQ played in cells relied on the antioxidant activity. As an antioxidant, UQ effectively scavenge ROS and prevent ROS-induced damage to membrane lipids, DNA, and proteins.

Metabolic Engineering of UQ

Plastoquinone plays significant roles in plants; however, it is not directly used in human life. Moreover, PQ derivatives are mainly synthesized by chemical methods. Metabolic engineering of PQ and its derivatives is rarely reported. Differing from PQ, metabolic engineering of UQ, particularly UQ_{10}, has been conducted in prokaryotes and eukaryotes, including bacterial, yeast and plants. Currently, the majority of commercially available UQ_{10} comes from yeast fermentation and chemical synthesis. Compared with microbial fermentation, chemical synthesis of UQ_{10} is more expensive and produces environmentally harmful waste products. Additionally, the scalability of both yeast fermentation and chemical synthesis is limited. Thus, plants are thought to be an attractive alternative source of UQ_{10}.

The natural UQ producers, such as *Agrobacterium tumefaciens, Paracoccus denitrificans, Rhodobacter sphaeroides*, and their chemical mutants have been successfully used for commercial production of UQ; while, with the increase of knowledge about enzymes involved in UQ biosynthesis and regulatory mechanisms modulating UQ production, opportunities have arisen for UQ metabolic engineering in other organisms. For instance, overexpression of some key genes, such as *ubiA* encoding *p*-hydroxybenzoate-polyprenyl pyrophosphate transferase, *ispB* encoding polyprenyl pyrophosphate synthetase and *ubiCA*, in *E. coli* may achieve the level of UQ content 3-4 times to that of wild-type cells (Zhu et al., 1995; Jiang et al., 2006). Even though, UQ production using these methods does not meet industrial needs, which require a yield of higher than 500 mg/L (Cluis et al., 2007). Therefore, in addition to the highly efficient microbial system, growth condition optimization and alteration of cellular regulatory mechanisms are important for UQ production (Sakato et al., 1992; Zhang et al., 2007a,b). Recently, multiple strategies have been employed in improving UQ production. One hundred and eighty percent increase of UQ_8 content is achieved in *E. coli* (*menA*) through a comprehensive approach, including blocking menaquinone pathway, coexpressing *dxs-ubiA*, and supplementing PYR and pHBA (Xu et al., 2014). The highest UQ10 titer and yield, 433 mg/L, is obtained in engineered *E. coli* through integrating *dps* into chromosome of *E. coli* ATCC8739, modulating *dxs* and *idi* genes of the MEP pathway and *ubiCA* genes, and recruiting the glucose facilitator protein of *Zymomonas mobilis* was to replace native phosphoenolpyruvate: carbohydrate phosphotransferase systems (PTS) (Dai et al., 2015).

Metabolic engineering of UQ in plants mainly concentrated on UQ_{10} production. Although UQ widely exists in plant cells, most cereal crops produce mainly UQ_9. Tomato, *Datura tatula* and tobacco BY-2 cells can produce UQ_{10} naturally; however, its yield is very limited (Ikeda and Kagei, 1979; Ikeda et al., 1981; Matsumoto et al., 1981). It has been shown that overexpression of rate-limiting genes and increase of UQ precursors can improve UQ production in plants. Expression of *ddsA* from *Gluconobacter suboxydans* in rice leads to efficient production of UQ_{10} in rice seeds (Sakiko et al., 2006, 2010). Since PPT plays the catalyzing role in a rate-limiting step of the UQ biosynthesis pathway, it is a significant target for UQ metabolic engineering in plants (Ohara et al., 2004; Stiff, 2010; Parmar et al., 2015). For instance, expression of yeast *coq2* gene resulted in a sixfold increase of UQ_{10} levels in transgenic tobacco plants (Ohara et al., 2004). Compared with wild type plants, *coq2* transgenic tobacco with high UQ_{10} level are more resistant to oxidative stresses caused by methyl viologen or high salinity (Ohara et al., 2004). Similarly, overexpression of *AtPPT1* in tobacco increases UQ_{10} content and enhances oxidative stress tolerance caused by high NaCl (Stiff, 2010). Increase of UQ precursors, such as 4HB and/or PPS, may potentially improve UQ production in plants (Sommer and Heide, 1998; Viitanen et al., 2004). However, due to the complex relationship among precursors, UQ production and many other intersecting metabolic pathways, the expected goal of improving UQ production is difficult to achieve. Further improvement of UQ content in plant cells may be expected using comprehensive approaches (Kumar et al., 2012), such as improving the amount of UQ precursors combined with overexpression of rate-limiting genes.

CONCLUSION AND PERSPECTIVES

Plastoquinone and ubiquinone are two important compounds in plants. They function as electron transporters in the electron transport chain of oxygenic photosynthesis and the aerobic respiratory chain, respectively, and play indispensable roles in plant growth and development. UQ, particularly UQ_{10}, has also been widely used in people's life. Great efforts have been done to elucidate their biosynthetic pathways and genes associated with PQ and UQ production. As shown in Figure 2 and Table 1, significant achievements have been made. However, there are still several issues related to the biosynthetic pathways, regulatory mechanisms and metabolic engineering need to be addressed.

Although a great amount of studies have been done to the MEP and MVA pathways and various enzymes, such as PPT and HST, involved in the attachment of isoprenoid side chain to the benzoquinone ring, pathways and enzymes involved in isoprenoid side chain elongation, UQ benzoquinone ring biosynthesis, and PQ and UQ benoquinone ring modification are largely unknown. The key enzyme responsible for 4HB production has not been identified (Block et al., 2014). PPSs are a group of enzymes converting IPP and DMAPP to diphosphate precursors. Each PPS may play different role in isoprenoid side chain elongation, which needs to be clarified. Although various enzymes involved in benoqunone ring modifications have been identified in bacteria and yeast, few achievements have been made in plants. With more and more transcriptome and whole genome sequence available, gene network reconstruction becomes possible and can be used to address these problems.

Non-coding RNAs, including small RNAs and long non-coding RNAs, play significant regulatory roles in many aspects of plants (Lu et al., 2005, 2007, 2008; Lu S. et al., 2013; Li D. et al., 2015; Wang et al., 2015). Various microRNAs have been identified to be associated with secondary metabolism (Wu et al., 2012; Lu S. et al., 2013; Fan et al., 2015; Wei et al., 2015). However, the regulatory roles of non-coding RNAs in PQ and UQ biosynthesis have not been revealed. In addition to non-coding RNAs, transcription factors, such as MYB, WRKY and SPL, potentially play significant regulatory roles in PQ and UQ biosynthesis (Li and Lu, 2014a,b; Zhang et al., 2014; Li C. et al., 2015), which needs to be further demonstrated. Understanding the regulatory mechanisms of PQ and UQ are important for manipulating the content of PQ and UQ in plants.

Metabolic engineering of UQ_{10} has been successfully performed in bacteria and yeast. However, they were found to be low yield and high production cost. UQ_{10} metabolic engineering in plants has various advantages and great perspectives, whereas current efforts are limited to a few plant species (Parmar et al., 2015). Increasing UQ_{10} content in UQ_{10}-producing plant species and engineering UQ_{10} in non-UQ_{10}-producing plant species are two routes for UQ_{10} metabolic engineering in plants. With more and more genes involved in UQ_{10} biosynthesis and regulation to be identified, great achievements may be expected for UQ_{10} production in plants.

AUTHOR CONTRIBUTIONS

Both authors listed, have made substantial, direct and intellectual contribution to the work, and approved it for publication.

CONFLICT OF INTEREST STATEMENT

The authors declare that the research was conducted in the absence of any commercial or financial relationships that could be construed as a potential conflict of interest.

ACKNOWLEDGMENTS

This work was supported by the National Key Research and Development Program of China (2016YFD0600104); the Natural Science Foundation of China (31370327, 31570667); the Beijing Natural Science Foundation (5152021); the CAMS Innovation Fund for Medical Sciences (CIFMS) (2016-I2M-3-016); and the Fund for Postgraduate Innovation in Peking Union Medical College (2015-1007-14).

REFERENCES

1. Adamiec, M., Drath, M., and Jackowski, G. (2008). Redox state of plastoquionone pool regulates expression of *Arabidopsis thaliana*genes in response to elevated irradiance. *Acta Biochim. Pol.* 55, 161–173.
2. Amirsadeghi, S., Robson, C. A., and Vanlerberghe, G. C. (2007). The role of the mitochondrion in plant responses to biotic stress. *Physiol. Plant.* 129, 253–266. doi: 10.1111/j.1399-3054.2006.00775.x
3. Anisimov, V. N., Bakeeva, L. E., Egormin, P. A., Filenko, O. F., Isakova, E. F., Manskikh, V. N., et al. (2008). Mitochondria-targeted plastoquinone derivatives as tools to interrupt execution of the aging

program. 5. SkQ1 prolongs lifespan and prevents development of traits of senescence. *Biochemistry* 73, 1329–1342. doi: 10.1134/S0006297908120055

4. Anisimov, V. N., Egorov, M. V., Krasilshchikova, M. S., Lyamzaev, K. G., Manskikh, V. N., Moshkin, M. P., et al. (2011). Effects of the mitochondria-targeted antioxidant SkQ1 on lifespan of rodents. *Aging* 3, 1110–1119. doi: 10.18632/aging.1000404
5. Araújo, W. L., Ishizaki, K., Nunes-Nesi, A., Larson, T. R., Tohge, T., Krahnert, I., et al. (2010). Identification of the 2-hydroxyglutarate and isovaleryl-CoA dehydrogenases as alternative electron donors linking lysine catabolism to the electron transport chain of *Arabidopsis* mitochondria. *Plant Cell* 22, 1549–1563. doi: 10.1105/tpc.110.075630
6. Ashby, M. N., Kutsunai, S. Y., Ackerman, S., Tzagoloff, A., and Edwards, P. A. (1992). COQ2 is a candidate for the structural gene encoding para-hydroxybenzoate: polyprenyltransferase. *J. Biol. Chem.* 267, 4128–4136.
7. Aussel, L., Pierrel, F., Loiseau, L., Lombard, M., Fontecave, M., and Barras, F. (2014). Biosynthesis and physiology of coenzyme Q in bacteria. *Biochim. Biophys. Acta* 1837, 1004–1011. doi: 10.1016/j.bbabio.2014.01.015
8. Avelange-Macherel, M. H., and Joyard, J. (1998). Cloning and functional expression of AtCOQ3, the *Arabidopsis* homologue of the yeast COQ3 gene, encoding a methyltransferase from plant mitochondria involved in ubiquinone biosynthesis. *Plant J.* 14, 203–213. doi: 10.1046/j.1365-313X.1998.00109.x
9. Baba, S. W., Belogrudov, G. I., Lee, J. C., Lee, P. T., Strahan, J., Shepherd, J. N., et al. (2004). Yeast Coq5 C-methyltransferase is required for stability of other polypeptides involved in coenzyme Q biosynthesis. *J. Biol. Chem.* 279, 10052–10059. doi: 10.1074/jbc.M313712200
10. Balercia, G., Buldreghini, E., Vignini, A., Tiano, L., Paggi, F., Amoroso, S., et al. (2009). Coenzyme Q10 treatment in infertile men with idiopathic asthenozoospermia: a placebo-controlled, double-blind randomized trial. *Fertil. Steril.* 91, 1785–1792. doi: 10.1016/j.fertnstert.2008.02.119

11. Balercia, G., Mosca, F., Mantero, F., Boscaro, M., Mancini, A., Ricciardo-Lamonica, G., et al. (2004). Coenzyme Q(10) supplementation in infertile men with idiopathic asthenozoospermia: an open, uncontrolled pilot study. *Fertil. Steril.* 81, 93–98. doi: 10.1016/j.fertnstert.2003.05.009

12. Barkovich, R. J., Shtanko, A., Shepherd, J. A., Lee, P. T., Myles, D. C., Tzagoloff, A., et al. (1997). Characterization of the COQ5 gene from *Saccharomyces cerevisiae*. Evidence for a C-methyltransferase in ubiquinone biosynthesis. *J. Biol. Chem.* 272, 9182–9188. doi: 10.1074/jbc.272.14.9182

13. Bentinger, M., Tekle, M., and Dallner, G. (2010). Coenzyme Q–biosynthesis and functions. *Biochem. Biophys. Res. Commun.* 396, 74–79. doi: 10.1016/j.bbrc.2010.02.147

14. Blanc, G., Agarkova, I., Grimwood, J., Kuo, A., Brueggeman, A., Dunigan, D. D., et al. (2012). The genome of the polar eukaryotic microalga *Coccomyxa subellipsoidea* reveals traits of cold adaptation. *Genome Biol.* 13:R39. doi: 10.1186/gb-2012-13-5-r39

15. Blanc, G., Duncan, G., Agarakova, I., Borodovsky, M., Gumon, J., Kuo, A., et al. (2010). The Chlorella variabilis NC64A genome reveals adaptation to photosymbiosis, coevolution with viruses, and cryptic sex. *Plant Cell* 22, 2943–2955. doi: 10.1105/tpc.110.076406

16. Block, A., Fristedt, R., Rogers, S., Kumar, J., Barnes, B., Barnes, J., et al. (2013). Functional modeling identifies paralogous solanesyl-diphosphate synthases that assemble the side chain of plastoquinone-9 in plastids. *J. Biol. Chem.* 288, 27594–27606. doi: 10.1074/jbc.M113.492769

17. Block, A., Widhalm, J. R., Fatihi, A., Cahoon, R. E., Wamboldt, Y., Elowsky, C., et al. (2014). The origin and biosynthesis of the benzenoid moiety of ubiquinone (Coenzyme Q) in *Arabidopsis*. *Plant Cell* 26, 1938–1948. doi: 10.1105/tpc.114.125807

18. Bouvier, F., Suire, C., d'Harlingue, A., Backhaus, R. A., and Camara, B. (2000). Molecular cloning of geranyl diphosphate synthase and compartmentation of monoterpene synthesis in plant cells. *Plant J.* 24, 241–252. doi: 10.1046/j.1365-313x.2000.00875.x

19. Chai, W., Cooney, R. V., Franke, A. A., Shvetsov, Y. B., Caberto, C. P., Wilkens, L. R., et al. (2010). Plasma coenzyme Q10 levels

and postmenopausal breast cancer risk: the multiethnic cohort study. *Cancer Epidemiol. Biomarkers Prev.* 19, 2351–2356. doi: 10.1158/1055-9965.EPI-10-0396

20. Chang, J., Fu, X., An, L., and Chen, T. (2006). Properties of cellular ubiquinone and stress-resistance in suspension-cultured cells of *Chorispora bungeana* during early chilling. *Environ. Exp. Bot.* 57, 116–122. doi: 10.1016/j.envexpbot.2005.05.001
21. Chao, Y., Kang, J., Zhang, T., Yang, Q., Gruber, M. Y., and Sun, Y. (2014). Disruption of the homogentisate solanesyltransferase gene results in albino and dwarf phenotypes and root, trichome and stomata defects in *Arabidopsis thaliana*. *PLoS ONE* 9:e94031. doi: 10.1371/journal.pone.0094031
22. Chatterjee, A., Kandwal, A., Singh, N., and Singh, A. (2012). Evaluation of Co-Q10 anti-gingivitis effect on plaque induced gingivitis: a randomized controlled clinical trial. *J. Indian Soc. Periodontol.* 16, 539–542. doi: 10.4103/0972-124X.106902
23. Chen, S. W., Liu, C. C., and Yen, H. C. (2013). Detection of suppressed maturation of the human COQ5 protein in the mitochondria following mitochondrial uncoupling by an antibody recognizing both precursor and mature forms of COQ5. *Mitochondrion* 13, 143–152. doi: 10.1016/j.mito.2013.01.007
24. Cheng, Z., Sattler, S., Maeda, H., Sakuragi, Y., Bryant, D. A., and DellaPenna, D. (2003). Highly divergent methyltransferases catalyze a conserved reaction in tocopherol and plastoquinone synthesis in cyanobacteria and photosynthetic eukaryotes. *Plant Cell* 15, 2343–2356. doi: 10.1015/tpc.013656
25. Clarke, C. F. (2000). New advances in coenzyme Q biosynthesis. *Protoplasma* 213, 134–147. doi: 10.1007/BF01282151
26. Clarke, C. F., Williams, W., and Teruya, J. H. (1991). Ubiquinone biosynthesis in *Saccharomyces cerevisiae*. Isolation and sequence of COQ3, the 3,4-dihydroxy-5-hexaprenylbenzoate methyltransferase gene. *J. Biol. Chem.* 266, 16636–16644.
27. Cluis, C. P., Burja, A. M., and Martin, V. J. (2007). Current prospects for the production of coenzyme Q10 in microbes. *Trends Biotechnol.* 25, 514–521. doi: 10.1016/j.tibtech.2007.08.008
28. Collakova, E., and DellaPenna, D. (2001). Isolation and functional

analysis of homogentisate phytyltransferase from *Synechocystis* sp. PCC 6803 and *Arabidopsis*. *Plant Physiol.* 127, 1113–1124. doi: 10.1104/pp.127.3.1113

29. Cramer, W. A., Hasan, S. S., and Yamashita, E. (2011). The Q cycle of cytochrome bc complexes: a structure perspective. *Biochim. Biophys. Acta* 1807, 788–802. doi: 10.1016/j.bbabio.2011.02.006
30. Dai, D., Miao, L., Sun, T., Li, Q., Xiao, D., and Zhang, X. (2015). Production of coenzyme Q(10) by metabolically engineered *Escherichia coli*. *Chin. J. Biotechnol.* 31, 206–219. doi: 10.13345/j.cjb.140211
31. Dai, Y. N., Zhou, K., Cao, D. D., Jiang, Y. L., Meng, F., Chi, C. B., et al. (2014). Crystal structures and catalytic mechanism of the C-methyltransferase Coq5 provide insights into a key step of the yeast coenzyme Q synthesis pathway. *Acta Crystallogr. D Biol. Crystallogr.* 70(Pt 8), 2085–2092. doi: 10.1107/S1399004714011559
32. Darwish, M., Vidal, V., Lopez-Lauri, F., Alnaser, O., Junglee, S., El Maataoui, M., et al. (2015). Tolerance to clomazone herbicide is linked to the state of LHC, PQ-pool and ROS detoxification in tobacco (*Nicotiana tabacum* L.). *J. Plant Physiol.* 175, 122–130. doi: 10.1016/j.jplph.2014.11.009
33. de Dieu Ndikubwimana, J., and Lee, B. H. (2014). Enhanced production techniques, properties and uses of coenzyme Q10. *Biotechnol. Lett.* 36, 1917–1926. doi: 10.1007/s10529-014-1587-1
34. Devadas, S. K., Enyedi, A., and Raina, R. (2002). The *Arabidopsis* hrl1 mutation reveals novel overlapping roles for salicylic acid, jasmonic acid and ethylene signalling in cell death and defence against pathogens. *Plant J.* 30, 467–480. doi: 10.1046/j.1365-313X.2002.01300.x
35. Du, Q., Li, C., Li, D., and Lu, S. (2015). Genome-wide analysis, molecular cloning and expression profiling reveal tissue-specifically expressed, feedback-regulated, stress-responsive and alternatively spliced novel genes involved in gibberellin metabolism in *Salvia miltiorrhiza*. *BMC Genomics* 16:1087. doi: 10.1186/s12864-015-2315-5
36. Ducluzeau, A. L., Wamboldt, Y., Elowsky, C. G., Mackenzie,

S. A., Schuurink, R. C., and Basset, G. J. (2012). Gene network reconstruction identifies the authentic trans-prenyl diphosphate synthase that makes the solanesyl moiety of ubiquinone-9 in *Arabidopsis*. *Plant J.* 69, 366–375. doi: 10.1111/j.1365-313X.2011.04796.x

37. Dutta, A., Chan, S. H., Pauli, N. T., and Raina, R. (2015). HYPERSENSITIVE RESPONSE-LIKE LESIONS 1 codes for AtPPT1 and regulates accumulation of ROS and defense against bacterial pathogen *Pseudomonas syringae* in *Arabidopsis thaliana*. *Antioxid. Redox Signal.* 22, 785–796. doi: 10.1089/ars.2014.5963

38. Dzyubinskaya, E. V., Ionenko, I. F., Kiselevsky, D. B., Samuilov, V. D., and Samuilov, F. D. (2013). Mitochondria-addressed cations decelerate the leaf senescence and death in *Arabidopsis thaliana* and increase the vegetative period and improve crop structure of the wheat *Triticum aestivum*. *Biochemistry* 78, 68–74. doi: 10.1134/S0006297913010082

39. Ewbank, J. J., Barnes, T. M., Lakowski, B., Lussier, M., Bussey, H., and Hekimi, S. (1997). Structural and functional conservation of the *Caenorhabditis elegans* timing gene clk-1. *Science* 275, 980–983. doi: 10.1126/science.275.5302.980

40. Fan, R., Li, Y., Li, C., and Zhang, Y. (2015). Differential microRNA analysis of glandular trichomes and young leaves in *Xanthium strumarium* L. reveals their putative roles in regulating terpenoid biosynthesis. *PLoS ONE* 10:e0139002. doi: 10.1371/journal.pone.0139002

41. Folkers, K., Brown, R., Judy, W. V., and Morita, M. (1993). Survival of cancer patients on therapy with coenzyme Q10. *Biochem. Biophys. Res. Commun.* 192, 241–245. doi: 10.1006/bbrc.1993.1405

42. Folkers, K., Hanioka, T., Xia, L. J., McRee, J. T. Jr., and Langsjoen, P. (1991). Coenzyme Q10 increases T4/T8 ratios of lymphocytes in ordinary subjects and relevance to patients having the AIDS related complex. *Biochem. Biophys. Res. Commun.* 176, 786–791. doi: 10.1016/S0006-291X(05)80254-2

43. Folkers, K., Langsjoen, P., Nara, Y., Muratsu, K., Komorowski, J., Richardson, P. C., et al. (1988). Biochemical deficiencies of coenzyme

Q10 in HIV-infection and exploratory treatment. *Biochem. Biophys. Res. Commun.* 153, 888–896. doi: 10.1016/S0006-291X(88)81179-3

44. Forsgren, M., Attersand, A., Lake, S., Grünler, J., Swiezewska, E., Dallner, G., et al. (2004). Isolation and functional expression of human COQ2, a gene encoding a polyprenyl transferase involved in the synthesis of CoQ. *Biochem. J.* 382, 519–526. doi: 10.1042/BJ20040261
45. Frederickson Matika, D. E., and Loake, G. (2013). Redox regulation in plant immune function. *Antioxid. Redox Signal.* 21, 1373–1388. doi: 10.1089/ars.2013.5679
46. Goewert, R. R., Sippel, C. J., and Olson, R. E. (1977). The isolation and identification of a novel intermediate in ubiquinone-6 biosynthesis by *Saccharomyces cerevisiae*. *Biochem. Biophys. Res. Commun.* 77, 599–605. doi: 10.1021/bi00517a041
47. Gulmezian, M., Hyman, K. R., Marbois, B. N., Clarke, C. F., and Javor, G. T. (2007). The role of UbiX in *Escherichia coli* coenzyme Q biosynthesis. *Arch. Biochem. Biophys.* 467, 144–153. doi: 10.1016/j.abb.2007.08.009
48. Hajj Chehade, M., Loiseau, L., Lombard, M., Pecqueur, L., Ismail, A., Smadja, M., et al. (2013). ubiI, a new gene in *Escherichia coli*coenzyme Q biosynthesis, is involved in aerobic C5-hydroxylation. *J. Biol. Chem.* 288, 20085–20092. doi: 10.1074/jbc.M113.480368
49. Hans, M., Prakash, S., and Gupta, S. (2012). Clinical evaluation of topical application of perio-Q gel (Coenzyme Q10) in chronic periodontitis patients. *J. Indian Soc. Periodontol.* 16, 193–199. doi: 10.4103/0972-124X.99261
50. Hayashi, K., Ogiyama, Y., Yokomi, K., Nakagawa, T., Kaino, T., and Kawamukai, M. (2014). Functional conservation of coenzyme Q biosynthetic genes among yeasts, plants, and humans. *PLoS ONE* 9:e99038. doi: 10.1371/journal.pone.0099038
51. Heide, L., and Berger, U. (1989). Partial purification and properties of geranyl pyrophosphate synthase from *Lithospermum erythrorhizon* cell cultures. *Arch. Biochem. Biophys.* 273, 331–338. doi: 10.1016/0003-9861(89)90491-8
52. Hemmi, H., Ikejiri, S., Yamashita, S., and Nishino, T. (2002).

Novel medium-chain prenyl diphosphate synthase from the thermoacidophilic archaeon Sulfolobus solfataricus. *J. Bacteriol.* 184, 615–620. doi: 10.1128/JB.184.3.615-620.2002

53. Hirooka, K., Bamba, T., Fucusaki, E., and Kobayashi, A. (2003). Cloning and kinetic characterization of *Arabidopsis thaliana* solanesyl diphosphate synthase. *Biochem. J.* 370(Pt 2), 679–686. doi: 10.1042/BJ20021311
54. Hirooka, K., Izumi, Y., An, C. I., Nakazawa, Y., Fukusaki, E., and Kobayashi, A. (2005). Functional analysis of two solanesyl diphosphate synthases from *Arabidopsis thaliana*. *Biosci. Biotechnol. Biochem.* 69, 592–601. doi: 10.1271/bbb.69.592
55. Hsieh, F. L., Chang, T. H., Ko, T. P., and Wang, A. H. J. (2011). Structure and mechanism of an *Arabidopsis* medium/long-chain-length prenyl pyrophosphate synthase. *Plant Physiol.* 155, 1079–1090. doi: 10.1104/pp.110.168799
56. Hsu, A. Y., Do, T. Q., Lee, P. T., and Clarke, C. F. (2000). Genetic evidence for a multi-subunit complex in the O-methyltransferase steps of coenzyme Q biosynthesis. *Biochim. Biophys. Acta* 1484, 287–297. doi: 10.1016/S1388-1981(00)00019-6
57. Hsu, A. Y., Poon, W. W., Shepherd, J. A., Myles, D. C., and Clarke, C. F. (1996). Complementation of coq3 mutant yeast by mitochondrial targeting of the *Escherichia coli* UbiG polypeptide: evidence that UbiG catalyzes both O-methylation steps in ubiquinone biosynthesis. *Biochemistry* 35, 9797–9806. doi: 10.1021/bi9602932
58. Hundal, T., Forsmark-Andrée, P., Ernster, L., and Andersson, B. (1995). Antioxidant activity of reduced plastoquinone in chloroplast thylakoid membranes. *Arch. Biochem. Biophys.* 324, 117–122. doi: 10.1016/j.bbabio.2007.10.008
59. Ikeda, M., and Kagei, K. (1979). Ubiquinone content of eight plant species in cell culture. *Phytochemistry* 18, 1577–1578. doi: 10.1016/S0031-9422(00)98506-6
60. Ikeda, T., Matsumoto, T., Obi, Y., Kisaki, T., and Noguchi, M. (1981). Characteristics of cultured tobacco cell strains producing high levels of ubiquinone-10 selected by a cell cloning technique. *Agric. Biol. Chem.* 45, 2259–2263. doi: 10.1080/00021369.1981.10864865
61. Iomdina, E. N., Khoroshilovamaslova, I. P., Robustova, O.

V., Averina, O. A., Kovaleva, N. A., Aliev, G., et al. (2015). Mitochondria-targeted antioxidant SkQ1 reverses glaucomatous lesions in rabbits. *Front. Biosci. (Landmark Ed.)* 20, 892–901. doi: 10.2741/4343

62. Ishizaki, K., Schauer, N., Larson, T. R., Graham, I. A., Fernie, A. R., and Leaver, C. J. (2006). The mitochondrial electron transfer flavoprotein complex is essential for survival of *Arabidopsis* in extended darkness. *Plant J.* 47, 751–760. doi: 10.1111/j.1365-313X.2006.02826.x
63. Jiang, Y. E., Lin, M. A., Wu, H., Liu, X., and Zhang, H. (2006). Effect of overexpressing ubiCA genes responsible for ubiquinone biosynthesis on ubiquinone production in *Escherichia coli*. *J. East China Univ. Sci. Technol. (Nat. Sci. Ed.)* 2006, 269–273.
64. Jonassen, T., and Clarke, C. F. (2000). Isolation and functional expression of human COQ3, a gene encoding a methyltransferase required for ubiquinone biosynthesis. *J. Biol. Chem.* 275, 12381–12387. doi: 10.1074/jbc.275.17.12381
65. Jonassen, T., Marbois, B. N., Kim, L., Chin, A., Xia, Y. R., Lusis, A. J., et al. (1996). Isolation and sequencing of the rat Coq7 gene and the mapping of mouse Coq7 to chromosome 7. *Arch. Biochem. Biophys.* 330, 285–289. doi: 10.1006/abbi.1996.0255
66. Jones, M. O., Perez-Fons, L., Robertson, F. P., Bramley, P. M., and Fraser, P. D. (2013). Functional characterization of long-chain prenyl diphosphate synthases from tomato. *Biochem. J.* 449, 729–740. doi: 10.1042/BJ20120988
67. Jun, L., Saiki, R., Tatsumi, K., Nakagawa, T., and Kawamukai, M. (2004). Identification and subcellular localization of two solanesyl diphosphate synthases from *Arabidopsis thaliana*. *Plant Cell Physiol.* 45, 1882–1888. doi: 10.1093/pcp/pch211
68. Kawamukai, M. (2000). Biosynthesis, bioproduction and novel roles of ubiquinone. *J. Biosci. Bioeng.* 294, 511–517. doi: 10.1263/jbb.94.511
69. Kellogg, B. A., and Poulter, C. D. (1997). Chain elongation in the isoprenoid biosynthetic pathway. *Curr. Opin. Chem. Biol.* 1, 570–578. doi: 10.1016/S1367-5931(97)80054-3
70. Kim, E. H., Lee, Y., and Kim, H. U. (2015). Fibrillin 5 is essential for

plastoquinone-9 biosynthesis by binding to solanesyl diphosphate synthases in *Arabidopsis*. *Plant Cell* 27, 2956–2971. doi: 10.1105/tpc.15.00707

71. Kovtun, Y., Chiu, W. L., Tena, G., and Sheen, J. (2000). Functional analysis of oxidative stress-activated mitogen-activatedprotein kinase cascade in plants. *Proc. Natl. Acad. Sci. U.S.A.* 97, 2940–2945. doi: 10.1073/pnas.97.6.2940
72. Kruk, J., Holländer-Czytko, H., Oettmeier, W., and Trebst, A. (2005). Tocopherol as singlet oxygen scavenger in photosystem II. *J. Plant Physiol.* 162, 749–757. doi: 10.1016/j.jplph.2005.04.020
73. Kruk, J., and Trebst, A. (2008). Plastoquinol as a singlet oxygen scavenger in photosystem II. *Biochim. Biophys. Acta* 177, 154–162. doi: 10.1016/j.bbabio.2007.10.008
74. Ksas, B., Becuwe, N., Chevalier, A., and Havaux, M. (2015). Plant tolerance to excess light energy and photooxidative damage relies on plastoquinone biosynthesis. *Sci. Rep.* 5:10919. doi: 10.1038/srep10919
75. Kumar, S., Hahn, F. M., Baidoo, E., Kahlon, T. S., Wood, D. F., McMahan, C. M., et al. (2012). Remodeling the isoprenoid pathway in tobacco by expressing the cytoplasmic mevalonate pathway in chloroplasts. *Metab. Eng.* 14, 19–28. doi: 10.1016/j.ymben.2011.11.005
76. Lange, B. M., and Ghassemian, M. (2003). Genome organization in *Arabidopsis thaliana*: a survey for genes involved in isoprenoid and chlorophyll metabolism. *Plant Mol. Biol.* 51, 925–948. doi: 10.1023/A:1023005504702
77. Lee, B. J., Huang, Y. C., Chen, S. J., and Lin, P. T. (2012). Coenzyme Q10 supplementation reduces oxidative stress and increases antioxidant enzyme activity in patients with coronary artery disease. *Nutrition* 28, 250–255. doi: 10.1016/j.nut.2011.06.004
78. Lepetit, B., Sturm, S., Rogato, A., Gruber, A., Sachse, M., Falciatore, A., et al. (2013). High light acclimation in the secondary plastids containing diatom *Phaeodactylum tricornutum* is triggered by the redox state of the plastoquinone pool. *Plant Physiol.* 161, 853–865. doi: 10.1104/pp.112.207811
79. Li, C., Li, D., Shao, F., and Lu, S. (2015). Molecular cloning and

expression analysis of WRKY transcription factor genes in *Salvia miltiorrhiza*. *BMC Genomics* 16:200. doi: 10.1186/s12864-015-1411-x

80. Li, C., and Lu, S. (2014a). Genome-wide characterization and comparative analysis of R2R3-MYB transcription factors shows the complexity of MYB-associated regulatory networks in *Salvia miltiorrhiza*. *BMC Genomics* 15:277. doi: 10.1186/1471-2164-15-277
81. Li, C., and Lu, S. (2014b). Molecular characterization of the SPL gene family in *Populus trichocarpa*. *BMC Plant Biol.* 14:131. doi: 10.1186/1471-2229-14-131
82. Li, D., Shao, F., and Lu, S. (2015). Identification and characterization of mRNA-like noncoding RNAs in *Salvia miltiorrhiza*. *Planta* 241, 1131–1143. doi: 10.1007/s00425-015-2246-z
83. Lin, F., Ferguson, K. L., Boyer, D. R., Lin, X. N., and Marsh, E. N. (2015). Isofunctional enzymes PAD1 and UbiX catalyze formation of a novel cofactor required by ferulic acid decarboxylase and 4-hydroxy-3-polyprenylbenzoic acid decarboxylase. *ACS Chem. Biol.* 10, 1137–1144. doi: 10.1021/cb5008103
84. Littarru, G. P., and Tiano, L. (2007). Bioenergetic and antioxidant properties of coenzyme Q10: recent developments. *Mol. Biotechnol.*37, 31–37. doi: 10.1007/s12033-007-0052-y
85. Littarru, G. P., and Tiano, L. (2010). Clinical aspects of coenzyme Q10: an update. *Nutrition* 26, 250–254. doi: 10.1016/j.nut.2009.08.008
86. Liu, F. Z., Guo, A. Q., and Wan, Y. S. (2013). Cloning and polymorphism analysis of the 2-methyl-6-phytyl-1,4-benzoquinol methyltransferase gene (VTE3) in *Arachis hypogaea, A. duranensis*, and *A. ipaënsis*. *Genet. Mol. Res.* 12, 1859–1871. doi: 10.4238/2013.January.4.5
87. Lockwood, K., Moesgaard, S., and Folkers, K. (1994). Partial and complete regression of breast cancer in patients in relation to dosage of coenzyme Q10. *Biochem. Biophys. Res. Commun.* 199, 1504–1508. doi: 10.1006/bbrc.1994.1401
88. Lu, S., Li, Q., Wei, H., Chang, M. J., Tunlaya-Anukit, S., Kim, H., et al. (2013). Ptr-miR397a is a negative regulator of laccase genes

affecting lignin content in *Populus trichocarpa*. *Proc. Natl. Acad. Sci. U.S.A.* 110, 10848–10853. doi: 10.1073/pnas.1308936110

89. Lu, S., Sun, Y. H., Amerson, H., and Chiang, V. L. (2007). MicroRNAs in loblolly pine (*Pinus taeda* L.) and their association with fusiform rust gall development. *Plant J.* 51, 1077–1098. doi: 10.1111/j.1365-313X.2007.03208.x
90. Lu, S., Sun, Y. H., and Chiang, V. L. (2008). Stress-responsive microRNAs in *Populus*. *Plant J.* 55, 131–151. doi: 10.1111/j.1365-313X.2008.03497.x
91. Lu, S., Sun, Y. H., Shi, R., Clark, C., Li, L., and Chiang, V. L. (2005). Novel and mechanical stress-responsive microRNAs in *Populus trichocarpa* that are absent from *Arabidopsis*. *Plant Cell* 17, 2186–2203. doi: 10.1105/tpc.105.033456
92. Lu, T. T., Lee, S. J., Apfel, U. P., and Lippard, S. J. (2013). Aging-associated enzyme human clock-1: substrate-mediated reduction of the diiron center for 5-demethoxyubiquinone hydroxylation. *Biochemistry* 52, 2236–2244. doi: 10.1021/bi301674p
93. Lütke-Brinkhaus, F., and Kleinig, H. (1987). Ubiquinone biosynthesis in plant mitochondria. *Methods Enzymol.* 148, 486–490. doi: 10.1016/0076-6879(87)48047-6
94. Lütke-Brinkhaus, F., Liedvogel, B., and Kleinig, H. (1984). On the biosynthesis of ubiquinones in plant mitochondria. *Eur. J. Biochem.*141, 537–541. doi: 10.1111/j.1432-1033.1984.tb08226.x
95. Ma, W., Deng, Y., and Mi, H. (2008). Redox of plastoquinone pool regulates the expression and activity of NADPH dehydrogenase supercomplex in *Synechocystis* sp. strain PCC 6803. *Curr. Microbiol.* 56, 189–193. doi: 10.1007/s00284-007-9056-x
96. Ma, Y., Yuan, L., Wu, B., Li, X., Chen, S., and Lu, S. (2012). Genome-wide identification and characterization of novel genes involved in terpenoid biosynthesis in *Salvia miltiorrhiza*. *J. Exp. Bot.* 63, 2809–2823. doi: 10.1093/jxb/err466
97. Maciejewska, U., Polkowska-Kowalczyk, L., Swiezewska, E., and Szkopinska, A. (2002). Plastoquinone: possible involvement in plant disease resistance. *Acta Biochim. Pol.* 49, 775–780.
98. Mancini, A., and Balercia, G. (2011). Coenzyme Q(10) in male

infertility: physiopathology and therapy. *Biofactors* 37, 374–380. doi: 10.1002/biof.164

99. Mancini, A., De Marinis, L., Littarru, G. P., and Balercia, G. (2005). An update of Coenzyme Q10 implications in male infertility: biochemical and therapeutic aspects. *Biofactors* 25, 165–174. doi: 10.1002/biof.5520250119
100. Marbois, B. N., and Clarke, C. F. (1996). The COQ7 gene encodes a protein in *Saccharomyces cerevisiae* necessary for ubiquinone biosynthesis. *J. Biol. Chem.* 271, 2995–3004. doi: 10.1074/jbc.271.6.2995
101. Marbois, B. N., Hsu, A., Pillai, R., Colicelli, J., and Clarke, C. F. (1994a). Cloning of a rat cDNA encoding dihydroxypolyprenylbenzoate methyltransferase by functional complementation of a *Saccharomyces cerevisiae* mutant deficient in ubiquinone biosynthesis. *Gene* 138, 213–217. doi: 10.1016/0378-1119(94)90810-9
102. Marbois, B. N., Xia, Y. R., Lusis, A. J., and Clarke, C. F. (1994b). Ubiquinone biosynthesis in eukaryotic cells: tissue distribution of mRNA encoding 3,4-dihydroxy-5-polyprenylbenzoate methyltransferase in the rat and mapping of the COQ3 gene to mouse chromosome 4. *Arch. Biochem. Biophys.* 313, 83–88. doi: 10.1006/abbi.1994.1362
103. Matsumoto, T., Kanno, N., Ikeda, I., Obi, Y., Kisaki, T., and Noguchi, M. (1981). Selection of cultured tobacco cell strains producing high levels of ubiquinone 10 by a cell cloning technique. *Agric. Biol. Chem.* 45, 1627–1633. doi: 10.1271/bbb1961.45.2259
104. Mattila, P., and Kumpulainen, J. (2001). Coenzymes Q9 and Q10: contents in foods and dietary intake. *J. Food Comp. Anal.* 14, 409–417. doi: 10.1006/jfca.2000.0983
105. Mayer, M. P., Beyer, P., and Kleinig, H. (1990). Quinone compounds are able to replace molecular oxygen as terminal electron acceptor in phytoene desaturation in chromoplasts of *Narcissus pseudonarcissus* L. *Eur. J. Biochem.* 191, 359–363. doi: 10.1111/j.1432-1033.1990.tb19130.x
106. Meganathan, R. (2001). Ubiquinone biosynthesis in microorganisms. *FEMS Microbiol. Lett.* 203, 131–139. doi: 10.1016/

S0378-1097(01)00330-5

107. Moludi, J., Keshavara, S., Javad, H. M., Frooshani, A. R., Sadeghpour, A., Salakia, S., et al. (2015). Coenzyme Q10 effect in prevention of atrial fibrillation after coronnary artery bypass graft: double-blind randomized clinical trial. *Tehran Univ. Med. J.* 73, 79–85.

108. Motohashi, R., Ito, T., Kobayashi, M., Taji, T., Nagata, N., Asami, T., et al. (2003). Functional analysis of the 37 kDa inner envelope membrane polypeptide in chloroplast biogenesis using a Ds-tagged *Arabidopsis* pale-green mutant. *Plant J.* 34, 719–731. doi: 10.1046/j.1365-313X.2003.01763.x

109. Mubarakshina, M. M., and Ivanov, B. N. (2010). The production and scavenging of reactive oxygen speciesin the plastoquinone pool of chloroplast thylakoid membranes. *Physiol. Plant.* 140, 103–110. doi: 10.1111/j.1399-3054.2010.01391.x

110. Mühlenbock, P., Szechynska-Hebda, M., Plaszczyca, M., Baudo, M., Mateo, A., Mullineaux, P. M., et al. (2008). Chloroplast signaling and LESION SIMULATING DISEASE1 regulate crosstalk between light acclimation and immunity in *Arabidopsis*. *Plant Cell* 20, 2339–2356. doi: 10.1105/tpc.108.059618

111. Mukai, N., Masaki, K., Fujii, T., Kawamukai, M., and Iefuji, H. (2010). PAD1 and FDC1 are essential for the decarboxylation of phenylacrylic acids in *Saccharomyces cerevisiae*. *J. Biosci. Bioeng.* 109, 564–569. doi: 10.1016/j.jbiosc.2009.11.011

112. Mur, L. A., Kenton, P., Lloyd, A. J., Ougham, H., and Prats, E. (2008). The hypersensitive response; the centenary is upon us but how much do we know? *J. Exp. Bot.* 59, 501–520. doi: 10.1093/jxb/erm239

113. Naqvi, S., Farré, G., Zhu, C., Sandmann, G., Capell, T., and Christou, P. (2011). Simultaneous expression of *Arabidopsis* ρ-hydroxyphenylpyruvate dioxygenase and MPBQ methyltransferase in transgenic corn kernels triples the tocopherol content. *Transgenic Res.* 20, 177–181. doi: 10.1007/s11248-010-9393-6

114. Nguyen, T. P., Casarin, A., Desbats, M. A., Doimo, M., Trevisson, E., Santos-Ocaña, C., et al. (2014). Molecular characterization

of the human COQ5 C-methyltransferase in coenzyme Q10 biosynthesis. *Biochim. Biophys. Acta* 1841, 1628–1638. doi: 10.1016/j.bbalip.2014.08.007

115. Nievelstein, V., Vandekerckhove, J., Tadros, M. H., Lintig, J. V., Nitschke, W., and Beyer, P. (1995). Carotene desaturation is linked to a respiratory redox pathway in *Narcissus pseudonarcissus* chromoplast membranes. *Eur. J. Biochem.* 233, 864–872. doi: 10.1111/j.1432-1033.1995.864_3.x
116. Norris, S. R., Barrette, T. R., and DellaPenna, D. (1995). Genetic dissection of carotenoid synthesis in *Arabidopsis* defines plastoquinone as an essential component of phytoene desaturation. *Plant Cell* 7, 2139–2149. doi: 10.1105/tpc.7.12.2139
117. Norris, S. R., Shen, X., and DellaPenna, D. (1998). Complementation of the *Arabidopsis* pds1 mutation with the gene encoding p-hydroxyphenylpyruvate dioxygenase. *Plant Physiol.* 117, 1317–1323. doi: 10.1104/pp.117.4.1317
118. Nosek, M., Kornaš, A., Kuzniak, E., and Miszalski, Z. (2015). Plastiquinone redox state modified plant response to pathogen. *Plant Physiol. Biochem.* 96, 163–170. doi: 10.1016/j.plaphy.2015.07.028
119. Obukhova, L. A., Skulachev, V. P., and Kolosova, N. G. (2009). Mitochondria-targeted antioxidant SkQ1 inhibits age-dependent involution of the thymus in normal and senescence-prone rat. *Aging* 1, 389–401. doi: 10.18632/aging.100043
120. Ohara, K., Kokado, Y., Yamamoto, H., Sato, F., and Yazaki, K. (2004). Engineering of ubiquinone biosynthesis using the yeast coq2 gene confers oxidative stress tolerance in transgenic tobacco. *Plant J.* 40, 734–743. doi: 10.1111/j.1365-313X.2004.02246.x
121. Ohara, K., Muroya, A., Fukushima, N., and Yazaki, K. (2009). Functional characterization of LePGT1, a membrane-bound prenyltransferase involved in the geranylation of p-hydroxybenzoic acid. *Biochem. J.* 421, 231–241. doi: 10.1042/BJ20081968
122. Ohara, K., Sasaki, K., and Yazaki, K. (2010). Two solanesyl diphosphate synthases with different subcellular localizations and their respective physiological roles in *Oryza sativa*. *J. Exp. Bot.* 61, 2683–2692. doi: 10.1093/jxb/erq103
123. Ohara, K., Yamamoto, K., Hamamoto, M., Sasaki, K., and Yazaki,

K. (2006). Functional characterization of OsPPT1, which encodes p-hydroxybenzoate polyprenyltransferase involved in ubiquinone biosynthesis in *Oryza sativa*. *Plant Cell Physiol.* 47, 581–590. doi: 10.1093/pcp/pcj025

124. Okada, K., Ohara, K., Yazaki, K., Nozaki, K., Uchida, N., Kawamukai, M., et al. (2004). The AtPPT1 gene encoding 4-hydroxybenzoate polyprenyl diphosphate transferase in ubiquinone biosynthesis is required for embryo development in *Arabidopsis thaliana*. *Plant Mol. Biol.* 57, 567–577. doi: 10.1007/s11103-004-1298-4

125. Olson, R. E., and Rudney, H. (1983). Biosynthesis of ubiquinone. *Vitam. Horm.* 40, 1–43. doi: 10.1016/S0083-6729(08)60431-8

126. Ozeir, M., Mühlenhoff, U., Webert, H., Lill, R., Fontecave, M., and Pierrel, F. (2011). Coenzyme Q biosynthesis: Coq6 is required for the C5-hydroxylation reaction and substrate analogs rescue Coq6 deficiency. *Chem. Biol.* 18, 1134–1142. doi: 10.1016/j.chembiol.2011.07.008

127. Palan, P. R., Shaban, D. W., Martino, T., and Mikhail, M. S. (2004). Lipid-soluble antioxidants and pregnancy: maternal serum levels of coenzyme Q10, α-tocopherol and γ-tocopherol in preeclampsia and normal pregnancy. *Gynecol. Obstet. Invest.* 58, 8–13. doi: 10.1159/000077011

128. Parmar, S. S., Jaiwal, A., Dhankher, O. P., and Jaiwal, P. K. (2015). Coenzyme Q10 production in plants: current status and future prospects. *Crit. Rev. Biotechnol.* 35, 152–164. doi: 10.3109/07388551.2013.823594

129. Payne, K. A., White, M. D., Fisher, K., Khara, B., Bailey, S. S., Parker, D., et al. (2015). New cofactor supports α,β-unsaturated acid decarboxylation via 1,3-dipolar cycloaddition. *Nature* 522, 497–501. doi: 10.1038/nature14560

130. Petrov, A., Perekhvatova, N., Skulachev, M., Stein, L., and Ousler, G. (2016). SkQ1 Ophthalmic solution for dry eye treatment: results of a phase 2 safety and efficacy clinical study in the environment and during challenge in the controlled adverse environment model. *Adv. Ther.* 33, 96–115. doi: 10.1007/s12325-015-0274-5

131. Pfander, H. (1992). Carotenoids: an overview. *Methods*

Enzymol. 213, 3–13. doi: 10.1016/0076-6879(92)13105-7

132. Phatthiya, A., Takahashi, S., Chareonthiphakorn, N., Koyama, T., Wititsuwannakul, D., Wititsuwannakul, R., et al. (2007). Cloning and expression of the gene encoding solanesyl diphosphate synthase from *Hevea brasiliensis. Plant Sci.* 172, 824–831. doi: 10.1016/j.plantsci.2006.12.015
133. Poon, W. W., Barkovich, R. J., Hsu, A. Y., Frankel, A., Lee, P. T., Shepherd, J. N., et al. (1999). Yeast and rat Coq3 and *Escherichia coli*UbiG polypeptides catalyze both O-methyltransferase steps in coenzyme Q biosynthesis. *J. Biol. Chem.* 274, 21665–21672. doi: 10.1074/jbc.274.31.21665
134. Premkumar, V. G., Yuvaraj, S., Sathish, S., Shanthi, P., and Sachdanandam, P. (2008). Anti-angiogenic potential of coenzyme Q10, riboflavin and niacin in breast cancer patients undergoing tamoxifen therapy. *Vascul. Pharmacol.* 48, 191–201. doi: 10.1016/j.vph.2008.02.003
135. Qin, G., Gu, H., Ma, L., Peng, Y., Deng, X. W., Chen, Z., et al. (2007). Disruption of phytoene desaturase gene results in albino and dwarf phenotypes in *Arabidopsis* by impairing chlorophyll, carotenoid, and gibberellin biosynthesis. *Cell Res.* 17, 471–482. doi: 10.1038/cr.2007.40
136. Reape, T. J., Molony, E. M., and McCabe, P. F. (2007). Programmed cell death in plants: distinguishing between different modes. *J. Exp. Bot.* 59, 435–444. doi: 10.1093/jxb/erm258
137. Rock, C. D., and Zeevaart, J. (1991). The aba mutant of *Arabidopsis thaliana* is impaired in epoxy-carotenoid biosynthesis. *Proc. Natl. Acad. Sci. U.S.A.* 88, 7496–7499. doi: 10.1073/pnas.88.17.7496
138. Rozhnova, N. A., and Gerashchenkov, G. A. (2006). Hormonal status of tobacco variety Samsun NN exposed to synthetic coenzyme Q10 (ubiquinone 50) and TMV infection. *Izv. Akad. Nauk Ser. Biol.* 33, 581–590. doi: 10.1134/S1062359006050074
139. Rozhnova, N. A., and Gerashchenkov, G. A. (2008). Effect of ubiquinone 50 and viral infection on phytohemagglutinin activity in development of induced resistance in tobacco plants. *Izv. Akad. Nauk Ser. Biol.* 35, 442–447. doi: 10.1134/S1062359008040080
140. Sadre, R., Frentzen, M., Saeed, M., and Hawkes, T. (2010).

Catalytic reactions of the homogentisate prenyl transferase involved in plastoquinone-9 biosynthesis. *J. Biol. Chem.* 285, 18191–18198. doi: 10.1074/jbc.M110.117929

141. Sadre, R., Gruber, J., and Frentzen, F. (2006). Characterization of homogentisate prenyltransferases involved in plastoquinone-9 and tocochromanol biosynthesis. *FEBS Lett.* 580, 5357–5362. doi: 10.1016/j.febslet.2006.09.002
142. Sadre, R., Pfaff, C., and Buchkremer, S. (2012). Plastoquinone-9 biosynthesis in cyanobacteria differs from that in plants and involves a novel 4-hydroxybenzoate solanesyltransferase. *Biocheml J.* 442, 621–629. doi: 10.1042/BJ20111796
143. Safarinejad, M. R. (2012). The effect of coenzyme Q10 supplementation on partner pregnancy rate in infertile men with idiopathic oligoasthenoteratozoospermia: an open-label prospective study. *Int. Urol. Nephrol.* 44, 689–700. doi: 10.1007/s11255-011-0081-0
144. Sakato, K., Tanaka, H., Shibata, S., and Kuratsu, Y. (1992). Agitation-aeration studies on coenzyme Q10 production using *Rhodopseudomonas spheroides*. *Biotechnol. Appl. Biochem.* 16, 19–22.
145. Sakiko, T., Ohtani, T., Saton, H., Nakamura, Y., Kawamukai, M., and Kadowaki, K. (2006). Metabolic engineering of Coenzyme Q by modification of isoprenoid side chain in plant. *FEBS Lett.* 580, 955–959. doi: 10.1016/j.febslet.1006.01.023
146. Sakiko, T., Ohtani, T., Saton, H., Nakamura, Y., Kawamukai, M., and Kadowaki, K. (2010). Development of coenzyme Q10-enriched rice using sugary and shrunken mutants. *Biosci. Biotechnol. Biochem.* 74, 182–184. doi: 10.1271/bbb.90562
147. Samuilov, V. D., and Kiselevsky, D. B. (2015). Effect of cationic plastoquinone SkQ1 on electron transfer reactions in chloroplasts and mitochondria from pea seedlings. *Biochemistry* 80, 417–423. doi: 10.1134/S0006297915040045
148. Savidge, B., Weiss, J. D., Wong, Y. H., Lassner, M. W., Mitsky, T. A., Shewmaker, C. K., et al. (2002). Isolation and characterization of tocopherol phytyltransferase genes from *Synechocystis* sp. PCC 6803 and *Arabidopsis*. *Plant Physiol.* 129, 321–332. doi: 10.1104/

pp.010747

149. Schmelzer, C., Lindner, I., Vock, C., Fujii, K., and Döring, F. (2007). Functional connections and pathways of coenzyme Q10-inducible genes: an in-silico study. *IUBMB Life* 59, 628–633. doi: 10.1080/15216540701545991

150. Schmelzer, C., Niklowitz, P., Jürgen, G., Okun, J. G., Haas, D., Menke, T., et al. (2011). Ubiquinol-induced gene expression signatures are translated into altered parameters of erythropoiesis and reduced low density lipoprotein cholesterol levels in humans. *IUBMB Life* 63, 42–48. doi: 10.1002/iub.413

151. Shen, W., Wei, Y., Dauk, M., Tan, Y., Taylor, D. C., Selvaraj, G., et al. (2007). Involvement of a glycerol-3-phosphate dehydrogenase in modulating the NADH/NAD+ ratio provides evidence of a mitochondrial glycerol-3-phosphate shuttle in *Arabidopsis*. *Plant Cell* 18, 422–441. doi: 10.1105/tpc.105.039750

152. Shintani, D. K., Cheng, Z., and DellaPenna, D. (2002). The role of 2-methyl-6-phytylbenzoquinone methyltransferase in determining tocopherol composition in *Synechocystis* sp. *FEBS Lett.* 511, 1–5. doi: 10.1016/S0014-5793(01)0323-9

153. Siebert, M., Bechthold, A., Melzer, M., May, U., Berger, U., Schröder, G., et al. (1992). Ubiquinone biosynthesis: cloning of the genes coding for chorismate pyruvate-lyase and 4-hydroxybenzoate octaprenyl transferase from *Escherichia coli*. *FEBS Lett.* 307, 347–350. doi: 10.1016/0014-5793(92)80710-X

154. Skulachev, M. V., Antonenko, Y. N., Anisimov, V. N., Chernyak, B. V., Cherepanov, D. A., Chistyakov, V. A., et al. (2011). Mitochondrial-targeted plastoquinone derivatives. Effect on senescence and acute age-related pathologies. *Curr. Drug Targets* 12, 800–826. doi: 10.2174/138945011795528859

155. Solovieva, A. D., Frolova, O. Y., Solovyev, A. G., Morozov, S. Y., and Zamyatnin, A. A. (2013). Effect of mitochondria-targeted antioxidant SkQ1on programmed cell death induced by viral proteins in tobacco plants. *Biochemistry* 78, 1006–1012. doi: 10.1134/S000629791309006X

156. Sommer, S., and Heide, L. (1998). Expression of bacterial chorismate pyruvate-lyase in tobacco: evidence of the presence of chorismate in

the plant cytosol. *Plant Cell Physiol.* 39, 1240–1244. doi: 10.1093/oxfordjournals.pcp.a029326

157. Stiff, M. R. (2010). *Coenzyme Q10 Biosynthesis in Plants: Is the Polyprenyltransferase an Appropriate Gene Target for the Increased Production of CoQ?*. Ph.D. thesis, North Carolina State University, Raleigh, NC.
158. Sweetlove, L. J., Fait, A., Nunes-Nesi, A., Williams, T., and Fernie, A. R. (2007). The mitochondrion: an integration point of cellular metabolism and signalling. *Crit. Rev. Plant Sci.* 26, 17–43. doi: 10.1080/07352680601147919
159. Swiezewska, E. (2004). Ubiquinone and plastoquinone metabolism in plants. *Methods Enzymol.* 378, 124–131. doi: 10.1016/S0076-6879(04)78007-6
160. Tang, Y., Fu, X., Shen, Q., and Tang, K. (2016). Roles of MPBQ-MT in promoting α/γ-tocopherol production and photosynthesis under high light in lettuce. *PLoS ONE* 11:e0148490. doi: 10.1371/journal.pone.0148490
161. Teran, E., Hernandez, I., Nieto, B., Tavara, R., Ocampo, J. E., and Calle, A. (2009). Coenzyme Q10 supplementation during pregnancy reduces the risk of pre-eclampsia. *Int. J. Gynecol. Obstet.* 105, 43–45. doi: 10.1016/j.ijgo.2008.11.033
162. Thoma, I., Loeffler, C., Sinha, A. K., Gupta, M., Krischke, M., Steffan, B., et al. (2003). Cyclopentenoneisoprostanes induced by reactive oxygen species triggerdefense gene activation and phytoalexin accumulation in plants. *Plant J.* 34, 363–375. doi: 10.1046/j.1365-313X.2003.01730.x
163. Tian, L., DellaPenna, D., and Dixon, R. A. (2007). The pds2 mutation is a lesion in the *Arabidopsis* homogentisate solanesyltransferase gene involved in plastoquinone biosynthesis. *Planta* 226, 1067–1073. doi: 10.1007/s00425-007-0564-5
164. Torres, M. A., Jones, J. D., and Dangl, J. L. (2006). Reactive oxygen species signaling in response to pathogens. *Plant Physiol.* 141, 373–378. doi: 10.1104/pp.106.079467
165. Tran, M. T., Mitchell, T. M., Kennedy, D. T., and Giles, J. T. (2001). Role of coenzyme Q10 in chronic heart failure, angina, and hypertension. *Pharmacotherapy* 21, 797–806. doi: 10.1592/

phco.21.9.797.34564

166. Tran, U. C., and Clarke, C. F. (2007). Endogenous synthesis of coenzyme Q in eukaryotes. *Mitochondrion* 7, S62–S71. doi: 10.1016/j.mito.2007.03.007

167. Tripathy, B. C., and Oelmuller, R. (2012). Reactive oxygen species generation and signaling in plants. *Plant Signal. Behav.* 7, 1621–1633. doi: 10.4161/psb.22455

168. Uchida, N., Suzuki, K., Saiki, R., Kainou, T., Tanaka, K., Matsuda, H., et al. (2000). Phenotypes of fission yeast defective in ubiquinone production due to disruption of the gene for p-hydroxybenzoate polyprenyl diphosphate transferase. *J. Bacteriol.* 182, 6933–6939. doi: 10.1128/JB.182.24.6933-6939.2000

169. Vajo, Z., King, L. M., Jonassen, T., Wilkin, D. J., Ho, N., Munnich, A., et al. (1999). Conservation of the *Caenorhabditis elegans* timing gene clk-1 from yeast to human: a gene required for ubiquinone biosynthesis with potential implications for aging. *Mamm. Genome* 10, 1000–1004. doi: 10.1007/s003359901147

170. Van Eenennaam, A. L., Lincoln, K., Durrett, T. P., Valentin, H. E., Shewmaker, C. K., Thorne, G. M., et al. (2003). Engineering vitamin E content: from *Arabidopsis* mutant to soy oil. *Plant Cell* 15, 3007–3019. doi: 10.1105/tpc.015875

171. Venkatesh, T. V., Karunanandaa, B., Free, D. L., Rottnek, J. M., Baszis, S. R., and Valentin, H. E. (2006). Identification and characterization of an *Arabidopsis* homogentisate phytyltransferase paralog. *Planta* 223, 1134–1144. doi: 10.1007/s00425-005-0180-1

172. Viitanen, P. V., Devine, A. L., Khan, M. S., Deuel, D. L., Dyk, D. E. V., and Daniell, H. (2004). Metabolic engineering of the chloroplast genome using the *Escherichia coli* ubiC gene reveals that chorismate is a readily abundant plant precursor for p-hydroxybenzoic acid biosynthesis. *Plant Physiol.* 136, 4048–4060. doi: 10.1104/pp.104.050054

173. Wang, K., and Ohnuma, S. (1999). Chain-length determination mechanism of isoprenyl diphosphate synthases and implications for molecular evolution. *Trends Biochem. Sci.* 24, 445–451. doi: 10.1016/S0968-0004(99)01464-4

174. Wang, M., Wu, B., Chen, C., and Lu, S. (2015). Identification of

mRNA-like non-coding RNAs and validation of a mighty one named MAR in Panax ginseng. *J. Integr. Plant Biol.* 57, 256–270. doi: 10.1111/jipb.12239

175. Wang, X., Wang, H., Wang, J., Sun, R., Wu, J., Liu, S., et al. (2011). The genome of the mesopolyploid crop species *Brassica rapa*. *Nat. Genet.* 43, 1035–1039. doi: 10.1038/ng.919
176. Wanke, M., Swiezewska, E., and Dallner, G. (2000). Half-life of ubiquinone and plastoquinone in spinach cells. *Plant Sci.* 154, 183–187. doi: 10.1016/S0168-9452(00)00200-4
177. Wei, R., Qiu, D., Wilson, I. W., Zhao, H., Lu, S., Miao, J., et al. (2015). Identification of novel and conserved microRNAs in Panax notoginseng roots by high-throughput sequencing. *BMC Genomics* 16:835. doi: 10.1186/s12864-015-2010-6
178. White, M. D., Payne, K. A., Fisher, K., Marshall, S. A., Parker, D., Rattray, N. J., et al. (2015). UbiX is a flavin prenyltransferase required for bacterial ubiquinone biosynthesis. *Nature* 522, 502–506. doi: 10.1038/nature14559
179. Wu, B., Li, Y., Yan, H., Ma, Y., Luo, H., Yuan, L., et al. (2012). Comprehensive transcriptome analysis reveals novel genes involved in cardiac glycoside biosynthesis and mlncRNAs associated with secondary metabolism and stress response in *Digitalis purpurea*. *BMC Genomics* 13:15. doi: 10.1186/1471-2164-13-15
180. Xu, W., Yang, S., Zhao, J., Su, T., Zhao, L., and Liu, J. (2014). Improving coenzyme Q8 production in *Escherichia coli* employing multiple strategies. *J. Ind. Microbiol. Biotechnol.* 41, 1297–1303. doi: 10.1007/s10295-014-1458-8
181. Yadav, D. K., Kruk, J., Sinha, R. K., and Pospíšil, P. (2010). Singlet oxygen scavenging activity of plastoquinol in photosystem II of higher plants: electron paramagnetic resonance spin-trapping study. *Biochim. Biophys. Acta* 1797, 1807–1811. doi: 10.1016/j.bbabio.2010.07.003
182. Yang, D. H., Andersson, B., Aro, E. M., and Ohad, I. (2001). The redox state of the plastoquinone pool controls the level of the light-harvesting chlorophyll a/b binding protein complex II (LHC II) during photoacclimation. *Photosynth. Res.* 68, 163–174. doi: 10.1023/A:1011849919438

183. Yang, R., Jarvis, D. E., Chen, H., Beilstein, M. A., Grimwood, J., Jenkins, J., et al. (2013). The reference genome of halophytic plant Eutrema salsugineum. *Front. Plant Sci.* 4:46. doi: 10.3389/fpls.2013.00046

184. Yazaki, K., Kunihisa, M., Fujisaki, T., and Sato, F. (2002). Geranyl diphosphate:4-hydroxybenzoate geranyltransferase from *Lithospermum erythrorhizon*. Cloning and characterization of a ket enzyme in shikonin biosynthesis. *J. Biol. Chem.* 277, 6240–6246. doi: 10.1016/S0014-5793(01)03223-9

185. Zhang, D., Li, Z., Wang, F., Shrestha, B., Tian, P., and Tan, T. (2007a). Expression of various genes to enhance ubiquinone metabolic pathway in *Agrobacterium tumefaciens*. *Enzyme Microb. Technol.* 41, 772–779. doi: 10.1016/j.enzmictec.2007.06.014

186. Zhang, D., Shrestha, B., Niu, W., Tian, P., and Tan, T. (2007b). Phenotypes and fed-batch fermentation of ubiquinone-overproducing fission yeast using ppt1 gene. *J. Biotechnol.* 128, 120–131. doi: 10.1016/j.jbiotec.2006.09.012

187. Zhang, L., and Lu, S. (2016). Overview of medicinally important diterpenoids derived from plastids. *Mini Rev. Med. Chem.* 16. doi: 10.2174/1389557516666160614005244 [Epub ahead of print].

188. Zhang, L., Wu, B., Zhao, D., Li, C., Shao, F., and Lu, S. (2014). Genome-wide analysis and molecular dissection of the SPL gene family in *Salvia miltiorrhiza*. *J. Integr. Plant Biol.* 56, 38–50. doi: 10.1111/jipb.12111

189. Zhou, X., Yuan, Y., Yang, Y., Rutzke, M., Thannhauser, T. W., Kochian, L. V., et al. (2009). Involvement of a broccoli COQ5 methyltransferase in the production of volatile selenium compounds. *Plant Physiol.* 151, 528–540. doi: 10.1104/pp.109.142521

190. Zhu, X., Yuasa, M., Okada, K., Suzuki, K., and Nakagawa, T. (1995). Production of ubiquinone in *Escherichia coli* by expression of various genes responsible for ubiquinone biosynthesis. *J. Ferment. Bioeng.* 79, 493–495. doi: 10.1016/0922-338X(95)91268-A

CITATION

CHAPTER 2

Koichi Kobayashi and Tatsuru Masuda; Transcriptional Regulation of Tetrapyrrole Biosynthesis in Arabidopsis thaliana; https://doi.org/10.3389/fpls.2016.01811.

CHAPTER 3

Robert M. Larkin; tetrapyrrole signaling in plants; https://doi.org/10.3389/fpls.2016.01586.

CHAPTER 4

María del Rocío Gómez-García and Neftalí Ochoa-Alejo; Biochemistry and Molecular Biology of Carotenoid Biosynthesis in Chili Peppers (Capsicum spp.); doi:10.3390/ijms140919025.

CHAPTER 5

Henry Daniell, Choun-Sea Lin, Ming Yu and Wan-Jung Chang; Chloroplast genomes: diversity, evolution, and applications in genetic engineering; https://doi.org/10.1186/s13059-016-1004-2.

CHAPTER 6

Krzysztof Bobik and Tessa M. Burch-Smith; Chloroplast signaling within, between and beyond cells; https://doi.org/10.3389/fpls.2015.00781.

CHAPTER 7

Roberta Croce and Herbert van Amerongen, Light-harvesting in photosystem I; DOI 10.1007/s11120-013-9838-x.

CHAPTER 8

Herbert van Amerongen and Roberta Croce; light harvesting in photosystem II; DOI 10.1007/s11120-013-9824-3.

CHAPTER 9

Sari Järvi Peter J. Gollan and Eva-Mari Aro; Understanding the roles of the thylakoid lumen in photosynthesis regulation; https://doi.org/10.3389/fpls.2013.00434.

CHAPTER 10

Miaomiao Liu and Shanfa Lu; Plastoquinone and Ubiquinone in Plants: Biosynthesis, Physiological Function and Metabolic Engineering; https://doi.org/10.3389/fpls.2016.01898.

INDEX